DESIGN AND ENGINEERING OF INTELLIGENT COMMUNICATION SYSTEMS

DESIGN AND ENGINEERING OF INTELLIGENT COMMUNICATION SYSTEMS

Syed V. Ahamed
Professor of Computer Science
City University of New York
&
Bell Labs Innovations, Lucent Technologies
Holmdel, New Jersey

and

Victor B. Lawrence
Bell Labs Innovations, Lucent Technologies
Holmdel, New Jersey

Kluwer Academic Publishers
Boston / London / Dordrecht

Distributors for North America:
Kluwer Academic Publishers
101 Philip Drive
Assinippi Park
Norwell, Massachusetts 02061 USA

Distributors for all other countries:
Kluwer Academic Publishers Group
Distribution Centre
Post Office Box 322
3300 AH Dordrecht, THE NETHERLANDS

Library of Congress Cataloging-in-Publication Data

A C.I.P. Catalogue record for this book is available
from the Library of Congress.

Edited and prepared by Lucent Technologies, Bell Labs Innovations.

ESS™ is a trademark of AT&T.
5ESS® is a registered trademark of AT&T
4ESS™ is a trademark of AT&T.
EWSD ® is a registered trademark of Siemens.
AXE® is a registered trademark of Ericsson.
CLASS™ is a trademark of Bell Communications Research
Sprint™ is a trademark of MCI
UNIX ® is a registered trademark in the United States and other countries,
  licensed exclusively through X/Open Corporation
Vnet ™ is a trademark of MCI.
MAC® is a registered trademark.

Printed on acid-free paper.

Printed in the United States of America

dedicated to

our families

in

Canada, Ghana,

India and the United States.

DESIGN AND ENGINEERING OF

INTELLIGENT COMMUNICATION SYSTEMS

Syed V. Ahamed and Victor B. Lawrence

TABLE OF CONTENTS

LIST OF FIGURES

3.0 VIDEO AND TV ENVIRONMENT 63

4.0 CURRENT DIGITAL NETWORKS 85

PART III THE METALLIC MEDIA: THE COMPUTATIONAL

7.0 SIMULATION AND CAD ASPECTS

10.0 COMPUTER BASED OPTIMIZATION TECHNIQUES FOR HDSL DESIGN.. 369

13.0 PERFORMANCE OF TRELLIS CODING 421

PART V RECENT HIGH-SPEED NETWORK ENVIRONMENTS

15.0 KNOWLEDGE HIGHWAYS

16.0 IMPACT OF FIBER OPTIC TECHNOLOGY 503

17.0 OPTICAL LIGHTWAVE SYSTEMS IN EXISTING NETWORKS ... 559

18.0 A PC BASED FIBER OPTIC CAD ENVIRONMENT 595

PREFACE

The emerging networks in our society will touch upon the life of everyone. These networks have started to bring about an immense information revolution. The revolution within our intellectual life will be similar to the materialistic revolution that followed the invention of the steam and the internal combustion engines. From the perspective of the 1980s, the information networks are indeed evolving and their influence can only be gradual. However, the strides of progress are accelerating in the 1990s. Networks in our society offer the most candid area of convergence for the computer and the communication technologies.

The two technologies are mature in their own right. However, there are a few major factors that prevent network engineers from constructing modern communication systems from components borrowed from each of these two technologies:

- Major innovations are happening.
- Specialized components evolve in synergistic patterns.
- New technologies emerge.
- Inquisitive minds cross disciplinary barriers.

Examples of these factors are plentiful. Researchers have demonstrated super conducting materials (at low temperatures) for new devices and are actively investigating them. Fiber optics for communication has impacted the network technology in a major way. Fiberonics integrates optical devices and integrated circuits. Quantum well devices offer new challenges. Optical computing is a distinct possibility.

The explosive growth of networks that has occurred is likely to continue. It is also very likely to leave society behind in ignorance because of the multiplicity of options that are created and the opportunities that follow. It is impractical, if not impossible, to write one book that can cover the recent developments in the field of intelligent communication systems. For this reason, we have focused upon the topic in the 1990s to provide the scientific and factual material that will help graduate students in computer science, electrical engineering, or telecommunications become familiar with telecommunication sciences of the twenty-first century.

Graduate or undergraduate knowledge of computer architecture makes the study of network architecture easier to comprehend. Intelligent communication systems come about from the blending of the computational process with the switching and transmission of information. Digital switching and digital transmission also get unified into network design. Stored program control concepts, which founded the architectural framework of the third generation computers, play a vital part in the architecture of modern switching systems. For these reasons of unique, yet similar, functionality between the two architectures, we look to the computer architecture and the newer network designs to help extrapolate the implications intelligent communication systems might have for our society. Computer networks simply become precursors of the evolving intelligent communications digital networks.

Two graduate level books are presented, which cover the area of broadband telecommuncation networks:

- Volume 1   *"Intelligent Broadband Multimedia Networks"*
- Volume 2   *"Design and Engineering of Intelligent Communication Systems"*

In this volume, *"Design and Engineering of Intelligent Communication Systems"*, the engineering and computer-aided design aspects of the components that constitute the intelligent telecommunications systems are discussed. The primary focus is on the transmission aspects of digital information through the network because the switching systems are supplied by a few well-established global vendors who adhere strictly to the interface requirements, CCITT (now ITU) protocols, and standards. The transmission media can vary drastically from country to country, and the practicing telecommunication engineers are charged with the responsibility of building novel, general purpose, digital networks to suit the particular communication and social needs of the country.

This book has *five* major parts. The *first* part has *four* Chapters and primarily focuses on the switching and transfer of information in traditional and intelligent networks. In Chapter 2, the various FM techniques for long haul transmission are discussed. The FM and the channel bank hierarchy is introduced. In the third Chapter, video and TV environment is discussed in context to the evolving hybrid fiber coaxial system that takes information to the customers. In the fourth Chapter, the current digital networks and the functions of digital channel banks are completely explored.

In the *second* part of this book, the metallic media for the physical layer is discussed in great detail. The digital subscriber loop (DSL) environment and the basic building blocks of high-speed digital subscriber lines (HDSL) are presented as well as the limits of the technology and the constraints of the environment. Methodologies for interconnection and exchange of information are discussed in some detail because the existing telecommunication network of any country is the basic foundation upon which additional intelligent services are going to be offered. In this part, the students and practicing engineers are expected to become aware of the capability and limitations of their own networks before expecting them to serve as providers of intelligent services.

The *third* part of this book has *five* Chapters. The first two Chapters are dedicated to the digital transmission in the subscriber loop environment. Metallic media data transmission systems for the high-speed digital subscriber line (HDSL) are explored and discussed in considerable detail. The CAD techniques and optimization strategies for basic rate and primary rate ISDN systems are also explored in detail. The later Chapters explore the design of fiber optic systems in public and private domain networks. In this part, most of the significant aspects of computation for the design and engineering of the DSL, HDSL and the asymmetric digital subscriber line (ADSL) are presented. The computer-aided design aspects of these new lines are covered in complete detail to help the DSL, HDSL and ADSL designers build and operate a computational environment for the metallic media based data distribution systems. Most aspects of design and component optimization are also presented.

The *fourth* part of the book deals with the design questions for retrofitting the existing telephone plants for low speed to very high speed data through the loop plant to reach the individual customers. This covers the entire range from 64 kbps (in the loop plant) to 125 Mbps (in the premises distribution systems). Typical case studies of the loop environments and their ultimate digital capacity are also included.

The *fifth* part of the book deals with the transition to fiber. Such a transition has occurred drastically in most of the trunk environments of the major telecommunication networks. Long-haul facilities are experiencing a major overhaul to become more competitive in carrying multimedia information over longer distances. With the recent trend of the repeaterless fiber links to carry multigigabits of information over hundreds of miles, or even across the street for that matter, this underlying trend is explored and presented in this part of the book. In this part, recent high-speed network environments dealing with SONET, ATM, broadband networks and most aspects of fiber optic networks and their initial design are discussed. Most of the recent global optical networks are also covered in the last three Chapters, 16 through 18.

The inherent transmission capabilities and design issues for the copper and fiber media is also discussed. These two very essential media (metallic and optical fiber) will carry vast amounts of data for the intelligent network functions in the immediate future. Other media, such as free space, microwave, laser beams, etc., are firmly established and are tightly monitored by regulatory commissions.

This focuses on switching and information transfer in the traditional and intelligent networks, methodologies for interconnection and exchange of information, and transmission media and associated technology.

In an overall sense, the questions addressed here deal with issues pertaining to the types of networks and their environments, the local and global aspects in communication, the interfaces, the sources and users of information communicated, and the medium for communication. The issues that are also addressed are the types of intelligence that the network should possess to be functional in an effective, economic, smooth, and error-free way. The second set of issues is influenced by two distinct patterns of emerging hardware and the evolving network architecture. The hardware for implementation is computer based and encompasses the conventional processor units, memories, input/output interfaces, control memories, and software/firmware driving the hardware, which contains the algorithms that carry the framework of intelligence. Both the massive parallelism of call processing in the electronic switching centers and the generic features of the software make the slow transition of the current networks toward being more and more intelligent, rather than simply algorithmic.

In the companion volume, *"Intelligent Broadband Multimedia Networks"*, most of the concepts that are generic to the intelligent networks and their functions are presented. The material ranges from defining intelligent networks to specialized educational and medical networks.

There are *three* major parts in Volume 1. In the *first* part (with *four* Chapters), the transition from traditional communication environments to the intelligent networks is presented. In Chapter 1, an overview of communications networks and computer systems is presented. The remaining three Chapters in Part 1 deal with the primary characteristics of Intelligent Networks (INs), the complexities of information transfer within INs and between INs and users, and the application of the OSI model to the development of INs. The influence of the two major American research organizations, are investigated and documented. The enhancements resulting from the guidance and ITU standards that indicate the current status of the evolution of the intelligent network are presented. Each Chapter addresses issues to help students make a transition from computer design and the underlying concepts of traditional telecommunication systems to considerations essential for intelligent network design.

The *second* part of the book has *nine* Chapters and is dedicated to the current architectures and proposals for the evolving intelligent networks. In the first two Chapters, wireless communication networks are discussed and the impact of major data transport carriers and recent American INs are described. In the following Chapters, the proposals of AT&T's Universal Intelligent Services Network and Bellcore's three versions of IN/1, IN/1+, and IN/2 are discussed in detail and cross-compared. Finally in Part 2, Advanced Intelligent Networks (AINs) that can offer technology and services beyond the IN/2 architecture are described, and other INs located outside the United States, where topologies sometimes differ, are also addressed. The hardware and software considerations in these machines are explored and linked to their functional specifications.

The *third* part of the book deals with networks that are specialized to the needs of the educational and medical communities. This part has *seven* Chapters, and the first two focus on the needs of the local, national, or global universities. Specialized components and techniques are discussed in these Chapters. In the remaining Chapters, numerous architectures and configurations for INs, whose functionality extend far beyond the original uses, are explored. These include: knowledge machines, knowledge processing systems and personal computer-based intelligent home networks. Finally, there is a discussion of the social and cultural impact of intelligent broadband multimedia networks, their accessibility, application and potential for misuse.

We thank our friends and colleagues at AT&T Bell Laboratories and Lucent Technologies for their collaboration and their long association with us for the last thirty years. Much of the seminal work dates back to the numerous inventions and innovations at Bell Laboratories. We also thank our colleagues at the City University of New York for their continued support for the project, which has culminated in these two graduate level text books for computer scientists and electrical engineers.

Syed V. Ahamed
Victor B. Lawrence

Holmdel, New Jersey.
November 1996

PART 1

OVERVIEW OF THE COMMUNICATION FACILITIES

Communications networks consist of terminal devices, transmission facilities, signaling systems, and switching systems. Almost invariably, the human recipient and source are implied. The older trend has been to tailor the communicating systems to serve as an error free medium which could be invoked at will to let any one "talk" to any one else at any time. More recently, video, computer, and other data communication have started to exert their own particular needs and protocol. In this part, the infrastructure of most of the existing communication networks that permit exchange of information is presented. The evolutionary nature of these systems in view of the changing social needs and expectations from the communication systems is also presented. The oldest of these plain old telephone service networks is the POTS network. Whereas it existed earlier to serve as a circuit switched voice transport facility, the POTS network now carries significant amounts of data and is turning all digital, in spite of ample analog systems and devices within its infrastructure. The existing, the evolving, and the transitory aspects of the communication networks are present in this part.

CHAPTER 1

BASIC COMMUNICATION NETWORKS

A handful of scientists (Boole), experimentalists (Tyndell, Marconi, Bell), mathematicians (Fourier, Maxwell and Shannon) and technologists (Bardeen, von Neumann, Pierce, Patel and Tarbox) dominate the arena of seminal communication *systems*. *Networks* are dominated by corporations rather than individuals, and the collective effort becomes the driving force in the realization of gigantic information grids that perform mostly without grid locks. From the perspective of the 1990s, communication networks have an economic, political, and social context. Information being the power, and communication being its transport, they can be catalysts towards progress and change. In this chapter, the basic technical milestones in the Western cultures that have accelerated progress and the economic growth are surveyed.

1.1

EVOLVING TELEPHONE NETWORKS

During the last three decades, communication networks have undergone major reconstitution and restructuring. Being the beneficiaries of *four* dominant forces altering the social and business perspectives within the society, communication systems are experiencing a slow and tedious, but

definite pattern of evolution and modernization. These *four* major forces are
presented in this section.

The *first* of these forces is the digital revolution. The trend towards
making all aspects (control, storage, processing, switching and transmission)
of data accurate and programmable is well received in the communication
field. Data representation is quantized and digital. All functions are precise,
verifiable and repeatable. Information and its entropy are both maintained at
every stage. Not all these events in nature are digital but over a small
enough time slice, all aspects of data can be accurately quantified. This calls
for all digital networks that perform as if computers functioning in a widely
dispersed cooperative role. Making it directed towards the accurate
communication of information from one point to any other point in a
dependable and predictable way.

The *second* of these forces is the computer revolution in the precision and
logical execution of most scientific, business and knowledge related
functions, however simple or however complex and/or however distributed
they may be. Hardware, software and the firmware are at play. The major
contribution comes from the portability of the hardware devices, especially
the laptops and palmtop systems. Making them portable and yet functional
with other such systems demands a networking facility.

The *third* of these forces comes from the major breakthrough in
transmission technology (satellite, wireless, fiber, and to some extent copper
wire-pairs). The major impact was felt during the sixties when satellites
started to participate in the long-haul of information (see Section 2.6).
Indirectly, they assumed the role of repeaters in the sky and opened a
detailed study of outer space as a viable medium for long-haul and then later
as ground controlled switching systems. Wireless communications in radio
bands (see companion book *"Intelligent Broadband Multimedia Networks"*,
Chapter 5) had once been considered a viable means of two-way
communication. On a commercial basis, it started out as Telepoint services in
Europe. Cordless telephone 1 (from the seventies) became cordless telephone
of the second generation (CT2 in the eighties) and third generation (CT3)
with additional frequency allocation for these devices (see the companion
book *"Intelligent Broadband Multimedia Networks"*, Sections 5.2 and 5.3).
Cellular radio has also made serious inroads in the paging, messaging and
possible low rate data services since early nineties. Fiber optic transmission
(Chapters 15 and 17) started another major phase in the study of a self-
contained closed medium of communication. The overall impact of all such
systems (coherent and noncoherent, land and underwater, fiber coaxial and

fiber to the curb, etc.) is yet to be felt in all demographic and geographic terrains of the world.

Twisted wire-pairs, originally intended for voice communication, are indeed capable of carrying data to the OC1 or OC3 rates under special line circumstances. The very first impact was initially felt in the seventies when the viability of the repeaterless digital subscriber line (see Chapter 12, Reference 1) for data up to 320 kbps was proved. Next came the era of the high-speed digital subscriber line (HDSL) for the 384, 768 and 1,554 kbps in the eighties (see Chapters 12 and 14). In the early nineties, the feasibility of the asymmetric digital subscriber line (ADSL) was established, even though the concept was known through the mid eighties. Finally, the feasibility of premises distribution systems (PDS) using specially fabricated wire-pairs, with random twists carrying very high-speed data ranging from 30 to 155 Mbps is also documented.

The *fourth* of these forces is derived from the switching technology (private branch exchange, portable, satellite and mobile switching, and wireless controlled cell site switching systems). This technology is also in the process of realigning its architecture and its interface to permit all types of transmission devices and systems (wireless, microwave, fiber optic and satellite) to connect to the network infrastructure that itself is going digital.

The net result of these four major trends is *three*fold:

- the change of network architecture in the subscriber side, customer signaling and End Offices;

- the change of architecture at interoffice signaling, interoffice trunking and middle level of Central Offices;

- the change of architecture at international networks, standardized ITU (formerly CCITT) global signaling and regional, national and international gateways.

The change of the network architecture in the distribution of voice and data to the customers is depicted in Figures 1.1 and 1.2. Figure 1.1, depicting the architecture of the early nineties, uses the distribution plant of telephone networks to carry voice frequency signals to and from the Central Offices. The typical carrier systems most commonly used were the T1 (E1 in Europe) and the 90 Mbps optical carrier. Both of these systems terminated in remote terminals (RT) and the final distribution is at voice frequency (VF) with or without feeder distribution interface (FDI). Figure 1.2, depicting the current architectures of the mid-nineties, uses the distribution part of the telephone network to carry T1/E1, optical carriers, and SONET OC3 carriers (with

standardized digital signaling) and much more fiber media in the
distribution plant.

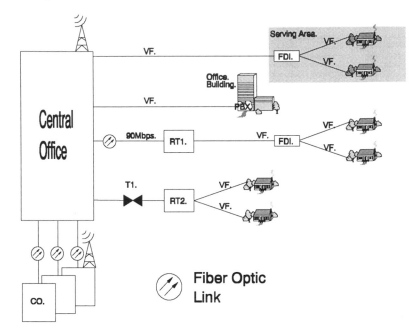

**FIGURE 1.1. Older Network Configuration Using Voice Frequency
(VF), Optical Carrier, and T1/E1 Carrier System.** CO - Central Office;
FDI - Feeder Distribution Interface; RTn - Remote Terminal; Tn -
Standard North American Carrier Lines Used to Transmit a Formatted
Digital Signal

The extensive use of fiber occurs at the End Offices to reach the customer
via RT, and then via distribution terminal (DT). The transition to fiber is
strong and eminent at all classes of Central Offices from the End Office, the
middle tier of Central Offices and the gateways.

1.2

EXISTING TELEPHONE ENVIRONMENTS

Telephone networks are generally the most dispersed networks reaching
most every home in the developed countries. Typically, telephone service
extends to remote areas and also to every user in a metropolitan area. For

this reason, immense geographic dispersion and branching occur. Individual users are generally served by twisted wire-pair subscriber lines or loops.

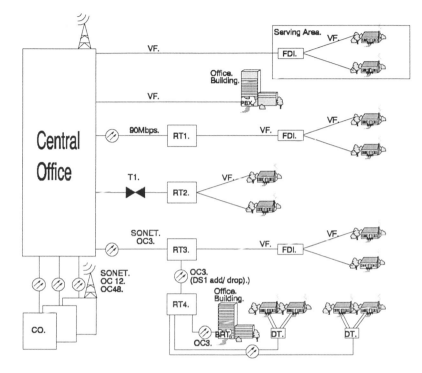

FIGURE 1.2. Recent Configuration Using the OC3 Carrier (155 Mbps) together with T1 and Optical Carrier System and Fewer Voice Frequency (VF) Systems. BRT - Broadband Radio Terminal ; CO - Central Office; DS-n - A Formatted Digital Signal; DT - Distribution Terminal; FDI - Feeder Distribution Interface; OCn - Optical Carrier; PBX - Private Branch Exchange; RTn - Remote Terminal; SONET - Synchronous Optical Network; Tn - Standard North American Carrier Lines Used to Transmit a Formatted Digital Signal

Each loop has a unique individual identifying number at the Central Offices and the subscriber is thus identified by the wire-pair number in the wire-center of the Central Office. When a party line (i.e., two or more subscribers on the same line) exits, individual subscribers are assigned distinct ringing and on-hook/off-hook signals and are thus correctly identified at the Central Office by the loop number and the distinct signaling pattern. The economic viability of the earlier telephone networks is based upon sharing its expensive resources such as switching systems, transmission facilities, administrative and billing procedures, etc.. In the same context, computing environments have traditionally shared their resources, such as central processors, array processors, memories, disk space, and even their I/O devices. Time sharing of expensive computing systems is also common. The expansion of the concept to the intelligent communication systems is evident by the use of SCP, SSP, STP facilities and also the IN-core CCS network. In perspective of the conventional telephone networks, there are three concepts implemented; concentration, distribution and expansion. These concepts are presented next.

Concentration permits the individual twisted wire-pairs from a large number of users to share a smaller number of trunk lines to widely dispersed Central Offices in a region. Subscriber loops are far more numerous than the trunk lines. Time and usage diversity occur. This principle is valid since all the subscribers are not likely to be using or for that matter trying to reach the same distant Central Office at the same time. Hence the Central Office responds to the called number and assigns one of the numerous communication paths (trunks and carriers) to the distant Central Office via its own switching facility.

Distribution permits the Central Office to access trunk lines to different regions by sharing these lines, whereas individual lines tend to become dedicated to the users. Trunk lines are considered as a shared network resource being switched into service for the duration of its use and then being released for reuse.

Expansion permits the Central Office to redistribute or reconnect the trunk lines to any one of the subscriber lines. This permits the establishment of a temporary communication path between the various subscribers by sharing the more expensive switching system and the trunk facilities among many users and allocating the cheaper twisted wire-pair subscriber lines to each user.

1.3

TRUNK NETWORKS

Distribution in telephone networks is done from Central Offices where numerous trunk lines converge, switching systems reside, and wire centers exist. All geographical areas do not require the same communication facilities and thus high capacity trunk lines do not have to link every Central Office. For this reason, an orderly concentration to very high capacity lines exists. The Central Offices thus have varying capabilities depending upon their hierarchical function. The type of Central Offices is thus identified by its capacity, function and the trunk lines terminating at the office. Trunk telephone networks thus become essential around cities, regions and nations to carry high books of traffic from one city to another, from one region to another, or from one country to another.

In larger networks, *four* trunk categories have been identified. *First*, the direct trunks connect End Offices. Covering very short distances, they operate within a given toll area. The length of these trunks is governed by the population and phone density in the given area. The average length of such trunks in the United States ranges between 8 to 10 miles with 8 to 20 trunks per group. The customer loop-to-trunk concentration ranges between 10 to 20. *Second*, the toll trunks provide access from an End Office to a Toll Center or any higher level office. Traffic on these trunks is monitored for billing and administration. Number (caller and called party) information, duration and statistics are passed to the OA&M software. Typically, the average length is about 8 to 10 miles and the customer loop-to-toll trunk concentration ratio can be 40 to 70. *Third*, the operator trunks provide operator access for call completion or operator assistance. In a topological sense, they are also grouped as toll trunks and provide the office interconnect facility. *Fourth*, the metropolitan intertoll trunks provide access among Toll Centers, and Toll Centers to other (Primary, Sectional or Regional) Central Offices placed higher in the hierarchy (see Section 1.5).

1.4

COAXIAL CABLES IN NETWORKS

Coaxial cables are used for carrying analog and digital data effectively. Cable networks servicing a large sector of customers (see Chapter 3) prevail in some nations. Cable networks between Central Offices to carry high density interoffice traffic exist in most nations. In LANs, WANs and MANs coaxial cables serve very dependably as digital pathways from 10 Mbps to 100 Mbps (10BT and 100BT systems). However, their capabilities and costs are continuously challenged by the emerging fiber optic networks discussed in Chapter 15 and 17. Bringing the fiber to home has become a serious concern for many telephone operating companies. Bringing the fiber to desks has become a serious concern for most designers of premises distribution systems.

Coaxial cables are also used for distributing analog cable TV signals. The societal demand for carrying data has been discouragingly low. However, the potential of these CATV systems for carrying high capacity digital information exists and needs serious exploration. Interfaced to intelligent networks, they can distribute highly customized data to a large number of subscribers bringing in new services generally not available in the cable TV industry. The FM hierarchy of the CATV network (see Chapter 3) that permits the transfer of small quantities of upstream data to control large streams of downstream data makes these networks potential contenders for very personalized and individual intelligent video network services.

In the telephone environment, coaxial cable is frequently used to distribute high rate information/data (ranging between a few analog channels to many hundreds of encoded speech channels) between the Central Offices. They are also used to provide access between Central offices and remote terminals. Such remote terminals (RTs) act as miniaturized Central Offices that concentrate and distribute the voice channels to and from a digital carrier system such as DS-1 or DS-3. In fact, these coaxial cables also served the needs of the long-haul trunk facilities to cover distances over 250 miles in the forties and fifties.

Coaxial cables for interoffice toll facilities have been used in analog carrier systems extensively. Frequency modulation aspects of such coaxial carrier systems are discussed in Sections 2.1 and 2.5. Here the physical and transmission aspects of coaxial cables while carrying FM and data signals are presented. The physical design of the coaxial cables can vary significantly. However, the serrated seam coaxial unit was the de facto standard for these cables. The inside conductor is typically solid 0.1003-inch axial copper centered within 0.369-inch (internal diameter) copper tube. Polyethylene insulating disks with one inch spacing hold the inside conductor in the tube. The outer tube is formed with a 0.012-inch copper strip. Steel reinforcement is also used to hold the outer copper tube intact. Such designs can be altered provided the four primary electrical constants (R, L, G, and C defined as Resistance in Ohms, Inductance in Henries, Conductance in mhos, and C in Farads; all designated per unit length) of the composite structure is held within the tight tolerance. Consistency of the transmission characteristics of the toll trunk lines carrying interoffice data can thus be guaranteed. With suitable design of the coaxial structure, two secondary constants (the characteristic impedance and the propagation constant, derived from the four primary constants) can be tuned to the carrier requirements. For typical coaxial cables, the first constant α known as the attenuation in dB/mile is roughly proportional to the square root of the frequency. The second constant β in radians per mile is nearly proportional to the frequency. For the third and fourth constants, DELAY and Z_0 are almost insensitive to frequency. All these four constants, which can indeed be derived from the four primary constants (see Chapter 7, Section 7.4) are tabulated for the serrated-seam coaxial units in Table 1-1.

TABLE 1-1 Transmission Characteristic Constants of a Typical 0.369" Cable

Frequency in MHz	α dB/mile	β Rad/mi.	DELAY μs/mi.	Z_o Ohms
0.1	1.217	3.660	5.83	77.5
1.0	3.845	35.70	5.69	75.5
10.0	2.15	54.0	5.64	74.9
100.	8.68	533.	5.62	74.7

At higher frequencies (above 1 MHz) the coaxial cables offer significantly lower loss. For example the 0.375 inch coaxial cable displays a 3.9 dB loss per mile compared to the 18.3 dB loss of a 19 AWG, or the 24 dB of the 22 AWG, or the 32.6 dB of the 24 AWG or the 41.2 dB of the 26 AWG wire-pairs at 1 MHz. A corresponding set of secondary characteristics for the 24 AWG cable at 70 degrees F is shown in Table 1-2. The frequency range is restricted because wire-pairs are generally not used at higher frequencies except in the possible premises distribution systems.

The attenuation and dispersion characteristics differ radically for the two media. The spectral band of possible usage is also significantly different. For this reason, the pre-fiber era used the coaxial medium for trunk routes between Central Offices where the carrier band spectral characteristics are most emphasized.

TABLE 1-2 Transmission Characteristic Constants of a Typical 24 AWG Wire-Pair

Frequency in MHz	α dB/mile	β Rad/mi.	DELAY μs/mi.	Z_o Ohms
0.1	12.235	5.710	9.088	112.79
1.0	32.639	51.80	8.244	99.59
2.0	46.267	101.41	8.072	97.36
3.0	56.753	150.60	7.989	96.31
5.0	73.405	248.39	7.906	96.31

In the distribution of telephone services, the older systems used the wire-pair medium for subscriber plants where voiceband spectral characteristics are most emphasized. The United States loop plant has some primitive design rules (see Section 6.3.1) incoporated within it. For higher speed data, these design rules have no significance.

1.5

CARRIER CONCEPTS

When high communication traffic is encountered, the use of individual physical medium for each path is futile if not impossible. High volume traffic is generally handled by high bandwidth media. Numerous communication paths have logical identifiers but are merged unto one physical or a larger logical entity. This principle of being able to freely merge (multiplex), travel long(er) distances, and be separated at the far-end by the logical identity (demultiplex) gives rise to the concept of carrier systems. Communication paths are thus carried through the telephone network by the appropriate carriers whose electrical, electromagnetic or optical properties are matched to the medium and the distance.

Electrical properties (typically the primary constants i.e., the R, L, G and C, discussed in Chapter 7) are important to twisted wire-pairs, coaxial cables and other metallic media that carry voltage/current signals. Electromagnetic (EM) wave propagation properties are important to wave-guides. Open or free space media carry EM signals. Optical properties are important in all types of fiber carrying photonic information. Specially designed media, such as coaxial cables, though metallic in nature use the wave guide structure and thus have both the electromagnetic and electrostatic fields within the inner coaxial structure. Twinaxial TV cables also display an interesting blend of electrical and EM properties. The components that make the overall carrier system feasible are optimally designed for the purpose of dependably accomplishing the signal transfer function from an electrical, EM, or optical consideration. The mechanism of all the (electrical, EM or photonic) energy transport from sender to the receiver is formulated by the generalized Maxwell's equations. Carriers in the existing networks can be analog and/or digital.

1.5.1 ANALOG CARRIERS

Analog carriers have a carrier signal or tone of optimally chosen characteristics. These characteristics are modulated from their predefined value according to the signal that they are carrying. The extents of deviation (after correcting for the characteristics of the medium) at the received end indicate the signal modulated upon the carrier. Analog carriers can operate in electrical wires, coaxial cables, terrestrial space via microwave radio, or outer space via satellites. The characteristics modulated could be amplitude, frequency or phase of the carrier even though certain media are matched to the modulation techniques to optimally exploit the technology.

In the United States, the telephone network analog carriers have been used as far back as 1925 to carry four voice circuits over one open wire-pair (C-Carrier System) by modulating the amplitude of the carrier and thus using single sideband amplitude modulation (SSBAM; discussed further in Chapter 2) technique. Wire-pair carriers have been developed and deployed in the fifties and sixties extensively. Accurate equalization and temperature compensation (in the analog sense) permitted these types of carriers for both short and long-haul of voice channels. The limit of open wire-pairs/cabled pairs is already reached with a very few (about 12) voice channels for the long-haul (250-4000 miles) systems. For the short-haul systems (under 250 miles) the open wire-pair systems provide 16 channels with double sideband transmitted carrier (DSBTC) modulation and cabled pairs can provide up to 24 channels with SSBAM modulation.

Coaxial cables (0.375" diameter), also using SSBAM modulation techniques, have been used to carry from 600 (L1 system) to 13200 (L5 system) voice circuits unidirectionally for the long-haul systems. For the short-haul systems (under 250 miles), the N carrier system provides 24 channels. Other systems also exist for lower (50 to 4000 channels) capacity voice trunk applications.

Over the terrestrial free space frequency modulation (FM) technique with 6 GHz radio carrier (TM system) yields the capacity to carry 1800 voice circuits over one channel for the long-haul facilities. For the short-haul systems (under 250 miles) the TJ (11 GHz) and TM (6 GHz) carrier system provides 600 to 1800 voice channels. Frequency modulation is generally used since the carrier frequency is much higher than the carrier frequencies of copper-based carrier systems.

Over the satellite link (see Section 2.6), the carrier is also at 6 GHz and capable of carrying 1500 voice circuits on each channel with FM modulation. Typically, the 4 and 6 GHz radio satellite route has 12 two-way channels. In a generic framework, each voice circuit has a logical address and can be coupled to a subscriber line to complete the connection throughout the network. Numerous other systems prevail with variations in different countries. In the United States, as many as 28 analog carrier systems using different modulation techniques (SSBAM, DSBTC, and FM; see Chapter 2) have prevailed until a few years back when fiber made serious inroads into the emerging all-digital telecommunication networks. Some of these systems (such as satellite and remote location radio systems) are here to stay, since the equipment cannot be economically replaced. Numerous capacities and systems are feasible for the analog carrier systems deploying wire-pairs. These systems and their capacities differ from vendor to vendor. The regulations on the use of microwave frequency space is an important consideration in the use of radio and satellite carrier systems.

1.5.2 DIGITAL CARRIERS

Digital carrier systems carry the information in the digital format. These systems carry binary data as raw pulses (with appropriate line code) or as symbols (i.e., bits grouped as discrete in-phase and quadrature components of a carrier). In the digital carrier systems, many lower speed digital channels are multiplexed onto one system and the resulting high-speed digital channel is used to carry transport data from one place to another. This data may also be used to modulate very high-speed carrier systems. In this case the digits of a digital carrier system are used to modulate analog radio signals, thus giving rise to digital radio systems (e.g., now outdated 11 GHz digital radio links in the United States). Fortunately the digital hierarchy in most of the telephone environments around the world is well established. Three such hierarchies are well documented as the PCM hierarchies and are consistent with the 64 kbps speech encoding (using the μ- or the A- law for companding). These hierarchical relations are tabulated for the United States, Canada, the United Kingdom, (Table 1-3); for CEPT or Conference of European Postal and Telecommunications Administration (Table 1-4); and for Japan (Table 1-5).

Digital carriers in conventional telephone networks can have highly variable capacities. Typically these carriers can range between 1.544 Mbps (DS-1 rate in the United States) to 274.176 Mbps (DS-4 rate in the United States). In Table 1-6, the specifications for the T carrier digital transmission system are presented. The applications, medium, and the line code generally found are presented.

At higher optical rates, such carrier systems are been envisioned to take advantage of the 2 to 8 (or higher) gigabit-per-second capacity of optical fiber, discussed further in Chapters 15, 16 and 17. Digital systems operate under the time division multiplex concept. Bytes of data from numerous users are interleaved to make up one higher rate into a more densely packed data stream. This stream is then communicated from one city to the next, from one region to the next, or one country to the next. At the destination, the carrier is unpacked and distributed to the individual recipients. The logical address identifies the channel and the physical address identifies the recipient.

TABLE 1-3 Digital Hierarchy in the United States, Canada, and the United Kingdom

Level 0: DS-0	Level 1: DS-1	Level 2: DS-2	Level 3: DS-3	Level 4: DS-4	Level 5: DS-5
64 kbps X 1* x **	1.544 Mbps X 24 x 24	6.312 Mbps X 96 x 4	44.736 Mbps X 672 x 7	274.176 Mbps X 4032 x 6	560.16 Mbps X 8064 x 2

* indicates the number of 64 kbps channels at each level.
** indicates the number of previous level channel at each level.
Note: The older T1C & T1D systems operate at 3.152 Mbps;
T1G systems operate at 6.443 Mbps in the United States.

TABLE 1-4 Digital Hierarchy in European (CEPT) Systems

Level 0:	Level 1:	Level 2:	Level 3:	Level 4:	Level 5:
64 kbps X 1* x**	2.048 Mbps X 30 x 30	8.448 Mbps X 120 x 4	34.368 Mbps X 480 x 4	139.264 Mbps X 1920 x 4	565.148 Mbps X 7680 x 4

*   indicates the number of 64 kbps channels at each level.
** indicates the number of previous level channel at each level.

TABLE 1-5 Digital Hierarchy in Japan

Level 0:	Level 1:	Level 2:	Level 3:	Level 4:	Level 5:
64 kbps X 1 * x **	1.544 Mbps X 24 x 24	6.312 Mbps X 96 x 4	32.064 Mbps X 672 x 5	97.728 Mbps X 4032 x 3	397.2 Mbps X 5780 x 4

*   indicates the number of 64 kbps channels at each level.
** indicates the number of previous level channel at each level

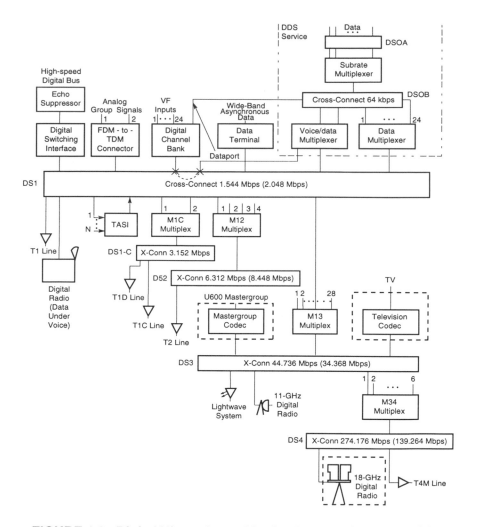

FIGURE 1.3. Digital Hierarchy and Its Implementation in Pre-SONET, Pre-Divestiture Bell System. This arrangement is modified by the worldwide acceptance of SONET hierarchy in the ATM switch. Older switches have additional units to interface with the evolving optical networks. ATM - Asynchronous Transfer Mode; DS-*n* - A Formatted Digital Signal; DS0A - Digital Signal of a Specified Rate (64 kbs); DS0B - Digital Signal of a Specified Rate (64 kbs); FDM - Frequency-Division Multiplexing; M - Multiplexer; N - Number; SONET - Synchronous Optical Network; TASI - Time Assignment Speech Interpolation; TDM - Time-Division Multiplexing; T*n* - Standard North American Carrier Lines Used to Transmit a Formatted Digital Signal; VF - Voice Frequency

The implementation of the digital hierarchy within the pre-divestiture Bell System (the United States) is depicted in Figure 1.3. Additional information on the type of systems (such as type of digital radio or the light-wave systems) that deploy the various carrier systems is also provided. At the DS1 cross connect, bus access to other systems (such as other digital channel banks, trunk lines, T1 lines, etc.) is also shown.

TABLE 1-6 Specifications for the North American Digital Hierarchy

Level 0: DS-0	Level 1: DS-1	Level 2: DS-2	Level 3: DS-3	Level 4: DS-4	Level 5: DS-5
1 PCM	24 PCM	96 PCM	672/ 1-TV	4032/ 6-TV	8064/ 12-TV
Wire-Pair	Wire-Pair	Wire-Pair	Coaxial	Coaxial	Coaxial
2 -level Binary	Bipolar RZ	B6ZS RZ	B3SZ RZ	Polar NRZ	Polar NRZ
--	1-mile Repeater	2.3-mile Repeater	within office	1-mile Repeater	1-mile Repeat
--	50 mile max.	200 mi max.	--	500 mi max.	500 mi max.
--	10^{-6} BER	10^{-7} BER	---	10^{-6} BER	4×10^{-7} BER

Note: T1C uses bipolar RZ, T1D uses duobinary and T1G uses 4-level NRZ code. The bit error rate (BER) is 10^{-6} for all the intermediate rates.

This architectural integration of the digital hierarchy, the hardware of the switch, and the transmission facilities prevails within most of the existing Central Offices. The practice of using T1C and T1D carriers at 3.152 Mbps is also shown by combining two of the DS1 channels. The intermediate level between levels 1 and 2 is sometimes used when the traffic demands do not justify the use of the T2 line at 6.312 Mbps. The ATM technology has recently brought about newer interfaces and standards for interfacing with the optical networks and ATM switches. Such architectures for taking the traditional DS2 and DS3 channels and absorbing them within the ATM framework are presented in Chapter 16. Both the signaling and data traffic aspects have been addressed.

Both analog and digital systems can operate as baseband systems or broadband systems. In the baseband systems, the information bearing signal is directly transmitted into the communication channel. In the broadband system a carrier is modulated and it is the carrier that flows through the communication medium. The capacity of both these baseband and broadband systems can vary dramatically.

1.5.3 TWISTED WIRE-PAIR DIGITAL CARRIERS

Rates in the DS-1, DS-2 (or E1, E2) range can be well supported by twisted wire-pairs in numerous ways. Subscriber loop carrier (SLC) such as SLC-40® were implemented as far back as the seventies in the United States. This system simply used the T1 line and used lower bit rate (37.6 kbps) to accommodate as many as 40 (rather than 24 channels at 64 kbps) voice channels in remote areas. Signaling and remote switching have been an integral part of the SLC40 system. The channel units are self-contained. This permits the operation, maintenance and fault detection on individual lines to reach the 40 subscribers. Also, these channel units use a specially designed (but non-standardized speech encoder using 4 bit sequential companding) codec.

With the SLC-96® systems for 96 voice channels, multiplicity of wires is used and the system is feasible with some flexibility and growth potential. In these systems, the encoders conform to the standard PCM speech encoding algorithms. A fiber version of the SLC-96 system has been designed. From the current perspective, the capacity of such systems is towards accommodating many thousands, or even tens of thousands, of voice channels (the current SLC-2000® systems) on a single or dual (up to six) fiber in each direction. These systems, though not popular in all the countries around the world, still hold an enormous potential.

Subscriber multiplexers in the European environment also serve an important role in carrying the 64 kbps PCM encoded speech for up to 30 subscribers over the E1 carrier. The Central Office units for a typical Swedish environment connects the remote subscriber multiplexer (RSM) to the Central Office subscriber multiplexer (CSM) over a paired cable (4 wire systems), an optical cable or a radio link. These systems have been in use since the mid 1980s in the Swedish Telecommunication Administration. Ericsson's AXE exchange usually serves as a digital office via a 2 Mbps interface. When the analog Central Offices are involved, then a CSM unit with its own D/A converters is placed between the carrier system and the exchange. Typically,

60 line rural exchanges can be served with 8-14 circuits and the lines can be better utilized for increased demand for voice/data services.

In the twisted wire-pair arena, HDSL and ADSL technologies have also been considered. Even though the concepts and coding methodologies have evolved in the HDSL and ADSL technology, they are still applicable in the loop plant carrier system. The fundamental difference is that HDSL and ADSL are repeaterless systems. The digital loop carriers may be repeatered, thus increasing the range and application. In the duplex system, a single wire-pair (HDSL and ADSL) carries the data at a higher rate using echo reduction and echo cancellation techniques. In the simplex systems (loop carriers), the directionality is isolated in the two physical wire-pairs, thus eliminating the need for the expensive echo reduction and cancellation circuitry from the system. In most of the application, the rate is confined to T1 or E1 rates. When the rate is even higher for the duplex system, the bidirectional data rate is halved, but two wire-pairs are used to reach a longer range of data communication. In the dual simplex system, four wire-pairs are used in half-rate simplex data. Different costs and blends of serving areas are possible by using these systems. The fiber-optic subscriber loop carrier (FSLC-96® and SLC-2000®) are logical extensions of the older copper-based loop carrier systems.

1.6

CENTRAL OFFICE HIERARCHY

Trunks and Central Office carriers have varying capacities to match their individual traffic patterns. Every Central Office is not provided with the same capability, instead its capacity is tailored to its localized needs. A hierarchy thus results with the End Office at the lowest level and a Regional Office at the highest level. Switching capacity and terminating trunk capacities vary. Additional trunking between any two levels is freely interspersed to accommodate growing communication needs of various geographical areas.

Fortunately, the hierarchical format of the Central Offices is well structured and is well documented in most national networks. In Figure 1.4, a five level Central Office hierarchy is depicted. The Central Offices' organization, patterns of growth, and trunks have been intertwined in the systematized evolution of the telephone network in the United States. For this reason the types of Central Offices/Centers (1.6.1 to 1.6.5 and 1.7) and the trunks for the interconnecting facilities (1.8) are introduced in the next few sections. In other countries, the Central Office hierarchy does not reach as high. For example, in most of the Middle Eastern, South American, and Pacific Rim countries, two-level hierarchy exists and suffices to serve the current needs of those socio-economic environments.

1.6.1 LOCAL CENTRAL OFFICES OR END OFFICE

Local Central Office (LCO, sometimes referred as the End Office or EO) communicates information with the subscribers over the customer loops. These loops (referred to as the subscriber loops in the United States) generally consist of twisted copper wire-pairs of a single or multisection between the LCO and the customer (See Section 2.1).

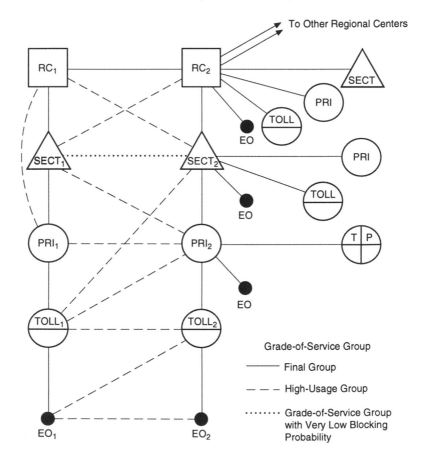

FIGURE 1.4. The Central Office Hierarchy and the Trunk Routing Patterns in Major National Telecommunication Networks. EO - End Office; PRI - Primary Center; RC - Regional Center; SECT - Sectional Center; TOLL - Toll Center; T/P - Toll Point

Each loop carries a permanent identification number associated with it. These loops may terminate at the customer premises with an FCC-approved digital and/or voice interface or an instrument such as a modem or a telephone. They may also terminate in a private branch exchange (PBX). The private exchange then forwards or transmits the information into the appropriate link. The LCOs perform the function of switching within the local area they serve. Concentration is used here because a fractional percentage of the customers are actively using their terminal facilities at any given time. The LCOs are also connected together with interoffice trunks over which information generally is multiplexed and transmitted at higher speed, if high volume is encountered. LCOs may also be connected with others through Tandem Offices via tandem trunks. These trunks may operate by repeated voice frequency systems or by any other type of multiplexed carrier system.

1.6.2 TOLL CENTERS

The next higher level in the switching hierarchy consists of the Toll Centers (TC) or Toll Office. Switching appropriate to the intertoll trunks takes place here. The trunks connect to the other Toll Offices or other End Offices. These trunks employ four wire transmission rather than the two wire transmission between the customer and the End Office. Concentration may also be employed because only a smaller fraction of the lines need to be connected outside the toll area. Intertoll office communication generally occurs over a multiplexed high capacity carrier system, with severe requirements on the extent of degradation that can be tolerated over these trunks.

1.6.3 PRIMARY CENTERS

These centers provide the third level of hierarchy providing communication of information between various TCs and other offices higher up in the hierarchy. These centers communicate with each other over high capacity carrier systems. Coaxial cables and microwave facilities are generally used for these interconnecting facilities between the primary offices. The requirements on these facilities are stringent. The distance between the centers, the type of trunk facility already connecting similar or neighboring areas, the type of terrain, the trunk capacity requirements play an important part in the choice of the carrier system between the primary centers. The physical separation between the centers is generally large, but the ultimate capacity to communicate to and from the primary center dictates the switching facility used within the primary centers.

1.6.4 SECTIONAL CENTERS

At the top of the switching center hierarchy, within the confines of section designated by the network boundaries, reside the Sectional Centers. These centers are distributed throughout the region they serve and have the facility to interconnect different sections with high capacity specialized trunk lines (mostly fiber) as indicated in Figure 1.4. This grade of service trunks carries an enormous amount of traffic. They are used to limit the hierarchical routing of sectional calls and to prevent sectional calls reaching Regional Centers. Typically, they are the final routing chain and designed for high usage with little/very little blocking probability.

1.6.5 REGIONAL CENTERS

Numerous regions span a country. Regional Centers become an integral part of the telephone network. Very high capacity carriers are necessary if these centers span large population centers. The distance between these centers determines the type of carrier system used. Radio, L-series cable, satellite, fiber optic facilities are generally used at the higher levels of inter-center communication facilities.

1.7

NODES WITHIN THE NETWORK

From a topographical consideration, the five switching centers (presented in Section 1.6) with numerous links connecting them, constitute a traditional regional telephone network. A large number of network patterns exist in a nation. The particular pattern that evolves in a geographical area is governed by the need of services in that area, the location, and the trunking facilities around the population center(s), the state of technology and the forecasted need for services. In a typical telephone organization, the eventual justification is still based upon the laws of microeconomics governing the expenditure of capital to establish a center, a trunk, a link, or the deployment of a new or an additional carrier facility. The integration of the facilities into one well construed long range plan for the nation is being constantly challenged by the recent deregulation of the telecommunication industry and the need to remain solvent in the near future.

The mode designation for the Central Offices also depends upon the traffic pattern. The LCOs generally have the function of tapping into the network for receiving and transmitting. Through-traffic is generally curtailed except in very small LCOs. The Tandem Office allows for through traffic and generally curtails the tapping function. These functions may be permanently blocked or programmable depending upon the flexibility of the particular office. Hence the switching centers have their own characteristics associated with the particular type of technology used in switching. At the low end of the flexibility and functional spectrum is the step-by-step mode (invented 1889, deployed 1919). At the high end of the spectrum, the gigantic Electronic Switching System is in an active phase of evolution. The machines incorporate the large computer design and fabrication procedures to control and switch the information carrying channels of the customers either collectively or as a single information carrying entity. Being strictly program controlled, the microcode, the firmware and the software control their functions. Flexibility in accommodating new technologies are an inherent requirements for these newer types of switching centers. The compatibility issues are discussed in Chapters 16-18 of the companion volume *"Intelligent Broadband Multimedia Networks"*, for fiber, and in Chapter 5 for wireless and VSAT. The interfacing issues of the new switches are discussed in Chapter 5 of the last reference for PCS mobile radio, in Section 2.6 for satellite, in Chapter 11 for ISDN; and in Chapter 16 for ATM and SONET.

Within the range of the two extremes of the functional flexibility, there are a large number of other switching centers in various national environments. However, the major vendors of these switching systems are trying to cooperate in making these nodes (switching centers) standard and replaceable.

In the North American network environment, the LCOs have employed the electromechanical system in small numbers for as many as 40,000 lines per Central Office. These systems using a mechanical ratchet was slow, error prone, expensive and have become obsolete. The panel system (early 1900s) used a flat layout of the contacts rather than a vertical and circular layout of the step-by-step system. Electric motors was used to drive the contact arm across the panels. This system could access larger number of terminals, but suffered from the same restrictions as the earlier system. The three crossbar systems were popular up to the early seventies. These electromechanical systems used series of horizontal and vertical bars (crossbars). Electromagnets between these bars activated by control signals establish contact for the duration of the call. These systems were gradually phased out by the introduction of the next generation of Electronic Switching Systems first introduced in 1965. In the 1990s, it is inconceivable that any switching system can be other than an electronic switching system, if not a photonic switching system.

1.8

LINKS WITHIN THE NETWORK

The interconnect rules within the network are very flexible. For example, any two centers in the top four (i.e., up to the Regional Office in Figure 1.4) of the hierarchical centers may be interconnected. The capacity of the links or the interconnect facility is highly variable. The capacity ranges from a few channels to DS4 rate of 274 Mbps over fiber optic links or the 140 Mbps optical fiber link currently being deployed in the network. This depends upon the type of physical layer (the type of media over which transmission occurs). As discussed earlier, the state of technology and the laws of economics determine the actual format of the network segment in any particular environment. More recently the move towards SONET OC-n rates is being investigated more and more seriously in view of the eminent ATM networks (see Chapter 16).

Newer subscriber links are becoming digital even at relatively low rates (28.8 kbps modems, 9.6 kbps videos and faxes, etc.). The evolution towards being all-digital is a well stated policy of most major national communication networks. This trend exists even within the subscriber loop environment with the proposed basic rate ISDN (144 kbps) and its precursor the dialed digital service (DDS) at 56 kbps available in the United States. There has been a conscious effort to make most telephone environments all-digital since the eighties, and to make most voice/data transport systems all-fiber during the nineties. The new trunk lines are generally all-fiber, all-digital and well-integrated into massive information grids to serve the medical, educational and the personalized needs of the society. Information highways are mere multimedia pathways within such information grids.

1.9

NEW ARCHITECTURES OF END AND INTERMEDIATE OFFICES

The newer technologies, namely wireless (Chapter 5 of the companion volume *"Intelligent Broadband Multimedia Networks"*), satellite (Section 2.6), high-speed twisted wire-pairs (Chapters 12 and 14), and fiber (Chapters 15 and 16), impacting upon the plain old telephone communication system, are altering its fundamental structure in *four* distinct ways discussed next.

First, consider wireless and the End Office systems. The new Central Offices now have to interface with the cellular switching systems and cooperate in call completion from all authorized mobile units. The switching information at the switching offices that traditionally arrives via twisted wire-pairs from individual subscribers now arrives over a wired/wireless medium from the cell sites or from mobile telephone switching offices (MTSOs, see companion volume *"Intelligent Broadband Multimedia Networks"*, Section 5.5). The traditional hardware interface and the OA&M (operations, administration and maintenance) software of these conventional End Offices gets substantially modified. Intelligent network services may also triggered from these MTSOs for the mobile roamers.

The typical enhancements to an electronic switching system to accommodate the digital cell sites is shown in Figure 1.5. A specific unit (personal communications services center) with its own hardware (i.e., the digital cellular switch, DCS), and its software module (PCS access manager), is incorporated to handle the voice traffic from the roamers in wireless contact with the radio ports. The DCS permits access to the public switch telephone network (PSTN) for the voice channel. Signaling and control information is handled directly by the SS7 network via the PCS access manager. OA&M information is handled by the administrative module of the switch. This architecture is suited to communications to and from high mobility fast moving traffic. For PCS dealing with low mobility traffic, a radio port control unit (RPCU) acts as an intermediary between radio ports and the ESS.

The digital format of the output data stream from RPCU is in the basic rate ISDN (see companion volume *"Intelligent Broadband Multimedia Networks,"* Chapter 7) format. In this particular case, the PCS center of Figure 1.5 is not necessary, and the switch itself provides data path to PSTN and signaling information to and from the network applications part of the ESS facility.

For handling message switching, a similar architecture is used and a more global framework is depicted in Figure 1.6. Three network layers, consisting of mobile terminals, access network, the existing backbone network, are identified. Typically, the access devices include mobile data terminals for telemetry, vehicle mounted data terminals, vehicle tagging devices, and portable laptops. Wireless local area networks (LANs), PCS and

cellular networks can thus be interfaced on to the backbone transport
network via the access network.

PCS CENTER

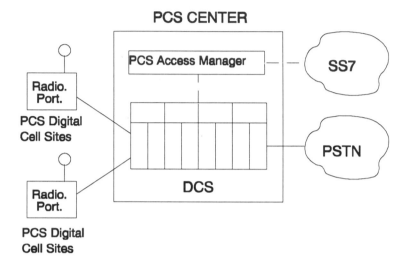

**FIGURE 1.5. Proposed Enhancement to the Switching Systems to
Accommodate the Growing PCS Networks and Mobile Units.** DCS -
Digital Cellular Switch; PCS - Personal Communication Services; PSTN -
Public Switched Telephone Network; SS7 - Signaling System 7

Second, consider the satellite based paging systems and the End Offices.
The paging information arrives as a packet from the customers or a
centralized paging dispatch center. The Central Office has to tie into the
remote location via its circuit-switched, packet- or message-switched
systems, then forward its packet with the caller identification. The proposed
low Earth orbit satellite provide excellent means to interconnect paging and
other value-added network customers. Typically, the access is provided via a
VSAT ground station in the switching systems of the PSTN and the access is
completed via the backbone transport network of Figure 1.6.

Third, consider the digital lines (DSL, HDSL, ADSL, or even ISDN
services). These lines now carry switching and service protocol in addition to
customer data. The End Office interface hardware and its processing
capability to separate protocol, control (bytes or even bits) and data bytes
becomes essential. This calls for new improvised channel units and new
software or firmware to perform specific channel functions. Data and signals
on these lines are both structured and standardized by the ITU-T (formerly

CCITT) committees. The access provisioning is handled by the ESS facility as it conventionally handles a BRISDN line for DSL application, a T1/E1 line for a high-speed DSL (HDSL). Channel banks (discussed in Section 4.5) in the Central Offices are designed to respond to these digital streams and provide localized switching and initiate additional signaling if the request for other network services are warranted.

A typical overview of a modern switch with digital networking units (DNUs) for DS3, STS-1, OC-3, OC 12, DS1, and the ISDN-type of Central Office terminations is shown in Figure 1.7 These DNUs are integrated with the access interface units (AIUs) to provide communication pathways to existing switching modules. Global signaling to other network elements is provided via the communications module.

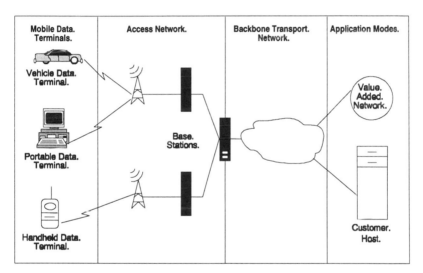

FIGURE 1.6. Provisioning of Access for Wireless Messaging Devices and Networks, such as Data Terminal Devices, Wireless LANs and Laptops into the Backbone Network with Electronic Switching Systems for Access into the Switch. (Also see Figure 1.5).

Further, consider the copper-based digital subscriber carrier systems (SLC) and the End Offices. These lines tie the Central Office to the remote terminals. They carry switching, OA&M, and customer data all on single or shared twisted wire-pairs (for T1, E1 carrier services) and optical fibers (for fiber subscriber loop carriers FSLC in the multiple T1/T2/T3, E1/E2/E3 systems). In a sense, the channel banks (interface, local software and ROMs) in the Central Offices have to be enhanced to perform the additional SLC and

FSLC functions. The net effect of the SLC systems is that the End Office to which these systems are connected, has generated a still lower level of a switching facility. It is an unattended mini-Central Office (called the "hut"), the remote terminal (RT) or collection of such terminals with their own localized trunks and distribution systems. The architecture of the subscriber plant is thus altered.

Fourth, consider the Central Office and SONET interface. The fiber media, which has already made dramatic inroads into the trunk facilities of a POTS network, is now connecting computer centers, business complexes, universities, hospitals, etc., and the Central Office at OC1 through OC192 rates for asynchronous transfer mode (ATM) of data communication. A typical scenario for the network evolution is shown in Figure 1.8.

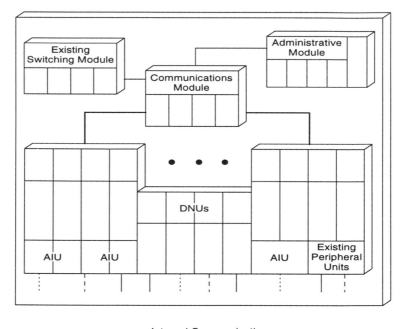

———————— Internal Communications
· · · · · · · · · · DS3, STS-1, OC3, OC12
———————— DS1
– – – – – – DS0, ISDN, Voice Frequency

FIGURE 1.7. Central Office Termination within Typical Modern ESS Providing Access and Signaling Facility via Digital Networking Units (DNUs) and Access Interface Units (AIUs). DS-*n* - A Formatted Digital Signal; ISDN - Integrated Services Digital Network; OC*n* - Optical Carrier; STS - Synchronous Transport System

The ATM hub/router receives data in the clear channel/ATM format from private branch exchanges (PBX), from front end processors (FEP) in the services network architecture/frame relay format, and from the brouter in the frame relay/ATM format. The output of the ATM hub in the DS3/OC3 format, is integrated into backbone network. Access to ATM backbone, software defined network (SDN), software defined data network (SDDN), ISDN or the private dedicated line is provided via the circuit switched capability of the backbone network. Packet switching, frame relay and ATM services are provided via the emerging ATM backbone. In this configuration, the network will provide separate access to data and voice networks, and also provide access to frame relay (FR) networks and to the emerging ATM backbone. Internetworking between X.25 to FR, FR to cell relay, and private line to cell relay is also accomplished at the ATM hub.

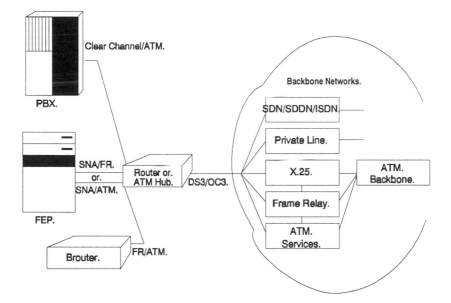

FIGURE 1.8. The Evolving Integration between the Existing SDN/SDDN/ISDN, Private Lines and Frame Relay/Cell Relay ATM Data Networks via the ATM Hub. ATM - Asynchronous Transfer Mode; DS-*n* - A Formatted Digital Signal; FEP - Front End Processors; FR - Frame Relay; ISDN - Integrated Services Digital Network; OC*n* - Optical Carrier; PBX - Private Branch Exchange; SDDN - Software Defined Data Network; SDN - Switched Digital Network; SNA - System Network Architecture (IBM)

The protocol for the widely accepted SONET network and the structure of the data bytes (or the playload) is radically different from the protocol and data structure of any other digital carrier or data transport system. For this reason, the ATM switch (a new switch for handling calls to/from another switch) for switching of SONET frames, and consequently ATM cells, now interfaces with the new digital telephone networks of the twenty-first century. It appears futile to improvise the ESS facilities to function as an ATM switch, even though these ESS switches have to interface with the ATM switches and vice versa.

1.10

REFERENCES

1.1   L.L. Birch 1994, "Network Characterization," *Private Communication*, AT&T Bell Laboratories.

1.2   Bell Telephone Laboratories 1982, *Transmission Systems for Communications*, Fifth Edition, Bell Telephone Laboratories Inc.

1.3   AT&T 1985, *Telecommunication Systems Engineering*, second printing, New York, NY.

1.4   M.J. Miller, and S.V. Ahamed 1988, *Digital Transmission and Systems*, Vol. 2, Computer Science Press, Rockville, MD.

1.5   R.L. Wilson and J.J. Mac Namara 1994, "Cellular and PCS Digital-Mobile Services Switching Center," *AT&T Technical Journal*, No. 6, November/December 55-61.

1.6   K. Rattray 1994, "Wireless Messaging," *AT&T Technical Journal*, May-June, 14-20.

CHAPTER 2

FM TECHNIQUES IN EXISTING NETWORKS

Frequency and phase modulation are the two major subsets of angle modulation techniques for encoding signals onto carriers. The instantaneous phase of the carrier is equally influenced by both of these analog coding methodologies. If the modulated signal can be represented as

$$M(t) = A_c \cos [w_c t + \phi (t)] \qquad\qquad (2.1)$$

then the instantaneous frequency, defined as the first time derivative of the instantaneous phase, $[w_c t + \phi(t)]$, in radians, is $(w_c + \phi'(t))$ in radians per second. In the phase modulated (PM) systems $\phi(t)$ *is directly proportional to the signal* and in the frequency modulated (FM) systems $\phi'(t)$ *is directly proportional to the signal.* Whereas for the PM systems, the signal is recovered by demodulating the instantaneous phase, for the FM systems, the signal is recovered by demodulating the frequency deviation $\phi'(t)$. Certain inherent difference arise between the two systems and the FM techniques offer more practical advantages (discussed next).

2.1

GENERIC CONCEPTS

Frequency modulation techniques are not restricted to any one type of carrier signal. In the older networks, the FM techniques have implied analog systems, but at high enough carrier frequencies, they encompass digital systems such as digital radio systems. The carrier penetrates terrestrial space, outer space, and even cable wave guides. Microwave radio frequency modulation techniques are also used in the 4 and 6 GHz band for satellite communications. In the earlier systems, each band carries 1500 voice (range) circuits per channel with 12 channels per route. However, the media character of the satellite paths can render them unsuitable for the telephone type of two-way communication. The FM techniques are also particularly applicable to the radio systems, to coaxial cable environments, and sometimes to twisted wire-pairs for analog subscriber carrier systems (such as SLC-1® systems sometimes deployed in the United States during the seventies). In many applications, the analog FM systems are gradually yielding to the more robust and integrated digital carrier systems.

Frequency modulation techniques in the radio telephony environment are particularly appropriate because of the dependable signal bearing quality of the carrier. Signal attenuation and dispersion are both encountered. The frequency range, antenna design (for wave-guides) and media characteristics influence the frequency dependent loss curve causing attenuation and dispersion. Suitable equalization techniques thus become essential. The repeatering of the carrier is specially designed with directional receiver and broadcast antennas. Due to the wide range of possible applications of the signal (voice, data, image and video) and of radio communication systems that broadcast in the frequency space, well designated carrier systems has been allocated for various applications. Largely, the extent of frequency modulation of the carrier permits many thousands of voice frequency channels to be multiplexed on one carrier. For example, the 6 GHz radio (TH-3 system used for long-haul in the United States) band carries 2340 voice circuits per channel with 8 channels per route. Other carriers operate in the 4 GHz and 11 GHz bands.

2.2

MULTIPLEXING OF FM SIGNALS

Much like the digital hierarchy (discussed in Section 1.4.2) for multiplexing numerous lower speed data channels into higher capacity digital channels, there exists a similar frequency multiplexing hierarchy. Frequency modulated signals covering lower bandwidths can be thus multiplexed into four groups: basic Group, Supergroup, Mastergroup, Jumbogroup, and onto a carrier. Five stages of multiplexing become necessary. The basic FDM hierarchical plan is shown in Figure 2.1. Typically in the United States, this basic frequency division multiplexing (FDM) hierarchy permits 12 voiceband (4 kHz) circuits into one basic Group at one channel bank. Five such basic groups are frequency division multiplexed to one Supergroup at the Group bank. Ten Supergroups map onto one Mastergroup at the Supergroup bank and so on (see Figure 2.1). Preallocated numbers (such as 2, 2.5, 3, 4 and 6) of these Mastergroups can be combined for different carriers. For example, with 2, 2.5, or 3 Mastergroups, the TD radio carrier systems are built. With four Mastergroups the TH3 carrier systems are built, and with six Mastergroups a Jumbogroup is formed. Three Jumbogroups give rise to the L5 (10,800 voice channel) carrier system.

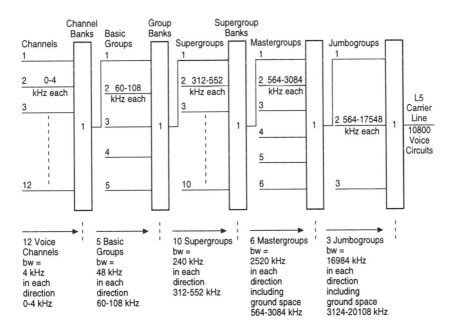

FIGURE 2.1. The Breakdown of the Analog L5 Carrier Systems into its Subgroups. The specific details of channel and guard space allocations are not shown. Also see Figures 2.5 to 2.10 and Tables 2-2 to 2-6. L5 - Very Wideband Analog Carrier System

Multiplexed groups in the FM hierarchy can be regrouped to yield other carrier systems. For example, when two Mastergroups are multiplexed onto one TD carrier, 1200 voice circuits can be carried, and when twenty-two Mastergroups are combined, the 13,200 voice circuit L5E carrier system results. The concept of frequency domain multiplexing has been used in these high bandwidth carrier systems, which serve as the major communication links. The basic FDM hierarchical plan in the United States is shown in Figure 2.1. Other plans that restructure the frequency space of carrier systems are also possible and can be used to advantage depending upon the telecommunication needs of different countries.

The flexibility that results from these FDM concepts and equipment is enormous. For example, the coaxial carrier system and the radio carrier systems can be intertwined along any short/long distance route; both perform according to the same signal-to-noise and transmission quality requirements. Thus, tracing a channel, a Group, Supergroup, Mastergroup, or Jumbogroup, from place to place (regardless of the carrier system that is being used) permits logical continuity of any telephone/communication channel. This concept of tagging built into analog carrier systems is greatly outstripped by the currently accepted concept of unique addressing of every logical digital channels in all digital systems.

One significant difference from the digital systems is that the spectral carriers cannot be packed as tightly as the digital carriers because of the imperfection of the filters that isolate the spectral bands. Counter-activated gates act as perfect cut off devices in the digital systems. Such devices cannot be as easily built with an analog system, because of the low level spill over of energy from one band to another at band edges. This problem grows acute in the higher frequency carrier systems (such as L5 systems discussed later) which have voiceband information channels at the band edge. Under such conditions, guard bands between high capacity groups are necessary to prevent the obliteration of the voice channels that populate the band edges. In addition, any non-linearity of the amplifiers used in repeatering the FM signals causes undesirable energy transfer from one band to another. This is a result of intermodulation, i.e., coupling of carrier systems via higher frequency signals generated in amplifiers or other decoupling devices. Carrier and signal energy get transferred (at low levels) from one system to another causing random noise (due to thermal noise of the systems, stray media signals, etc.) and signal correlated energy transfer (crosstalk due to varying signal strength on the other carrier bands). Thus the nature of signal distortion as it affects signal recovery and its reconstruction in the FM carrier systems is quite different from digital systems (discussed in Section 1.4.2)

In well designed systems, the extent of signal distortion is well within the range controlled by the carrier standards. Such distortions are not crucial in voice communications over short distances, but can become significant for video signals and for long distance communications. Hence some precaution is necessary in handling the media and carrier system characteristics in relation to the type of signal being carried.

2.3

HIGH FREQUENCY FM NETWORK LINKS

Broadly categorized, the major short-haul and long-haul (250 to 4000 miles) FM carriers were radio systems and coaxial cables systems. The generic impact of fiber on the communication networks is discussed in Chapter 15. Microwave radio systems for communication of speech dates back to the days of Marconi. Single channel radio communication networks have been long established for unidirectional transmission of wireless broadcasts. Television broadcast networks demand bandwidth and sophisticated encoding and synchronization. Initially used for black and white TV broadcasting, the carriers have subsequently been used for color TV networks. Typically, the major long-haul microwave TV links carry unidirectional information. These systems use the FM techniques effectively to carry the video information. Such networks have been firmly established in the communication environments of most countries. In essence, these specialized applications have called for specialized transceiver designs.

To enhance their bandwidth, these networks can only be exploited in context to the limitations that are inherent in their original design. For example, when voice telephony was expanding dramatically in the United States, the telephone system around the world (the United States, Canada, Australia, and Japan, etc.) permitted the use of bridged taps for party lines thus reducing the cost of a new line. In addition, some networks (the United States and Canada) also permitted loading coils for capacitive compensation of long loops. In the context of modern digital subscriber lines (DSL) discussed in Section 4.3, and asymmetric digital subscriber line discussed in Chapter 14, the taps are a major source of signal degradation and reflection of the digital signals. In addition, the loading coils completely choke off any longer distance (greater than 12,000 feet) high-speed digital signals, thus rendering them totally unsuitable for any very high-speed digital subscriber line applications based on any viable FM techniques. The loaded loops of the subscriber loop plant is thus out of bounds for any FM applications.

The conventional applications of various segments (such as subscriber loops, trunk, FM links, and CO networks, signaling and operations networks, private networks, etc.) of the communications networks as they exist now are consistent with their initial designs. It is still to be investigated to determine if the radio links (or any other segment, if it matters) that were designed for specific bands in the fifties and sixties have a latent potential as data link for yet a higher band (the 22, 28, 31, and 39 GHz yet to deployed) by re-engineering the transceivers. In the context of subscriber loops, such an opportunity remained dormant until the POTS lines of the thirties were re-engineered as the digital subscriber loops (DSL) of the eighties (see Reference 2.4) and then as high-speed digital subscriber loops (HDSL), and then as asymmetric digital subscriber lines (ADSL), and yet again as VDSL for very high-speed applications.

2.4

RADIO LINKS

Even though electromagnetic wave propagation equations are applicable to all media, and the media properties may be similar, the microwave radio networks are treated differently from the cable networks in the engineering of communication systems. For this reason, we differentiate between radio links and coaxial cable trunk links. We also differentiate between CATV networks and coaxial cable trunk networks. We reexamine the radio and cable carriers (for communication networks) in this section. We classify and review the CATV network (for distribution of TV signals to the home) in Section 2.5. Typically, the commercial TV broadcasts are dubbed over the communication carriers for long-haul, due to their very wide bandwidth and high analog capacities. The details of the 11 GHz digital radio, which is also an excellent carrier of video signals, is presented in Section 2.4.5.

There are *two* important aspects of radio carrier systems: *(a)* the carrier frequency and *(b)* the bandwidth for modulation. The carrier frequencies are selected and standardized by the terrestrial and outer space properties of the medium and to some extent by the terrain. These carrier frequencies have been well studied by the standards committees of individual countries and their properties documented in context to the location of the country on the globe, the influence of sun spot and solar activities, weather and storm patterns, and the terrain. These factors influence the carrier attenuation, dispersion and its fade, and thus influence its capacity to carry information dependably. The terrain influences the reflection of the signal and causes signal-dependent coherent echoes. Single and multiple reflections from self-signal cause most annoying picture and voice transmission. Signals caused by the reflection of neighboring band causes crosstalk. Coherent video and audible crosstalks both cause enormous signal degradation. Hence, the choice of carrier frequency in relation to its use and the terrain is a matter of serious deliberation, optimal matching and (perhaps) detailed simulation in the design of most radio systems.

The bandwidth of modulation directly influences the extent of variation of the carrier frequency and thus its ability to carry information. The factors influencing the bandwidth are based upon coding, amount of information and the extent of signal degradation that can be tolerated at band edges. The transceiver performances at the edges of the band becomes crucial. These devices have to exhibit uniform power spectrum and amplification. Linear delay distortion requirements also are stringent in order to minimize the signal dispersion at the receiver.

In the United States, the Federal Communications Commission (FCC) has classified the frequency space for carrier transmission at nominal frequencies of 2, 4, 6, 11, 18, 22, 28, 31, and 39 GHz. The spectral bands at lower frequencies of 2, 4, 6, 11, and 18 GHz are actively used for carrier systems studies. The upper ranges of carrier frequency (22-39 GHz) are yet to be explored for their possible applications. The bandwidth at each carrier that is actively modulated to carry information at each of the nominal frequencies is also established. The features of the carriers in each one of the bands are presented next.

2.4.1 THE 2 GHZ BAND

The 2 GHz common carrier system employs two 20 MHz bands ranging from 2.110 to 2.130 GHz and again from 2.160 to 2.180 GHz. These carrier systems have limited channel capacity for both analog and digital transmission. In addition, the larger intercity and metropolitan television distribution systems also share the upper segment of the 20 MHz frequency space. Limited digital signal transmission is feasible within the 20 MHz bands.

2.4.2 THE 4 GHZ BAND

The 4 GHz (referred to as TD-2 and TD-3 in the United States) system provides a wider band between 3.7 and 4.2 GHz for modulation with twelve 20 MHz bidirectional radio channels. The upper limit of each of the radio channels is 1500 message channels. A majority of these radio channels have the capability for 1200 channels. An intermediate frequency (IF) carrier of 70 MHz is used to modulate the radio frequency carrier between 3.710 GHz and

4.170 GHz (i.e., 12 two-way radio channels at 20 MHz spacing leading to a band spread of 0.460 GHz). Very accurate discrimination at the intermediate frequency and dual polarization are essential for 12 two-channels utilization. This keeps within the noise limits of long-haul systems. The TD-2 system precedes the TD-3 system even though the two systems are functionally the same. TD-3 system (introduced in the early sixties) uses solid-state technology whereas TD-2 (introduced in the late forties and been modified through the late sixties) uses electron tube technology. Both systems tend to have a noise performance of 43 dBrnc0* while carrying the ultimate capacity of 1500 message channels per each radio channel.

2.4.3 THE 6 GHZ BAND

The 6 GHz system operating between 5.925 to 6.425 GHz spanning the 500 MHz band encompasses eight bidirectional 30 MHz radio channels. The channel allocation has become standard for the American TH and TM systems using frequency modulation (FM) techniques. Each of the 30 MHz channels carry approximately 1800 message channels. The notable exception is the single sideband amplitude modulated AR 6A radio system all using heterodyne repeaters. The TM systems are designed to accommodate two types of repeater designs. The first type uses the standard heterodyne repeaters used in the TH systems. The second type uses the baseband remodulating repeaters used in the 1950s and 1960s in the 11 GHz TJ systems (next section) employing electron tube technology.

•   In the United States, dBrnc0 is chosen as a measure of noise power. This value is a measure of noise in any system relating to a 1000-Hz tone from noise-power at 10.0E-12 watts. This value of 10.0E-12 watts is arbitrary and corresponds to 10.0E-09 milliwatts or -90 dBm since 0 dBm corresponds to the power of 1 mV of signal.

However, there is a reference noise to consider. The -90 dBm level is referred to as dBrn to indicate the reference noise of the 1000-Hz signal at -90 dBm level. Any system noise expressed above this reference noise is called dBrn. There is one more consideration in the measure of noise as it relates to hearing capabilities of the human ear. From research, with respect to annoyance of noise tones at different frequencies, a weighting curve called the C-message curve has been deduced. This curve relates the power levels of signals at different tones that cause the same annoyance as the 1000-Hz tone. If the noise power of the system can now be measured with this C-message weighting curve, the noise level above the reference level is call dBrnc, to indicate the C-message weighting function.

Finally, the signal level is important in all measurements of power in dBm, since the measure is the ratio rather than the absolute noise power or the signal power. For this reason the noise measure is referred to as 0 dB of transmit level power or 0 TLP, thus the 0 is appended to dBrnc making the final noise measure to be dBrnc0. Thus, the 40 dBrnc0 means the noise measured at 0 TLP in the system is 40 dB higher than -90 dBm level, when the entire measured noise in the voice channel is C-message weighted.

For these FM systems the 70 MHz IF is located about halfway between the lower limit of 30 MHz (for most radio systems) [2.2], and the upper limit of about 100 MHz (to contend with excessive leakage effects in the repeaters). Compatibility with existing systems also makes the choice of the intermediate frequency critical. Only the TH-1 frequency allocation plan calls for a deviation of the 70 MHz IF to 74.1 MHz IF. All these systems are designed to operate in conjunction with the 6 GHz long-haul facilities with fairly tight performance requirements approaching 44-46 dBrnc0 noise performance for the worst case 4,000 mile TH-1 system carrying 11,160 message channels (i.e., six regular and two standby, 30 MHz radio channels each with 1860 voiceband circuits). The TH-3 systems use all solid-state technology except for the highly efficient low distortion traveling wave tube used as the final amplifier.

This particular system has 41 dBrnc0 noise performance with 1800 voiceband circuits. The more recent TH-3 systems carry up to 2400 voice channels on each of the 30 MHz radio channels. The 6 GHz band can also carry short-haul traffic under split frequency, staggered, to the co-channel plan. The TH and the TM systems are designed to carry television broadcasts.

2.4.4 THE 11 GHZ BAND

The 11 GHz system short-haul service has a radio frequency space ranging from 10.7 to 11.7 GHz. The older (now obsolete) TJ system using baseband remodulating repeaters carry 600 voice circuits in each of the six bidirectional 20 MHz radio frequency channels.

The TL system encompasses six radio frequency channels requiring 40 MHz band for each direction. Each channel can carry 1200 voice circuits. It can also carry one television channel. A more optimal utilization of the frequency space occurs in the solid-state TN-1 system for the 11 GHz band. The system can function in two modes, operating as a carrier of 12 two-way radio frequency channels each carrying 1800 voice circuits or as a carrier of 23 two-way radio frequency channels each carrying 1200 voice circuits. The TN system is repeated with heterodyne intermediate frequency of 70 MHz. Cross channel interference and interference into other existing carrier systems are the major bases for the selection of long-haul or short-haul systems with specialized frequency allocation basis. The system is also effectively employed to carry digital information to and from the 3A digital radio systems.

2.4.5 THE 18 GHZ BAND

The 18 GHz band spreads between 17.7 to 19.7 GHz. The 18A digital radio systems (now obsolete) carry digital information at the DS-4 rate of 274.176 Mbps over the allocated bandwidth. Eight two-way radio frequency channels occupy the 2.0 GHz band. Each channel carries one DS-4 rate signal and as many as seven channels may be in use with one standby (protection) channel. At the standard 64 kbps PCM encoding of each voice channel, 4,032 voice channels may be carried by each of the DS-4 channels. This makes each digital DS-4 channel capable of carrying 168 DS-1 rate digital channels at 1.544 Mbps.

2.4.6 DIGITAL MICROWAVE RADIO LINKS

The data services offered by networks have to also meet the more stringent service requirement for data transmission than for voice/speech transmission. When speech is encoded, transmitted and decoded, casual errors in the data handling through the network do not drastically degrade the reconstructed speech especially at the higher PCM bit rates. The intelligibility is not lost and the listener barely perceives the degradation. However, digital devices at the end of the network are not more susceptible to error. Hence, the established networks are obligated to offer digital services with a high dependability. Some of the major measures of the enhanced requirement on the digital services are: percentage error free second (typically 99.5%), percentage of service availability (typically 99.96%) and finally that service restoration be quick (typically under two hours).

Frequency shift-keying (i.e., frequency shift from one frequency to the next, for example, binary value of 0 goes to binary value of 1) has been deployed for digital radio carrier systems. The simplest application is the Morse code. When used in conjunction with message storage terminals simple message switched facilities may be built. Phase shift-keying shifts the phase of the carrier from one finite state to the next in a 0-360 degree phase plane. When both the amplitude and the phase of the carrier is shifted from one finite value to the next, the constellations in the X and Y coordinate space of the carrier are well placed. The simplest form has four states in which the amplitude is held constant and the carrier phase is shifted by 90 degrees. The four finite states are located on a unit circle with quadrant angular displacements. Sixteen point constellations may be generated by placing the discrete points equally spaced in the X and Y dimensions. Figure 2.2 depicts some of the possible constellations in the transmission of digital data. The 32 and 64 point constellation become more and more attractive in the new higher speed modem (V.bis and V.fast) ranging from 28.8 to 32 kbps and for the HDSL technology (see Chapter 12). Modulation efficiency and ease of decision threshold(s) become the secondary issues on the choice of these constellations, even though optimal spectral utilization of the medium remains to be the focal issue.

In practice, these constellations may be generated by amplitude modulators and phase shifters in various combinations depending upon the incoming data bits or symbols generated by a sequence of four bits for a 16 point cluster (or constellation) or six bits for a 64 point rectangular clusters. They may also be generated by the independent amplitude modulation of two coherent carrier signals displaced in phase by 90 degrees. Constellations are generally chosen to be symmetric around the X and Y axis to simplify the terminal devices and to exploit the symmetric media properties to positive and negative signals. Complex clusters also exist but are rarely used except in specialized applications. Predistorted transmit cluster may be used like predistorted transmit signals to counter nonsymmetric media properties. Most of the encoding and emphasis technique commonly used with simple codes are applicable here. Typical advantages associated with specially designed line codes and channel codes [2.3] have been exploited in this application.

At the receiver, demodulation of the carrier, recovery of the carrier, timing, and symbols are essential. Intersymbol interference is caused by nonlinearities of the amplifiers, the amplitude-to-phase converters, imperfect filters and equalizers. Phase jitters in the carrier recovery devices and hence the uncertainty of the timing recovery for sampling contributes to the possible errors in transmission. The freedom of signal path in free space also causes multipath (i.e., a non unique path for the signal to traverse the transceiver spatial separation) fading. Thus all the encoded signal energy is not recovered at the same instant of time because of the variation of the path length and because of variation of media characteristic along different paths. The net effect of these impairments in transmission does cause noise and jitter in the signal recovered at the receiver. The engineering design rules and the system optimization techniques attempt to maintain enough signal-to-noise ratio (in spite of the system imperfections), such that dependable transmission of information is achieved.

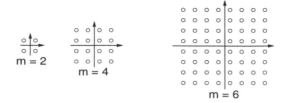

(a) Rectangular or QAM Constellations

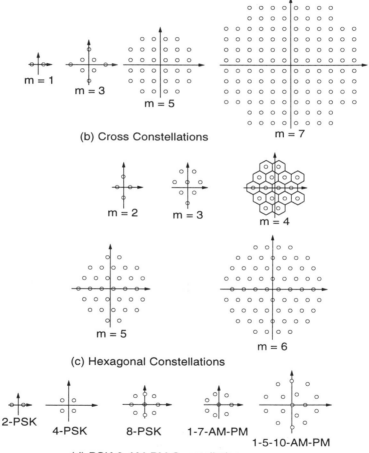

(b) Cross Constellations

(c) Hexagonal Constellations

(d) PSK & AM-PM Constellations

FIGURE 2.2. Configurations of 2^m Point Constellations. In (d) both phase shift-keying (PSK) and amplitude modulation (AM) are used in the last two constellations. The modulation efficiency (in view of the reduced limits for threshold detection) starts to decrease significantly after m reaches 6 for the $2^{**}m$ constellations.

2.5

COAXIAL LINKS

2.5.1 GENERAL FRAMEWORK OF COAXIAL SYSTEMS

Coaxial systems, though expensive at installation time, carry a large bandwidth of modulated carrier signal. The range is from 2.8 MHz for the L1 system to 61.5 MHz for the L5E system in the United States. This feature makes them well suited as long-haul communication links between distant cities as well as high density short-haul links within the metropolitan areas where high capacity is needed. These systems may be used for distances up to 4,000 miles for undersea coaxial cable systems (type SF and SG designs in the United States) where as many as four thousands voice circuits can be multiplexed and "hauled" from coast to coast. The coaxial carrier systems are generally used between major communication centers and use the four-wire or equivalent four-wire (single cable with spectral separation for the two directions of communication) mode for transmission. The major concern in designing these systems is that of powering the repeaters along the haul.

2.5.2 TRANSMISSION ASPECTS

The basic constraint in the transmission of coaxial cable systems for communication is that of noise at the receiver that degrades the signal. For voice grade channels at the maximum system range (i.e., 4,000 miles for the L5E systems), the objective is 40 dBrnc0. The cable design, the carrier frequency band, the maximum number of repeaters along the route, the repeater characteristics, and the extent of equalization all contribute to the system noise. In view of the inherent limitations, the design of the components that constitute the system, ranges and capacities are specified to maintain the quality of communication over the coaxial carriers. The two established systems in the United States are the L5 and the L5E systems with a spectral band of 3.124-60.556 MHz and 3.522-64.844 MHz. respectively. As presented in Section 2.2 (see Figure 2.1), *three* Jumbogroups each having six Mastergroups are combined to yield the L5 spectrum. In the` L5E systems, three sets of Mastergroups (with 7, 7 and 8 Mastergroups) are directly multiplexed onto the L5E spectrum.

This spectrum has *three* sub-bands in the ranges 3.252-21.900; 22.900-41.548; and 43.508-64.844 MHz to accommodate the 7, 7 and 8 Mastergroups each having a bandwidth from 0.564-3.084 MHz. The reference frequency

signal is at 20.480 MHz in the L5 system and it is shifted to 2.048 MHz in the
L5E systems. Likewise, the equalizing pilot at 20.992 MHz in L5 is shifted to
21.956 MHz in L5E. This frequency allocation of the L5E system makes it
consistent with other long-haul transmission systems that use 2.048 MHz
without any changes in the line design and layout of the two systems. The
voice channel capacity of the L5 system with eighteen Mastergroups is 10,800
voice circuits per pair of coaxial cables. For the L5E system with twenty-two
Mastergroups, the capacity is 13,200 voice circuits per coaxial cable pair.

System performance is maintained strictly by pilot tones, a hierarchical
combination of repeaters, transmission surveillance, and signal processing
functions throughout the route. The transmission layout for these two
systems is depicted in Figure 2.3 (discussed in Reference 1).

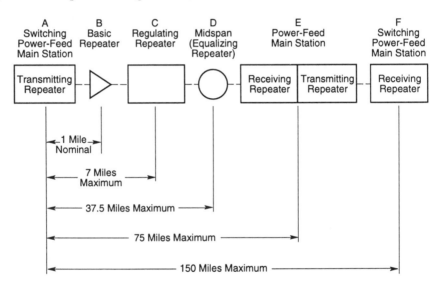

**FIGURE 2.3. Repeater Spacing and Transmission Layout for the L5
and L5E Systems (Very Wideband Analog Carrier Systems).**

The attenuation characteristics of the 0.375 coaxial cables used in these
carriers is depicted in Figure 2.4 (see Reference 2.1). There are four types of
repeaters used in the overland coaxial carrier system for the L carriers. The
basic repeater is designed for one mile of cable run (including an additional
31.6 feet for cable sag). The regulating repeater is designed to be placed at a
maximum distance of seven miles to regulate reduction on the third
harmonic buildup in the cable system, in addition to reducing
intermodulation signals cascading over the long run of the carrier. The

equalizing repeater reduces the loss of signal-to-noise ratio due to drift in the alignment of the frequency components of the modulated signal. When the spacing of these repeaters cannot be monitored accurately and errors are likely to cause any cumulative effects, a deviation equalizer can also be used with equalizing and main station equalizers. Finally, the main station repeater has the receiving and the transmitting components of earlier repeaters and two equalizing circuits to cover the entire band. The performance is monitored to maintain the transmission of the signal to lie in a range of plus or minus 0.4 dB over the 1.6 to 66 MHz spectral range of the carrier signal.

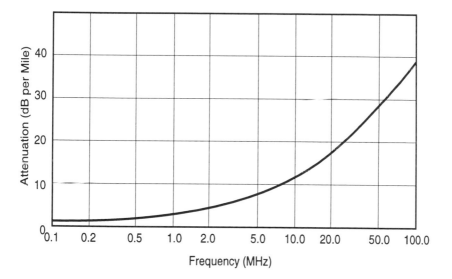

FIGURE 2.4. Attenuation/Frequency Characteristics of 0.375 Inch Diameter Coaxial Cable Unit Used for the L5 & L5E Very Wideband Analog Carrier Systems.

Three types of equalizers are also used to automatically maintain the carrier performance. The performance of these equalizers is done by four pilot tones at 2.976, 20.992, 42.88, and 66.048 MHz for the L5 carrier. For the L5E carrier, the second pilot tone is changed from 20.992 MHz to 21.956 MHz. The system performance is thus tightly controlled by a set of self-monitoring and regulating components throughout the route.

2.5.3 UNDERSEA CABLE CARRIERS

The transmission principles for overland and undersea are the same, but the implementation and design procedures are different. Installation, repair procedures, power feed and system equalization techniques vary. As the cable is being installed, measurements are made continuously and the equalizer characteristics are determined to switch in and out the equalizer networks. The equalizer is then sealed and placed on the ocean floor. Adapting the individual components is generally assisted by a computer aided procedure. The monitoring of the signal-to-noise ratio of the cable as it is being installed is unique to the undersea cable. The spacing of the voice channels in the undersea carriers is reduced to 3 kHz (from the conventional 4 kHz spacing in the land carrier systems) permitting more voice channels per carrier system. Certain undersea carrier systems also employ the time assignment speech interpolation or TASI equipment to use silent periods during the telephone conversations. This increases the capacity of channels carried by the same coaxial carrier system. The undersea cable system design and installation procedure thus becomes more complicated and cumbersome than similar land carrier systems. The modern transoceanic fiber optic systems (see Chapter 15) have now made the oceanic coaxial systems archaic.

2.6

SATELLITE LINKS

2.6.1 GENERAL CONCEPTS

Satellite links use an orbiting satellite as a repeater for the transmitted signal to amplify and rebroadcast to the receiver. Limited amount of channel switching is also accomplished. Communication satellites have a geostationary orbit that places it in the plane of the equator at an altitude of about 22,300 statute miles (35,784 km) above the earth. The direction of rotation is the same as that of the earth. The minimum spacing of different satellites is four degrees for the directional antennas. Geostationary satellites remain synchronous with the rotation of the earth thus being able to "view" a predefined area throughout the day without the need of the earth antennas to track the satellite. Station-keeping of the satellites is necessary since gravitation from the sun, the moon and other terrestrial causes slow drift from their ideal positions. Both north-south and east-west drifts are encountered and the angular drifts are regulated to be within ± 0.1 degrees by timely rocket thrusts. The intervals for the correction may be several weeks. Due to the slow wander of the satellite, the communication path length also changes slightly causing perturbation in the delay. Highly directional antennas to the earth stations are used and their stabilization becomes crucial. Two techniques (double spin configuration and three axis stabilization) can be used. In the former configuration the angular drift is much less than 0.1 degree and in the later stabilization the drift can be under 0.001 degree, thus the exterior of the satellite appears fixed with respect to any point on the surface of the earth.

Satellite design is generally considered from *two* distinct considerations: *one* for its tracking, telemetry and control, and the *other* for its role as communication units. Many designs separate the requirements, functions and the components that are isolated. Earlier non-geosynchronous low-orbit satellite systems that forced spin-stabilization about a fixed axis maintained a fixed angle with respect to the earth's axis. Tumbling was thus eliminated, however the antennas could not be properly aligned to the earth station.

To permit the use of highly directional antennas aboard, two techniques (double-spin configuration and three-axis stabilization) were later developed. In the first method, the satellite and antennas are spun in opposite directions at low speed (75 rpm with ± 25 rpm). The antennas would retain their orientation with respect to the earth station if the angular damping could be made accurate. In the second method, a flywheel rotating at high speed would provide stability in all three axes and the exterior of the satellite (including the antennas) would be stationary with respect to the earth station. Noteworthy, two specific directionalities, direction of the antenna and direction of the solar panel, are both necessary.

Earth stations monitor the satellite position and functions, the antenna, and solar panel orientations at regular intervals. These earth stations then relay appropriate control signals to the satellite. Most satellite systems have certain redundancy of critical components and subsystems, and rigorous preflight testing. Solar cells (including back-up batteries for earth eclipse) and thrust fuel restrict the weight. Solar cells die an exponential death and the minimum wattage at "end of life" is balanced against the weight of thrust fuel that satellites carry. Component failure (typically occurring over a span of seven to ten years) rather than satellite failure due to decay of solar cells or fuel exhaustion, is the chief cause of "dead" satellites.

Even though non-synchronous satellites may be used (e.g., three Soviet launched "Molniya" satellites in elliptical orbit with 12 hour periods), a 24 hour continuity of service is possible by a timed ritual of daily handover procedures from the setting satellite to the rising satellite. They are particularly important in order to reach the polar regions of the earth where geostationary satellites do not provide good coverage due to extreme latitudes of these areas.

The up-link from the transmitter and the down-link to the receiver are each a about thousand times longer than a corresponding terrestrial hop. The cumulative time and signal distortion of the multiple hops in the terrestrial systems are non-existent in the satellite systems. Selective fading of the signal, common in terrestrial microwave systems, do not exist in satellite links. A single satellite covers a wide area of the earth. Crossing rough terrain for humans and microwaves is thus resolved.

Due to limited satellite capabilities, *three* effects emerge: *(a)* longer delays, *(b)* more transmit power in the up-link and *(c)* amplification of the down-link. The weight, size and power of the satellite are major restrictions. The variation of media characteristics (especially the atmospheric region) and its loss place upper bounds upon what satellite links can accomplish as communication pathways.

Repeater satellite channels called transponders, are "visible" to the transmitters and receivers by specially designed antennas. Antennas covering very wide continuous areas cannot offer enough power and cause interference with other satellite systems. This effect dictates the fact that certain orbital positions more desirable than others to cover the Western Hemisphere. For example, COMSTAR-2 satellite occupies 95 degrees West Longitude. It uses a 3.2° X 8° elliptic antenna beam to cover the continental United States; for Alaska, a 2° X 4° elliptical beam, and two 3° circular beams for Hawaii and Puerto Rico. Other communication satellites have specialized and focused antennas to serve that particular region of the globe.

2.6.2 SPECTRAL GROUPS AND BANDS OF SATELLITE LINKS

Three major frequency groupings 4/6 GHz, 12/14 GHz, 19/29 GHz exist. The up-links and down-links are separated by about 2 GHz for the 4/6 and 12/14 GHz groups and about 10 GHz for the 19/29 GHz group. Frequency tolerance is also imposed by the Federal Communications Commission (FCC). For example, frequency tolerance is restricted to be 0.001 and 0.002 percent of nominal value, respectively for both the earth station transmitter and space station. The up-links use a higher frequency band in each group. Also, a spectral band is necessary to carry numerous conversations or channels in each satellite link. This spectral bandwidth depends upon the number of channels carried on each link. Typically the 4/6 and 12/14 groups each have a 500 MHz bandwidth and the highest group has a bandwidth of 3500 MHz. Largely, the bandwidth is too large for individual use and are subdivided into a series of lower bandwidth transponder channels. Each transponder channel is at least adequate to carry one standard television transmission requiring a 36 MHz band. These transponder channels need suitable (about 4 MHz) guard space, thus requiring a separation of about 40 MHz. These frequency groupings and their spectral bands are presented in Table 2-1.

TABLE 2-1 Stationary Satellite Antenna Coverage

Area Covered	Bandwidth (in degrees)	Gain* in dBi (#)	Aperture at 4 GHz (Meters)
Entire World	18° Circular	17	0.3
48 Cong. United States	8°X 4°(elliptical)	27	0.8 X 1.6
1 hour time zone	1.5°X 3°(elliptical)	36	4.2 X 2.1

* 3 dB down at band edge. # dBi = dB with respect to isotropic radiator.

2.6.3 INTERRUPTIONS IN SATELLITE COMMUNICATIONS

The satellite, having an orbit of its own, experiences the effects of eclipses. Of particular importance to communications via satellite are *two* events. *First*, during the spring and autumn seasons, the receiving antenna, satellite and sun align themselves, creating an eclipse. Any signal from the satellites is completely obliterated against thermal noise of the sun. These two annual eclipses are predictable and occur for about six successive days, for a few minutes each day, at noon at the location of the earth station. Another standby earth station takes over the communication from the satellite. At the end of the eclipse, the standby earth station hands back the function to the first earth station.

Second, the satellite also experiences an eclipse due to the earth's alignment between the sun and the satellite. During this time, the solar energy from the sun is blocked, making it necessary for storage batteries to power the transponders. The occurrence and duration (46 days) of these two annual eclipses are also predictable. The sudden change of temperature during the eclipse, lasting a maximum of about 72 minutes every day, can be of serious concern to the transponder components.

2.6.4 IMPAIRMENTS IN SATELLITE COMMUNICATIONS

The *four* major impairments to the propagation of the signal arise due to attenuation in the earth's atmosphere, effects of rain, ionosphere and low angles of satellite elevation from the earth station. Two important device limitations (antenna and system noise) also exist. Hence, it becomes essential to examine the design and operability of the satellite links in view of the propagation impairments and the device limitations. The effects of these have been documented [2.1] and the channel performance is crucial in maintaining its quality. These impairments are regulated by maintaining an acceptable ratio of the signal-to-noise ratio within the system.

Generally, the capacity quotient computed (in dB) as the ratio of the carrier power to the noise power density of the channel is calculated. The noise density is determined as the sum from all the noise components arising from thermal, intermodulation, interference, and distortion. Both up and down links need to be considered in the overall channel performance. Losses of the channel consist of space propagation loss, atmospheric loss, earth station antenna tracking loss and rain loss. Six system gains compensate the losses and reestablish the signal (and introduce some noise in the process). Three such gains on the up-link are earth station transmitter amplification, transmitting antenna, and satellite receiving antenna. The remaining three gains on the down-link are the satellite transmitting antenna, earth station receiving antenna, and receiver. The earth station gains are sometimes broken up into the IF (intermediate frequency) gain that amplifies the raw baseband signal and the main link amplifier that amplifies the entire satellite link. The channel quality is ascertained by monitoring the capacity quotient accurately.

To limit the weight of the satellite, the up-link receiver and down-link antenna gains at the satellite are roughly half of the earth stations up-link transmitter and down-link receiver antenna gains. The greatest loss in the satellite systems occurs due to signal propagation in space in the order of about 200 dB in either direction with the 30 meter diameter antennas. This value is for the 4/6 GHz system, 1200 voice channel transcontinental communication via COMSTAR satellite in clear weather. The atmospheric and antenna tracking losses are less than 0.25 percent in clear weather. The rain loss, being frequency dependent, is low for the 4/6 GHz systems [2.1] and can become significant for the higher frequency bands, causing a total system outage.

2.7

ANALOG CHANNEL BANKS

In Section 2.2, *five* stages (basic groups, Supergroups, Mastergroups, Jumbogroups, high capacity carriers) of multiplexing are presented for the FDM hierarchy in the United States. In essence, there are *five* analog (A-type) channel banks (see Sections 2.7.1 to 2.7.5). The flexibility in composing and decomposing the groups using analog channel banks is depicted in Figure 2.5.

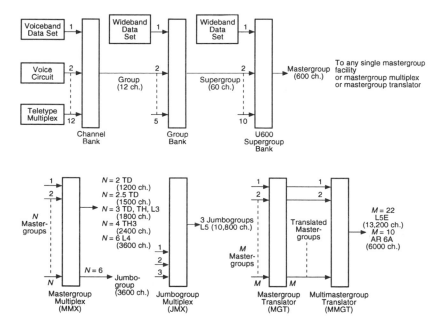

FIGURE 2.5. The Frequency Division Multiplexing Hierarchy Used in the Pre-Divestiture Bell System. It still exists in some remote systems. AR-6A - 6 GHz Radio SSBAM Analog Carrier System; L5E - Very Wideband Analog Carrier System.

2.7.1 THE A6 CHANNEL BANK

The A5 and A6 banks are dedicated to the generation of the "groups" from 12 voiceband circuits (typically voiceband data sets, voice circuits, teletype multiplexers, etc.). Each of the 12 channels is modulated over a frequency ranging from 8.140 to 8.184 MHz incremented by 4 kHz. The composite signal is filtered through a band pass filter and only the upper sideband is retained and combined into one band 8.140-8.188 MHz. Thus the 12 bands are mapped into 8.140-8.144, 8.144-8.148, ..., 8.184-8.188 MHz bands. Each of these 4 kHz bands contains the voiceband information/data (in the range of 100 Hz - 3400 Hz) from the information channel. When the composite signal is demodulated by a carrier of 8.280 MHz, and passed through a low-pass filter, a "Group" signal is generated between 60-108 kHz. This technique forces the twelfth information band (8.184-8.188 MHz) to reappear between 60-64 kHz generated as the difference between the 8.248 and the 8.184-8.188 MHz band. The other information bands are similarly mapped with the first information band appearing in the 104-108 kHz "group band". The frequency of operation for the various channels and bands is depicted in Figure 2.6 and represented in Table 2-2. The numbers in column four are generated by isolating the upper sideband of the modulated signal of the voiceband data in column two with the carrier signal. The group bands are generated by isolating the lower sideband of the product of the carrier signal at 8.248 MHz and the 12 upper sidebands.

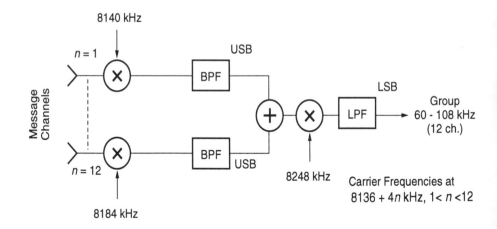

FIGURE 2.6. The A6 Channel Bank. Table 2-2 lists the carrier frequencies for each message channel. BPF - Band Pass Filter; LPF - Low Pass Filter; LSB - Lower Side Band; USB - Upper Side Band.

**TABLE 2-2 Frequency Allocation for the Message Channels
of the A6 FDM Carrier System**

Message Channel	Carrier kHz	USB Range kHz	LSB Range kHz
4 kHz $1 \leq n \leq 12$	f_c	$(f_c) - (f_c + 4)$	$(60 + n.4)$ kHz
1	8140	8140-8144	60-64
2	8144	8144-8148	64-68
- -	- - -	- - -	- - -
12	8184	8184-8188	104-108

2.7.2 THE GROUP BANK

Five of the groups are multiplexed onto one Supergroup. Each of the five groups has information bearing signals in the 60-108 kHz bands. In the receiving group banks, pilot tones at 104.08 kHz are inserted in each group to manage, monitor and control the net group loss as it traverses through the varying media and its dynamic losses. Group signals of standard and acceptable amplitude are thus regenerated in the group bank before mixing with carriers at five frequencies ranging from 420-612 kHz at 48 kHz intervals. The 48 kHz band now becomes the group bandwidth. Bandpass filtering and the extraction of the lower sideband now yield the Supergroup between 312-552 kHz accommodating the 60 original voice channels. The frequency of operation for the various channels and bands is depicted in Figure 2.7 and represented in Table 2-3.

Noteworthy, that the 60 kHz offset generated by the channel banks (Section 2.2.7.1) is completely utilized by the Group banks in generating the Supergroup band of 312-552 kHz. This frequency offset permits dependable function of the channel bank and the Group bank.

2.7.3 THE SUPERGROUP BANK

The function of these banks proceeds very much like the functioning of the Group banks. Ten Supergroups are multiplexed onto one basic Mastergroup.

This Mastergroup now contains 600 channels. The first six Supergroups are carried on carriers starting at 1.116 MHz and incremented by 248 kHz.

This frequency increment is more than 240 (= 552 - 312) kHz, thus providing some spectral isolation of the bands. The remaining four Supergroups are carried on carriers starting at 2.652 MHz thus leaving behind a gap of 56 (48+8) kHz in the spectral space between 2.044 and 2.100 MHz. The frequency of operation for the various channels and bands is depicted in Figure 2.8 and represented in Table 2-4. In Figure 2.9 the bidirectional mode of functionality for the Supergroup banks is depicted. It is essential to note that effective filtering is necessary after each carrier modulation. For example, only the lower sideband between 564 and 804 kHz is permitted to find access into the Mastergroup. By and large, all FM modulation techniques produce two sidebands and to isolate the lower sideband, accurate well-tuned and sensitive band pass filters are necessary. The U600 Supergroup analog bank used in the United States takes care of the filtering and multiplexing of the 600 voiceband channels onto one basic Mastergroup.

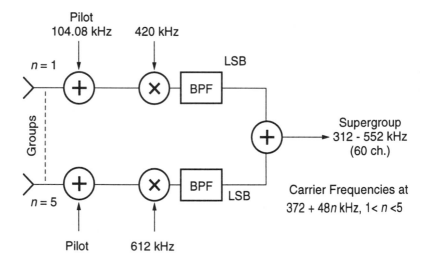

FIGURE 2.7. The Group Bank. BPF - Band Pass Filter; LSB - Lower Side Band. The Supergroup has 60, 4 kHz channels to cover 240 kHz for voice and voiceband data.

**TABLE 2-3 Frequency Allocation for the Channel Groups
of A6 FDM Carrier System**

Groups kHz	Carrier kHz	LSB Range kHz
60-108	f_c	$(f_c - 108) - (f_c -60)$ (see Figure 2.6)
1	420	312-360
2	468	360-408
3	516	408-456
4	564	456-504
5	612	504-552

This frequency increment is more than 240 (= 552 - 312) kHz, thus providing some spectral isolation of the bands. The remaining four Supergroups are carried on carriers starting at 2.652 MHz thus leaving behind a gap of 56 (48+8) kHz in the spectral space between 2.044 and 2.100 MHz. The frequency of operation for the various channels and bands is depicted in Figure 2.8 and represented in Table 2-4. In Figure 2.9 the bidirectional mode of functionality for the Supergroup banks is depicted. It is essential to note that effective filtering is necessary after each carrier modulation. For example, only the lower sideband between 564 and 804 kHz is permitted to find access into the Mastergroup. By and large, all FM modulation techniques produce two sidebands and to isolate the lower sideband, accurate well-tuned and sensitive band pass filters are necessary. The U600 Supergroup analog bank used in the United States takes care of the filtering and multiplexing of the 600 voiceband channels onto one basic Mastergroup.

2.7.4 THE MASTERGROUP BANK

Supergroups may or may not be multiplexed depending upon the capacity of the communication pathway. Some applications that need only one 600 (L1 system on TD-2 broadband channel) channel stop at the Supergroup. Some systems need two or more Supergroups. In this section, the multiplexing of the 3,600 voice circuit Mastergroup is presented.

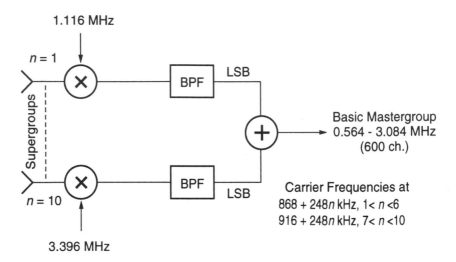

FIGURE 2.8. The U600 Supergroup Bank. The Mastergroup has 600, 4 kHz channels to cover 2400 kHz in two bands (0.564-2.044) MHz and (2.090-3.048) MHz. BPF - Band Pass Filter; LSB - Lower Side Band.

Six Mastergroups are multiplexed onto one Jumbogroup. Pilot tones are inserted in each of the Mastergroups to adjust for the variation of the media characteristics and to restore the modulated signal strength as discussed on the Group bank. Next individual carriers ranging in frequencies from 6.336 to 18.112 MHz are used for the five higher Mastergroups ranging from 2 to 6. The first master band proceeds onto the 0.564-3.084 MHz band of the Jumbogroup (see Section 2.7.3.).

TABLE 2-4 Frequency Allocation for the Supergroups

Supergroups Range MHz	Carrier MHz	LSB Range MHz
0.0-0.240	f_c	$(f_c - 562) - (f_c -312)$ (see Figure 2.7)
n=1 to 6	0.868+0.248*n	0.564 - 2.044
n=7 to 10	0.916+0.248*n	2.090 - 3.048

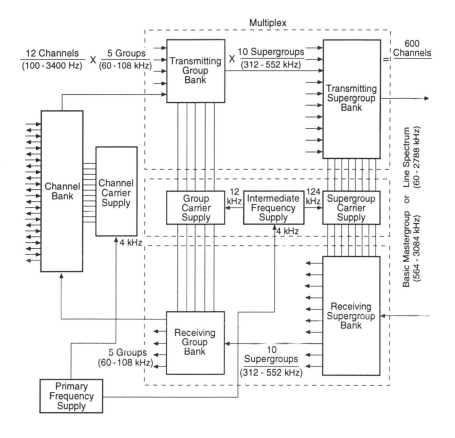

FIGURE 2.9. Broadband Mastergroup with the L Carrier Multiplexer Shown in the Enclosed Box (dashed line). L - Carrier System

The carrier frequencies for the various bands are separated by a idle band of frequencies to prevent excessive spill over of information from one master band to the next, thus retaining good quality of information transmitted and contained at the extreme ends of the master bands. This idle band, called the guard space, permits effective channel separation and filtering necessary for high quality carrier performance. The frequency of operation for the various channels and bands is depicted in Figure 2.10 and represented in Table 2-5.

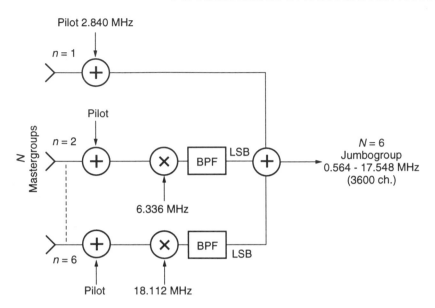

FIGURE 2.10. The Mastergroup Multiplex for Jumbogroup. The Jumbogroup has 3600, 4 kHz channels to cover 14,400 kHz in twelve bands. BPF - Band Pass Filter; LSB - Lower Side Band.

As in the case of the Supergroup, the lower sidebands are isolated by high quality band pass filters, thus providing a spectral space of 0.564 to 17.548 MHz for actually carrying the 3600 voiceband circuits.

TABLE 2-5 Frequency Allocation for the Jumbogroups

Mastergroup Number	Carrier MHz	LSB Range MHz (see Figure 2.8)
1	$f_c=0$,Direct input	0.564-3.084
2	f_c= 6.336	3.252-5.772
3	f_c= 18.112	15.028-17.548

2.7.5 THE JUMBOGROUP BANK (MULTIPLEXER)

Jumbogroups are "banked" or multiplexed into carrier signals only if such very high capacities are necessary and justified. There are two such systems for carrying 10,800 (i.e., three Jumbogroups) voiceband channels on the L5 carrier systems and for carrying 13,200 (i.e., 7, 7, and 8 Mastergroups) voicebands on the L5E carrier systems in the United States. In the second case the Mastergroups are directly multiplexed onto the L5E systems without resorting to the Jumbogroups. Technical design of the large systems can be tailored depending upon the individual requirements of the network and the geographical dispersion of the population centers. These two systems are presented next.

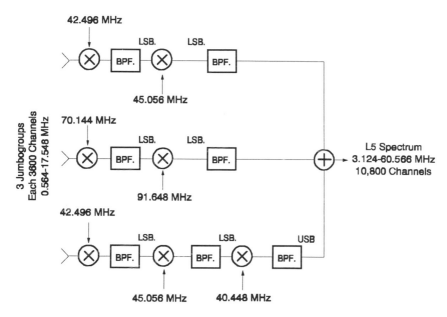

FIGURE 2.11. The Jumbogroup Multiplex for L5 System. The L5 has 10,800 of the 4 kHz channels to cover 57.432 MHz in 36 bands.

In the L5 or 10,800 voiceband carriers, the first two of the three Jumbogroups each carrying 3,600 channels in the range of 0.564 to 17.548 MHz are modulated twice, and the third Jumbogroup is modulated three times. The carrier frequency modulation for the three Jumbogroups is presented in Table 2-6. This is an elaborate system where the carrier frequencies are adjusted to yield the desired band in the L5 spectrum. The relation between the carriers frequencies and the L5 spectrum is depicted in Figure 2.11 and represented in Table 2-6. In the actual systems constructed, band pass filters are essential after each stage of modulation to ensure isolation of the appropriate band listed in the Table. In addition, these band pass filters ascertain that only the signal energy contained in the appropriate band is permitted to enter the next stage of modulation.

TABLE 2-6 Frequency Allocation for the L5 System

Jumbogroup Number	LSB;Range 1 MHz	LSB;Range 2 MHz	USB;Range 3 MHz (see Figure 2.10)
1	24.948-41.932	3.124-20.108	Same
2	52.596-69.580	22.068-39.052	Same
3	24.948-41.932	3.124-20.104	43.572-60.566 (see Figure 2.11)

2.8

CONCLUSIONS

Analog carrier systems can vary significantly from country to country. The design of these carriers is not unique nor universal. The use of such systems facilitates the short, medium and long-haul facilities of most national networks. In the earlier systems where analog facilities prevailed, the communication networks have invested heavily in building analog carrier "highways" for the then prevailing communication technology. Fiber and its advent has caused serious ramifications on the future use of analog carrier systems. But they do exist and it is to the economic advantage to exploit the ultimate capability of these prevailing analog systems.

For this reason, many a high capacity digital system simply "rides" upon these high capacity analog systems (e.g., the 274 Mbps digital radio for the DS-4 rate in the United States, even though it is rapidly getting archaic and antiquated).

The economic balance in other countries and the need for communication in remote areas will determine the slow global demise of analog carrier systems. Even though these systems are not the most effective means of conveying massive amounts of data totally error free, they offer an established methodology for some cable and most long-haul wireless systems. For this reason there is a niche for these analog systems in the dominance of digital carriers just as there is a niche for twisted wire-pair systems in the dominance of fiber optic systems.

2.9

REFERENCES

2.1.   Bell   Telephone   Laboratories   1982,   *Transmission   Systems   for Communications*, Fifth Edition, Bell Telephone Laboratories, Inc.

2.2.   AT&T 1985, *Telecommunication Systems Engineering*, second printing, New York, NY.

2.3.   E.A. Lee and D.G. Messerschmitt 1988, *Digital Communication*, Kluwer Academic Publishers, Boston, MA.

2.4.   S.V. Ahamed 1982, "Simulation and Design Studies of Digital Subscriber Lines," *Bell System Technical Journal*, Vol. 61, No.6, July-August, 1003-1077.

CHAPTER 3

VIDEO AND TV ENVIRONMENT

Information revolution has facilitated the movement of massive amounts of data over networks. Human expectations have also risen accordingly. Both audio and visual modes of communication are invoked in most human recipients and network transport providers. Multimedia communication is very much of a reality. Video camera and TV devices become the input and output devices of modern networks. In a sense, the entire video and TV environment now interacts with modern broadband networks. Traditionally, these environments had been influenced by non-network oriented CATV personnel and protocol (if it can be called protocol). However, the need to interface with them and utilize their technology now becomes more and more immediate for the networks' and systems' designers. This Chapter facilitates a graceful integration of the disciplines.

3.1

GENERAL CONCEPTS

Coaxial cables carry high frequency modulated signals extremely well. The primary characteristics of the cables are designed to suit the carrier frequency and are stringently controlled for any variations during the manufacturing process. For this reason, and by the inherent shielding provided by coaxial cables, they were used extensively from interoffice trunks to undersea intercontinental communication systems. Other applications range from high fidelity music carriers to wide area community TV programming.

The use of microwave for communication of speech information dates back to the work of Marconi. Radio communication systems have been used in military and civilian environments ever since. In fact, the entire radio frequency (RF) band is divided into *five* distinct bands (MF, HF, VHF, UHF and SFH), ranging from medium frequency (or MF from 0.3 to 3 MHz) to super high frequency (or SHF from 3 to 30 GHz). The utilization of this RF band for different applications is depicted in Figure 3.1. In the earlier applications, radio transmission had been long established for unidirectional wireless broadcasts.

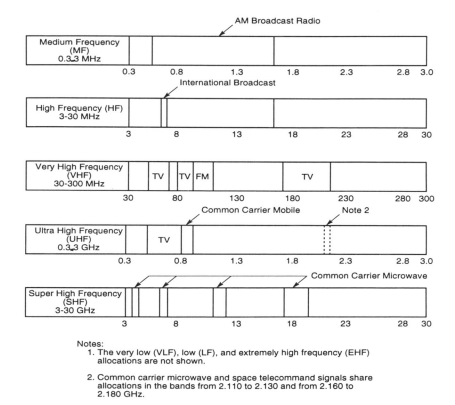

Notes:
1. The very low (VLF), low (LF), and extremely high frequency (EHF) allocations are not shown.

2. Common carrier microwave and space telecommand signals share allocations in the bands from 2.110 to 2.130 and from 2.160 to 2.180 GHz.

FIGURE 3.1. Radio Frequency Spectrum Allocations. Medium through super high frequency bands.

Television broadcast networks demand much more bandwidth and sophisticated encoding, even for black and white pictures. Subsequently, broadcast of color encoded video signals has emerged. Both these technologies are now mature. The more recent contributions to the TV industry comes as picture encoding and transmission of high definition television. Both analog and digital techniques of picture coding are significant.

Two aspects of transmission of the analog TV signals prevail; *first,* the transport (i.e., long/medium/short-haul) over the transmission media (terrestrial space, free space, coaxial cable or any other media) and *second,* the distribution via CATV networks (also coaxial cable systems for distribution). The signal transport systems are typically owned and operated by common carriers and are used for hauling any (voice, video, data, etc.) information consistent with the charters and permits issued to the common carriers. The transport quality aspects of the video signals is contingent upon the guidelines of the Networks Transmission Committee (NTC) presented in this section and the distribution aspects presented in the rest of the Chapter.

Typically, the major long-haul microwave TV links carry unidirectional information. These systems use frequency modulation techniques (Section 2.1) to carry the video information. The details of the microwave radio (Section 2.4) and satellite environment (Section 2.6) that carry the video signal over long distances (up to 4000 miles) have been presented in Chapter 2. Here, the intercity TV networks that carry video signals over shorter haul from city to city, from studios to broadcast stations, etc. are classified and briefly reviewed. A synopsis of this environment is presented in Section 3.3. Finally, the "distribution" aspects of TV signals by the CATV network are presented in Sections 3.4 through 3.6.

Over the last few decades, there has been an understanding between the TV broadcasters and the common carriers who deploy the transmission media (terrestrial and free space, co-axial cables, etc.). Considerable deliberation and expertise has been deployed to ascertain that the quality of the video signal is high before any distribution within the CATV network (see Section 3.4). Detailed test signals and test objectives have been published to assure that the transporters (i.e., the common carriers) do not degrade the video quality at the Head End station (see Section 3.5.2) from where CATV signal distribution starts. The acceptable quality of the picture is thus identified by a set of limits on *four* discernible parameters: distortions, chrominance-luminance inequalities and intermodulation, and noise in the received signal. These four parameters, the associated test signals and the

objectives are accepted by the NTC [3.2] and are used as guidelines in the
transport of video signals throughout the country.

First, there are *nine* (*a* through *i*) forms of distortions. The first *three* types
of waveform distortions are (*a*) field-time, (*b*) line-time, and (*c*) short-time.
The test associated with the waveform distortions are field-bar, line-bar and
2T pulse tests. The objectives for the first two distortions are that neither one
may exceed 4 IRE (Institute of Radio Engineers) units* and the last distortion
should lie within 100 + or - 6 IRE units. In addition, there are *two* (*d*) gain
and (*e*) phase, chrominance related nonlinear distortions. The nonlinear gain
distortion is measured by the three-level chrominance test and the limits for
the smallest subcarrier and the largest subcarrier should each lie within 20 +
or - 2, and 80 + or - 8 IRE units respectively. The non-linear phase distortion
is also measured by the same test, but the limit is at five degrees. The
differential (*f*) gain and (*g*) phase, tested by the modulated 5-riser staircase
test allows a limit of + or - 15 degrees for the former and + or - five degrees
for the later. Further, (*h*) the gain/frequency distortion is verified by the
color burst test limit to lie between 40 + or - 4 IRE units. Finally, the (*i*)
luminance non-linear distortion also measured by the modulated 5-riser
staircase test limits the smallest pulse to be over 90 IRE units. (See Table 3.1).

Second, there are *two* (*i* and *ii*) types of inequalities related to
chrominance-luminance balance. In dealing with both, (*i*) gain inequality
and (*ii*) the phase inequality, the chrominance pulse test is used. The limit for
the former inequality is set to lie within 100 + or - 3 IRE units and the upper
limit for the later is less than 75 nanoseconds.

Third, there is the chrominance to luminance intermodulation. In this
case a three level chrominance test signal is used and the limit for the
intermodulation is set at a maximum of 3 IRE units.

Fourth, there are *four* (*1* to *4*) types of noise that degrade the video signal.
In specifying (1) random noise, weighted noise measure is used and the
objective is to retain a signal-to-noise ratio (SNR) in excess of 53 dB. In
dealing with (2) impulse noise and (3) periodic noise, a flat (over the band)
field test is used. The limit for (2) is accepted as being less than 7 IRE with 1
V (p to p) or 140 IRE. This limits the SNR to be over 23 dB. The lower limit
for SNR for (3) is at 50 dB. For (4) crosstalk noise, the accepted lower limit
for the SNR is at 60 dB.

---

* When the maximum composite signal is 1V peak to peak, it corresponds to 140 IRE units.

TABLE 3-1 NTC Report No. 7 End to End Television Transmission Objectives

Parameter	Test Signal	Objective
Distortion Tests		
Field-time waveform	Field bar	≤ IRE units p-p
Line-time waveform	Line bar	≤ IRE units p-p
Short-time waveform	2T Pulse	100 ± 6 IRE units
Chrominance-luminance gain inequality	chrominance pulse	100 ± 3 IRE units (± 6%)
Chrominance-luminance delay inequality	chrominance pulse	±75 ns advanced or delayed
Gain/frequency distortion	color burst	40 IRE units ± 4 IRE units
Luminance nonlinear distortion	Modulated 5 riser staircase	Smallest pulse ≥ 90 IRE units
Chrominance nonlinear gain distortion	Three-level chrominance	Smallest subcarrier, 20 ± 2 IRE units
		Largest subcarrier, 80 ± 8 IRE units
Chrominance nonlinear phase distortion	Three-level chrominance	≤5 degrees
Differential gain	Modulated 5 riser staircase	≤15 %
Differential phase	Modulated 5 riser staircase	≤5 degrees
Chrominance to luminance intermodulation	Three-level chrominance	≤3 IRE units
Random noise weighted		SNR ≥ 53 dB
Impulse noise	Flat field	≤7 IRE units (23 dB)
Periodic noise	Flat field	≥50 dB
Crosstalk		≥60 dB

A representation of these tests and the objectives are presented in Table 3-1 taken from Reference 3.2 and published by NTC.

3.2

TV NETWORKS

TV networks have maintained an identity of their own distinct from the communication networks. In the past, [3.2a] the requirements, standards and the monitoring of the TV signals have been different from those of the voice and data communication networks. The common carriers have accommodated the specialized needs of the TV industry by tailoring the tests and modifying the measuring systems to cater to the TV industry. However, the collaboration between the TV broadcast industry and the common carriers (approved by the NTC) has provided a platform for the current analog TV systems as they have existed to date. However, as mentioned earlier, the two share the common media and facilities in the long/medium/short-haul arena, yet differ significantly in the distribution and broadcast techniques.

Digital networks do not differentiate between the signals they carry. The data rate of channels differs from application to application. Whereas a toll quality LPC (linear predictive code) coded speech signal can be "moved" at a modest rate of about 3 kbps [3.3], a subjectively well pleasing picture [3.4] requires 45 Mbps (44.736 Mbps or the DS3 rate) by using intrafield DPCM codec for NTSC color video. If the bandwidth is fiber cheap, stereo 3-D ultimate display to the limits of what the two human eyes can ultimately perceive can be conveyed at 810 Gbps [3.4].

The communication of television signal demands enhanced bandwidth from the network (about 0-4 or 6 MHz, analog TV; about 90 MHz, analog HDTV, without any sophisticated conservation of bandwidth; and 45 Mbps, HDTV quality). The final codec designs play a crucial role on the exact bandwidth/data capacity of the channel. Ultimately, most TV broadcasts are unidirectional to the end user or the subscriber. However, in the context of a network carrying TV signals, a network has to offer bidirectional capability (for programming functions) between selected television centers that monitor and manipulate TV signals from TV studios distributed in any major city and between the final broadcasters radio transmitter. The TV networks may be broadly classified as intercity networks and intracity networks.

3.3

INTRACITY TV NETWORKS

The television networks share the analog radio systems extensively for carrying information over the short, medium, and long-haul facilities that have nominal radio frequency ranging from 2 GHz to 39 GHz. Whereas metropolitan (intracity) networks may use systems that range up to 50 miles, including the common carrier facilities at 2 GHz for TV broadcasts, the intercity and cross country networks use short-haul systems for up to 250 miles and long-haul systems for up to 4000 miles. The TV transmission in the long-haul facility accounts for under 10 percent of the system usage. The two systems combined provide approximately 60 percent of the toll circuit mileage.

The data services offered by networks have to also meet the more stringent service requirement for data transmission. When speech is encoded, transmitted and decoded, casual errors in the data handling through the network do not drastically degrade the reconstructed speech. The intelligibility is not lost and the listener barely perceives the degradation. However, digital devices at the end of the network are not more susceptible to error. Hence, the established networks are obligated to offer digital services with a higher dependability and lower bit error rates. *Three* of the major measures of the enhanced requirement on the providers of digital services are: *(a)* percentage error free second (typically 99.5%); *(b)* percentage of service availability (typically 99.96%); and *(c)* quick service restoration (typically under 2 hours). In addition, the signal quality has to meet the stringent NTC requirements presented in Section 3.1.

3.4

INTERCITY TV NETWORKS

Within large cities, numerous and interspersed TV recording studios might exist. In addition, the TV signals from neighboring cities and areas may travel bidirectionally over coaxial cables. Generally, a TV operating center monitors the flow of the programming signals to and from the broadcasters central control facility. These links meet stringent quality requirements such that even multiple cascading does not produce notable degradation in the quality. In the United States, the National Television Systems Committee (NTSC) has specified the constraints on the scanning line and field synchronization pulses for the 525 line TV signals for the conventional broadcasts. This accounts for about 93 percent for the visible signal. The 7 percent blanking period carries the vertical synchronizing pulse and the equalizing pulse. The waveforms measurement for the relative amplitudes of TV signals has been standardized and well documented by the Institute of Electrical and Electronics Engineers (IEEE).

One of the common TV media for carrying the video signals is a 16-gauge insulated wire-pair shielded from the rest of the cable. The shielding consists of longitudinal and spiral wrapping of copper shields. Housed within the standard sheathing of local area telephone cables TV signal can be carried freely over limited distances. Special attention is necessary to limit the audio channels crosstalk in the high quality video channel within the shielded wire-pair. At the video frequency the wire-pair characteristic impedance* (see Chapter 7) is about 125 ohms. The variation of this impedance can be as high as ± 3.5 ohms in the older systems. At the high-end of the entire band (30 Hz to 4.5 MHz) the attenuation is about 19 dB per mile. The nominal value of the attenuation of the video pair is 17 dB per mile (4.0 MHz) at 67 degrees F. Accurate monitoring of the attenuation is necessary for maintaining the video quality.

The longer range (AT&T's) A2-type video system can span up to 15 miles with repeaters at a maximum spacing of about four and a half miles. This system has been used to link the Television Operating Center (TOC) and long-haul transmission facilities. A minimum signal level at 4.5 MHz to the repeaters in the A2- type video system is -60 dBV. The low frequency input level is -10 dBV. A pre-emphasis of 0 dB (low frequency of 30 Hz) to 32.5 dB at (high frequency of 4.5 MHz) is also employed at the transmitting terminal. The equalization is the inverse of the attenuation of the 16 gauge video cable. Both the attenuation and delay are controlled in this system.

Video cable loss of up to 80 dB at the high video frequency end can be compensated. The enhanced design of the A2AT system [3.1] can tackle distances of up to 30 miles with multiple repeaters. In this system 0/0 dBV (at low frequency/high frequency levels) at the input is pre-emphasized to the -10/22.5 dBV level and travels to a level of -10/-60 dBV level over the video link. The post equalization restores the levels to the input level of 0/0 dBV. If a next video link is necessary, then the pre-emphasis also becomes essential.

AT&T's A4 video system is generally used for short runs of about half a mile with the 16 gauge video cable. The bandwidth is 10 MHz and shorter

---

* $Z_0 = \sqrt{((R+jX)/(G+jB))}$, for details see Section 7.2.1

length (200 feet to 0.3 miles) systems are used for the A4 systems with other types of cable. Complex terminating impedances and different operating levels make this system incompatible with the A2 systems. Applications are limited to linking the TOC and with the microwave radio systems. Linking the video switch to the monitor and test positions also is done by the A4 system.

Video switching, testing, setting up various circuits, monitoring and maintaining is carried out from the TOCs or the television facility test positions (TFTP). TOCs also facilitate the switching between long-haul microwave systems after testing and adjusting the levels of the signals appropriately (both manual and remote controlled switching facilities). The functions of the TOCs become more elaborate compared to those of the TFTP that provides for intercity network and the local loops.

The allocation of the spectral bandwidth generally consists of *five* bands. *First,* the sub-VHF band accommodates 5 to 54 MHz signals and includes the T7 to T13 channels with 6 MHz spacing between the channels. Upconverting these channels yields the VHF channels 7 to 13. A single upconverter generally suffices the entire band conversion. *Second,* the low-VHF band accommodating 54 to 88 MHz sometimes includes the FM broadcast band of 88 to 108 MHz. *Third,* the mid-VHF band encompasses 120 to 174 MHz. *Fourth,* the high-VHF band covering the range of 174 to 216 MHz carries the channels 7 through 13. *Finally,* the super-VHF band covers a band of 174-216 MHz. All the bands and the channels are not entirely used in the cable environments. Enormous information capacity is still untapped in the frequency space of the cable TV network.

Other existing cable video channels include closed-circuit television (CCTV), community antenna television (CATV), educational television, (ETV), industrial television (ITV) and pay-TV. Each of these services has variations in the frequency bands and the number of channels they use. Accordingly, they are monitored differently and according to different tariff structures. At the present time, these channels and TV facilities are being considered for digital communication, switching and in the framework of digital networks. However, some of the TV systems and facilities are likely to compete with intelligent networks in providing specialized services in the modern digital network environments.

3.5

FEATURES OF EXISTING TV SYSTEMS

3.5.1 CABLE FREQUENCY ALLOCATION

The current video standard calls for a bandwidth of 4.2 MHz. The high definition TV (HDTV) requirement is about 30 MHz for each of the three (red, green and blue) signals without any sophisticated bandwidth reduction techniques. For conventional TV, the National Television Systems Committee (NTSC) has allocated 6 MHz band due to the amplitude modulation vestigial-sideband (generally referred to as AM-VSB), which transmits on complete sideband and only a visage of the other sideband. The frequency allocation [3.5] of the cable thus uses 54-72 MHz band for channels 2-4, 78-88 MHz for channels 5-6, 88-108 MHz for FM band, 108-174 MHz for the 11 midband channels, 174-210 MHz for channels 7-12, 210-300 MHz for the 15 superband channels, and 300-412 MHz for the 17 hyperband channels. The very low end of the frequency spectrum is sometimes used for upstream signals. For this reason the low end frequency band up to 30 MHz is called the upstream band. The currently deployed frequency allocation for the CATV systems is shown in Figure 3.2.

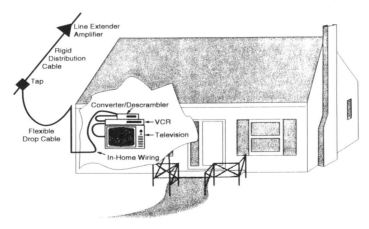

FIGURE 3.2. Typical Terminal Equipment and Cable Drop. VCR - Video Cassette Recorder

The actual use of the frequency space depends upon the channels and the type of coaxial cables used. The 0.5 inch diameter aluminum cable introduces a 10 dB loss/kft at 181 MHz. The corresponding loss is 5.9 dB for 1.0 inch diameter cable. The standard square root of frequency relation holds for the attenuation of these cables, thus making the channel 13 loss at 213 MHz, roughly twice the loss of channel 2 at approximately 57 MHz. The attenuation vs. frequency curve for a typical coaxial cable is shown in Figure 3.3. The loss depends on the type of cable (super-trunk, trunk or distribution; see next Section) and fairly rigorous monitoring of the signal-to-noise ratio is essential at every stage throughout the CATV network.

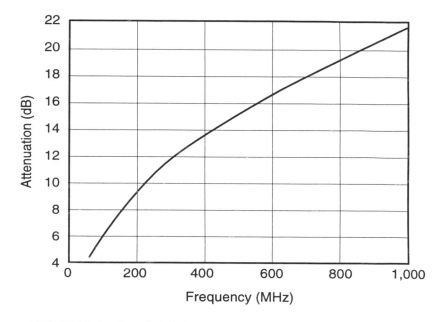

FIGURE 3.3. Coaxial Cable Attenuation (dB) vs. Frequency (MHz).

3.5.2 CATV NETWORK TOPOLOGY

The CATV uses a fairly standard topology for the distribution of the TV channels. The *five* elements of most CATV networks are depicted in Figure 3.4. *First*, the head end, consisting of a transmitter and multiplexer, combines signals from satellites, special long distance antennas, local signals, local studios for community access, video machines, etc., and other signal sources.

Second, the trunk cables and the broadband trunk amplifiers are every 2 kft, going to different population centers to service various communities.

Third, the distribution cable reaching the customer homes with bridged amplifiers boosts the signal for multiple or cluster homes with line extenders. The *fourth* element is the cable drop permitting TV signals to reach the customer premises. For the *fifth*, the internal in-home wiring goes to the converter/descramblers and TV sets. The distribution system and the flexible drop account for about 90 percent of the actual TV cable footage.

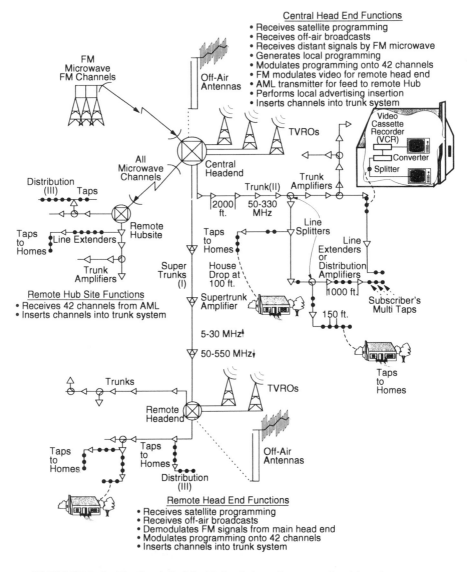

FIGURE 3.4. Typical Cable Television System Architecture.

3.5.3 CATV COAXIAL CABLES

Coaxial cables used for cable TV distribution to reach CATV subscribers have been standardized in the United States. *Five* sizes (0.412, 0.500, 0.625, 0.750 and 1.00 inches) specifying the outer diameter (without the cladding) are commonly used. The center conductor is generally copper-clad aluminum even though solid copper wire is also used. The dielectric material can be air, foam or solid polyethylene. For air core cables the support is provided by thermoplastic discs placed at regular intervals along the length of the cable. Loss considerations tend to limit the use of solid dielectric cable and cost considerations generally rule out the one inch diameter cables. Loss in the coaxial cables is of serious concern to maintain acceptable signal strength throughout the distribution plant reaching individual subscribers. Typically, for one-half inch cable losses, the air dielectric versus foam dielectric material at 5, 30, 50, 100, 200, 300 and 400 MHz are (0.16 vs. 0.16), (0.38 vs. 0.40), (0.50 vs. 0.54), (0.69 vs. 0.75), (1.00 vs. 1.08), (1.20 vs. 1.31) and (1.39 vs. 1.53) dB respectively for 100 feet of the cable.

For the CATV applications, the spacing of the amplifiers to compensate for the cable loss (without any taps) is typically about 30 dB for the 220 MHz applications and about 28 dB for the 300 MHz applications. A few tests to carry voice and digital data over the cable network have been attempted. The pulse code modulated carrier has been used within the existing CATV coaxial systems without any modifications. The quality for voice communication has been the same as that of paired wire telephone systems. For point-to-point digital communication the bit error rate (BER) is marginal due to a lack of component optimization. No digital repeaters were used since digital signals are carried at an RF frequency within the passband of conventional coaxial systems. With suitable design of components and reallocation of the usable bandwidth, the existing CATV distribution media can serve as dependable digital links in addition to serving the needs of the CATV industry. The bandwidth for communication over most of these systems can spread between 5 to 400 MHz reaching over distances of about 20 miles. The utilization of the communication capabilities of the coaxial systems already reaching individual subscribers is far from complete.

3.5.4 CATV SIGNAL QUALITY

The distribution quality is maintained by monitoring the carrier power to the noise of the system. Most cable systems aim to maintain the carrier to noise power ratio in the order of 46 dB and maintain the two composite (second order and triple beat) signals at about or below minus 53 dB. The signal level at TV is on the average of about 0 dBmV or 1 mV.

Sometimes it is necessary to boost the signal quality. Occasionally the digital video is transported over the cable. Adding fiber interconnects in the CATV network also enhances the signal quality. Analog-video fiber technology carrying 6 to 12 video channels is used, even though the on-going research is for 30-40 NTSC channels over one fiber for distances to 20-25 km. Currently this technology is used for studio quality transportation of TV signals from remote earth stations, studio-to-head end video transport, etc.. It is expected that when this technology matures, fibers running between nodes located at approximately the sixth trunk amplifier can directly insert broadband signals (combining 30-40 channels) into the coaxial network. Each node is thus capable of serving an area with about 1.5 mile radius. In essence, this is an inexpensive fiber backbone of the CATV network.

3.5.5 DIGITAL CAPABILITIES OF CATV SYSTEMS

The cable can carry digital traffic and special services based upon the data exchanges and two-way digital communication such as video, text, electronic publishing, home security and digital audio programming. However, to date, the public response to such data service offering is very low. Some CATV companies that had offered such services have retracted them because of the lack of demand. Thus, the CATV network is focusing most of the attention on efficiently and economically carrying video programs. For the present, it appears that the CATV networks are likely to remain entertainment-oriented rather than becoming competitive in data transport and communication type of services.

3.6

HYBRID FIBER COAX (HFC) SYSTEMS

Fiber capacity and its economic advantage to long-haul multiple gigabits of information are well documented in the telecommunications field. Fiber and its impact in the trunk and carrier systems are well suited, even in the older and more established POTS network. However, the capacity for the newer multimedia services (needing megabits per second data capacity) still remains to be completely exploited. The proposed hybrid fiber coax systems take advantage of providing a "fiber-backbone" network with ultra high capacity and quality. This carries data from video information providers, and broadband switches for entertainment, broadcast CATV, distance learning, etc., to "fiber nodes" and then span the "last couple of miles" via coaxial cables to reach the residences and business.

Coaxial cables for CATV reached over 50 million households in the United States in 1990, thus enabling 55 percent of households to access cable television. Eighty-seven percent of the households could have been connected to the existing cable services immediately. The rate of growth is sustained at one percent per year thus offering a high potential for services, (such as CATV, ATV, HDTV, video, picture, multimedia teleconferencing, interactive and participative video games, etc.) to grow in the residential markets. Providing access via cable and distribution via fiber combines the superb capabilities of both media for integrated broadband networks. The typical spectrum utilization for the existing CATV and proposed HFC networks are shown in Figures 3.5 a and b. The high end of the band (above 550 MHz) carries digital video signals.

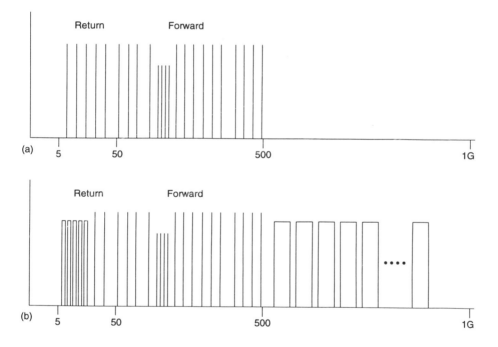

FIGURE 3.5. Cable Spectrum Utilization for the CATV Network. (a) Existing CATV networks. (b) HFC networks.

The HFC networks are not about to make a quantum leap towards becoming all digital all at once. There are no commercial all digital systems for HDTV applications as yet. A typical wait to 2002 or 2004 is expected for commercial all digital systems. However, in the interim, most HFC systems are likely to use analog video techniques through the end of the decade. The evolution of the analog CATV laser (typically known as distributed feedback laser or DAB) permits the direct modulation of AM multiplexed signals. In 1994, these lasers permitted 80-110 channel capacity, improved end of life sensitivity, analog/digital capability, higher splitting ratios, and optical redundancy. Laser technology for CATV applications is still being customized to the fibers of the HFC systems. In a laboratory environment at room temperature, QPSK techniques (see Section 6.6.3) have been used to demonstrate up to 60 channels through 24 km of fiber. Multiple quantum-well devices (see Chapters 16 and 17) are also being considered as possible contenders for CATV applications.

The deployment of fiber at the head end of a CATV distribution facility (see Figure 3.4) as fiber optic supertrunks (instead of coaxial supertrunks) brings a range of about 12 miles of passive transmission to the fiber optic hub. Typically, the maximum number of trunk cascades reduces from 32 to about 12. If a fiber is developed to provide a fiber backbone in the CATV environment, it is foreseen that the maximum number of trunk cascades reduces still further to 4. In the fiber to the feeder architecture, about 500 residences would be served by one optical hub receiving optical signals from the head end. The overall number of traditional supertrunk, trunk, and feeder amplifiers would be dramatically reduced, thus improving the video quality, the power requirement, and also the maintenance of the CATV plant. The replacement of trunks and supertrunks with fiber has been an ongoing process in the United States since 1988-89. A typical architecture of a HFC is shown in Figure 3.6. Here the fiber node uses an optical transceiver, and a linear bidirectional split band RF amplifier for the coaxial segment to handle the upstream/downstream signals.

The combined networks will become an integrated broadband (about 750 MHz) capability for most services needing increased bandwidth compared to the older POTS network or routine intelligent network (see companion volume, *"Intelligent Broadband Multimedia Networks"*). A typical architecture of this network to handle CATV, broadband data, and POTS services is shown in Figure 3.7. A broadband laser link to and from the head end permits CATV and broadband data traffic. Typically, the feeder fiber link (with five individual fibers) has to accommodate the following types of signals: phone transmit, phone receive, broadband transmit, broadband receive, and the traditional analog TV. Each of these nodes may support as many as 480 homes.

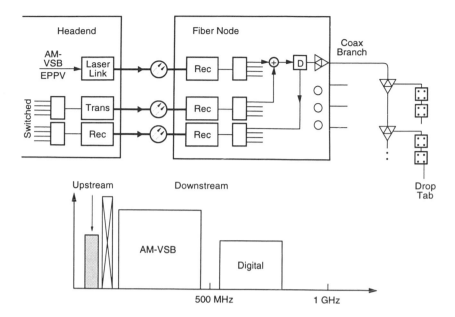

FIGURE 3.6. Typical Architecture of a Hybrid Fiber Coax Systems at Head End and Spectrum Utilization.

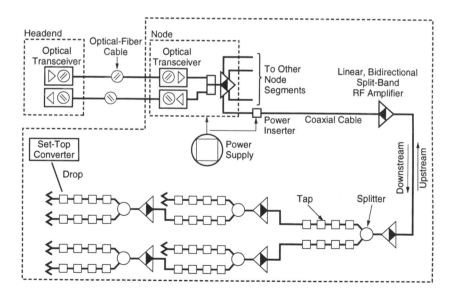

FIGURE 3.7. Typical Overview for a Hybrid Fiber Coax System. RF - Radio Frequency

The *five* fibers are deployed as follows: *one* fiber (the downstream video fiber) carries the same broadcast analog signal to all 480 homes; the remaining *four* fibers carry telephone/data traffic to 120 homes each. The downstream frequency allocation for each of the four fibers is 550-750 MHz, with 550-700 MHz for digital video services and 700-750 MHz for telephone. Each of the four fibers also carry upstream traffic in four equally spaced distinct 35 MHz bands within the 100-400 MHz band.

As can be seen, there is ample bandwidth for further application in the fiber part of the HFC network. A bottleneck exists in the cable part of the HFC network because the coaxial bus between the fiber node and the 120 homes has to accommodate an entire bandwidth from 5 to 750 MHz, since the 5 to 50 MHz can be used by conventional upstream and telephone signals. The overall broadband access architecture proposed [3.6] to/from the network service providers (with head end, broadband data and telephone services) from/to the individual home within the hybrid fiber coaxial network is depicted in Figure 3.8.

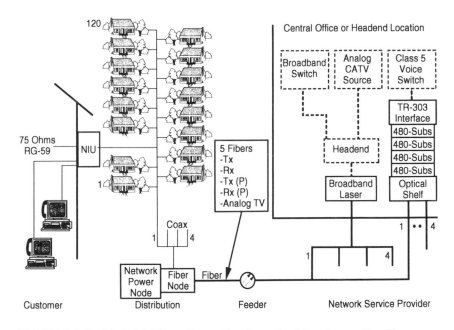

FIGURE 3.8. Hybrid Fiber Coax System Architecture. P - Phone; NIU - Network Interface Unit; Tx - Transmit Fiber; Rx - Receive Fiber; CATV - Cable Television; TR-303 - Transmit/Receive Interface in Class 5 Switch; RGn - Cable Type

Appropriate switching and software control in the newer network will permit symmetric and asymmetric flow of network information. Digital and programmed control (via stored program control in the switching systems) will permit high reliability and high quality of a mixture of customized telecommunication and entertainment services. It is the object of the fiber coax network to bring in the impact of digital technology and to carry a broad spectrum of multimedia services to the household rather than CATV or ATV services that plain old cable TV brings to the public. The bandwidth allocation and the multiplexing/demultiplexing for up to 120 homes is shown in Figure 3.9. The Figure compliments the layout shown in Figure 3.8.

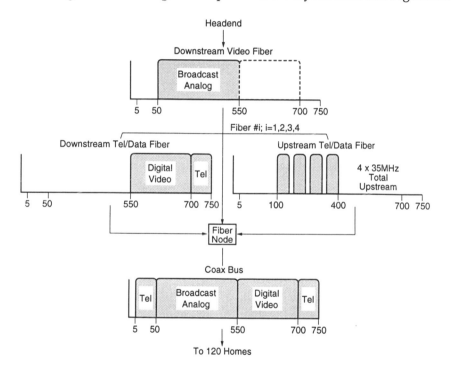

FIGURE 3.9. Schematic Representation of a Hybrid Fiber Coax Distribution System

In a sense, this network is likely to become the successor of the plain old telephone services (POTS) network, making the existing cables serve as access and/or distribution systems and making room at the top for fiber to carry much higher capacity data. However, the analogy stops here since CATV systems are not generally bidirectional to the same extent as POTS networks. Some of the services (such as distance learning, video-on-demand, enhanced pay-per-view, broadcast CATV, video-catalog rather than bidirectional interactive TV, video-conferencing, video-telephony, etc.) become more compatible with the new network. Converting the existing CATV plant for complete bidirectionality is expensive if not impossible. For services needing complete bidirectionality, other networks architectures incorporating microwave, all fiber networks with fiber-to-the-home (FTTH) or fiber-to-the-curb (FTTC) become more appealing for high bandwidth. For the very low bandwidth voice/facsimile application, the cellular radio network comes in as a strong contender.

It is the object of the newer fiber coax network to be rid of the major drawbacks of the existing POTS and CATV networks. There are *five* major factors that make the telephone network unsuitable for multimedia services: narrow bandwidth, labor intensive nature for changes or new services, relative inflexibility in the type of video/interactive services, high initial and operating cost, and its limited switching and interconnect capability. There are *three* major factors that make the CATV networks unsuitable for multimedia services: it is mostly analog and carrier modulated, unidirectional and less reliable than the POTS or other computer networks. The new fiber coax network offers the advantage of about 750 MHz of bandwidth, symmetric or asymmetric, highly reliable for voice and (analog and digital) video telephony, and relatively inexpensive. Present estimation is that such service may be feasible for about twice the CATV service rate.

3.7

CONCLUSIONS

The present day needs of the Cable TV subscribers are substantially different from the needs of the subscriber for information and intelligent network services. Some of the newer services (such as shop-at-home, video-on-demand, etc.) have features that overlap video and intelligent network services; they have not established a firm hold in the communication industry. It is a fond hope of the intelligent network service providers that such services can be met by intelligent wide-band networks (such as broadband ISDN, ATM, etc.). The CATV network and such broadband intelligent network appear doomed to become strange bedfellows in the emerging arena within the half glass-wired telecommunications scene. The possible entry of the fiber coaxial networks as viable carriers in the multimedia services to the home, makes the status quo of traditional analog CATV networks less and less likely.

The impact of digital encoding of video information and the HDTV technology are likely to have a strong impact on the TV broadcast and TV production industry. This effect is likely to persist throughout the decade and well into the 21st century. The slow and insidious trend in the TV environment is to go all-digital like most of the other communication systems. The rewards of an all-digital TV environment is already documented and the HDTV picture quality rates dramatically superior to the analog TV quality. When this technology reaches a viable commercial status, the emerging high capacity all-digital networks are well equipped to bear the digital video signals both for transport (typically the fiber-optic backbone and ATM technology) and for distribution (typically fiber to home facility, fiber coax facility or a perhaps new digitized CATV network environment)

3.8

REFERENCES

3.1 Bell Telephone Laboratories 1982, *Transmission Systems for Communications*, Fifth Edition, Bell Telephone Laboratories Inc.

3.2 Network Transmission Committee 1976, "Video Facility Testing, Technical Performance Objectives," *N.T.C.*, *Report No. 7*, The Public Broadcasting System, January. Also see Ciciora, W.S. 1990, W. S., "An Introduction to Cable Television in the United States," *IEEE LCS Magazine*, Vol. 1, No. 1, February.

3.3 AT&T 1985, *Telecommunication Systems Engineering*, second printing, New York, NY.

3.4 Netravali, A.N. and Haskell, B.G. 1988, *Digital Pictures: Representation and Compression*, Pelnum Press, New York, NY.

3.5 Ciciora, W.S. 1990, "An Overview of Cable Television in the United States," *Reprort of Cable Television Laboratories*, Boulder, CO.

3.6 Grout, S.C. 1994, "HFC-2000 Broadband Access System," *Private Communication*.

CHAPTER 4

CURRENT DIGITAL NETWORKS

Public domain networks and totally private networks constitute the extremes of network ownership and thus its organization and management. Whereas there is considerable freedom in the architecture and operation of the private network, the public domain network architecture and operation tend to be standardized and streamlined (see Section 4). In the discussion presented here, acknowledged international standards are used in context to the public domain networks and for ISDN. There are *two* major architectural variations (circuit-switched and packet-switched) in the networks used in the public domain.

4.1

MAJOR ARCHITECTURES

Circuit Switched Architectures: In these architectures (connection oriented), there are *five* stages for data transfer through a switched channel. The sequence may be summarized as follows: an idle channel is identified and tagged, a connection is established, the data is transferred, the channel is released, the channel resumes to be idle again. This sequence of stages parallels the well established network steps in a typical voice call which follows the network functions: call setup, alert, connect, disconnect and release. In reality, the detailed functions within the network far exceed the five major steps listed earlier. For example, at the outset, the line scanners detect the off-hook condition. The scanning rate is enhanced to collect the dialed digits in the older exchanges or the tone detectors are invoked to identify the touch-tone frequencies. The call setup procedure is initiated after the digit collection is complete. Such a sequence of well identified responses (network macros) are "assembled" until the call is completed as a successful call or is terminated by user or the network. Such responses (macros) are amenable to computer environments rather than to mechanized systems, and this is the basic reason for the "call processing" environments in most of the digital Central Offices.

In context to the ISDN, the information necessary to accomplish these individual steps are incorporated in the ITU-T (formerly CCITT) Q.931/931 protocol at the network layer used over the D (Delta) channel to set up the circuit switched B (Bearer) channel.

Packet-Switched Architectures: In these connectionless oriented architectures, individual packets of information are assembled with their own header block including its sender and destination addresses and the sequence number of the individual packet. Any one of the packets may transit through any variable number of nodes until it reaches its destination. The exact physical routing depends on the dynamic network conditions. The innate functioning of every participating node through the network facilitates the correct delivery of the packet. In context to ISDN packet-switching, ITU-T (formerly CCITT) has specified the widely accepted X.25, X.75, X.28, and X.29 protocol at the network layer (see companion volume *"Intelligent Broadband Multimedia Networks"*, Chapter 4).

At the network level, the two (circuit and packet) architectures function cooperatively. At the media level, the data appears as encoded signals. At the switch level, the routing and channeling functions occur in the (synchronous or the asynchronous) time division multiplex (TDM) framework. At the signaling level, the ITU-T (formerly CCITT) standards for both modes of data transfer are well accepted throughout the world. In fact the modern networks strive to be massively distributed, but well modularized computer systems. These networks are designed with the objectives of being as broadband as optical systems, of being as accurate as a computer system and of being as evolved as the OSI model with the ITU-T standards. This transition can only be a slow and insidious process because of the expanse of most national networks.

At the media level, wired, coaxial, wireless (cellular and microwave), satellite and optical facilities are most common. Frequency division (for the cable, wireless, satellite, etc.) and time division multiplexing exist for the synchronous (such as frame relaying in fiber-oriented SONET) and asynchronous (such as cell relaying also in fiber-oriented ATM) networks. Differences in concept and methodology do exist but the emerging all-digital networks tend to be broadband, architecturally well planned and designed to cope with congestion, well oriented to the OSI model, and operate with the worldwide ITU-T standards and protocol.

The network switch designers also strive to successfully switch all customer (circuit-switched, packet-switched, message-switched, private virtual, frame-relayed and cell-relayed) channels over all (optical, wired, cellular, wireless, cable) media with appropriate interfacing and signaling. In a sense, this assures the compatibility and cooperation between any of the major networks' ideologies. The integrated network can thus be all-digital and all-services oriented rather than be only POTS oriented as it used to be.

Private and Semi-Private Network Architectures: The networks (sometimes called channel-switched networks) may also use the standard components and associated interfaces. Generally, private branch exchanges (PBXs), private and dedicated lines, and digital distribution facilities with computerized information handling capacity may be encountered in private (intelligent) networks. These networks are emerging with alarming amounts of sophistication and intelligence. Their function can be as flexible and adaptive as circuit/packet/virtual networks. Typically, such networks are found in the scientific community (e.g., National Science Foundations NSFNET, Defense Advanced Research Projects Agency's ARPANET, Carnegie Mellon & Bell of Pennsylvania, Metropolitan Campus Network (MCN), etc.). Other examples from industry (e.g., VLSI vendors' networks) and national laboratories (e.g., Lawrence Livermore National Laboratory's private network for their employees, spanning 500 buildings) also prevail.

INTERNET: A Computer Network: Another example of the modern all-digital networks is the INTERNET with global access and variable (dynamically adaptive) bandwidth. Designed to be a computer network, it far exceeds its original target by being an equally effective communication network (in a rather limited sense of its user and access capabilities). But for users being proficient in both (personal) computer and communication systems, this network offers extensive access, services and dynamic adaptation. Occasional abuse and encroachment of services remains to be problem for these (very) inexpensive communication facilities. Though not a direct threat to the modern communication systems, the role of networks such as the INTERNET cannot be ignored. In a sense, these networks invoke almost the real-time response that communication systems offer and are almost as data-transparent as the evolving networks (ISDN, SONET and ATM) are striving to be.

The INTERNET packets (and thus the user messages) get through the network fast enough (depending upon the network congestion) most of the time. Fast packet technologies and strategic location of the switching nodes within these networks permit the user to get the "almost" real-time response but at a small fraction of the cost of a regular/dedicated communication-network channel. INTERNET has its special protocol (the INTERNET Protocol), four number (each less than 256 i.e., 2^8) address and 1-1500 byte packet-size. The transmission control protocol (TCP) deploys checksum parity to detect errors in the received packets and retransmission of error ridden packets. User datagram protocol (UDP) may also be used to convey short messages rather than the packet or piecemeal delivery of the long messages.

The SS7 Network: This supporting network has emerged because of necessity. This network is crucial for global, regional and local communication networks. The need to control and monitor the flow of customer information is vital in view of the volume, diversity, routing and destinations of information handled. A detailed discussion of this network is presented in the companion volume, *"Intelligent Broadband Multimedia Networks"*, Chapter 3). An information explosion has accompanied the information revolution. The older signaling methodologies (single-frequency, multi-frequency, in-band, and even the CCISS number 6 systems) have gracefully given way to the SS7. The SS7 all-digital network which handles the signaling (to control and monitor most of the information bearing network) is sometimes referred to as the backbone signaling SS7 network since signaling is essential to the functionality of any modern network. Signals in networks are as important as the opcodes in computers and signaling network is as important as the control (memory and its associated) circuits in computers.

4.2

ISDN AND DIGITAL TELEPHONE ENVIRONMENTS

Digital telephone environments allow for transmission, switching, routing and storage to be accomplished in the digital mode. The channel control of voice and speech channels is also digital. The encoding and decoding of speech are both done at the telephone set. Even low bit rate speech (such as stored messages and announcements) is reconverted to analog form at the telephone set. All the inherent advantages of digital computing and processing techniques can be imported in the digital telephone networks. In a sense, the ISDN framework blends digital telephone network features with other digital services such as packet-switching, message storage, forwarding, automated message selection, etc.

ISDN as a Universal All-Purpose Digital Services Network: Signaling is accomplished over the SS7 network fairly well evolved and adopted in the United States and Europe. This network uses the out-of-band information (i.e., the information on the D channel), to control and signal the various switches to complete and monitor the B channels. Figure 4.1 depicts the role of the signaling and control of the information bearing B channels.

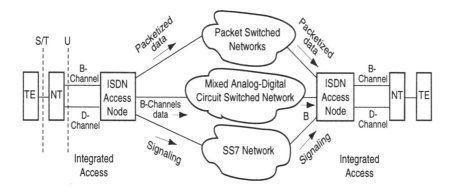

FIGURE 4.1. ISDN Architecture Showing the Separation of Signaling Packet and Circuit Switched Data. ISDN - Integrated Services Digital Network; NT - Network Termination; SS7 - Signaling System 7; TE - Terminal Equipment

This network becomes essential in the circuit-switched context because the B channels provide transparent end-to-end digital connectivity for the network users. If and when a transition to the packet mode on the 'B2' channel of the BRISDN is to occur, then the SS7 signaling information is provided for the transition and vice versa. BISDN and N-ISDN functions also depend upon the separation of the customer data in the 'B' channel from the control and packet data in the 'D' channel. Figure 4.2 depicts the functional details of an ISDN node including customer interfacing, channel routing, and D channel splitting to separate control signals from packetized/localized control and telemetering information (if any).

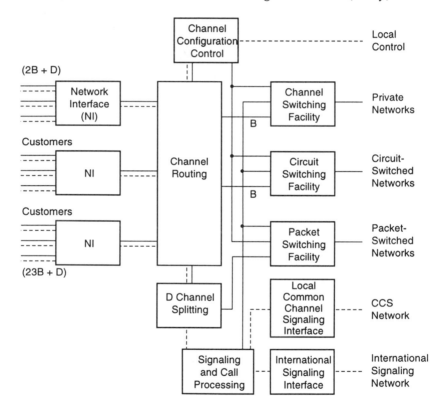

FIGURE 4.2. Architecture of a Gateway Node. The need for Common Channel Signaling Network to control and monitor the B channels via the D Channel in reference to the ISDN environment is depicted.

B (Bearer) Channel and D (Delta) Channel of ISDN: The B channels are switched throughout the network to carry data from its source to the destination in a bidirectional mode. The signaling information on the D channel carried by the SS7 network conveys the address and network control data at their appropriate nodes. The data on the D channel is demultiplexed from signaling information for the B channels. This D channel signaling information may be used for line functions, and for packets assembly functions for user packet switching. The telemetry channels (if any, for use in the United States), are separated out and granted local access within the localized area of the Central Office, telemetry services, security offices, etc..

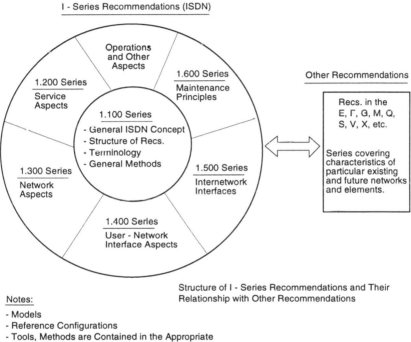

I - Series Recommendations (ISDN)

Other Recommendations

Structure of I - Series Recommendations and Their Relationship with Other Recommendations

Notes:
- Models
- Reference Configurations
- Tools, Methods are Contained in the Appropriate
  I - Series Recommendations.

FIGURE 4.3. ITU-T (CCITT) I-Series Recommendations (ISDN).
CCITT - International Telegraph and Telephone Consultative Committee; ISDN - Integrated Services Digital Network; ITU-T - International Telecommunications Union - Telecommunications Standardization Sector (formerly CCITT)

The network transition to its final ITU-T (formerly CCITT) defined ISDN standards (I-Series Recommendations as shown in Figure 4.3) is a slow and tedious process in the United States. For instance, the circuit-switched digital capability (CSDC) at 56 kbps introduced in the early 1980s uses a 64 kbps (DS0) channel, but uses the extra 8 kbps for signaling and other line functions, which would have been done via the D channel in the standard

ISDN service for controlling either of the two 'B' channels. In fact, the CSDC transmission scheme [4.1, 4.2], switching and signaling concept is itself redundant, if either of the two major ISDN services, BRISDN, i.e., 2B+D, or the PRISDN i.e., 23B+D, in the United States is available. For both these ISDN services (depicted in Figure 4.4), the channel banks (see network interfaces in Figure 4.2) and signaling facilities need to be updated to comply with the ITU-T recommendations (see Figure 4.3). In this instance, the CSDC service has not been accepted as a global service throughout the rest of the world.

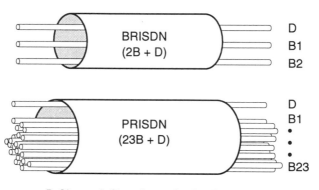

D Channel: Signaling + Packet Data
B Channel: Clear 64-kbps Digital Access

FIGURE 4.4. Two of the Major ISDN Access in the United States.
The Primary Rate ISDN in Europe has (30B+D) channel configuration.
BRISDN - Basic Rate Integrated Services Digital Network; ISDN -
Integrated Services Digital Network; PRISDN - Primary Rate Integrated
Services Digital Network.

The emerging digital networks specialize in transporting any amount of data to any user defined rate over the high capacity links, and to switch the user channels by the innate time-division multiplexing capability within the digital switches. The most evident realization of these networks for data are the ISDN (basic-rate, primary-rate and broadband), the SONET and ATM networks. It is also to be appreciated that some media is more suited to the rather specialized network encoding and switching techniques. For example, the fiber optic systems suit the implementation of the SONET and ATM techniques rather than twisted wires, whereas the twisted wire-pair media suits the digital subscriber loop (DSL), the high-speed digital subscriber loop (HDSL), ADSL and the recent modem techniques, etc. Force-fitting technologies is sometimes uneconomical and sometimes impossible because of the differences in media characteristics, but some broadband technologies (such as ATM) can be implemented in the fiber, coaxial, and even special wire-pair (premises distribution systems or PDS) environments. It is the vision of modern network designers to provide all all-digital network services to anybody, at anytime, at any (data) rate, anywhere in the world, provided the customer is willing to pay for such a service. This is the beginning of global information highways in earnest because of the capacity, the range of services, the flexibility, and most of all the economic viability in the global arena.

4.3

CAPACITY OF EXISTING TELEPHONE NETWORK FOR ISDN

Whereas ISDN brings the digital revolution to the POTS environment, ISDN is itself a precursor for the more intelligent broadband multimedia communication services to follow. The telephone network, generally the most dispersed form of existing networks, is the most viable means for introducing new services brought about by the digital and network revolution in the public domain. Cellular radio networks are also rapidly contending to reach a large cross section of the population through the deployment of a shared, pre-allocated radio frequency spectrum. From a hierarchical consideration, the trunk facilities are capable of carrying higher capacity speed data and a larger number of channels. On the other hand, the subscriber loop environment and the cellular radio, being the last links over which the information has to be communicated to the customer, are restricted by inherent range-bandwidth constraints.

Two major conclusions regarding transmission (Section 4.3.1) and switching (Section 4.3.2) are emphasized here to stress the compromise between cost of the digital subscriber lines and digital capacity of the loop plant. The discussion is limited to rates below the 144 kbps. Rates substantially higher than the 144 kbps and T1 rate are also possible with the current (mid 1990s) technology. These limits are investigated and presented in Part IV.

4.3.1 TRANSMISSION ASPECTS

Since the early days (1950s and 1960s) within the twisted wire-pair environment, repeaters and specially selected wire-pairs have been used to carry data at the T1 rate between the switching centers and large customers. More recently, (early 1980s) the cost of conditioning the line, repeating the data, and the costs incurred in making the line traffic worthy are rapidly making this T1 type of system to reach every customer an uneconomic if not a futile proposition. The high labor costs in conditioning the line are a prime factor in rejecting conditioning of lines. In addition, T1/E1 rate data (i.e., data at PRISDN in the United States/Europe) is generally not a requirement for every customer. Between these two extremes of data rates, voice-band modem data (at a high rate of 28.8 kbps) and the 1.544/2.048 Mbps T1/E1 rate data, other rates such as B+D at 80 kbps, 2B+D at 144 kbps, 5B+D at 384 kbps, 11B+D at 768 kbps also exist. These options at intermediate rates are not seriously considered in Europe.

One of the options to enhance the data rate is to use the existing wire-pair connections with suitably designed line termination for the data traffic on these twisted-wire-pair channels. This direction of investigation was initiated as early as the mid 1970s and led to viable and robust systems for data ranging from 56 kbps to 324 kbps over the twisted wire-pair telephone line subscriber loops. The results of the system design have been presented in detail in Part II (also see Reference 4.3). *Two* factors that had been foreseen in the past are listed with a modern perspective.

First, the basic ISDN rate at 144 kbps over the existing subscriber plant can be economically communicated over most of the loops to 18,000 ft. (5.48 km) in the United States. The 2B1Q code presently chosen by the Bell Operating Companies (BOC) provides extra protection against loop plant noise contamination of the digital signal, especially over long customer loops. The carrier serving area (CSA) concept proposed by the BOC further facilitates the communication of data at the basic rate and brings the H_0 (i.e., 5B+D) rate, at 384 kbps in the ISDN context, closer to reality. The introduction of the fiber optic technology further enhances the viability of the true broadband ISDN (BISDN) directly up to the customer premises.

4.3.2 SWITCHING ASPECTS

Second, the switching technology that had been singled out as the major bottleneck [4.4] in the final implementation of the broadband ISDN is undergoing closer scrutiny. The Electronic Switching Systems (ESSs) currently in use utilize semiconductor devices for switching and routing digital channels. Wire centers of the past are being replaced by the digital channel banks that offer flexible and programmable architectures. The component limitations associated with ICs severely restricts their speed and capacity. The eventual limit on the functioning of these devices is due to the distributed capacitance and resistance within the chip layout. The finite capacitance in conjunction with the finite resistance ultimately dampens the rise time and the fall time of pulses. The capacity necessary for switching becomes too enormous if every customer demands broadband ISDN services all at once.

Optical switching (though not commercially realized just as yet) has promise beyond this limitation.   The already introduced fiber optic technology (fiber to the home, fiber to the curb, SLC-2000®, presented in Chapter 1 and in companion volume *"Intelligent Broadband Multimedia Networks"*) in the subscriber loop plant CSA systems is designed to carry high-speed data. The optical switching is yet to be realized in the switching centers. Together, these technologies, compliment each other in the evolution toward ultra high capacity digital facilities (in multigigabits per second). In harmony, they form the next generation communication systems, if the society has need for such networks. A technological evolution of the communication systems from the electronic domain to the photonic domain will be complete. In this totally optic environment, information highways will be as common as telephone channels and the information explosion would have reached every home, if the subscriber is willing to pay for it. Fiber optic transmission technology per se, is current and routine, but optical switching is yet to be viable in the switch and in the remote terminals.

4.4

DIGITAL FEATURES OF THE EXISTING TELEPHONE NETWORKS

The advantages of digital processing over analog processing are well documented and widely accepted.   This is especially the case in the telecommunications   arena.   Similar   advantages   exist   in   digital communication of information. Digital control of communication networks and channels (by digital switches and the D-channel banks) offers more dramatic improvements in speed and accuracy. Such features was conceived very early in the switching of the telephone channels. For example, the early electro-mechanical type [4.5] switch operated by "clicks" or electrical pluses at a rate of 10 pulses per second generated by a rotary dial phone.   The switch would then respond by a vertical movement and then, a rotation of a rotary wiper to correspond to two digits of a call. Such numerous banks would be cascaded to complete a multi-digit call. Transformations from the two digits to connections were simple but slow due to the number of vertical steps for the plunger and the discrete angular movements of the wiper. Thus, the discrete number of vertical "plunges" and discrete number of "angular rotations" simply correspond to the two digit numbers dialed by the users or operators of the forties. A typical 10X10 switch matrix of a step by step system (from the 1920s) is depicted in Figure 4.5.

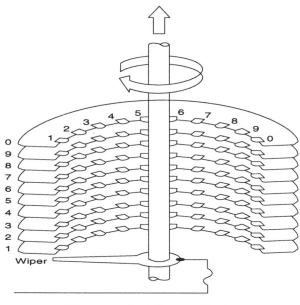

Wiper and Fixed Contact Matrix
for One Set of Conductors
(Tip, Ring, or Sleeve)

FIGURE 4.5. A 10x10 Matrix Switch in the Older Step by Step Switching Systems.

In a sense, these types of switches though discrete are not digital. They correspond to the era of mechanical devices that made up the first few mechanical calculators built by Schickhard in 1623, or by Pascal in 1642, or by Leibniz in 1671. The move to digital data processing was completed by Aiken's design in 1937. Such machines seemed to have reached a point of saturation at the third generation electronic computers and are still awaiting the implementation with optical computing devices.

In the telecommunication environment, the move to digital control of the switch occurred in 1965 by the introduction of the first ESS. Such systems are still being updated to be compatible with each and every type of communication channel. Much like computing systems, communication systems are still waiting to be completely optical!

In the more recent past, digitization of telephone networks is considered in *two stages:* digitization of switching systems and digitization of the communication facilities. Digitization of the switching systems is discussed in detail in companion volume, *"Intelligent Broadband Multimedia Networks"*, Chapters 9 to 13. Digital communication is also considered in *two* steps: digital communication between the user and Central Office (i.e., the digital subscriber loop), and between other Central Offices (i.e., all digital trunking facilities). Digitization is a well conceived move in almost all aspects of telecommunications; it is the strategy of implementation that is different. Differences between the interoffice (i.e., between offices) and local loops media and facilities account for the differences in strategy for digitization. This creates a telephone network ready to become all-digital, such as the proposed ISDN, and be able to feature any of the IN services. Digitization of the communication systems, including the localized channel switching (of channel banks) and small offices, are discussed here.

4.4.1 DATA FLOW TO/FROM THE CENTRAL OFFICE

A large amount of freedom exists in communicating data between the local Central Offices and the end user. Traditionally, the subscriber lines have been open/twisted wire-pairs or microwave in some remote areas. More recently, microwave-cellular, VSAT, plain wireless, coaxial, hybrid-coaxial, and even fiber links are also becoming feasible. Each of these links have distinctive features and the network designers exploit their features to provide the best combination of service(s) to the customers.

Numerous types of Central Offices exist. Traditionally the classification has been Local/End Office, Toll, Primary, Sectional and Regional Offices (see Section 1.5). The functions and the access of these Central Offices vary. Numerous modes of communication over the physical media also exist. Typically, two-wire (i.e. full-duplex or one wire-pair for both directions of communication), four-wire (one wire-pair for each direction) and carrier systems (analog and/or digital systems) carry voice/data. The mode of operation could be simplex (i.e., one direction), duplex (both directions simultaneously), dual duplex (two sets of lines each carrying half the rate in the duplex mode), TCM (time compression multiplex, i.e., one direction in one time slot and the other direction in the next time slot) or even FDM (frequency division multiplex).

The current United States environment also permits BRIDSN and 56 kbps service in addition to the various subrate (2.4, 4.8, 9.6. 14.4 and 28.8 kbps) services. Basic rate ISDN service is routinely offered in large sections of the country and uses the a two wire, full duplex service at a customer rate of 144 kbps. Adaptive hybrid-echo-cancellation is used to isolate the effects of the reflected echo from the received signal.

The 56 kbps service is offered under the older four-wire at the 56 kbps under the Digital Data System (DDS) with ITU-T (formerly CCITT) V.35 standard interface or the more recent CSDC service offered under the two-wire TCM mode [4.3] throughout the country. Finally, subrate data capability (discussed next) is the most prevalent telephone network data service over two-wire systems. In some of these services line conditioning was a primary requirement. The recent high grade modems permit data rates to 19.6 kbps over unconditioned lines. In a majority of cases, these modems permit the highest possible rate depending upon the subscriber line. Some adaptive features are also incorporated in the V.bis and V.fast modem technology.

4.4.1.1 Lower Data Rates

Amplitude modulation of signals in existing networks is not always practical because of the varying attenuation of the medium and variable switching at different local levels. However, the frequencies do retain the integrity in most of the existing voice networks. Hence, the early very low rates (300-1200 bps) for communicating data resorted to frequency shift-keying techniques for asynchronous data transmission. The condition of the line in use ultimately restricts the rate at which data can be exchanged. Higher rates are possible with specialized lines and/or special channel units at the end of the line. Four phase shift-keying (PSK) provides synchronous rates of 2400 bps with the end unit or the data set generating the clock. Eight phase PSK offered rates at 4800 bps over public and/or private networks. For higher rates, quadrature amplitude modulation (QAM) provides 4800 bps on public networks or 7200 bps on private networks in a synchronous mode of data transmission. At 9600 bps on private lines, QAM techniques are used in conjunction with automatic equalization. Some form of equalization may also become necessary at 4800 bps over public networks.

The bit error rate (BER) can vary depending upon the rate and line condition. Most of the random tests at the 1200 bps offer better than 10^{-05}

BER. At the 4800 bps, the BER may become 10^{-04}. The measure of burst error also offers medium dependability with these systems. Typically, a 1000 bit block faces a 5^{-3} probability or less of being error-prone in about eighty to ninety percent of the calls at the lower bit rate.

When the subrate data signals (at 2.4, 4.8, or 9.6 kbps, also classified as the DDS rates) are received at the Central Office, they have to be repeated or multiplexed (see DS0-A or DS0-B formats in Section 4.7) onto a standard 64 kbps channel and two basic functions (line monitoring and channel identification) become necessary. The customer terminal for these three subrates is the standard RS-232-C interface. The combined effect of these functions essentially reduces the capacity to 48 kbps. Thus each byte in the successive T1 frame gets assigned to five 9.6 kbps, or to ten 4.8 kbps, or twenty 2.4 kbps channels. The consistency between the subrates and the overall 64 kbps rate is maintained. The subscriber digits at the lower rates between 2.4 through 9.6 kbps use a RZ (return to zero) fifty percent duty cycle bipolar (AMI) coding to reach the serving area Central Office. Here the 64 kbps (DS0) rate also uses the bipolar code with 100 duty cycle and a NRZ (non return to zero) coding to reach the D channel bank converting the rate to the T1 rate or 1.544 Mbps.

The capacity to achieve a rate of 9.6 kbps over plain old telephone loops is achieved by using the more efficient line codes. Many such codes have been designed and two very efficient and closely related codes in the communication of data are the Ungerboeck and/or trellis codes discussed in further detail in [4.6]. When the information bits are mapped in signal-space (i.e., each symbol having unique amplitude and phase coordinates), then efficiently encoded symbols at a lower rate of 2400 symbols per second are communicated. Thus, four bits can be mapped onto one symbol to offer a 16 point symbol constellation of the quadrature amplitude space. In practice, four bits are mapped in a 32 point constellation. This increases the Hamming distance between the symbols thus reducing the probability of error at the receiver. The receiver is thus capable of placing each symbol accurately and regains the bits used when the symbol is received.

4.4.1.2 The 56 kbps Data

For the four wire 56 kbps digital data service (DDS), one repeater may be used. Data is transmitted in the bipolar (AMI) form. Bipolar violations are used to signal special conditions. Specially coded signals are used to indicate *four* special conditions: *(a)* Distant data service unit (DSU) is idle, *(b)* error condition in the receive path, *(c)* DSU under test, and/or *(d)* there are seven or more zeros in the customer data stream. ITU-T (formerly CCITT) V.35

standard interface with specialized line units, DSU at the Central Office end and channel service unit (CSU) at the subscriber end need careful installation. Loop selection is also stringent. No loading coils may exist, and the maximum allowable insertion loss (at half-rate frequency) is 31 dB, thus defining the range of pure 19, 22, 24, and 26 AWG cables at 40.8 kft (12.44 km), 24.2 kft (7.38 km), 17.2 kft (5.24 km) and 12.9 kft (3.93 km), respectively. The permissible loop topology also restricts the bridged tap configurations. Similar restrictions also apply to the sub-rate data facilities.

For the newer 56 kbps CSDC applications over repeaterless two wire subscriber lines, the 8-bit byte of 64 kbps channel has to contain one reserved bit (signaling) for network control. To carry the minimum, one's density for timing recovery of the T1 channel is eventually going to contain the standard 64 kbps channel. Hence, seven of the eight bits may carry information relevant to the user channel. The actual bit rate reduces to 56 kbps. In addition, if the information at the subrate has to be channeled into an appropriate port of the receiver, and then the data clock (of the submultiplexer) at the receiver has to be synchronized, one additional bit is called for, thus reducing the effective rate to 48 kbps.

4.4.2 CENTRALIZED SWITCHING OF DIGITAL CHANNELS

The switching of digital channels is accomplished in Central Offices or switching centers. These facilities may be large, switching many hundreds of thousand channels, or quite small private branch exchanges (PBXs) serving a limited number of subscribers. A typical example of a large switching facility is the 5ESS facility capable of handling up to about 200,000 lines, with compatibility for the emerging services (multimedia, hybrid-fiber coax, integrated ATM and the full range of packet switching) at the switching modules. Other examples of such switches are Siemens' EWSD, Alcatel's 8300 System, Ericsson's AXE-10, Northern Telecom's DMS, etc. In this section, discussion is confined to large exchanges that may serve larger regional areas.

Toll switching (i.e., switching between hub offices located in different toll areas) has been in use since 1926 in the United States. The demand for switching capacity at the Toll Offices has been met by a steady introduction of the more sophisticated switching systems. The evolving technology brought about the crossbar architecture in 1941 In 1965, technology brought electronic switching systems (ESS) to the local switching systems, and to toll switching in 1970. The last of the non-ESS, i.e., the crossbar No. 4 introduced for toll systems in 1969 uses common control with markers to identify the sender, connectors, etc. The ESS use the stored program control to monitor and complete the switching in the network. These switching systems become focal points in carrying large amounts of circuit data within the network. The progression towards the ATM environment in the 1990s is forcing new architectures for these switching systems that will permit them to handle "frame" and "cell" relaying through the older circuit switches.

4.4.3 ELECTRONIC TOLL SWITCHING AND DIGITAL CHANNELS

The notable ESS designs for larger switching centers are the #1 ESS (up to 65,000 lines) introduced in 1965 and the #1A ESS (up to 128,000 lines) introduced in 1976. These two ESS facilities have largely been used for the larger metropolitan areas throughout the world. Smaller systems, such as #2 ESS introduced in 1970, #3 ESS introduced in 1976 with remreed, #2B ESS, also introduced in 1976, but with ferreed switching elements, have also been built for rural and suburban areas. The #1 ESS has been built for Toll Central Offices for a maximum of 180,000 lines. These systems use the ferreed (shown in Figure 4.6) or remreed switch (a more compact electronically controlled version of the ferreed switch) for establishing the actual cross point connection [4.7].

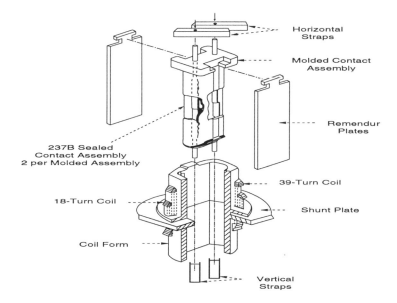

FIGURE 4.6. A Cross Section of the Two Wire Ferreed Switch to Connect/Release the Call. The stored program controlled currents in the two coils activate the make, hold and break connections at the contact assembly.

Both ferreed and remreed switches make the cross connections in a switching matrix by sealed, miniaturized relay contacts activated magnetically by short duration control pulses to establish or release the contact. Two control currents (row and column) are both necessary to activate a switch. The current-in row or column alone opens all the contacts in that specific row or column. In principle, the operation is similar to the functioning of a ferrite core memory to store or retrieve bits of information.

Elaborate controls are necessary to ascertain the cross point in any row or column that can be operated at a time. These connections may be made for two-wire duplex systems or for the four-wire circuits in the office. The cross connect switches are generally arranged in groups of 64 (8 rows and 8 columns).

Current surges to operate the relays (and other causes such as lightning, power line switching and dial pulses) cause impulse noise surges in the transmission paths. Differential delays due varying path lengths can cause pulse transpositions especially when these devices are controlled by the currents to perform the time division multiplexing (TDM) function with 4ESSs.

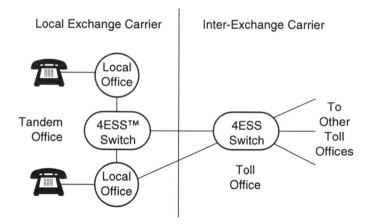

FIGURE 4.7. Role of the Tandem/Toll Office Switch in the Central Office Hierarchy. ESS - Electronic Switching System

The 4ESS was introduced in 1976 with an initial capacity to receive 107,520 trunk lines and was capable of handling 500,000 calls per hour. The key component of this toll switch is the 1A processor with busy-hour call-attempt rates of about 800,000 per hour. The switch fits in the network Central Office hierarchy elegantly (see Figure 4.7).

Since 1994, [4.8] the 1B processor, with about 2.4 times the throughput of the 1A processor, has been deployed. The busy hour call attempts (BHCA) with this processor are demonstrated to be well over one million. The architecture of this switch (see Figure 4.8) accommodates a flexible switching fabrics platform. It is the switching platform that contains space-division (different trunk termination), time multiplex switch which interconnects numerous time slot interchange units. At the trunk, the T1 [i.e., D1 or 24 (voice/data) 64 kbps, DS0 rate] channel bank interfaces with the core switch fabric. A network clock keeps the network and the switching fabric synchronized, and also retains the synchronization between various signal processors for in-band signaling, maintenance and control.

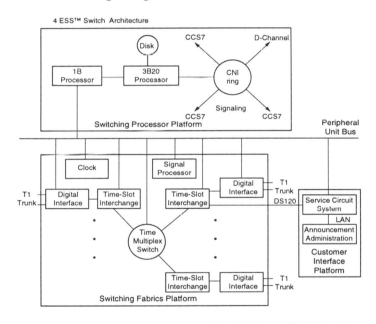

FIGURE 4.8. Generic Architecture of a Modern ESS Tandem/Toll Office Switch. 1B - A Type of Switching Processor; 3B20 - Type of Switching Processor; CCS7 - Common Channel Signaling 7; CNI - Common Network Interface; DS-120 - A Formatted Digital Signal; ESS - Electronic Switching System; Tn - Standard North American Carrier Lines Used to Transmit a Formatted Digital Signal

This switch serves as a computerized software controlled structure that also uses solid-state switches (rather than the ferreed or the remreed switches) for making and releasing connections. The noteworthy feature of these systems is the time multiplexing feature implemented to make multiplexing and demultiplexing of numerous channels (individually or collectively) extremely easy and programmable. Gates and logical devices, readily available computer hardware, can be used to perform the channel switching functions in the ESS environment.

In fact, the modern ESS perform very much like the modern computer systems. All the hardware, software and firmware concepts are applicable in this environment. However, it has its own operating system, its own specialized version of the office data administration system, and its own special data bases. Basic call processing capability is an integral part of the high level modular software needed to operate the switch. Service related functions can also be provided by the switch by invoking the service circuit system shown in Figure 4.8.

Typically, at a high level of software control, the switching platform can be instructed to create a 64 kbps (DS0) voice/data path through the switching fabric to complete a call. Call processing macros can also be implemented. The channel banks provide an interface between the switching fabric and the lines that also carry network information in analog or digital form. The switch, at its highest capacity (as of 1994), can provide 131,072 - DS0 paths to 32 expanded time-slot interchanges as shown in Figure 4.8. There is no line concentration at the edges of the switching fabrics platform. The channel banks are physically separated out to serve the individual trunks. In addition, there is one channel unit for each of the 24 channels of the D1 bank serving a T1 line.

4.5

CHANNEL BANK FUNCTIONS

For the voice circuit, the channel bank provides the D to A (digital-to-analog) and A to D (analog-to-digital) conversion. The signaling bit (which informs the channel units about the signaling/non-signaling state of that particular channel), and the framing bit (to verify that the T1 frame is maintaining synchronization at the transmitter and at the receiver) is also added in the channel bank.

The carrier systems transfer digital data to and from the Central Offices. The composite train of these three signals [4.9] constitutes the T1 carrier digital signal. The lines terminate in the D-type channel banks. The analog-to-digital converter is included for the voice channels during the multiplexing phase for data entering the carrier systems at 1.544 Mbps. Converse processes occur for the received signal. The digital signals after decoding, demultiplexing and digital-to-analog conversion go to the line units for traversing to their appropriate destination via subscriber loop or other toll circuits. The filtering, the channel selection, the multiplexing and demultiplexing of the signals from and to the various customers is carried out by channel banks. The encoding and decoding of the AMI code to the unipolar code respectively, for the T1 carrier system also takes place in the channel banks.

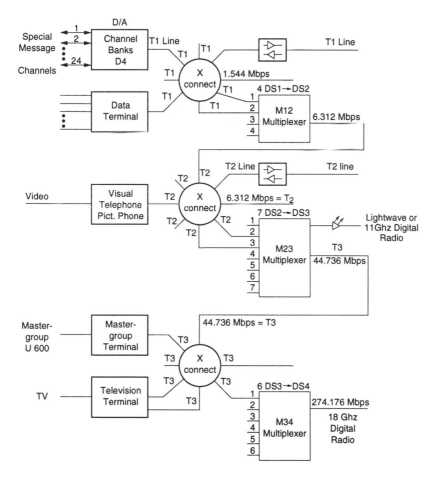

FIGURE 4.9. Implementation of the Pre-SONET Digital Hierarchy with Up-and-Down Multiplexers. Also see Figure 4.17. American Digital Hierarchy (DS0-DS4). DS4 - A Formatted Digital Signal; FDM - Frequency-Division Multiplexing; TASI - Time Assignment Speech Interpolation; TDM - Time-Division Multiplexing; Tn - Standard North American Carrier Lines Used to Transmit a Formatted Digital Signal; VF - Voice Frequency

The D channel banks also work in conjunction with the digital multiplexer units. These multiplexers (in reality mux-demux units) are essential to traverse the digital hierarchy up and down the *six* levels in the North American digital hierarchical layers: DS0, DS1 (frequently known as T1), DS1C (sometimes known as T1C), DS2 (known as T2), DS3 (also known as T3), and DS4 (occasionally known as T4).

Figure 1.3 depicts the *six* layers and shows the M1C, M12 M13, M34 multiplexers in some detail. The CEPT hierarchy is also shown. In Figure 4.9, the layers are redepicted, and the D channel banks and the cross connects are also shown to illustrate the switching aspect in the digital communication capability of the network.

4.6

D CHANNEL BANKS IN THE UNITED STATES

There are *four* distinct D channel facilities. *First*, the D1 channel bank facilitates the termination of the T1 lines and the functions associated with the 24 voice channels that they carry in the earlier usage of this digital carrier system. Different versions (D1A, D1B, D1C, and D1D) of the D1 bank perform the specific functions slightly differently. From a network point of view, these variations have insignificant effect except in the final stages of digital data or the voice distribution.

Second, the D2 facility serves 96 channels at 64 kbps interfacing with four DS-1 (T1 rate signals). The output of this bank is at the DS-2 rate of 6.312 Mbps and in the DS-2 format. Four DS-1 rate signals at 1.544 Mbps can be demultiplexed from this signal. A certain amount of digital interfacing is necessary to sort out the signaling bits and the framing bits before the 96 individual 64 kbps are recovered from the composite signal. Interfacing between the four type of D1 channels and D2 channel need special care, since the differences in companding, channel identification, and subgroups of the individual channels can be significantly different.

Third, the D3 bank introduced during the early seventies incorporated the newer technologies that evolved after the introduction of the D1A and D1B banks. The increased number of options, smaller size, lower costs, and improved performance has made this channel bank a more viable option over the D1 bank in the newer installation. Since all the factors relevant to the successful introduction of the newer generation of minicomputers are applicable in D channel bank design and architecture, the newer D3 channel banks are more optimal choice in Central Offices and switching centers. The

D3 bank can process the information (digital and digitized voice) with considerably improved dependability at reduced cost and with greater flexibility. Typical of the features as far back as the 1970s are the interfacing with the common channel interoffice signaling (CCIS), capacity for diagnostics and telemetry, and master/slave relationship to internal or external clocking signals. Flexibility in repair and maintenance, fast service restoration and parallel operation of these banks are the result of the IC technology in the design of the D3 bank.

Finally, the D4 bank operates with a maximum amount of flexibility. The functional mode of these banks can be changed by the use of appropriate plug-in hardware modules. Designed to generate the DS-1, DS1-C (rate of 3.152 Mbps), or the DS-2 signaling needs, the D4 channel bank can serve up to a maximum of 48 DS0 or 64 kbps channels. One T1 C-carrier or two T1 carriers may be interfaced to a switching system. Two of the D4 banks (in coupled pairs) can also serve one T2 carrier system. Increased power efficiency, and reduced size makes these systems attractive compared to the earlier D banks. The line interface units and multiplexers need special consideration when the D4 system is to be used for terminating the carrier systems. The D4 channel bank can also be used as an interface between the carrier systems and the digroup terminal (an input/output port) for the 4ESS. Since this switching system uses the TDM principle extensively, it is feasible to interface the D4 system directly or through multiplexers depending upon the rate of the carrier systems.

The D4 system has distinct modularity. The processing is controlled by the seven common control plug-in units. There are two additional units for power conversion and power distribution [4.5]. Processing in the D4 bank consists of coding and decoding sampled data to PCM pulses. Adaptability of the D4 bank permits it to function in any given type of Central Office; to interface processing required for line and far-end transmitter banks; to protect switching office and calls in progress against transmission errors; to process emergency conditions and to invoke response; and finally to process and buffer data streams when minor differences in carrier rates are detected. This last function is important and is carried out by the Syndes (synchronizer and desynchronizer) plug-in unit.

There are *four* operating modes (see Figure 4.10) for the D4 channel banks. *First*, D4 banks may be directly tied together with a T1C line at 3.154 Mbps. *Second* a D4 bank may be tied to a M1C (which multiplexes two D1 channel banks) with one T1C line. The function of the SU or the Syndes becomes necessary to maintain synchronism between the data in the D1 channel banks. *Third*, two D4 banks may be coupled to one D4 and two D1 banks via four T1 lines and *finally*, two D4 banks may be coupled to four D1 banks via two M12 multiplexers and four T1 banks. The need for the SUs and the line interface units (LIU) is also shown.

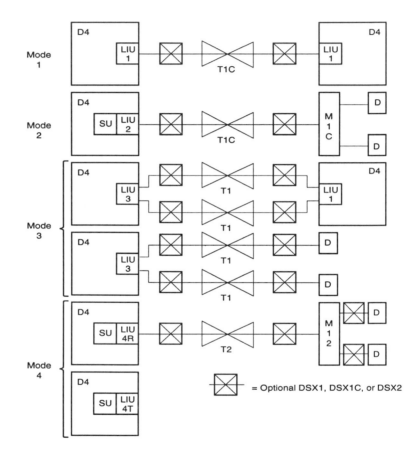

FIGURE 4.10. Four Modes for the Deployment of the D4 Channel Subrates and Digital Channel Banks. D4 - Digital Channel Bank for Switching each of 24 Voiceband Signals into 8-bit PCM Words; DS - A Formatted Digital Signal; ESS - Electronic Switching System; LIU - Line Interface Unit; M -Multiplexer; R - Receive; SU - Syndes Unit; T - Transmit; T*n* - Standard North American Carrier Lines Used to Transmit Formatted Digital Signal

It is important to note the use of digital computer-oriented hardware and software features in the D4 channel bank development. The four designs of channel banks (D1 through D4) are also affected by the IC technology and architectural enhancements. This effect also persists in the last two generations of computers. The concept of specialized hardware is more dominant in the telecommunications environment than in the computer technology where program control and general purpose microprocessors accomplish more generic and more modular functions.

4.7

SUBRATES AND DIGITAL CHANNEL BANKS

In the United States, the T1 systems are the major carriers of data traffic. Attempts to use this facility more and more at the Central Offices resulted in the introduction of the digital data system (DDS, see Section 4.4.12) to the subscribers. Digital data signals at 2.4, 4.8, 9.6 and at 56 kbps are converted to a 64 kbps format by synchronous data multiplexers [4.9] at the Central Office. The two formats (DS0-A and DS0-B at 64 kbps) contain the customer data at the lower rates. Data bytes in the DS0-A and DS0-B formats are carried on the T1 carrier system quite effectively in 23 out the 24 bytes within the T1 frame.

For the DS0-A format and for the subrates, 6 bit customer data bits (bits 2-7 of each byte) and one network control bit make seven bits of the byte, while a one in the eighth bit flags a customer data byte and a zero in the eighth bit indicates a control byte. Byte stuffing (i.e., repeating the byte 5 times for 9.6 kbps rate, 10 times for 4.8 kbps rate and 20 times for 2.4 kbps rate) then brings up actual data rate to the DS0 or 64 kbps rate. For the 56 kbps rate, the 7 bit-customer data and one network control bit make up the byte. No byte stuff is necessary. The clock rate is at 64 kHz or at every 15.625 milli-seconds.

For the DS0-B format and for subrates, bytes of data from other customers are substituted (as need be) for repeated bytes from the same customer. This flexibility to block repetition and reuse the bytes for other customers channels permits ample flexibility at the Central Offices. In large metropolitan areas, the use of a second multiplexer to convert the single user DS0-A format to multi-user DS0-B format may be justified. Both these multiplexers meet the DDS requirements. The later type of multiplexer, largely used in metropolitan areas, is designed for business and computer data traffic. Twenty-three of the twenty-four 64 kbps channels are actively used (see Figure 4.11) for DDS services in a synchronous mode and one byte monitors the line to the carrier system.

Digitized PCM voice and data from the other D (D3 and D1D) channel banks are multiplexed onto the synchronous 64 kbps channels of T1 carrier system. Some of the features of these multiplexers include the flexibility to function between Central Office to Central Office as data carrier exclusively, or to carry up to 12 data channels, or to transmit the data over the partial length of the T1 line and drop channels at the appropriate hub offices.

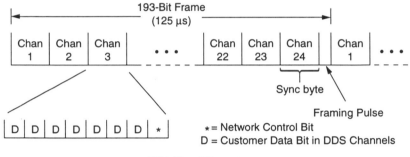

DS1 Signal Format

FIGURE 4.11. Arrangement of the DS0-A and DS0-B Data Bytes in the T1 Frames. Note the 24th byte contains the synchronizing byte for the customer data and the 193rd bit contains the control framing bit. DS0A - Digital Signal of a Specified Rate (64 kbps); DS0B - Digital Signal of a Specified Rate (64 kbps); DS1 - Digital Signal

Digitized PCM voice and data from the other D (D3 and D1D) channel banks are multiplexed onto the synchronous 64 kbps channels of T1 carrier system. Some of the features of these multiplexers include the flexibility to function between Central Office to Central Office as data carrier exclusively, or to carry up to 12 data channels, or to transmit the data over the partial length of the T1 line and drop channels at the appropriate hub offices. When the hub offices are properly distributed, data at fractional T1 rates can be brought to the individual users effectively and economically. From the perspective of the 1990s, it is to appreciated that the emerging ISDN and/or HDSL/SDSL technology will replace most of the archaic systems.

4.8

THE ROLE OF NEWER ELECTRONIC SWITCHES

At the main switching facility, the lines and trunks are handled as input and output ports. Channels simply are carried in the TDM mode over these lines and trunks. The operating system (known as the operating system for distributed switching, OSDS) is unique to the architecture of the 5ESS mainframe that houses its main switch module. Terminal processes govern the functioning of each line or trunk. These subprocesses, activated as the trunks and lines are deployed, monitor the local functions of the ports. The processes are initiated and killed as any other process in a mainframe environment. The OSDS is thus the main centralized command software supporting multiple processes monitoring the ports. It also supports the interprocess communication to facilitate the cross channeling of different ports.

Electronic Switching Systems (ESS) have diagnostic and fault locating capability typically found in most sophisticated computational systems. In addition, it has the operation, administration and the maintenance (OA&M) capability for the telephone network. This feature is greatly enhanced versions of 5ESS. The maintenance functions are divided for the system, trunks, loops and special services. Diagnostic checks are routinely performed to monitor system functions. Part of the local network administration can also be accomplished at these main switching facilities.

The additional support that these newer switching systems provide is that they can effectively interface the different carrier systems, such as the SLC 96® [4.10] (from the 1980s) and the fiber SLC® carrier systems (also from 1980s) at the DS-2 rate of 6.314 Mbps handling 96, 64 kbps 'B' channels. The SLC-2000® which is the more recent vintage (mid 1990s) of the subscriber loop systems, is presented in Chapter 1. These subscriber carriers need to be interfaced to and serviced by the switching systems.

4.9

FAST PACKET TECHNOLOGIES

For the transport of data within the LAN, MAN or WAN arena, as many as *thirteen* methodologies prevail ranging from (a) Ethernet, Isochronous

Ethernet or IsoEthernet, Fast Ethernet; (b) Token Ring passing, FDDI, FDDI II, FFOL; (c) DQDB, PS, MDS to (d) N-ISDN, Frame Relay, IEEE Standard 802.9, ATM/BISDN, SDH/SONET. Star topologies also exist, and some of the key contenders are AT&T StarLAN with rates from 1 to 10 Mbps (or much higher depending upon the switching capability of the centralized and focal node). StarLAN and CO-LAN with CENTREX systems both serve to distribute data rather than to transport it, and for this reason they are not presented in this section. In addition to the thirteen methodologies listed above, the ISO/IEC 8802.3 (1993 ANSI/IEEE 802.3) Standard includes and defines the twisted-pair medium attachment unit (MAU). A summary overview of the data transport networks and their features are provided next for the sake of completeness.

Ethernet (IEEE Standard 802.3 CSMA/CD Bus): This rather old bus-oriented packet-switched technology (see Figure 4.12) is characterized by having no priority of service/packets. The rate is typically at 10 Mbps and operates under CSMA/CD (carrier sense multiple access/collision detect) principles suited to LAN environments. It uses a bus topology ranging up to 4,920 ft or 1,500 meters with passive taps for each station. The frame size varies between 64 to 1,518 octets.

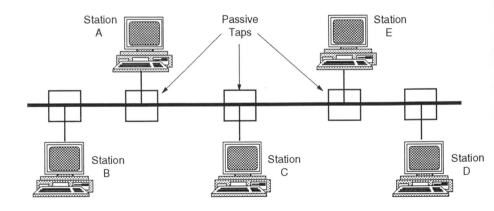

FIGURE 4.12. Ethernet Passive Bus Topology from IEEE 802.3 Specification for CSMA/CD. CSMA/CD - Carrier Sense Multiple Access/Collision Detect

The transmission mode is neither synchronous nor isochronous, giving rise to low delays at low loading, but the performance rapidly deteriorates with increased loading. It is inexpensive and can use the copper and/or wireless media.

IsoEthernet: This provides an inexpensive short-term intermediate isochronous services capability with CSMA/CD. A 16.384 Mbps physical layer is specified, and a new physical layer medium dependent (PMD) sublayer modification to the hybrid multiplexer sublayer is shown in Figure 4.13. IsoEthernet is based on the IEEE Standard 802.9 TDM frame format (discussed later in this section). The IsoEthernet also specifies the rates as 10 Mbps for synchronous/asynchronous data traffic and 6 Mbps (or 96 ISDN bearer channels) for isochronous traffic.

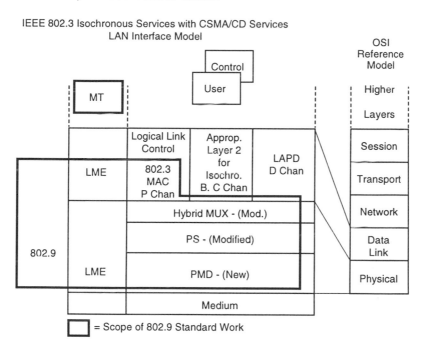

FIGURE 4.13. CSMA/CD LAN Model for Isochronous Service Provisioning with IsoEthernet. CSMA/CD - Carrier Sense Multiple Access/Collision Detect; LAN - Local Area Network; LAPD - Link Access Protocol for the D Channel; LME - Layer Management Entites; MAC - Media Access Control; MT - Management; MUX - Multiplexer; OSI - Open System Interconnection; PMD - Packet Mode Data; PS - Packet Service

As can be seen, the IsoEthernet can be classified as a retrofit to the new developments and demands for higher data rates. In the proposed network, the rate may be increased to 25 Mbps, but the media access control (MAC) layer is detailed in the IEEE Standard 802.3. All the specifications need to be finalized for this retrofit technology.

Twisted-Pair Medium Attachment Unit (MAU) for the Baseband Medium, Type 10Base-T: The purpose of the MAU is to provide a simple, inexpensive and flexible means of attaching devices to the medium. The standard enables the coupling of the physical signaling sublayer to the baseband twisted pair link via an attachment unit interface (AUI).

The data rate is 10 Mbps and the data transfer takes place based on the ISO/IEC 8802.3 CSMA/CD LAN specification discussed earlier. It provides for operation to at least 10 m (or 328 ft) of twisted pairs (0.4 to 0.6 mm or 26 to 24 AWG wire-pairs) without the use of any repeaters. It also provides for a DTE (data terminal unit) or a repeater to interface and function with the MAU. The access method is the CSMA/CD defined in 1993 (8802.3 Standard) with baseband signaling. This provision supports point to point access between MAUs, and if used with repeaters, it supports a star topology with multiple ports.

Fiber Distributed Data Interface, FDDI II and FFOL (FDDI Follow-On LAN): The configuration is derived from the family of ring network architectures. FDDI II (sometimes referred to as FDDI/2) partitions the 100 Mbps bandwidth into an isochronous band and token passing packet LAN. FFOL, like IsoEthernet or IsoEnet, also combines isochronous and packet traffic. The rate is scalable to over 2 Gbps. The nodes and interfaces on the ring(s) are active. Specifically, the FDDI architecture is based on two fiber rings carrying information in opposite directions with typical speeds of 100 Mbps ranging up to 62 (100 KM) miles per ring. As many as 500 stations may by connected to each ring. The access is provided by token passing and token is released immediately after packet transmission. Duality of rings, bypass switch, concentration and fault isolation provide reasonable reliability of FDDI networks. The FDDI protocol conforms to the two lowest layers (Data Link and Physical) of the OSI model (see Figure 4.14) via Data Link - Media Access Control (MAC), Physical Layer (PHY) - Physical Layer Medium Dependent (PMD), and finally the Station Management (SMT) protocols.

The FFOL Reference Model preserves the FDDI infrastructure, but enhances the speed to SDH/SONET rates well into the Gbps. The data link functions in the FFOL environment (see Figure 4.15) include the provisioning for logical link control and circuit-switched multiplexers. FDDI-II offers all the features of the IsoEthernet and is suited for all voice/data/video environments, except that it offers a 100 Mbps bandwidth (rather than 16 Mbps). Although it provides fixed bandwidth capability it is not compatible with the FDDI standard developed by the ANSI X3T9.5 committee. The differences between the first generation FDDI and FDDI-II can be seen by comparing the FFOL Reference Model shown in Figure 4.15 with the less specific FDDI Reference Model shown in Figure 4.14.

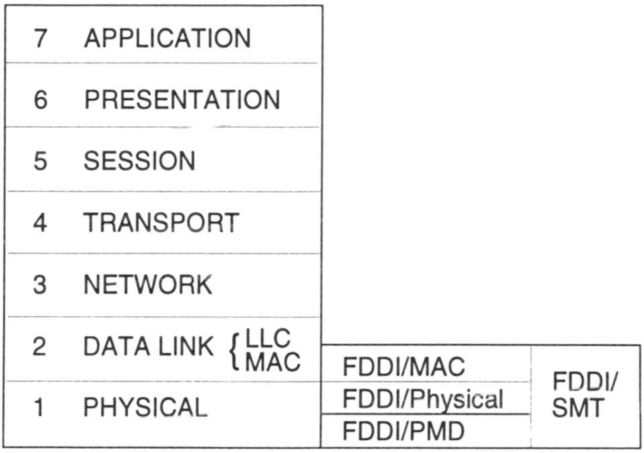

7	APPLICATION		
6	PRESENTATION		
5	SESSION		
4	TRANSPORT		
3	NETWORK		
2	DATA LINK { LLC / MAC	FDDI/MAC	FDDI/ SMT
1	PHYSICAL	FDDI/Physical FDDI/PMD	

FIGURE 4.14. FDDI and its Functional Mapping with Reference to the OSI Model. FDDI - Fiber Distributed Data Interface; LLC - Logical Link Control; MAC - Media Access Control; OSI - Open System Interconnection; PMD - Packet Mode Data; SMT - Station Management

Distributed Queue Dual Bus DQDB: Based on IEEE Standard 802.6 recommendations, two shared buses carrying information in opposite directions dominate this architecture. There are two head ends for the two directions of data flow. Both may physically be located at one place or node. Each node connects with both buses. The topology may also be altered to permit a star configuration similar to that of an 802.5 token-passing ring. The emphasis is on packet-switched voice, data and/or video applications.

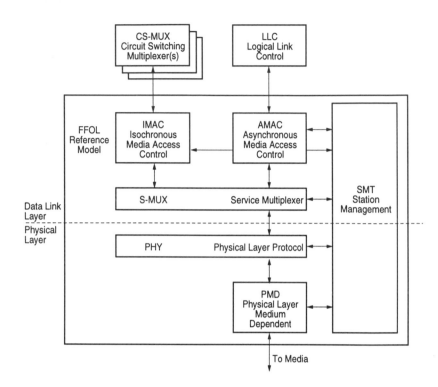

**FIGURE 4.15. The Two Lowest Layers for the 100 Mbps
Isochronous and the Asynchronous Capabilities of the Fiber Loops.**
Isochronous and asynchronous capability on the fiber laps. AMAC -
Asynchronous Media Access Control; CS-MUX - Circuit Switching
Multiplexer; FDDI - Fiber Distributed Data Interface; FFOL- FDDI Follow-
On LAN; IMAC - Isochronous Media Access Control; LAN - Local Area
Network; LLC- Logical Link Control; PHY - Physical Layer Protocol; SMT
- Station Management; S-MUX - Service Multiplexer

The architecture for DQDB allows for asynchronous (data) and
isochronous (fixed bandwidth, voice/video) access with switched
multimegabit data service (or SMDS, standardized in late 1989 and appended
to the original standard) capability.   This allows for high-speed large
bandwidth depending on the media, coax, fiber, unshielded twisted wire-
pair (UTP), by deploying a fixed length link layer entity with 53 octets (i.e.
cell). The medium to carry data is initially targeted as DS3. The range and
speeds make DQDB well suited for MAN applications. DQDB establishes
distributed queues for packets of different priorities thus facilitating highest

priority packets to get through immediately. The Reference Model for DQDB is shown in Figure 4.16.

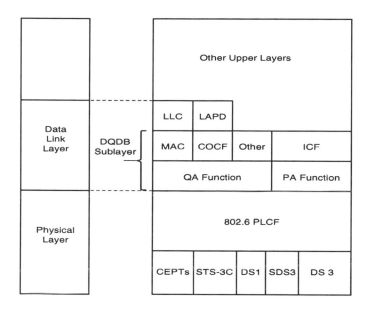

FIGURE 4.16. DQDB Functions and OSI Reference Model. CEPT - European Rates; CO CF - Connection Oriented Convergence Function; DQDB - Distributed Queue Dual Bus; DS3 - Synchronized Digital System #3; ICF - Isochronous Convergence Function; LAPD - Link Access Protocol for the D Channel; LLC - Logical Link Control; MAC - Media Access Control; OSI - Open System Interconnection; PA - Physical Access; PLCF - Physical Layer Convergence Function; STS - Synchronous Transport System

IEEE Standard 802.9 Specifications: The TDM frame format specifying a 125 microsecond cycle period repeats with N (depending upon the rate) octets. Within this N octet frame, the first eight octets (0-7) have a specific arrangement. The eight octet-based fields are Synchronization (SYN), TDM maintenance field (TDM-MTN), Hybrid Multiplexer Control (HMC), Reserved field, D-channel, B1-channel, B2-channel and Access Control (AC). The playload is carried from octet number eight to the (N-1) octet.

ATM: This offers all the advantages of fast packet-switching and fiber optic transmission technology. The ATM cross connect is geared towards switching speeds of about 2 Gbps (if not more) and will carry SMDS data quite efficiently and economically, acting to switch information highways between large local and metropolitan area networks. ATM is discussed in detail in Chapter 15 and also in Chapter 7 of the companion volume *"Intelligent Broadband Multimedia Networks"*.

4.10

IMPACT OF SONET AND ATM

The impact of the SONET and ATM on the older modes of digital communication is profound. In a sense, the newer fiber optic networks and the ATM switches break into the architectures shown in Figures 1.3 and 4.9. Typically, the DSn (n=2, 3 or 4) information is absorbed into 48 byte ATM cells and the DSn signaling in re-encoded into the 5-octect header of each ATM cell. Similar remapping occurs from the DSn signals to the SONET frames. The cheaper more versatile ATM switches replace the bulkier cross connects of Figure 4.9. Extreme care and attention is necessary to determine the viability of such conversion in view of the application at hand. Chaos is likely to clutter the customer information especially if real-time multimedia or video information is being carried by the DS-n signals.

Ideally, the composite architecture of the telephone network would evolve as the all-digital broadband facilities and the circuit/packet switches would become integrated as fast packet relay technology frame. Whereas the private line, software defined networks, ISDN and switched digital networks need to keep their identity for reasons of service requirements, it is possible to see other packet-oriented digital and computer networks being unified into the ATM backbone network. The scenario is depicted in Figure 4.17. Internetworking options [4.11] permit conversions from ATM or cell relay mode to the synchronous and frame relay mode as DS3/OC3. These functions are well accomplished by routers or by ATM hubs. The ATM cross connect (AXC) switch permits the network both to switch at very high rates (in the order of tens of Gbps) and to route the cells to the appropriate address. It is also feasible to foresee the ESS facilities being integrated with AXC to makeup one universal switch architecture.

A new broadband network will evolve with both network service capabilities: traditional digital hierarchical (DS0-DS4) network and the SONET synchronous digital hierarchy WAN. The switch scenario is depicted in Figure 4.18 and the network scenario is depicted in Figure 4.19. This broadband switch would be the key integrator of almost all digital network services with one unified voice/data/video services network.

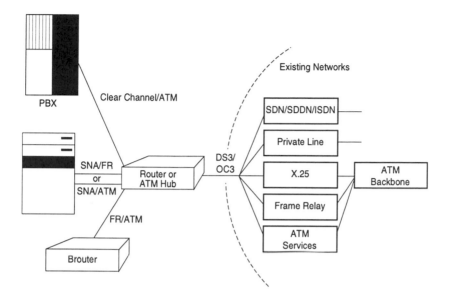

FIGURE 4.17. Schematic of an Evolving Broadband Network.

ATM to the desktop is more than a proposition. ATM adapters at 155 Mbps full duplex campus networks bring in more bandwidth and flexibility. The other alternatives such as Ethernet, full-duplex Ethernet or token ring, fast Ethernet (i.e., 100 Mbps half duplex mode), or FDDI can only become obsolete in a few years after the ATM cross connects and adapters, routers and brouters become common and inexpensive.

The OC-3c, 155 Mbps adapters uses category-5 unshielded twisted wire-pair (UTP-5) media compared to the more frequently used fiber media for ATM LANs. Complete standardization can only bring down the cost and point-to point ATM is likely to bring the information highway to the desktop in the next few years.

But more than that, wireless personal communication services (PCS) networks permits portables laptops to exchange data at rates ranging from 2 to 20 Mbps. They may be also connected via a server or peer-to-peer to a "portable base station" (PBS).

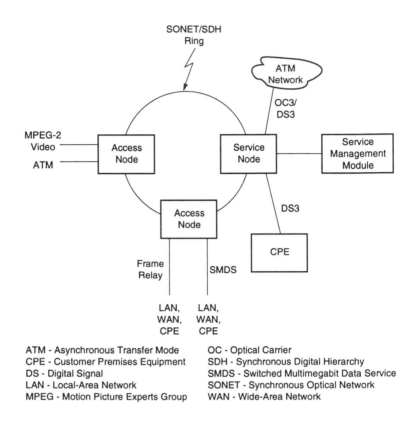

FIGURE 4.18. A Typical Node of a Service Provider on a SONET Ring. Services include both the conventional circuit switched and ATM. Routing and bridging services are also provided.

The link rate between these PBSs is in the order of a few Gbps by using the cell relay ATM technology. Typical rates of 1.2 Gbps over radio links or free-space optic (line of sight optical) links have been functional in the laboratory environment. Finally, a gateway-PBS permits the access to-and-from the wireless ATM LAN to the wired/fibered/cabled LAN, MAN or WAN. Personal and mobile communication networks (PCNs) at multi-megabits can be elegantly switched via wireless ATM networks since ATM supports the SMDS capability. Indoor wireless ATM LANs complete the promise of multimedia services to portable "plug-n-play" laptops.

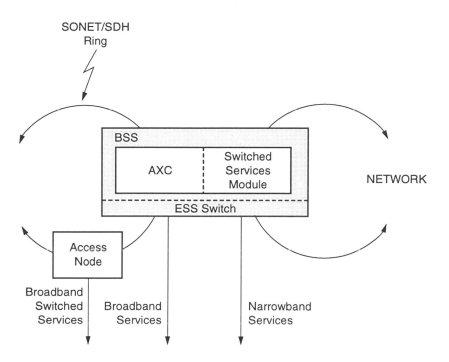

FIGURE 4.19. A Typical ATM Node to Provide Circuit and Packet Switched Services. It also serves as an ATM cross connect to the SDH ring, thus providing both frame/cell relay functions. AXC - ATM cross connect; BSS - Broadband Switched Services; ESS - Electronic Switching System

On a realistic note, most of rate requirements are modest. For example, a word processing application with 6400 byte text object retrieval needing 30% transmission time, will need a peak data rate of 0.7 Mbps to still "get" the object in less than 250 (i.e., 244) msecs. CAD/CAM applications with 50 kB manufacturing objects need about 5 to 6 Mbps, and a fairly detailed 144 kB office image needs 15 to 16 Mbps to meet similar object retrieval objectives over LAN networks. In computing environments the requirements are easily estimated. For example, a 200 byte message (1,600 bits) is furnished in one-tenth msec (10^{-04}) over a 16 Mbps link. Time for file transfers may be calculated quite readily at different rates but the slowest element is generally the human response.

Very detailed X-ray images for real-time diagnostic analysis need about 45 Mbps. Transmission of HDTV quality imaging applications with MPEG++ video compression techniques need about 18 Mbps. With standard NTSC TV quality of pictures (3 to 4 Mbps rate) on personal computers, video conferencing with 6 coworkers needs a 25 Mbps ATM LAN.

4.11

REFERENCES

4.1   S.V. Ahamed, P.P. Bohn, and N.L. Gottfried 1981, "A Tutorial on Two-Wire Digital Transmission in the Loop Plant," *IEEE Transactions on Communications*, COMM, Vol. 29, 1554-1564.

4.2   B.S. Bosik and S.V. Kartalopoulos 1982, "Time Compression Multiplexing for a Circuit Switched Digital Capability," *IEEE Transactions on Communications*, COMM, Vol. 30, 2046-2052.

4.3   M.J. Miller and S.V. Ahamed 1988, *Digital Transmission Systems and Networks, Volume II Applications*, Computer Science Press, Chapters 5 and 6.

4.4   F.T. Andrews and R.J. Haas 1984, "The Role of Digital Switching in the Evolution of the Subscriber Loop," *Proceedings of the Sixth International Symposium on Subscribers and Services*, 29-33.

4.5   AT&T 1985, *Telecommunications Transmission Engineering, Volume 2 Facilities*, New York, NY.

4.6   E.A. Lee and D.G. Messerschmitt 1988, *Digital Communication*, Kluwer Academic Publishers, Boston, MA.

4.7   Bell Laboratories 1977, *Engineering and Operations in the Bell System*, Chapter 9, Western Electric Company.

4.8   T.W. Anderson, P.D. Carestia, J.H. Foster, M.N. Meyers 1994, "The Evolution of the 4ESS™ Switch," *AT&T Technical Journal*, November/December, 93-100.

4.9   Bell Telephone Laboratories 1982, *Transmission Systems for Communications*, Fifth Edition.

4.10   AT&T Bell Laboratories Technical Journal 1984, *SLC 96 Carrier*, Vol. 63, No. 10, Part 2, December.

4.11   K.M. Ryan 1994, "ATM and Data Networking," AT&T Bell Laboratories, *Private Communication*.

PART II

THE METALLIC MEDIA: THE PHYSICAL ENVIRONMENT

The physical media for the exchange of information over communication networks varies from free space to the most tightly shielded electromagnetic wave guides. Every one of these highly variable media for the transport of information follows Maxwell's Laws for the propagation and distribution of the electromagnetic energy that carries the information bearing signal, however weak it might be. From an analytic consideration, these generic and universal equations assume numerous formats ranging from Pyonting theorem (in its numerous formats for different media) to Snell's Law (in fiber optics). All the same, from an engineering and design consideration, the computer aided design environment needs the program platforms and media characteristic data bases to execute and optimize. As an implementational consideration, the optimized designs need the economic, social and political seal of approval before they can become a reality. In this part of the book, only the first two links in the metallic-network life cycle ar emphasized, with most of the attention focused upon CAD based optimization techniques.

Chapter 5

SUBSCRIBER LOOPS

Over the evolutionary phase of most voice networks, a large amount of copper in the form of wires has been used to carry electrically encoded information bearing signals. Copper is a very good conductor (resistivity 1.712x10-8 Ohms/m, 25° C) of electrical signals, being next to silver (resistivity 1.617x10-8 Ohms/m, 25°C). It is cheap, readily available, durable and ductile. For these reasons, it has been used for overhead wires, twisted wire-pairs, coaxial cables, CATV distribution, and for TV broadcast cable (twin-cable). In the context of communication networks, the copper medium has played a significant role and still retains a sizable (but diminishing) foothold in most established networks. Two well-established networks that depend heavily upon copper for distribution of information are the traditional telephone networks and the CATV networks. More recently, the impact of fiber has been dominant and resounding. Fiber as a viable alternate both for long-haul and for local distribution of information is firmly established. The impending hybrid-fiber/coax technology and cellular wireless "last mile" both pose a serious threat to the continued role of copper as an electrical signal carrier beyond the year 2020.

The limited bandwidth of the copper media, especially the twisted wire-pairs, has been a topic of serious contention, but the transmission engineers have been able to push the limit further and further by improvement of the terminal devices that "couple" with the media. However, in this era, it is not the characterization (attenuation and dispersion) of the medium that is the challenge; instead it is the contamination (crosstalk and stray electromagnetic coupling) within the medium. Hence, the scientists attempt to isolate the effects of contaminants that cannot be accurately characterized. There is a statistical estimation of the damage they cause, but not a scientific methodology to eliminate them. The scientific breakthrough will result when and if it becomes possible to force the copper medium innately to "couple" the electromagnetic fields for the signal and "uncouple" them for the noise. New transmission materials that perform signal processing at the atomic/molecular levels may result. Such fabricated compound semiconductor materials [5.1] for optical devices are being investigated. However, a true parallel cannot be drawn between the two modus operandi.

The current integrated optical technologies for transmission (see Part V) and switching [5.2] can leave behind some of the older communication technology in a state of premature obsolescence [5.3]. In this part of the book, the utilization of copper already in place as a medium to carry voice and data in the subscriber loop environments is presented. Here the older communication networks are heavily invested with established metallic connections to almost every household. Deploying copper (by itself) for new communication services needs considerable deliberation.

In most older telephone networks, the subscriber loops (Figures 5.1 and 5.2) are used in star topology to connect individual subscribers to the nodal Central Office or to a remote terminal. These remote terminals (RTs) are in turn served by a higher capacity carrier system from a Central Office. Figure 5.3 depicts a more recent subscriber plant layout. Here, the carrier systems are integrated with the optical systems and carry data at SONET (OC3 [at 155.520 Mbps] or at 12 [at 622.080 Mbps]) rates over optical fiber. These Central Offices themselves have trunks or other high capacity lines (OC-12 or OC-24) for interconnections to other Central Offices in widely dispersed geographic areas. The subscriber loop still remains as the last link for distributing information to the end user and also becomes the first link to receive the information from the user. This dual requirement from the loop to serve as both the first and last link of a conventional telephone network is due to the interchangeability of the caller and the called party, or the talking and the listening party.

The design of the subscriber loops in the conventional telephone environment has been essentially to carry voice in a bidirectional mode. It also carries the minimum ringing current at 20 Hz to ring the older distant telephone sets.

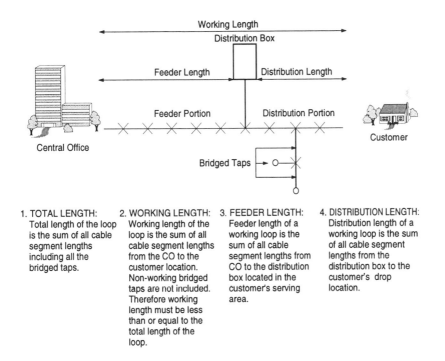

1. TOTAL LENGTH: Total length of the loop is the sum of all cable segment lengths including all the bridged taps.

2. WORKING LENGTH: Working length of the loop is the sum of all cable segment lengths from the CO to the customer location. Non-working bridged taps are not included. Therefore working length must be less than or equal to the total length of the loop.

3. FEEDER LENGTH: Feeder length of a working loop is the sum of all cable segment lengths from CO to the distribution box located in the customer's serving area.

4. DISTRIBUTION LENGTH: Distribution length of a working loop is the sum of all cable segment lengths from the distribution box to the customer's drop location.

FIGURE 5.1. Typical Subscriber Loop Configuration in the United States.

The ringing current is superimposed on dc current components used for ring tripping. The Central Office battery power is typically at 48 V dc for loops up to 15 kft, or at 72 V dc (with range extension) for unigauge (26 AWG) loops to 24 kft. Longer loops would usually have loading coils (see Section 6.3.1). The newer ISDN services do not suffer from such restrictions, but they cannot tolerate any load coils.

5.1

WIRE-PAIRS

Wire-pair environments differ significantly from country to country and from one region to another. A large number of factors influence the nature of the telephone loop plant environment where wire-pairs are used extensively. Two (open and twisted) types of wire-pairs are used in the loop plant.

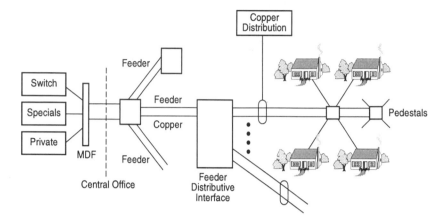

FIGURE 5.2. Typical Metallic Access to and from the Telephone Central Office. MDF - Main Distribution Frame

5.1.1 OPEN WIRE-PAIRS

Open wires strung on top of poles still exist in some areas even though they are no longer actively installed in most countries. The existing open wire-pairs still used in the United States for metallic connection to the Central Office, usually consists of bare copper wire of 0.080, 0.104, 0.128, or 0.165 inches. The wire-pairs are generally mounted on poles spaced at about 130 feet. Frequent transposing and close spacing between the pairs tends to provide immunity to noise and power frequency electromagnetic induction. It also helps reduce crosstalk from other wire-pairs. Some older carrier systems ranging from 15 to 150 miles (O-Type for 16 single-sideband two-way voice grade channels on open-wire) were implemented. However, these carrier systems were also quickly adapted to twisted wire-pairs, coaxial cables, and even microwave radio systems in order to deal with problems of terrain, population density, pole-line congestion, and also environmental conditions including rain, ice, sleet, frost and high wind.

5.1.2 TWISTED WIRE-PAIRS

Twisted wire-pairs to carry voice and data generally exist in most national networks. These connections to the customer premises were originally provided for voice traffic with a limited bandwidth of 200 Hz to 3200 Hz. Since the inception of POTS, the terminating devices have become more and more adaptive and IC oriented. The cost of these components has plummeted

initiated the investigation of complex circuitry at the subscriber line termination to enhance the data rate.

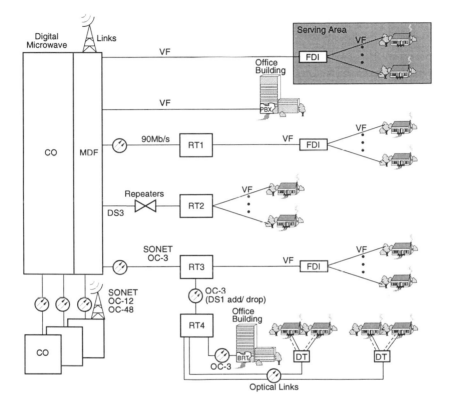

FIGURE 5.3. Typical Enhancements to the Subscriber Loop Plant.
BRT - Broadband Radio Terminal; CO - Central Office; DS-n - A Formatted Digital Signal; DS3 - Standard Third Level Digital Carrier System Used in the US; DT - Distribution Terminal; FDI - Feeder Distribution Interface; MDF; Main Distribution Frame; OCn - Optical Carrier; PBX - Private Branch Exchange; RTn - Remote Terminal; VF - Voice Frequency

Rate enhancement over the existing twisted wire-pairs and extending the range are the two main objectives. The reasoning behind this approach is to design an optimal system that can maximize the exchange of information with almost all the customers.

This effort has resulted in optimally designed inexpensive termination of twisted wire-pairs without the large expense of reinstalling new transmission medium. Bandwidth expansion of this primary medium of transmission has been target of research by adaptive digital signal processing. Analog, digital and hybrid systems have resulted. In the context of digital carriers and their transmission, channel units with delicate balancing circuits, equalizers, echo cancelers, timing recovery circuits, etc., are necessary as the bit rate and distance for transmission start to increase.

5.1.3 UNSHIELDED TWISTED WIRE-PAIRS (UTP)

Much of the metallic signal carrying color coded, bound in bundles, and enclosed in sheaths. Typically, media consists of the twisted metallic wire-pair. Twisted wire-pairs are typically waterproof coating, corrugated aluminum and steel shields and/or polyethylene jackets are used outside the core insulation of the twisted wire-pairs. The choice of the outside material depends upon the type of application and the environmental protection most desirable. Immunity against electrical interference and physical protection are offered by metal clad cables, whereas resistance to corrosion is higher with polyethylene cladding.

Being inexpensive and relatively indestructible, UTP is used within Central Offices, sometimes between Central Offices, almost always within the loop plant to reach the subscribers, and almost always within the customer premises. Mostly the wire-pairs have a random twist along its length. They have been manufactured to range between being six wire-pair units to several thousand wire-pair units. Physically, individual wire-pairs are bundled to occupy either the inner most bundle or one of the outer bundles within the physical cable.

Figure 5.4 depicts the distribution of a typical 600 pair cable. The number of pairs per unit can vary between 12 to 100 depending on the gauge of the wire and the size of the cable. Sometimes concentric layered structure of wire-pairs is also used. The purpose of the random twist is an attempt to undo the stray electromagnetic coupling with the neighboring wire-pairs. This crosstalk coupling between these wire-pairs can differ substantially depending on the location of the individual wire-pair, due to differences in the spatial distribution of the electromagnetic fields within the cable. In fact it has been suggested that appropriate choice of wire-pairs can make the high-speed services to a few businesses without any modification to the loop plant.

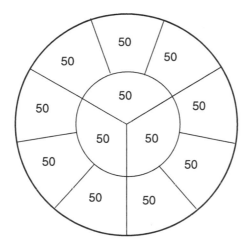

FIGURE 5.4. Distribution of the Twelve 50 Wire-Pair Bundles in a 600 Pair Cable.

Numerous layers of such wire-pair bundles may be placed within one physical cable. In the subscriber environment, the Central Office side's wire-pairs (referred to as main feeder going onto branch feeder cables) usually have a finer gauge [5.4] than the subscriber side (distribution cables). In view of their extensive usage, twisted wire-pairs are manufactured in numerous sizes and gauges. Typically, the coarser i.e., the 19 AWG copper and 17 AWG aluminum gauges are used for special services and carrier systems, and the 22 AWG copper and 20 AWG aluminum gauges serve special services, trunks, carrier systems, and long loops. By and large, the polyethylene insulated cable (PIC) in 19 and 22 gauges are manufactured in smaller sizes ranging from 6 to 600 wire-pairs. On the other hand, the pulp insulated 26 AWG cables are manufactured in the largest of the size ranging between 1200 to 3600 wire-pairs.

The gauge of the copper and that of the corresponding aluminum cables is adjusted to offer the same resistance per mile of cable. In the past, this may have been a valid justification to use the cheaper aluminum for the subscriber plant, but at higher rates for DSL and HDSL applications the increase in skin effect (and thus its increased attenuation) at T1 (E1), T1C or T2 (E2) symbol rates makes the aluminum cables unsuitable for wide use. For this reason, the primary characteristics (R, L, G, and C) in the cable data bases for simulation studies have not been included. Only a detailed simulation study will provide information on the capacity of these cables (aluminum in the United Kingdom, PEUIT in Australia, quads in Germany, etc.) that are occasionally found in the loop plants around the world.

5.1.4 SHIELDED TWISTED PAIRS (STP)

Shielded wire-pairs have the same type of makeup as the new untwisted wire-pairs. In addition, there is a metallic shield between cable pairs. This metallic shield modifies the field distribution substantially to the extent that the propagation properties of the cable are themselves altered or even degraded. Due to eddy currents, any loss in the shield shows up as additional copper loss. To prevent any build up of voltage in the shield, a single drain wire is connected between the shield (or foil in most cases) and the outside sheath. Such cables are sometimes referred to as screened cables. Shielding itself takes two forms: braided shield and foil shield, each offering different electrical properties. Sometimes the braided shield find use in applications below 100 MHz and foils shields find use in applications above 100 MHz. The grounding connections of the sheath at either end of the cable, also influence the net reduction of the electromagnetic pickup in shielded wire-pairs, in addition to the connectors and cross connects.

The shielded wire-pairs can become slightly more bulky and less flexible, thus crowding the conduits in which the cables are buried. These shielded wire-pairs are also heavier, which makes the installation more labor intensive. The process of including the STP in the networks becomes difficult and expensive. Even though these wire-pairs are used in some premises distribution systems (PDS, see Section 5.10), they are not as popular as the unshielded wire-pairs. The Electrical Fast Transient Tests based upon the IEC 801-4 recommendation has indicated that UTP and STP perform only as long as adequate care is executed in the balancing and grounding of the shields. With proper care, the tests have indicated that the UTP can perform as well as the STP in the electromagnetic compatibility studies.

5.2

TELEPHONE SUBSCRIBER LINES

A vast range of variation exists in wire-pair environments. Numerous countries and even some Bell Operating Companies have embarked on surveys to generate statistically significant parameters of the individual loop plants. Twisted wire-pairs have been studied extensively for their electrical characteristics and performance. For conveying voiceband (300-2300 Hz) signals, twisted wire-pairs offer adequate performance for distances up to 18,000 feet (in the United States) without any special devices (or load coils) to compensate for the high frequency signal attenuation. However, *three* issues need special consideration.

First, as a part of the information carried to/from the subscriber premises from/to the Central Office, signaling (such as ringing at the subscriber, Central Office scanning of the station equipment like on-hook or off-hook, etc.) is also necessary. For this reason, in most of the existing telephone networks, the electrical resistance of the loop was limited to about 1300 Ohms. This limit assured the older networks to supply enough ringing current for the telephones to ring as a call was coming in. Over many decades, the subscriber loop plant used twisted wire-pairs of numerous gauges (or sizes) especially as the loops became longer. This effect shows in long loops that have coarser wire gauges (and thus lower resistance) at the subscriber end of the loop.

Second, in the nations that experienced a large growth, the demand of telephone services during their infancy of the network has prompted two side effects. In the first case, the records of the subscriber loop configuration were very sloppy. The second case, a more serious nature, when a new request for service was received and no line connections to the Central Office was available, the connection was established by tapping into an existing line that ran into a Central Office. This resulted in multi-party lines that have been phased out gradually, but the taps linger on. In the telephone environment, these open circuited wires (that used to serve a subscriber in older days) are called bridged taps. Even though they are of very little importance for voice communication, these taps cause signal reflection at higher frequencies. For digital transmission, they are a source of serious concern in the design of equalizers and echo canceler.

Third, in an effort to carry voice to distant subscriber locations, the transmission engineers in the past have used load coils along long subscriber loops. These coils compensate for the distributed capacitance of the twisted wire-pairs. The value of the inductance of each of the coils is typically 88 milliHenries (in United States) and when placed at regular spacing (typically every 6000 feet between coils and 3000 feet to the first coil from the Central Office) tends to cancel out the largely capacitive nature of the loop. The capacitive nature of the typical loop does not become significant until the loop length is greater than 18,000 feet, and for this reason the load coils are generally not encountered in shorter (i.e., < 18,000 feet) loops, even though some of the shorter loops (in the 1983 Loop Survey) still had these coils. The combined effect is to offer a uniform impedance in the voice frequency band (200-3200 Hz), but significantly increase it beyond the voice frequency. The telephone speech over long loops thus sounds more natural.

United States Loop Surveys: The subscriber loops in the pre-divestiture Bell System has been surveyed *three* times for their important physical and electrical characteristics. The *first* survey [5.5], focusing on physical and audio frequency characteristics, was undertaken in 1964 and intended to provide the extended use of the telephone line to carry telephone messages over the limited bandwidth, ranging from about 200 Hz to about 3200 Hz. At that time, the extended use of the facilities to carry digital data, encoded speech, and other high-speed digital services were not foreseen. In this section, a subset of the published data is presented in order to provide an evolutionary basis of the lines that connect the Central Office to the residences of individual users.

The initial intent of the *second* survey [5.6], undertaken in 1973, was essentially to update the earlier results and detect any major changes. However, during the late seventies, the study of the loop plant became essential because of the high-speed subscriber digital lines. The 1973 data base was reanalyzed in order to evaluate the potential of the plant to carry data in the range of 64 to 384 kbps. The results [5.4], published in 1982, established the relatively high-speed digital capabilities of that era and the limitations of the American loop plant.

The *third* survey [5.7] was undertaken in 1983 with the intent of updating the earlier results and also characterizing the American loop plant in view of the basic rate ISDN services and high-speed data communication from Central Offices to the carrier serving centers. This Chapter reports the results [5.8, 5.9] from the last and perhaps the most exhaustive survey of the pre-divestiture Bell System in Sections 5.1.1 and 5.1.2. *Two* major tends evident from these three Surveys (1963, 1973, to 1983) are that *(a)* the working loop length has increased from 10,613; to 11,413; to 11,723 ft. for all the general loop population and *(b)* the bridged tap length has reduced from 2,478; to 1,821; to 1,490 ft. for all the loops. Business loops are shorter and have fewer, shorter taps.

Some of the Bell Operating Companies have undertaken the survey of their own loops during the eighties for transmitting BRISDN data. A notable example of such a survey is the NYNEX Loop Survey in 1985. Comparison with the national survey has indicated that the loops in the Manhattan and Boston (typically metropolitan) areas have loop lengths about two to three kft shorter than the average length (7,535 feet) computed from the entire 1983 pre-divestiture Bell System Loop Survey. Data specific to the bridged taps and their distribution is not available in the public domain.

A survey to evaluate the digital capability of the subscriber loops in the GTE (General Telephone and Electronics) loop plant was undertaken in 1982 (see Section 5.6). Partial results indicate that the average loop length for all loops to be about 8,796 ft. vs. 10,787 ft. for the pre-divestiture Bell System loops. Also, the average bridged tap length is about 1,300 ft. vs. 1,299 ft.. For the digital subscriber environments, the average loop length is about 7,708 ft. vs. 7,535 ft. with a bridged tap length of 1,352 ft. vs. 1,100 ft. Published results do not disclose all the electrical characteristics such as the image impedances, loop input loss, and input impedances from the Central Office side and the subscriber side.

5.3

DIGITAL SUBSCRIBER LINES (DSL)

Data rate over twisted wire-pairs can vary widely (300 bps in the early modems of the fifties to 155 Mbps in the PDS of the nineties). Various modes of transmission, and subsequently the methodology for signal separation (simplex, duplex, half-duplex, dual-duplex, etc.) also exist. Numerous line codes used for transmission (from AMI/bipolar to carrierless amplitude phase modulation) are also deployed. A vast array of the DSL configurations can thus be derived. Voiceband modems use the limited spectral (200-2300 Hz) capabilities of the twisted wire-pairs. The most rudimentary codecs use frequency modulations of the zero and one states of the bit transmitted. More sophisticated codecs use bandwidth efficient cluster codes to increase the throughput to as much as 14,400 bits per second (bps), while being confounded to the usual spectral band of 2400 Hz for the telephone channels. Line condition also becomes a significant factor in the capability to carry data through the loop plant.

5.3.1 WIRE-PAIRS AS HIGH-SPEED DATA SERVICES

From the very modest rate of 32 kbps in the late seventies [5.10], the DSL rates in the subscriber loop environments have climbed to E1 rates at 2.048 Mbps in the early nineties. In the United States, H_0 (5B+D; 384 kbps), H_1 (11B+D; 768 kbps) and T1 (23B+D; 1.544 Mbps) rates are being actively considered. In other countries, 1.024 Mbps or 15B+D and 2.048 Mbps or 30B+D are being considered. These rates also play a part in mode of operation. In the full-duplex mode, full-rate bidirectional data is transmitted over one wire-pair. In the half-rate duplex, bidirectional half-rate is transmitted over two wire-pairs. In the simplex modes, twice the wire-pairs are used, but the directionality is isolated. Hence, the frequency at which loops should be characterized depends upon the HDSL rate and mode of operation. A series of scatter plots for all the loops in the 1983 Loop Survey data base at discrete frequencies of 40, 60, 80, 120, 160, and 200 kHz is presented in Reference 5.7.

The potential use of twisted wire-pairs as high-speed digital lines reaching every subscriber at medium-rates is high. If this digital capability may be economically realized, then wire-pairs become one of the most viable carriers of digital data to the homes of millions of subscribers. The limits of such wire-pairs as carriers of very high-speed data over very short distances may also considered. Accordingly, the discussion is focused in *two* major areas in Section 2.4.7. *First*, in the general subscriber loop environment, rates to the T1 and E1, are currently being considered. *Second*, for very short loops in the tight local area environment that may include premises distribution systems (PDS) rates, up to 155 Mbps is feasible over very short distances (typically about 300 feet) with extremely high quality wire-pairs (such as category 5 UTP) where the twist and insulation quality are both tightly monitored. This application of the UTP in the PDS environment is presented further in Section 5.9. The quality of echo cancelers and near-end crosstalk cancelers eventually govern the implementation and the realization of these multi-megabit and possibly fractional gigabit per second twisted wire-pair data carrier systems.

5.3.1.1 Historical Perspective

The use of twisted wire-pairs in the subscriber loop plant to carry data at rates significantly higher than the subrates (up to 19.6 kbps) have been investigated since 1978 [5.11]. Even the modem technology with V.fast and V.bis standards facilitates data transmission at rates up to 28.8 (if not 32) kbps. From the perspective of digital carrier systems, the advent and success of the four-wire repeatered T1 and E1 carrier systems (during the sixties) at 1.544 Mbps in America and at 2.048 Mbps in Europe, were the precursors to the possibility of high rate data over twisted wire-pairs. However, during the mid seventies, the investigations were directed towards combining the repeaterless features of the modem driven lines with the high-speed capability of the T1 and E1 carrier systems. The era of the digital subscriber line (DSL) was thus started in 1978 [5.11] when it was decided to investigate the limit of the subscriber-line wire-pairs to carry encoded data rather than voice.

5.3.1.2 Twisted Wire-Pairs for BRISDN

Basic rate (BRISDN) applications are based upon the use of one non-repeatered single wire-pair to provide two 'B' (bearer) channels at 64 kbps and one 'D' channel at 16 kbps to every subscriber. It is also standardized to use the Adaptive Echo Canceler technology. However, for primary rate (PRISDN) services, 4-wire simplex technique offers freedom from very high-speed digital signal processor (DSP) based echo cancelers. In the final analysis, space-division multiplexing is still a losing proposition since full-duplex half-rate HDSL (also needing two wire-pairs) offers superior performance.

Wire-pairs display distinct electrical characteristics in different bands of spectral utilization. Most digital transmission systems attempt to counter the degradation that is introduced and maintain transparency between the sender and receiver of data. Numerous rates of transmission ranging from about 300 bps to T1/E1 rates exist in loop plant.

Numerous codes for representing the bits as discrete levels (bipolar, 2B1Q, etc.), phase and frequency encoding (PSK and FSK) also exist. The two most widely used codes (bipolar code and the 2B1Q code) prevail in the present ISDN twisted wire-pair transmission schemes. There is a wide choice of codes available and certain codes offer unique properties that may prove to be specially suited to specific environments.

Media restrictions (attenuation, dispersion and residual echo) and line impairments (crosstalk and impulse noise) have been identified as the constraint to the data rate in subscriber systems. More recently, with better devices, line impairments are identified as the limitations. During the late seventies, media restrictions caused the more serious limitations to transmission of high-speed data. These later limitations were studied in context to the subscriber loop population in the 1973 Loop Survey data base. This data base was the most recent one at that time and the results were encouraging and consistent with the design rules of the T1 carrier systems. For these reasons, an elaborate series of simulation studies was undertaken to investigate the capability of the loop plant for rates from 32 kbps to 324 kbps. Both time-compression multiplexing (TCM) and adaptive echo cancellation (AEC) techniques were investigated in detail. These results have been documented in Reference 5.12. The individual components of the digital subscriber line have been designed, and their performance has been streamlined; BRISDN transceivers have been available since the late eighties. The simulation and design of these BRISDN transceivers are presented in References 5.4, 5.12, and 5.13.

5.3.1.3 Recent Perspective

The commercial viability of the repeaterless digital subscriber line was established by the offering of the 56 kbps line in the early eighties. Time compression mode, with a line rate of 144 kbps using 3 msec bursts in each direction, was adapted. Later the viability of the adaptive echo canceler technology was introduced, proven and accepted for BRISDN transmission with the 2B1Q code. The interfaces have been standardized since the late eighties and the BRISDN systems and its components are firmly in place for the proverbial digital revolution to reach every home via the existing twisted wire-pair telephone line. A more ambitious and realistic application of the twisted wire-pair is for high-speed LANs at rates up to 155 Mbps over a very short distance of about 30 ft or 100 m. The devices and chips for these applications have been designed and fabricated (see Section 4.8). Copper distributed data interface during the nineties is a distinct possibility alongside FDDI. However, the other prospect of having all high-speed wireless networks, such as wireless ATM, is just around the corner (see Sections 4.9 and 4.10).

5.3.2 GENERAL TOPOLOGY OF THE SUBSCRIBER LOOP PLANT

Figures 5.1 and 5.2, show different variations of the subscriber network. The physical configuration of the subscriber loop generally consists of star topology with several thousands of copper wires from the Central Office (with different lengths and diameters in cables) installed overhead on poles or buried underground in ducts.

The wire-pair density tapers off gradually as the lines get more distant from Central Office towards the customer premises. The wire-pairs also form a star topology at Remote Terminals (RTs). One or more such RTs may form a carrier serving area or a business location. The RTs are usually served from Central Offices by digital carriers such as T1, SLC-96™ or any other loop carrier systems [5.14]. The number of the cables from the Central Office tapers because of the branching that occurs in both the feeder and the distribution plant. The distance from the customer premises to the Central Office ranges from a few hundred feet (less than a 100 meters to about 210 kft. (64 km). Figure 5.3 shows the star-tree transmission media and electronic apparatus (repeaters and multiplexers) that interface between the feeders and distribution and the interconnections between drops in the distribution cable to individual customer premises.

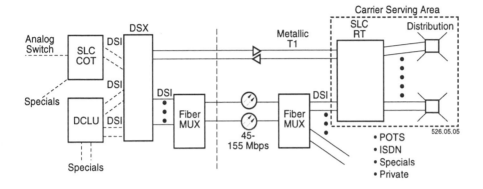

FIGURE 5.5. Stages of Integration of Fiber in the Loop Plant by the Use of Fiber Multiplexers and Subscriber Loop Carrier Systems to the Remote Terminals. DCLU - Digital Channel Line Unit; DSX - Digital Signal Cross Connect; ISDN - Integrated Services Digital Network; MUX-Multiplexer; POTS - Plain Old Telephone Service; RT - Remote Terminal; SLC - Subscriber Loop Carrier

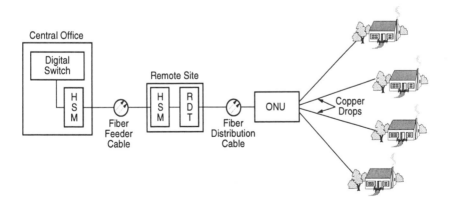

FIGURE 5.6. Fiber to the Curb Architecture with the Remote Optical Terminals with Fiber Feeder and Distribution. HSM -High Speed Multiplexer; ONU - Off Network Unit; RDT-Remote Distribution Terminal

It also shows the role of the main distribution frame at the Central Office that connects customer loops to various services within and out of the Central Office. From the pedestal to the customer, the pairs are dedicated and relatively inexpensive. Generally enough pairs are installed to meet future needs. From the distribution to the feeder, the number of pairs per customer is fewer. The taping of the wires produces bridges taps which allow parallel connections. In certain cases, there may be bridged taps on branches that have already been bridged. Bridged taps increase the flexibility of customer connections, because it allows the same line to be connected to customers in different geographical areas.

However, when subscribers are connected, the other parallel paths remain open circuited, except for party lines where customers share the same line. Figure 5.5 shows the architecture and the use of fiber in the feeder, while Figure 5.6 shows the transition to fiber in the distribution, which brings the fiber to the curb in the FTTC architectures.

5.3.3 ELECTRICAL CHARACTERISTICS OF LOOPS (1973 LOOP SURVEY)

Enhanced transmission rates envisioned for the DSL services range from data rates from 56 kbps, 64 kbps (or B rate), 80 kbps or B+D rate, and 144 kbps or 2B+D rate kbps in the late seventies and early eighties. The line rate envisioned could range from 144 kbps (TCM rate for 56 kbps) to 400 kbps (TCM rate 144 kbps). In the AEC mode of transmission the line rate is the data rate and the signaling rate. The line code influences the actual symbol rate. Thus the range of frequencies at which loop plant characteristics become significant depends upon the mode and the code chosen for this application. In the United States, Regional Bell Operating Companies have chosen the 2B1Q code for the basic rate access. The range of spectral interest ranges between direct current (DC) to about 80 kHz for the early DSL applications. In a global sense, the loop plant has to be investigated in view of the national and need characteristics of the particular environment. Codes, rates, and access methodologies have to be individually tailored to the specific region (or country) under consideration.

During the late seventies, time compression multiplexing (TCM) and adaptive echo cancellation (AEC) techniques were both investigated in detail. These results have been documented in References 5.9, 5.10 and 5.15. The individual components of the older versions of the digital subscriber line have been designed, simulated and optimized. The prototypes were constructed with accomplished field tests. In Figure 5.7, the technical feasibility of the TCM system in United States loop environment is depicted.

Each point on the scatter plot indicates the extent to which the AMI eye-diagram is open. Since the AMI eye-diagram has a positive and a negative part, each loop is identified by two points on the plot. The Y-axis depicts the extent to which the eye (positive and negative) is open and the X-axis depicts the length of the loop. Only the Subscriber side scatter plot is shown, since the Central Office side usually has better eye-openings. The length and gauge of the bridged taps present a more serious degradation of the eye-diagrams rather than the length of the loop at this circuit-switched digital capability (CSDC) rate. The documentation is complete in the references.

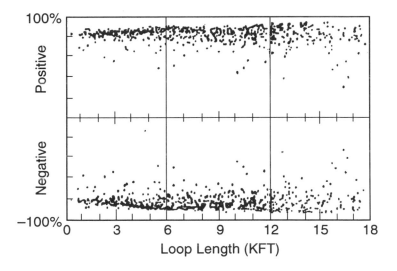

FIGURE 5.7. Eye Opening Scatter Plot of the TCM-CSDC at 56 kbps in the Loop Plant with Bridged Taps. CSDC - Circuit Switched Digital Capability; TCM- Time Compression Multiplexing

The viability of the digital subscriber line was established by the offering of the 56 kbps CSDC access [5.16] through the eighties. Time compression mode with a line rate of 144 kbps uses 3 ms burst repetition rate with 1.39 ms bursts in each direction. Two hundred bits (1.39 times 144) per burst are sent in *each* direction with a silent time after each burst of 0.111˙ ms every 3 ms cycle. The equalization is for all loops with a loss up to 45 dB at 72 kHz, since the bipolar or the AMI code is used for the line code. The timing recovery circuit is to retain the recovered clock through the two silent times and the transmit duration at the subscriber end with a pulse jitter within ±150 ns. The AMI pulses have 50% duty cycle and have with a pulse width of (3.4722 µs ± 150 ns) with an pulse height of (5.2 ± 5 %) V. The transmit power considerations are presented in Reference 5.13 and the penetration consideration are presented in Reference 5.4. These specifications are based upon the characteristic of the loop plant for the band of spectral interest for the AMI code at 144 kHz. These characteristics are presented next.

Bridged taps have a profound influence on the overall loop plant capability to carry higher data rates. The statistical average of loop lengths is about 7,750 ft. or 2.36 km and the equivalent length (about 12,900 ft. or 3.93 km) of 22 AWG cable. The equivalent length is based on the average length of 26, 24, 22, and 19 AWG cables present in a average loop. The loss of such a loop is about 16.75 dB at 60° F and at 32 kHz. However, the average loss computed for the entire loop population based upon the bridged taps being intact is 18.25 dB at 60° F and at 32 kHz. Thus the average loss of all the loops with bridged taps intact is about 1.75 dB higher than the corresponding loss with the bridged taps stripped at 60° F and at 32 kHz. At this frequency the equivalent length of 22 AWG with the same average loss increases by about 1400 ft. or 0.427 km.

The effects of the bridged taps on the image impedances are also significant. The subscriber side impedance at 32 kHz, changes from 134.36 - j 78.15 Ohms for loops without bridged taps to 99.8 - j 68.73 Ohms whereas the Central Office impedance experiences a smaller change from 140.4 -j 83.9 Ohms to 139.9 -j 82.0 Ohms due to the fact that fewer of the bridged taps are located nearer to the Central Office than to the subscriber side. A summary of these impedances at the six discrete frequencies (32, 40, 72, 96, 108, 162 kHz) is tabulated in Table 5-1 at 140°, 70° and 0° F. A summary of loop attenuations at the same temperatures is also presented in Table 5-2. The variations of the impedances and the attenuations at the different temperatures are too significant, and for this reason it is expected that the adaptation built in the digital subscriber line devices will accommodate the effects of temperature fluctuations.

At 32 kHz the average loss of all the 831 loops (for possible DSL applications) in the data base is 18.5 dB. This loss climbs to 23.2 dB, 25.2 dB, and 29.7 dB at 72 kHz, 96 kHz and 162 kHz, respectively. The image impedance at 32 kHz, 72 kHz, 96 kHz, and 162 kHz is computed to 140.4-j82.0 Ohms, 119.4-j62.4 Ohms, 115.4-j49.6 Ohms and 110.7-j27.9 Ohms, respectively at the Central Office side. The corresponding numbers at the subscriber side are 99.8-j68.7 Ohms, 84.6-j35.9 Ohms, 82.5-j28.1 Ohms and 79.2-j20.3 Ohms, respectively. The loops display a much greater variance at the subscriber side because more bridged taps occur near the subscriber. The average propagation delay is about 5.51062 µs/km at 162 kHz for 24 AWG cable.

TABLE 5-1 MAXIMUM AND AVERAGE LOOP IMAGE IMPEDANCES (WITH BRIDGED TAPS) AT 140°, 70°, AND 0° F

kb/KHz	Temperature °F	Central Office Side of Loops (Z_{in})						Subscriber Side of Loops (Z_{in})					
		Maximum		Minimum		Average		Maximum		Minimum		Average	
		Re	Im	Re	Im	Re	Im	Re	Im	Re	Im	Re	Im
64/32	140	196.0	-120.1	47.1	-88.9	144.3	-99.1	193.2	-121.0	28.9	-4.3	102.3	-74.7
	60	194.7	-112.2	43.8	-79.6	139.9	-82.0	188.3	-112.2	29.1	-0.9	99.8	-68.7
	0	192.9	-105.5	40.3	-71.9	136.4	-67.0	181.3	-103.2	28.6	+2.6	98.0	-63.7
80/40	140	187.8	-108.9	44.7	-76.1	135.7	-95.2	177.9	-105.0	25.5	+3.1	96.1	-63.2
	60	189.8	-99.4	43.9	-67.7	132.3	-82.2	181.3	-100.5	24.9	+8.9	94.2	-57.8
	0	189.6	-94.8	41.8	-60.8	129.7	-70.2	183.1	-96.6	24.8	+15.3	92.9	-53.4
144/72	140	187.1	-101.7	40.3	-49.3	120.5	-67.5	157.6	-84.3	21.5	+27.8	85.2	-39.7
	60	190.5	-98.4	40.1	-43.2	119.4	-62.4	164.1	-84.5	21.3	+43.7	84.6	-35.9
	0	192.0	-93.0	40.1	-38.3	118.7	-57.6	175.3	-85.7	21.2	+54.6	84.2	-32.8
192/96	140	206.7	-79.7	38.6	-39.5	115.8	-49.6	180.3	-84.1	19.4	+49.7	82.6	-31.5
	60	233.0	-80.2	83.6	-34.5	115.4	-49.6	186.8	-83.5	20.3	+54.4	82.5	-28.1
	0	253.3	-81.1	38.7	-30.5	115.2	-50.6	200.7	-85.1	18.9	+56.8	82.4	-25.4
216/108	140	212.2	-105.5	37.9	-36.1	114.4	-38.0	180.1	-82.7	16.3	+59.4	81.8	-29.0
	60	198.8	-111.5	38.0	-31.5	114.2	-36.8	194.0	-81.4	16.5	+68.5	82.0	-25.9
	0	201.8	-114.2	38.1	-27.8	114.2	-37.5	203.1	-84.4	16.8	+74.9	82.1	-23.3
324/162	140	216.0	-119.7	35.1	-26.7	110.2	-34.6	184.8	-100.6	15.2	+74.5	78.5	-22.4
	60	233.4	-133.4	35.4	-23.4	110.7	-27.9	213.6	-105.3	14.0	+81.4	79.2	-20.3
	0	261.5	-145.5	35.6	-20.8	111.2	-19.8	242.0	-108.1	13.3	+88.0	79.8	-18.6

TABLE 5-2 Maximum and Average Attenuation of Loops

| kbps/kHz | Temperature | Attenuation in dB | |
		Maximum	Average
2.4/1.2	140	11.7	4.93
	60	10.7	4.49
	0	9.9	4.13
4.8/2.4	140	15.6	6.92
	60	14.1	6.29
	0	13.3	5.78
9.6/4.8	140	21.5	9.63
	60	19.5	8.73
	0	18.0	8.01
56/28	140	45.5	19.93
	60	41.0	17.75
	0	37.4	16.02
64/32	140	47.6	20.82
	60	42.9	18.50
	0	39.1	16.66
80/40	140	50.6	22.27
	60	45.6	19.72
	0	41.7	17.72
144/72	140	60.0	26.30
	60	53.3	23.22
	0	47.9	20.84
192/96	140	66.2	28.46
	60	60.5	25.21
	0	56.4	22.73
216/108	140	69.1	29.44
	60	61.9	26.13
	0	56.4	23.62
324/162	140	76.0	33.15
	60	69.0	29.68
	0	63.5	27.08

From Table 6-2 in Section 6.4.3, this number (average propagation delay) is essentially the same for the three gauges (26, 22, and 19 AWG). The number also corresponds to 1.68 µs/kft at 162 kHz.   This value is approximated as 2 µs/kft at 72 kHz for the CSDC applications. The design of the bidirectional data transmission has to accommodate the wide disparity of loop characteristics such as its image and input impedances, highly variable frequency dependent losses looking from either direction of the loop. The range and statistics of these variations were also been presented in Reference 5.10.

The signal delay (derived from the imaginary term in the secondary constant $[\Gamma = \alpha + j\beta]$ of cables) is about 5.51 µs/km at 162 kHz. The wavelength thus becomes 1.12017151 km (6.1728 µs divided by 5.51062 µs/km). The quarter wave $\lambda/4$ length (see footnote on the next page) is about 280.004 m (918.77 ft.) and this number is just about equals 281 m (922.42 ft. from the 1973 truncated Loop Survey of 831 loops and about the same number for the 1983 truncated Loop Survey of 1520 loops), which happens to be an average bridged tap length. Hence, if the TCM mode of transmission is selected at 144 kbps, then the TCM rate of 324 kbps will suffer severe inter-symbol interference. This inference is documented by the simulation studies of the loop plant eye opening scatter plots discussed in References 5.4 and 5.13. The extent of such inter-symbol interference at other frequencies in other networks and other codes can be estimated.

Three frequencies chosen to span the spectral band are at 40, 60, and 80 kHz. This band is appropriate for rates below the BRISDN with the 2B1Q code in the duplex mode of transmission. Ninety percent of the loops display signal attenuation under 33, 37, and 40 dB loss at the respective frequencies. Ninety-five percent of the loops have signal attenuation less than 35, 40, and 43 dB at 40, 60, and 80 kHz. The mean loss of the forty-five percent (i.e., <1.83 km or 6.0 kft.), sixty-five percent (i.e., <2.74 km or 9 kft.), eighty-two percent (i.e., <3.65 km or 12 kft.), ninety-five percent (i.e., <4.57 km or 15 kft.) and 100 percent of the non-loaded loops have a mean loss of about 11.4, 15.1, 18.6, 21.1, and 22.2 dB at 80 kHz respectively. Ninety percent of the loops under 1.83 km (6 kft.), under 2.74 km (9 kft.), under 3.65 km (12 kft.), under 4.57 km (15 kft.) and under 5.49 km (18 kft.), display a loss of 18, 25, 33, 37, and 40 dB, respectively at 80 kHz. An estimation of the penetration of services may thus be derived from these attenuation loss values.

5.4

1983 LOOP SURVEY FINDINGS

5.4.1 LOOP ENVIRONMENTS

The loop environment suffers severe discontinuities between the Central Office and the subscriber. The three major impairments arising from the structural properties of the cable are: gauge changes, bridged taps, and loading coils. Any design of a bidirectional high-speed data transmission facility has to accommodate the wide disparity of cable characteristics resulting from the different composition of cables and multiplicity of bridged taps. Load coils make them unsuitable for any HDSL applications. The 1983 Loop Survey consists of a random sample of 2290 loops from the participating Bell Operating Companies.

Gauge Changes: In the United States and most Western European countries, telephone networks have a number of different diameters of the copper and aluminum wires. This produces gauge changes in the subscriber loop plant. The junction points between wires of different diameters are sources of reflection, and constitute non-uniformity in the cable characteristics. There are four dominant wire sizes used. The finest diameter wire generally encountered is 26 AWG, roughly equivalent to the 0.4039 mm wire used in European countries. The coarsest wire is 19 AWG with a diameter of 0.9119 mm. The two intermediate wire sizes used in the loop plant are 24 AWG (0.5105 mm) and 22 AWG (0.6426 mm). About eighty percent of the loops in the United States have gauge changes with an average of 2.3 gauge changes per loop. The percentage distribution of gauges are fifty-eight percent of the 26 AWG, thirty-one percent of the 24 AWG, ten percent of the 22 AWG, and less than 0.5% of the 19 AWG. PIC (Polyethylene Insulated Cables) dominate the loop plant. Paper insulated cables are rare.

Bridged Taps*: Open circuited cable sections tapped from the main loop between the Central Office and the subscriber known as bridged taps (BTs) are also abundant in the United States, Canada, Japan, Italy, and Australia.

---

* Quarter wavelength bridged taps cause the maximum inter-symbol interference since the interval for the pulse to travel up and down the open bridged tap is one half the period at half bit rate frequency. A binary +1 and a binary 0 (which is coded as -1 in the bipolar code) would be separated by this interval and thus the -1 pulse "sees" the highest amplitude of the reflected pulse at the instant it is being scanned. When decision feed back or other sophisticated equalization strategy is used, the effects of such bridged taps can be compensated.

The primary reason for these BTs is traced back to the multi-party lines that prevailed in the early days of telephone networks. Also, in areas where the demographic growth was uncertain, the cable installers used to leave behind open ended cables in the expectation that these lines would be used somewhere, sometime. Later, segments of the cable were used to connect new subscribers elsewhere, thus leaving behind the remaining sections of unused cables. In the United States, these are the two major reasons for BTs in the subscriber loop plant. Eighty-seven percent of the loops have at least one bridged tap, and the average bridged tap length is 281 meters. The percentage gauge distribution in the bridged taps is about the same as that in the loops. In this survey, the average length of a sampled loop is 10,787 feet, and about eighty percent of the sampled loops have a working length less than or equal to 15,000 feet.

Load Coils: Load coils of about 88 mH were also placed along long loops. Such coils are used to compensate for the predominantly capacitive character of the wire-pairs. At voice frequencies, these coils restored the proper tone of the telephone voice. The survey indicates that about 23.7% of the sampled loops are loaded, and thus unfit for high-speed data. This removes 543 loops from the survey with 1,747 loops remaining for high-speed data. However, 227 loop configurations either do not carry accurate descriptions for the analysis, or are longer than 18,000 feet, or are loops that have non-standard cable make-up, thus further reducing the subset to 1,520 loops (about sixty-six percent of the total sampled loops) considered for the higher data rates. These loops are non-loaded and less than 18,000 feet, and have standard cable sections.

Four dominant wire sizes are used. The finest diameter wire generally encountered is the 26 AWG and is roughly equivalent of the 0.4039 mm wire used in the European countries. The coarse wire is the 19 (with a diameter of 0.9119 mm) AWG. The two intermediate wire sizes used in the loop plant are the 24 (0.5105 mm) and 22 (0.6426 mm) AWG.

About forty-five percent of the loops under 5.48 km (18 kft.) in the 1983 survey have a working length less than 1.83 km (6 kft.); sixty-five percent of the loops have a working length less than 2.74 km (9 kft.); eighty-two percent have the length less than 3.65 km (12 kft.) and ninety-five percent are under 4.57 km (15 kft.). Shorter loops have the finer 26 AWG polyvinyl chloride (PVC) cable. Longer loops tend to have more sections of the cable with coarser (24, 22, and 19 AWG) wire sizes towards the Subscriber. The 19 AWG has an insignificant proportion in the ISDN loops, even though the 24 and 22

the loops at 5.33 KM (17.5 kft.)   At this loop length, the 26 and 19 AWG
cables share the remaining fourteen percent equally, at seven percent each.

5.4.2 1983 LOOP SURVEY (PHYSICAL CHARACTERISTICS)

Out of all the 2,290 loops investigated in the 1983 Loop Survey, about thirty-
four percent have load coils, or are longer than 18 kft. (5.486 km), or have
nonstandard cable sections, or are otherwise unsuitable for ISDN data
transmission. Eliminating these loops, about sixty-six percent (1,520 loops) of
the entire loop population surveyed constitutes the truncated data base over
which the digital data transmission is readily feasible. Twenty-four percent
of the all the loops surveyed had load coils to selectively strengthen the
audio frequency signal in the voice frequency applications, thus rendering
them unsuitable for the ISDN considerations. Only ten percent of these non-
loaded loops were longer than 4.57 km (15 kft.). Also only ten percent of the
non-loaded loops had bridged taps longer than 0.91 km (3 kft.). About fifty
percent of the loaded loops were shorter than 5.48 km (18 kft.) partially
because of voice special services and/or errors as the working length was
reduced from over 5.48 km (18 kft.) while rearranging the cable sections for
new customers.

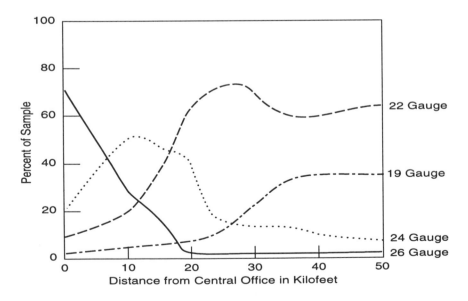

FIGURE 5.8. Gauge Distribution in the 1983 Loop Survey. Sample
includes bridged taps.

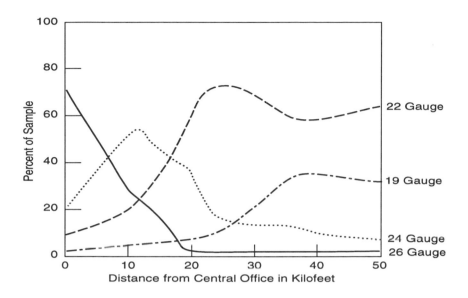

FIGURE 5.9. Gauge Distribution in the 1983 Loop Survey. Sample does not include bridged taps.

The gauge distribution is depicted in Figures 5.8 and 5.9 for the loop populations with bridged taps and for the same population without taps. From these two figures, it is evident that the coarser gauge (22 AWG vs. 26 AWG) become dominant as the loop length increases and that the inclusion bridged taps does not significantly change the overall distribution of the gauges in the loops. Thus the loop to any subscriber may be expected to have the same gauge distributions. Further, the influence of the 19 AWG cable is almost insignificant for the HDSL service.

Aerial, underground, and buried cables prevail in the HDSL loops and the percentage distribution is depicted in Figure 5.10. The buried cable structure of the plant dominates as the loop length increases, and any loop is likely to have all three structures roughly equally distributed with about the same probability in the 14 to 18 kft. (4.27 to 5.48 km) band.

Thirty-two percent of the loops constitute business loops. The average length of all business loops is 8.816 kft. (2.688 km) with an average bridged tap length of 894 ft.(272.5 m). Business loops are about 25 percent shorter than residential loops (based upon average length in each segment of the loop population). Ninety percent of the loops are less than 18 kft. (5.49 km) and have a bridged tap length of less than 2,000 ft. (609.7 m). Seventy-five percent of the loops are under 12 kft. (3.66 km) and have a bridged tap length of less than 1,200 feet (366 m).

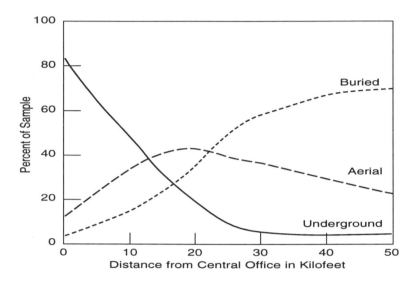

FIGURE 5.10. Cable Structure for the Loops in the 1983 Loop Survey Data Base.

Seven percent of the loops surveyed handle special services (i.e., services other than POTS). The average length of these loops is 9,059 feet with an average bridged tap of 736 feet. These loops are about 3 percent longer than business loops and about 30 percent shorter than residential loops. About seventy-one percent of these loops are shorter than 12,000 feet and have a bridged tap length of about 800 feet.

The 1983 Loop Survey also includes a cross section of all the loops that may be expected to carry the HDSL data traffic. The mean length of *these loops* surveyed is about 3.29 km (10.78 kft.) and the mean bridged tap length is 0.57 km (1.87 kft.). The presence of loading coils in loops beyond 4.57 km (15 kft.) or beyond 5.48 km (18 kft.) makes such loops unsuitable for basic rate repeaterless ISDN data transmission. (see Section 6.3.1 on Design Impairments.)

5.4.3 1983 LOOP SURVEY (ELECTRICAL CHARACTERISTICS)

Cable and conductor (metallic) access exists between the Central Office and the subscribers in most conventional telephone plants. These metallic paths have characteristic electrical properties. Four such parameters (the primary constants R, L, G, and C) uniquely characterize a uniform section of the cable. Many such sections may constitute an individual loop, but the characterization of the entire loop may be derived if the make-up of the loop defining these sections and the four primary constants of each section are known.

The primary constants vary considerably with frequency, temperature, and cable size. For example, a set of values for these constants for the 26 AWG polyethylene insulated cable (PIC) at 70 kHz and 70° F is 453.03 Ohms per mile, 0.9546 milliHenries (mH) per mile, 4.63 micromhos per mile, and 0.083 microFarads (µF) per mile. At 1 MHz and 70° F, the same constants become 956.65 Ohms per mile, 0.8381 mH per mile, 46.85 micromhos per mile, and 0.083 µF per mile. At 70 kHz and 120° F, these constants become 502.51 Ohms per mile, 0.9617 mH per mile, 4.643 micromhos per mile, and 0.083 µF per mile. These constants are published in Reference 5.14 for frequencies up to 5 MHz at 0°, 70°, and 120° F. Bare overhead open-wires have different primary constants associated with them, depending upon their size, pole spacing, etc. The ISDN loop plant may include a wide variety of transmission media. Special consideration is necessary to evaluate the type of media over which data can be transmitted at different rates [5.17].

Cable pairs in the loop plant also have two secondary electrical constants: the characteristic impedance, and the propagation constant. These two constants are derived from the primary constants (R, L, G and C). The two secondary constants are essential for computing the circuit performance of the cable pair in carrying any electrical signals such as line codes carrying high-speed data. During the manufacture of the cables, strict control of the conductor diameter, thickness, quality, and the consistency of the insulation are all essential to ascertain the uniformity of these four primary constants. A five percent manufacturing tolerance is still accepted as the standard range for the primary constants in the cable manufacturing facilities in the United States. The secondary constants thus have an implicit range of variation.

The cumulative loss characteristics of loops in the 1983 Loop Survey data base are depicted in Figure 5.11 for the HDSL frequencies. We have also investigated the loss characteristic of the loops for all possible bandwidth efficient codes for VHDSL and the ADSL frequencies and they are presented in Chapter 14. These families of curves become important in evaluating the penetration and range of new digital services over the telephone subscriber lines.

For example, consider the CSDC system discussed in Section 6.4. A 2.5 V peak signal in the HDSL frequency range of 80 kHz from 90% of the subscribers (Figure 5.11) suffers signal-loss less than 40 dB, thus attenuating the signal to 25 mV in the worst case. Alternatively stated, 90% of the subscribers will have a 80 kHz signal delivery to and from the Central Office at a level above 25 mV. If the Central Office noise floor is identified at 10 mV, then the SNR due to Central Office noise (by itself) is about 8 dB. In practice, the crosstalk, tail echo (from the burst), phase jitter effects, impulse noises, and other electrical disturbances, also degrade this SNR. However, since the business customers (requesting the CSDC service) are likely to be closer to the normal population, the SNR is likely to be higher and the CSDC service quality is correspondingly higher. A family of such curves at different frequencies is outlined in Figure 5.11 and also addressed in Chapter 14 provide an insight into the digital service viable in any given environment.

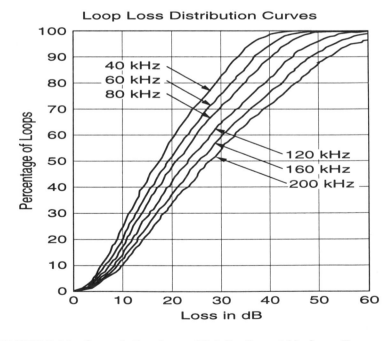

FIGURE 5.11. Cumulative Loss Distribution at Various Frequencies for the Loops in the 1983 Loop Survey Data Base.

Figure 5.12 presents a distribution of the cumulative loss in five different distance bands ranging from 6 kft. to 18 kft. Continuing the CSDC case, if the businesses requesting the 56 kbps service are within 12 kft. from the Central Offices, then almost 99% of the subscribers will have the same quality of service (SNR about 8 dB due to noise floor alone). The example presented here is for a first step approximation on the penetration of services, but in reality, more detailed computations are necessary and are presented in Reference 5.13.

In Figures 5.13 and 5.14, the line impedances of the loops in the data base at 40 kHz are presented, consistent with the BRISDN data rate of 160 kbps and the 2B1Q code. The symbol rate is 80 kbaud and the mid-band frequency is 40 kHz. The purpose of these two figures is to illustrate the electrical disparity of the subscriber loop plant when it is viewed from the subscriber side or from the Central Office.

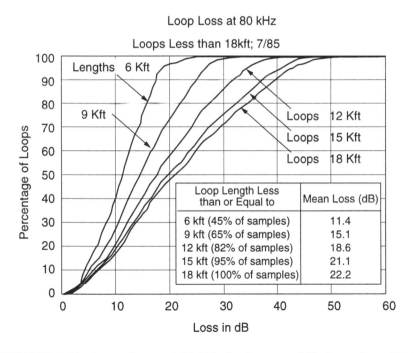

Loop Loss at 80 kHz

Loops Less than 18kft; 7/85

Loop Length Less than or Equal to	Mean Loss (dB)
6 kft (45% of samples)	11.4
9 kft (65% of samples)	15.1
12 kft (82% of samples)	18.6
15 kft (95% of samples)	21.1
18 kft (100% of samples)	22.2

FIGURE 5.12. Loop Loss at 80 kHz for Loops of Various Lengths in the 1983 Loop Survey Data Base.

These two clusters assume a much higher dispersion and variance at HDSL frequencies (see Chapter 13). The signals entering the loop plant do suffer reflections due to impedance mismatches at the entry and the exit from the loop plant. Figures 5.13 and 5.14 indicate the extent variation of these echoes from loop to loop. Typically, when the terminating impedance is 120 Ohms (i.e., the average value of the real components of the line impedances (see Figures 5.13 and 5.14) then the echo return loss (ERL) can be computed from the equation 3.2 of Reference 5.18:

$$ERL = 20^* \, Log \, [(Z_B + Z_L)/(Z_B - Z_L)] \quad dB \tag{5.1}$$

Indicating that Z_B (i.e., the terminating impedance in this case) and Z_L (the input impedance of the line) should be as close to each other to reduce the echo reflected signal (i.e., to enhance the loss of echo return signal to infinity). At the subscriber end (see Figure 5.14) the impedance matching is a serious problem and the demands on the echo canceler are going to be quite intense.

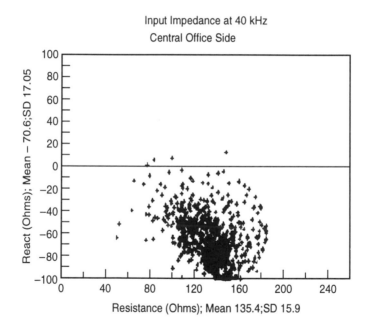

FIGURE 5.13. Loop Image Impedance at 40 kHz Viewed from the Central Office Side for the Loops in the 1983 Loop Survey Data Base.

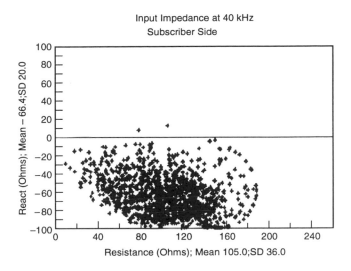

Input Impedance at 40 kHz
Subscriber Side

FIGURE 5.14. Loop Image Impedance at 40 kHz Viewed from the Subscriber Side for the Loops in the 1983 Loop Survey Data Base.

For the duplex mode, the residual echo is computed from the mismatch effects and added to the crosstalk, stray impulse noise and Central Office/subscriber noise to compute the final SNR and thus evaluate the quality of received signal. The scatter plots depicted in Figures 5.13 and 5.14 differ from one loop environments to the next, and a computational environment provides the extent and nature of the typical echoes in the loop plant. The device vendors and the chip designers for the transceivers of HDSL and VHDSL systems receive this requirement for the FIR filters embedded in the echo cancelers.

A large family of such scatter plots are depicted in Reference 5.7 for the 1983 Loop Survey for different HDSL frequencies. Such families of plots has not been generated for all the loop in the data base for the VHDSL and ADSL frequency bands, but their values can be computed for the ANSI and ETSI loop topologies presented in Section 5.7.1.

The image and input impedances at five frequencies in the DSL and HDSL range (40.0-200 kHz) are tabulated in Tables 5.3 and 5.4 at the Central Office and at the subscriber ends, respectively. The standard deviation values, indicating the range of variations for both the real and imaginary components of the two impedances are highest at the subscriber end and tend to get enhanced as the frequency increases.

TABLE 5-3 Central Office Side Image and Input Impedances

Frequency kHz	Aver.Image Impedance Ohms	Standard Develop. Re. Im.		Aver.Input Impedance* Ohms	Standard Develop. Re. Im.	
40.0	133.7-j68.7	15.1	18.0	135.4-j70.6	15.9	17.05
60.0	124.0-j50.6	13.7	16.1	124.2-j51.8	13.4	14.7
80.0	119.2-j40.8	14.1	14.8	119.0-j41.5	13.2	13.4
120.0	114.2-j30.0	13.1	14.3	114.0-j30.2	12.6	13.2
160.0	111.9-j24.3	12.4	12.4	111.7-j24.4	11.7	12.9
200.00	110.1-j20.6	12.6	12.6	110.1-j20.5	12.0	11.9

*The terminating impedance is 120 Ω.

TABLE 5-4 Subscriber Side Image and Input Impedances

Frequency kHz	Aver.Image Impedance Ohms	Standard Develop. Re. Im.		Aver.Input Impedance* Ohms	Standard Develop. Re. Im.	
40.0	103.2-j65.0	31.0	18.2	105.0-j66.4	36.0	20.0
60.0	94.2-j47.6	31.0	18.4	94.6-j48.5	32.1	19.3
80.0	90.0-j38.0	31.8	19.5	90.0-j38.5	32.3	20.0
120.0	85.7-j27.6	33.7	22.8	85.6-j27.7	33.9	23.0
160.0	83.9-j22.2	36.4	25.4	83.9-j22.3	36.4	25.6
200.0	82.2-j18.7	37.3	28.4	82.1-j18.7	37.3	28.4

*The terminating impedance is 120 Ω.

5.5

CARRIER SERVING AREA (CSA) CONCEPTS

A Carrier Serving Area (CSA) is a plant engineering entity as a distinct geographic area capable of being served by a single Remote Terminal (RT). The CSA maximum permissible [5.19] outer bounds are determined by the serving distance over copper of unrepeated 64 kbps service, and by POTS (Plain Old Telephone Services). However, the acceptance of CSA loops to carry most of the higher data rates have made them the prime target loops around which the system optimization studies are based. The market study for the need of the T1 rate data services and the location of the customers suggests that these high data rate customers are located well within the CSA, or close to the Central Offices. The loop length is thus restricted to a maximum length of 12 kft (3.6 km). Equivalently, loss is limited to the average loss of 9 kft (2.7 km) of 26 AWG twisted wire-pair in the traditional telephone cable network. Moreover, the loop configuration is likely to be simpler with zero, one or two bridged taps (BTs), as defined by the CSA loop selection criteria. From the studies in 1981 [5.15], the CSA loops have to meet the *five* following requirements:

1.   There are no loading coils anywhere in the loop;

2.   Any loop containing 26 AWG is restricted to a total length of 9 kft (2.7 km), including BT length;

3.   If there is no 26 AWG in the loop, the total length of all 24, 22, and 19 AWG cables in the loop (including all BTs) may be as much as 12 kft (3.6 km.) beyond the theoretical RT site on non-loaded coarse gauge loops;

4.   Total BTs length is limited to 2.5 kft, with no single tap greater than 2.0 kft;

5.   The number of gauges is limited to two (exclusive of BTs) along the loop [5.16].

The CSA loop selection is based on an existing plant design process, which results in the creation of distribution areas (DA), and allocation areas (AA). The DA is a planning entity with defined boundaries, 200-600 ultimate living units based on expected land usage, and uniform transmission requirements. An AA is a grouping of contiguous DAs (preferably two-four, although five is allowed), that is administered as a single unit from a feeder standpoint. To establish the carrier serving areas, the engineer should locate in each AA a single optimal Remote Terminal (RT) site, so that, if possible, all subscribers are within 9 or 12 kft of the RT site. Studies have shown that in most cases, having one site per CSA is more economical than dispersed sites, where the cable cost from the single RT site rarely prevails over the expense involved in establishing additional sites. The *three* elements to be considered in the CSA allocation are:

(a) The cost of placing the cable from the remote terminal site should be examined against that of establishing multiple RT sites;

(b) Routine maintenance and emergency restorable are critical items. Digital loop carrier systems are usually equipped with batteries capable of providing service for several hours in the event of commercial power loss. Continuity of service is also important in case of extended power outage, which requires the placement of portable generators at each site until commercial power is restored;

(c) Merging of RT sites will enable more effective maintenance and provisioning efforts.

5.5.1 WIRE-PAIRS IN THE CARRIER SERVING AREA (CSA)

A CSA composed of more than one AA is generally uneconomic when the costs of extra cabling required to reach the feeder-distribution interfaces of the more distant AAs are compared to additional site costs. Bandwidth efficient codes (see Reference 5.20 as well as Chapters 12 and 13). offer higher bit rates over subscriber loops without repeaters (and loading coils). For this reason, it is envisioned that the PRISDN services can be made available to most CSA subscribers.

In context to the 1983 Loop Survey of 2,501 loops, one of the RBOCs did not wish to participate in the characterization of the loops, thus reducing the number of loops in the data base to 2,290. Only 1,520 of the remaining loops satisfied the DSL criteria further reducing the possible DSL/HDSL in the data base. When the CSA loop selection rules are enforced on the remaining loops, then there are about 347 loops in the CSA loop data base that satisfy these criteria, thus eliminating about eighty-five percent of the loops.

Due to loop asymmetry, the two ends of the loop exhibit entirely different responses. The bridged taps in the CSA data base also tend to exist closer to the subscriber. Thus, the need for adaptation becomes essential to cope with the loop, its gauges, its discontinuities in the wire gauges, and the presence of bridged taps [5.4], especially at the subscriber end. The loop reflections become stronger (due to increased number of taps) but the crosstalk NEXT noise power becomes weaker (due to fewer disturbers) near the subscriber end. On the other hand, the echo becomes weaker (due to fewer taps) but the NEXT crosstalk becomes stronger (due to more disturbers) near the Central Office end.

The highlights of the loop characteristics in both environments have been reported and the feasibility of the HDSL project at E1 rates has been evaluated. We have also differentiated between these two loop environments and correlated the performance in view of the differences between American and European loop plants. In Europe, the loop plant statistics less systematically compiled published. Hence, loop composition and structure data is not readily available.

5.5.2 PHYSICAL CHARACTERISTICS OF CSA LOOPS

The American CSA loop environment consists of about sixteen percent (364 loops from the 2,290 loops) of the loops surveyed in 1983 Bell System Loop Survey. About forty-five percent of the CSA loop population (164) have no bridged taps, and about the same percentage (166) has one bridged tap. The remaining loops have two taps and about 34 loops have two taps along the working length.

The average length of the CSA loop is about 4,250 feet with an average bridged tap length of about 600 feet. The average length of zero bridged tap loop is about 4,183 feet and the corresponding lengths of one and two bridged tap loops are 4,365 and 4,082 feet. The average bridged tap lengths in the later two groups of loops are 615 and 581 feet, respectively. The loop length scatter plot is shown in Figure 5.15 with the number of taps (0,1 or 2) on the X-axis and loop length on the Y-axis. The scatter plot of the number of sections in the loop with 0, 1 or 2 bridged taps arc shown in Figure 5.16. Finally, the scatter plot relating the tap length (Y-axis) and the loop length (X-axis) for the CSA loop sample is shown in Figure 5.17.

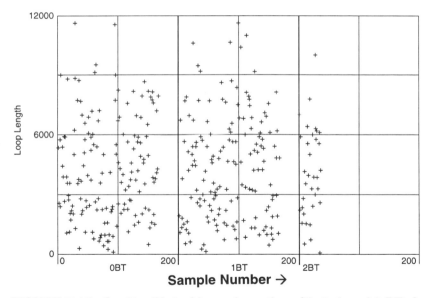

FIGURE 5.15. Scatter Plot of Loop Lengths with 0, 1 and 2 BTs in the CSA Data Base. BT - Bridged Taps; CSA - Carrier Serving Area

When the CSA loops are subdivided by their length in twelve 1000 ft bands, the average loop lengths is roughly at 500 feet mark in each band. The only notable exception occurs in the 9,000 to 10,000 feet band where the average length occurs at about 9,300 rather than at 9,500 feet. Longest bridged taps are concentrated in the 9,000-10,000 band and the shortest tap are concentrated in the 3,000-4,000 band.

5.5.3 ELECTRICAL CHARACTERISTICS OF CSA LOOPS

The image impedance of the CSA cables are depicted in two scatter plots (see Figures 5.18 and 5.19) at 200 kHz. The real and imaginary components vary in a much tighter range than the corresponding ranges of the general loop population (also see Figures 5.13 and 5.14, but note that the X and Y axis are interchanged in the two sets).

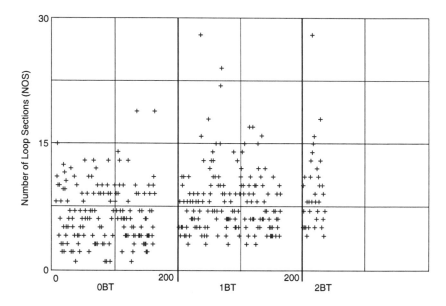

FIGURE 5.16. Scatter Plots of the Number of Loop Sections (NOS) for Loops with O, 1 and 2 BTs in the CSA Data Base. BT - Bridged Taps; CSA - Carrier Serving Area

The rules for CSA loop selection do not consider the inherent loop asymmetry regarding the concentration of taps nearer the subscriber. For this reason both loop populations (CSA and general) depict the same generic pattern; the Central Office side scatter plot is much tighter than the subscriber side scatter plot.

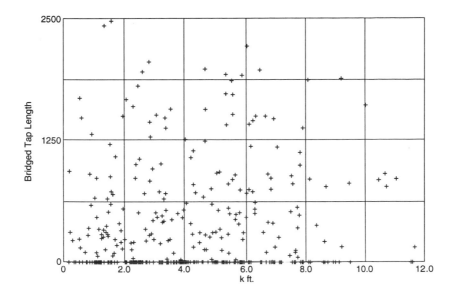

FIGURE 5.17. Scatter Plot of CSA Loops Depicting the Relation Between Length of BTs with the Length of the Loops in the CSA Data Base. BT - Bridged Taps; CSA - Carrier Serving Area

The focal point of the loop-impedance on the Central Office side (Figure 5.18) gets centered at the image impedance (which is also the characteristic impedance) of 107.8-j 22.8 Ohms for the 26 AWG cable at 200 kHz, which corresponds to the general loop design procedure of concentrating the finer gauge wire nearer the Central Office. The image impedance (Figure 5.19) on the subscriber side shows a rather weak focus at this complex value of the impedance due to the presence of 40 percent of the loops without any bridged taps.

The shape of the scatter plot on the subscriber side (Figure 5.19) also exhibits an additional feature: the presence of two overlapping segments of elliptical curves (in fact, spiral segments). These elliptical segments are common (see the scatter plots published in Reference 5.7) to a large number of the images (and to a lesser extent, the input) impedance scatter plots. The generic cause for these elliptical orbits is the presence of the taps and the closeness of the taps to the subscriber.

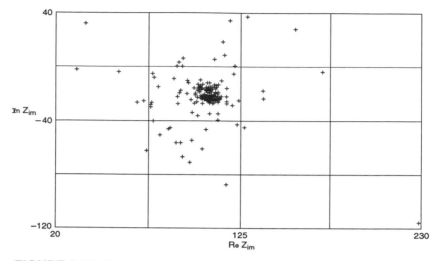

FIGURE 5.18. Image Impedance of the CSA Loops from the Central Office Side. CSA - Carrier Serving Area; Im Z_{im} - Imaginary Part; Re Z_{im} - Real Part

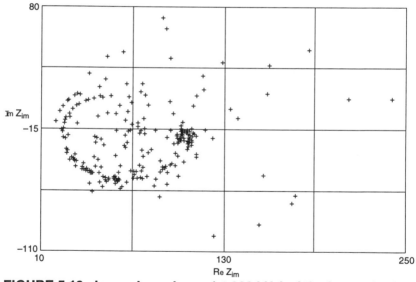

FIGURE 5.19. Image Impedance (at 200 kHz) of the Loops in the CSA Data Base from the Subscriber End. The formation of the spiral is starting in this figures. CSA - Carrier Serving Area; Im Z_{im} - Imaginary Part; Re Z_{im} - Real Part

The shape of the scatter plot on the subscriber side (Figure 5.19) also exhibits an additional feature: the presence of two overlapping segments of elliptical curves (in fact, spiral segments). These elliptical segments are common (see the scatter plots published in Reference 5.7) to a large number of the images (and to a lesser extent, the input) impedance scatter plots. The generic cause for these elliptical orbits is the presence of the taps and the closeness of the taps to the subscriber.

Taps of different lengths give rise to an elliptical segment. When the length of the loop gets limited to any definite value, say 12 kft. or 18 kft., then the two dominant gauges (26 AWG and 24 AWG) in the bridged taps force the impedances to traverse these elliptical paths. And loops with no taps near the subscriber termination will be similar to the loops at the Central Office termination. Loops without taps (e.g., the Swiss loops [see Section 5.6.1]) do not exhibit such elliptical segments in their electrical characteristics and also display a much tighter cluster of points. For the sake of completeness, the scatter plot for the entire band of HDSL frequencies (40-400 kHz is depicted in Figure 5.20.   Im

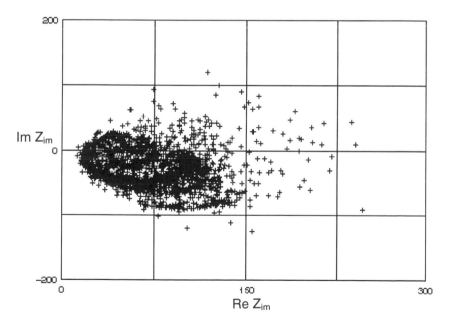

FIGURE 5.20. Image Impedance of the Loops at 40, 80, 120, 160, 200, 280, 360 and 400 kHz for the Loops in the CSA Data Base from the Subscriber Side. The formation of the spiral is almost complete due to the presence of the bridged taps nearer to the subscriber end. CSA - Carrier Serving Area; Im Z_{im} - Imaginary Part; Re Z_{im} - Real Part

The signal wavelength starts to become shorter with respect to the bridged tap length at the higher band edge. The spiral formation is evident in this Figure. Whereas a loss-less tap gives rise to an elliptical trajectory, a lossy tap gives rise to a spiral trajectory. When the lossy-tap (i.e. with finite values of R and G in the primary constants) length becomes infinitely long, the reflected signal never returns, thus forcing only a signal loss rather than a phase perturbation in the received signal. Under these conditions, the real component of the image impedance gets reduced but the imaginary component holds its own value. This effect is clearly starting to happen in Figure 5.20.

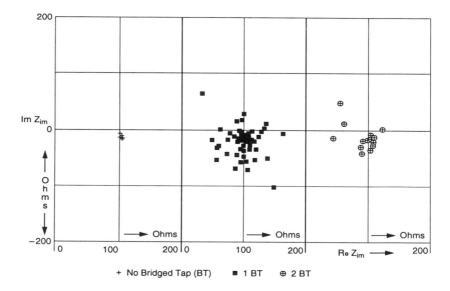

FIGURE 5.21. The Effect of Bridged Taps on the Image Impedance at 400 kHz at the Central Office, Re Z_{im} (Real Part) vs. Im Z_{im} (Imaginary Part).

In Figures 5.21 and 5.22, the image impedances of the CSA loops at 400 kHz are depicted. Here the loops are separated by the number of taps (0,1, or 2). Loops with no taps form the tightest cluster. Criticality of tap length in relation to the wavelength of signal propagation, rather than the number of taps, influences the variation of the image/input impedances.

One of the typical loss characteristics of the CSA loop population is shown in Figure 5.23 at 200 kHz. Some very short loops with critical bridged taps are identified to display little or no loss at certain frequencies due to quarter wavelength line-resonance effects discussed in conventional transmission line theory. In Figure 5.23, this effect is not present at 200 kHz, but appears at 240 kHz and 360 kHz for a 500 ft. loop with one tap.

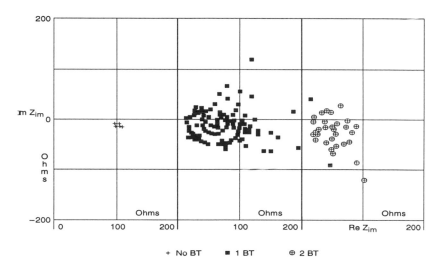

+ No BT   ▪ 1 BT   ⊕ 2 BT

FIGURE 5.22. The Effect of Bridged Taps on the Image Impedance at 400 kHz at the Subscriber Side, Re Z_{im} (Real Part) vs. Im Z_{im} (Imaginary Part). BT - Bridged Taps

5.5.4. PENETRATION OF T1 AND E1 SERVICES

Modulation and Encoding: Encoding of 'n' bits of binary data as one symbol reduces the symbol rate by a factor on 'n' and the ratio of bits per second to baud is sometimes referred to as bits per baud ('n' in this case). The concept of encoding blocks of binary information to the phase of a carrier is known as the phase shift-keying (PSK). The concept of encoding blocks of information to the amplitude of the carrier is also known as the amplitude shift-keying (ASK). In the context of the twisted wire-pairs, both the features of phase and amplitude keying have been recombined as carrierless amplitude and phase (CAP) modulation. Digital signals whose clocks are separated by 90 degrees are used to carry in-phase and out-of-phase signals. The amplitude of the two signals is also modulated. It becomes possible to address any single point in two dimensional (16 point discrete) space by assigning discrete amplitude levels (of -3, -1, +1, -3) to each of the two (in-phase and out-of-phase) orthogonal signals. Sixteen discrete points are feasible in with four-level encoding of the two orthogonal signals. The collection of such points is called a constellation and numerous other (4 point, 8 point, 32 point, 64 point, 128 point, etc.) constellations are also feasible.

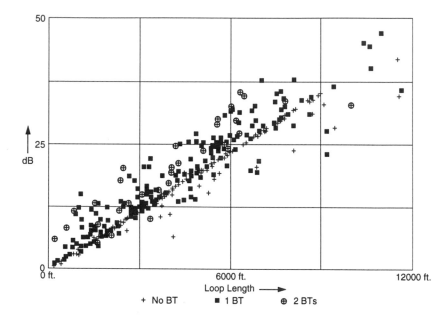

FIGURE 5.23. Typical Loss Scatter Plot for the CSA Loops at 200 kHz. BT - Bridged Taps; CSA - Carrier Serving Area

A systematic study of CAP modulation techniques also has been undertaken in the HDSL environment. It is also necessary to match the bandwidth requirement for each of these constellations with the overall loss and dispersion characteristics of each subscriber loop plant. The crosstalk and noise characteristics of the particular loop plant also becomes crucial because it is the signal-to-noise (S/N) ratio that influences the average probability of error. Recent studies have indicated that both T1 and E1 rates (i.e., two-wire full-duplex mode) are feasible over the typical average CSA ranges of about 5,000 feet.

Transmission Limitations: Crosstalk rather than equalization and residual echo are taken as the constraints to the range of HDSL data transmission. The use of bandwidth efficient codes becomes essential to the success of any high-speed data system between the Central Office and the subscriber. These coding algorithms can offer 6 bits/baud coding gain with 64 point clusters or even 7 bits/baud with 128 point clusters. There are many realizations of such codes from 4 point clusters to 256 or even 1024 point clusters. However, the realistic choices are usually confined to 32 point, 64 point or 128, since the Euclidean distance between symbols starts to fall rapidly thus making the symbol recovery more and more difficult at the receiver.

For the estimation of the T1 and E1 services in the CSA environment, two basic steps become necessary. On the one hand, the crosstalk power can be estimated from the published data [5.15, 5.20] for the CSA loop environment at various loop lengths. On the other hand, the signal power in the received constellation (cluster) can also be accurately estimated for the CSA environment from the topology of loops and the primary constants of the cables. The signal-to-noise ratio that is acceptable for the T1 and E1 quality of service is also well documented for various bit-error-rates (BER).

Coding Alternatives: *Three* types of coding algorithms (uncoded, trellis, and concatenated Reed-Solomon with trellis) are used for the 64 point constellation. In effect, the SNR required (for 10^7 BER) for the 16, 32, 64, and 128 point clusters is computed as 21.5 dB, 24.5 dB, 27.5 dB and 30.5 dB respectively.

TABLE 5-5 Theoretical Requirement for the Signal-to-Noise Ratios in the Presence of Gaussian Noise for Probability of Error P_e at 10^7

Constellation→ Code ↓	16 pt	32 pt	64 pt	128 pt
Uncoded	21.5	24.5	27.5	30.5
Trellis	18.0	21.0	24.0	27.0
Concatenated Reed-Solomon Trellis Code	15.5	18.5	21.5	24.5

Note: Trellis code offers an $\cong 3.5$ dB gain over the uncoded system and concatenated code offers an $\cong 2.5$ dB over the trellis for the clusters only.

Coding adds robustness against errors (Table 5-5). The SNR required can thus be relaxed in the presence of coding and the modified SNR with plain trellis code can thus be relaxed to 18 dB, 21 dB, 24 dB, and 27 dB from the earlier numbers. By taking the code one step further, the SNR requirement can be relaxed still further to 15.5 dB, 18.5 dB, 21.5 dB, and 24.5 dB for the concatenated Reed-Solomon and trellis codes. These results are summarized in Table 5-5.

In addition, an excess of 0.6 dB, 1.09 dB, and 1.55 dB reduces the probability of error P_e from 10^{-7}, to 10^{-8}, 10^{-9} and 10^{-10}, respectively. Conversely a reduction by 0.67 dB from these numbers in the Table increase the P_e to 10^{-6}.

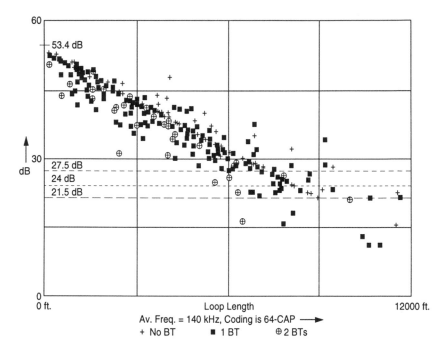

FIGURE 5.24. Scatter Plot of Noise Margins of the CSA Loop for 1.544 Mbps, 64 CAP Constellation. .... Uncoded; - - - - - Trellis, _ _ _ Concatenated Code. BT - Bridged Taps; CAP - Carrierless Amplitude and Phase Modulation; CSA - Carrier Serving Area

Simulation Results: We present the results of this approach in determining the loop failure to carry T1 and E1 rate data for the CSA loop population in Figures 5.24 and 5.25 respectively.

The effect of the three codes is depicted by three horizontal lines in the two Figures at 27.5 dB, 24 dB and 21.5 dB (see Table 5-5 Column 3). Individual points on the scatter plot indicate the actual SNR for each loop.

Note that the zero length loop has a SNR over the crosstalk of 53.4 dB indicating that this is indeed the crosstalk level in the de facto Bellcore Standard (discussed in detail in Chapter 14) for Central Office (crosstalk) power consistent with the NEXT loss of 57 dB at 80 kHz. The 53.4 dB NEXT loss value is obtained at the half-baud frequency of 140 kHz for the 64 point constellation with 8.8% excess bandwidth for the T1 rate of 1.544 Mbps.

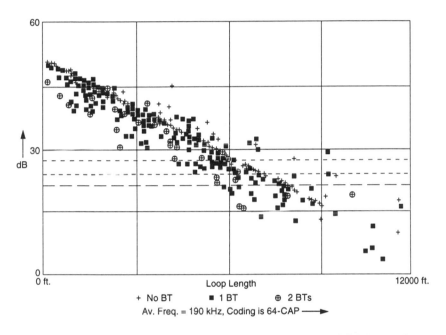

FIGURE 5.25. Scatter Plot of Noise Margins of the CSA Loop for 2.048 Mbps, 64 CAP Constellation. .... Uncoded; - - - - - Trellis, _ _ _ Concatenated Code. BT - Bridged Taps; CAP - Carrierless Amplitude and Phase Modulation; CSA - Carrier Serving Area

The SNR for each loop (i.e., the ordinate value of the points each depicting one of the 364 CSA loops) is thus derived as 53.4 - loop loss at 140 kHz. Loops that fall below the three horizontal line indicate their failure to carry the data. These computations indicate that at the full duplex T1 rate 11.8% (uncoded system), 7.4% (trellis encoded system) and 0.5% (concatenated code) of the CSA loops cannot carry T1 rate data. These percentages of loop failure increase to 23.4%, 20.5% and 10.0% for the E1 rate.

A family of such results for different constellations (16 point, 32 point, 64 point and 128 point) and for dual duplex T1, and dual duplex E1 has been generated. These results are summarized in Table 5-6. As can be seen, the two-wire PRISDN (i.e. T1 rate data or E1 rate data) has serious noise problems even with perfect devices, whereas the dual duplex systems do offer good potential with more and more sophisticated devices (equalizers, echo cancelers, timing recovery circuits, etc.).

TABLE 5-6 Percentage of Loops with Negative Margins for Various Transceiver Designs

Data Rate	Signal Constellation	Percentage of US CSA Loops		
		Uncoded	Coded	Concatenated
0.772 mpbs	16-CAP	1.1	0.3	0.0
1.024 mpbs	16-CAP	1.9	1.7	< 0.1
1.024 mpbs	32-CAP	2.5	1.1	< 0.1
1.024 mpbs	64-CAP	3.6	1.1	< 0.1
1.544 mpbs	64-CAP	11.8	7.4	0.5
2.048 mpbs	32-CAP	22.3	15.7	8.8
2.048 mpbs	64-CAP	23.4	20.5	10.0
2.048 mpbs	128-CAP	21.7	17.5	11.0

In summary, the CSA loops presented in this section have an approximate 88% penetration and satisfactorily carry T1 rate data with the 64 point CAP constellation without any special coding. With trellis encoding, the penetration increases to about 92%, and with the additional margin provided with concatenated Reed-Solomon codes and trellis encoding algorithms, the digital penetration of these loops is about 99.5%. These rates are based upon simulation studies using single wire-pair full-duplex, full T1 rate data transmission. The expected bit error rate (BER) is 1.E-07 and the quality of echo cancellation is very high (>>60 dB). With poorer quality cancelers we expect the BER to be worse than 1.E-07. Additionally, full-duplex, half-rate (two-2 wire-pair system), PRISDN offers a 98.8% penetration without trellis or concatenation, a 99.7% penetration with trellis only, and one 100% with trellis and concatenation. The topic is discussed in greater detail in Chapter 13.

5.6

THE GTE LOOP ENVIRONMENT

In 1985, about 11.5 million telephone lines were served by GTE with 2,250 Central Offices. In 1982, the subscriber loop data base was created with 2,002 loops to identify the significant characteristics of the GTE loop plant. Physical and a subset of electrical characteristics have been published [5.21 and 5.22]

The loop length and bridged tap distributions are summarized in Table 5-7. The average length of the loop is reported as 8, 796 ft. for all loops with two bridged taps with an average length of about 1,300 ft. The "average" gauge is reported as 24 AWG (see Table 5-8 from Reference 5.23 for the DSL environment).

TABLE 5-7 Appropriate Statistics of GTE Loops
(from References 5.21, 5.22 and 5.23)

Table A		Table B	
Loop Length kft.	Percentage * %	Number of BTs	Percentage* %
0- 2	3.5	0	21.3
2- 4	10.3	1	36.4
4- 6	15.	2	23.2
6- 8	14.	3	10.0
8-10	18.	4	4.9
10-12	14.	5	2.1
12-14	11.2	≥ 6	2.0
14-16	7.		
16-18	5.		
18-20	2.		

* Approximate values

The loops suitable for BRISDN access (i.e., non-loaded, single party lines) were separated from the general data base and loop statistics of the subset is reported in Table 5-8 [5.23]. The number of loops in this data base is 1,272 (approximately 63.53% of the loops) with an average length of 7,708 ft. with an average of two bridged taps per loop. The mean length of the bridge tap is 1,352 ft. The 1 kHz insertion loss of all loops is reported at 4.22 dB and a 100 kHz loss is reported at about 30.5 dB for loops in the 9-13 kft. band.

TABLE 5-8 Percentage Gauge Distribution of the GTE Loops for BRISDN Considerations (Reference 5.23)

Gauge kft	26 AWG %	24 AWG %	22 AWG %	19 AWG %	Others %
1- 3	59	28.5	9.2	2.3	2.3
.3- 6	49	33.6	12.3	2.9	2.9
6- 9	38.4	45.4	12.9	1.3	1.4
9-13	20.2	59.7	17.7	0.4	0.4
13-20	3.9	69.8	24.9	0.0	0.0
20+	0.0	0.0	100	0.0	0.0

5.6.1 OTHER LOOP ENVIRONMENTS

National loop surveys were also undertaken in Canada, Italy, France, Switzerland, Germany (to a limited extent), Australia, Japan, Korea, Taiwan and other countries. The results have not been systematically generated and published in open literature to be able to globally optimize system components. However, some of the important characteristics are available to justify the possibility of higher data rate in the subscriber loop plants throughout the world. The mode and the approach in these countries can vary significantly from one country to the next. Further, the range of operation and the maximum rates that can be expected can vary from one country to the next. An individual study of each country with its own subscriber loop environment can provide the realistic basis of the digital capabilities and the best approach (fiber, cable, twisted wire-pair, fiber coaxial cables, fiber-twisted wire-pair, etc.) to bring high-speed data to the customers. In general, it is observed that combined gauge discontinuities and bridged taps are prevalent in the United States, Japan, Canada, Italy, and Australia. Hence, the HDSL system tailored for the bridged taps should also accommodate discontinuities that result from gauge changes.

A limited amount of data is available from other countries. The information presented here is to provide an overview of the other national networks in the context of ISDN implementation. It should be noted that statistics can vary between the urban and suburban loops, domestic and business loops, metropolitan and rural loops.

The **Canadian** loops have the same gauges (26, 24, 22, and 19 AWG) found in the American loops, but only fifty-five percent have gauge discontinuities. The two finer gauges (26 and 24) account for sixty-five percent of the loop length, and eighty percent of the loops have bridged taps. Ninety-nine percent of the bridged taps are less than 1.6 km (5.25 kft.). PIC and paper insulated cables may be expected in the loops. About ninety percent of the Canadian loops are under 4.2 km (13.8 kft.), and the average length of the non-loaded loops is slightly longer than the Bell System loops. The average length is about 1.7 km (5.58 kft.), with eighty-eight percent of the loops being shorter than 2.7 km (8.86 kft.). The Canadian loops have an approximate average attenuation of about 22 dB at 80 kHz. The resistive component of the matching impedance lies between 110 to 120 Ohms.

In European countries, the physical characteristics of the **Italian** loops are available, where 0.4 and 0.6 mm PE (polyethylene) insulated cables are used. In the main population of the loops, eighty percent of the loop length consists of 0.4 mm cables, and the remaining twenty percent are of 0.6 mm cables. Bridged taps (BTs) are also available with an average tap length about 75 m. In the secondary population, 0.6 mm cables are used. Twenty percent of all loops have bridged taps with the average tap length being about 38 m. The Italian network has an average loss of 19.5 dB at 128 kHz with a resistive component of the matching impedance at about 120 Ohms and the average delay per km at 128 kHz is about 5 μs.

In **France**, 0.4 and 0.6 mm cable sizes are used. Eighty-eight percent of the French loops are less than 5 km (or 13.1 kft.). In **Sweden**, 0.4, 0.5 and 0.6 mm cables exist. The average loop length is about 1.7 km (5.58 kft.). Size discontinuity also occurs in most of these networks. The bridged tap information for these countries is not readily available. The maximum loss for the surveyed loops in France is about 40 dB at 88 kHz.

The loops in the **Swiss** environment are relatively free of bridged taps. A detailed survey is available within the PTT, but limited investigation [5.24], indicated the 0.4, 0.6, 0.8, and 1.0 mm copper cables are used. The typical loop is about 2.259 km (7.4 kft.) long [875 m (2,870 ft.) of 0.40 mm, 34 m (111 ft.) of 0.5 mm, 175 m (574 ft.) of 0.6 mm, and 1175 m (3855 ft.) of 0.8 mm cables]. The characteristics of these cable were published in 1982 [5.25]. Loops from **Norway** have an average loss of about 20 dB at 128 kHz with a resistive component of about 135 Ohms and the duration is 4.6 μs/km. The loop plant is devoid of any taps and the longer loops tend to be about 6 to 7 km (19.5 to 23 kft.) long composed of the coarser gauge (0.60 and 0.80 mm), but have relatively fewer discontinuities. Simulation of the limited number of loops [5.26] for E1 rate data has offered excellent results for this environment.

In **Germany** the loop environment has no bridged taps but the finest of the cable gauges is 0.35 mm. Three other sizes in the loop plant are 0.4, 0.5 and 0.6 mm. The skin effect is dominant over 2 MHz. Most of the characterization of the plant is based on the algebraic representation of the R,L,G and C, rather than on measured data especially in the 2-30 MHz band. The crosstalk coupling coefficients indicate that the three coarser gauges may be classified as category 5 cable standardized for premises distribution systems in the United States. The terminating impedance is fixed at 135 Ohms based upon a statistical evaluation of the mean impedance to be between 129 and 142 Ohms [5.27]. The measured data for the loops is sparse and the German telephone network has a loss in the range of 8.6 to 23.2 dB (frequency omitted) for the surveyed loops with a 5.44 µs/km, 5.24 µs/km, 4.87 µs/km, 4.84 µs/km for 0.35mm, 0.4 mm, 0.5mm, 0.6mm, cables respectively for the transit time (Singnallaufzeit).

In **Japan**, *five* cable sizes (0.32, 0.4, 0.5, 0.65, and 0.9 mm) exist. Forty-eight percent of the loops have gauge discontinuities and only fourteen percent have more than one bridged tap. The average bridged tap length is about 200 m (656 ft.). An average length of about 2 km (6.56 kft.) is estimated for the ISDN services, and ninety percent of the loops are less than about 4.1 km (13.45 kft.). Other published data [5.28] from NTT summarize that 50%, 90%, and 99% of the loops are less than 1.9 km (6.23 kft.), 4.1 km (13.45 kft.), and 6.5 km (21.32 kft.), respectively. No bridged taps exist on 58% of the loops, and one, two and over three taps exist of 30%, 9% and 3% respectively, of the loops. In Japan, the average loop loss at 100 kHz is 17 dB and about 90 percent of the loops have under 12 dB loss. The resistive component of the impedance is about 110 Ohms. The propagation duration is about 5.6 µs/km at 100 kHz.

NTT served about 43 million customers in the early 1980s and gathered the statistics from a sample of 5,500 loops. The electrical characteristics are summarized as follows: the 100 kHz loss is less than 17 dB and 42 dB for 50% and 99% of the loops. The resistance (for ringing current computation) is less than 420 and 1020 Ohms for 50% and 99% of the loops. The range is estimated at 22 km (72.2 kft.), 11 km (36 kft.) and 7.5 km (24.6 kft.) at 100 kbps, 200 kbps and 300 kbps, respectively. Bandwidth efficient codes should offer greater penetration at HDSL and VHDSL rates.

Compared to the United States, the Japanese networks offer fewer loops with discontinuities and fewer bridged taps. The average length is shorter and the average attenuation is also about 8 dB lower at about 100 kHz.

The **Australian** subscriber loop survey [5.29] is based upon a 1979 Survey. The data base contained only 423 randomly selected loops (within the metropolitan and country urban areas within the street lighting) with 271 loops in Victoria, 84 samples in South Australia and 68 samples in Western Australia. The plant uses 0.32, 0.4, 0.51, 0.64, 0.9 and 1.2 mm copper cable sizes, and 0.52, 0.81, 1.15 mm aluminum cable sizes. Both polyethylene insulated unit-twin copper (PEUIT) and paper insulated unit-twin copper (PIUT) cables are used. Polyethylene-insulated quad copper cable is used for local (PEIQL) services. Installation methods range from being aerial drop-wire, buried and/or duct installed. The 0.40 mm PIUT copper cable is the most dominant at 42% (46.3% in Victoria [VIC.], 28.8% in South Australia [S.A.] and 44.5% in Western Australia [W.A.]) in the plant.

Ninety eight percent of the loops are within 5 km (16.4 kft.) of the Central Office. These statistics differ from one area to another (99% loops from VIC, 94% from S.A. and 97% from W.A.). The average length of the loop is 2.1 km (6.89 kft.) for the entire population; (2.0 km [6.56 kft.] for VIC, and 2.4 km [7.87 kft.] each for S.A. and W.A.).

Out of the 423 loops in the data base, 421 (270 loops out of 271 in VIC., 83 out of 84 in S.A., and 68 out of 68 in W.A.) have size discontinuities, with a weighted average number of 3.4 (3.3 VIC., 3.9 S.A., 3.4 W.A.) discontinuities per loop.

Fifty-four (54%) percent having at least one "multiple" or tap (61% VIC., 54% W.A., 35% S.A.) and ninety percent are located very close to the subscriber. Only 4.9% (of the 54%) had two or more taps. Sixty-six percent of the taps use 0.64 mm PEIUT copper cable and 9.4% contain cable gauge changes. The average length of the tap is about 171 m (561 ft.) [169 m (554 ft.) VIC., 264 m (866 ft.) S.A., and 104 m (341 ft.) W.A.). The Australian loops suffer an average loss of 17 dB at 72 kHz.

According to a 1988 Survey, in the **Korean** environment, there were 17 million subscriber loops available, from which fifty-eight percent were in use, and the remaining forty-two percent were for future use. Of the loops in use, sixty-seven percent were 26 AWG, five percent were 24 AWG, one percent was 22 AWG, and the remaining twenty-seven percent have a mixture of 22, 24, and 26 AWG. The average loop length was 2.2 km (7.21 kft.). Two percent of the loops in use have bridged taps. The average BT length was 413 m (1355 ft.). There were no loaded loops in the Korean environment. About 337,000 loops have bridged taps, of which seventy-four percent have one BT, seventeen percent have two BTs, and nine percent have three BTs. The average bridged tap length was not more than 200 m (656 ft.) for fifty-seven percent of the 337,000 loops. Of this figure, seventeen percent have BTs length ranging from 201 m to 400 m, nine percent have BTs length ranging from 401 m to 600 m, seven percent have BTs length ranging from 601 m to 800 m, and ten percent have BTs length that was greater than 800 m [5.30].

In the **Taiwan** area, there are about 8.5 million subscriber lines [5.31]. From this population, 2805 lines were sampled, and 121 sample pairs (4.3%) were found loaded. The estimated overall mean working length from the Central Office to the subscriber end is 1.850 km (6.07 kft.). The loop plant in Taiwan is composed of two types of cables: the paper cable, and the Foam-Skin (FS) PIC cable. Both types of cables have three different wire gauges: 0.4, 0.5, and 0.65 mm. In this survey, it has been found that sixty-five percent of all sample subscriber lines had BTs. The maximum number of BTs in a Taiwan local loop is five. Statistics indicate that sixty-nine percent of BTs in the overall Taiwan area has a length less than 350 m, but sixteen percent of the BTs are longer than 609 m. Most BTs are far away from the Central Office end and they tend to be more concentrated near the subscriber end, indicating [5.15] that the more severe echoes will appear at the subscriber end.

In summary, an overview of the environments in the various countries in comparison with that in the United States is provided to indicate the limits of device performance necessary. The American loop plant has the most severe loop conditions with 2.3 gauge changes and 1.64 bridged taps per loop. The Japanese networks offer fewer loops with discontinuities and fewer bridged taps. The average length is shorter and the average attenuation is about 8 dB lower at about 100 kHz. Canadian loops have a fractional proportion of paper insulated cable and the percentage is unavailable, hence the impact of loop discontinuities is not clearly evident. The properties of loops that do have one or more bridged taps is about 80% as against 87% in the United States and there is considerable penetration of 22 AWG and 19 AWG cables in the network (35% vs. 10.7%) as compared to the loops in regional operating companies. The average bridged tap length is not available and hence the combined severity of higher proportions of 22 and 19 AWG cables with bridged taps cannot be easily compared against the severity in the Bell System loops. Average loop attenuations are only slightly lower (22 dB at 80 kHz vs. 23.2 dB at 72 kHz).

Australian loops have more variety of cables, more gauge discontinuities, but fewer bridged taps. The average length of the taps (or multiple) is also shorter. The effect of cable discontinuities between the dominant gauges (.4 mm PIUT, .51 mm PIQL, .64 mm PIQL, .4 mm PEIUT, and .64 mm PIUT) depends on the differences between their primary characteristics (R, L, G & C), and the effect of these differences in the spectral band of ISDN frequencies is not evident, even though the necessary simulations can be easily performed. The average attenuation at 72 kHz (Nyquist frequency for 144 kbps) is also about 8 dB less than that for the loops from the American operating companies. The impact of this loop environment, as far as the image impedance (for hybrid echo cancellation techniques, if it is considered in this environment) matching, should be studied in detail. In a general sense, the loop environment appears considerably less hostile than to the environment in the American operating companies in spite of gauge discontinuities. The extent of the gain from the introduction of the ISDN services for the more friendly loop environmental conditions can be computed.

Italy also offers a more loop friendly network with only two gauge discontinuities, shorter loops, fewer loops with bridged taps (20% vs. 87%), shorter bridged taps (38-75m vs. 281m) than the American telephone network. The average loop attenuation (19.35 dB at 128 kHz vs. 26.13 dB at 108 kHz) is also considerably less than that for the loops from the American loop environment.

The British environment offers a larger variety of gauge sizes, but few or no bridged taps. The average attenuation is also about 8 dB (17.5 dB at 100 kHz vs. 25.2 dB at 96 kHz) lower than the American loops. Norway has an average loop loss about 7-8 dB (20 dB at 128 kHz vs. 26.13 dB at 108 kHz) lower. The loop lengths for both the United Kingdom, and Norway are also about 0.75 km (2.46 kft.) shorter.

5.6.2 THE BRITISH LOOP ENVIRONMENT

5.6.2.1 Physical Characteristics

The British telephone environment has both copper and aluminum cables. The aluminum cables are being phased out of the network and for this reason only copper loops are included in the study reported here. A collection of eighty-one loops has been collected from five typical exchanges in the British network. The loop makeup is stored in a data base that can be accessed by the simulation programs to generate the physical and electrical characteristics of these loops. A summary of these characteristics are presented here for comparison to similar loops in the United States.

The distribution in this environment (see Figure 5.26) is not unlike the distribution in the United States environment (see Figures 5.1 and 5.2). The subscriber loops in the United Kingdom can be characterized as being shorter than those in the United States. In the United Kingdom, a wide mixture of cable sizes and insulation types may be expected, where copper and aluminum cable sizes of 0.32 and 0.4 mm may also be used. The average loop length can be expected to be from 1.7 to 2 km (5.58 to 6.56 kft.). The British loops surveyed do not have any bridged taps or the loading coils that are prevalent to the United States. Generally, the number of cable sections in the loop population is smaller than that in corresponding loops in the United States.

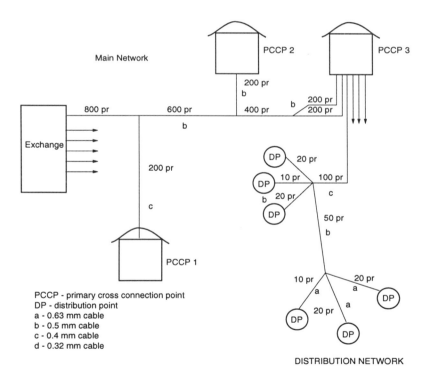

FIGURE 5.26. Typical POTS Distribution Facility and the Local Subscriber Loops in the European Environments. DP - Distribution Point; PCCP - Primary Cross Connection Point; POTS - Plain Old Telephone Service

The maximum number of cable sections in one of the loops surveyed in the United Kingdom is twenty-three compared to a maximum number of over sixty sections in United States. The loop plant has 2.4 percent of 28 gauge (0.32 mm) and very little of 19 gauge (0.9 mm) wire. These statistics are summarized in Table 5-9. In addition, the length statistics are presented in Table 5-10.

TABLE 5-9 Typical Cable Gauges and Electrical Characteristics in the United Kingdom Loop Environment

Diameter (mm)	AWG Equivalent	Gauge Distribution	Resistance Ω/km	Length in km for 1 $k\Omega$	Length for 40 dB Loss (100 kHz)	Resistance at 100kHz ($k\Omega$)
0.90	19	-	-	55	-	-
0.63	22	13.9%	110	9.1	7.3	0.80
0.50	24	47.5%	170	5.9	5.6	0.95
0.40	26	33.2%	270	3.7	3.8	1.02
0.32	28	2.4%	430	2.3	2.7	1.19

About eighty-six percent of the loops surveyed are under 12,000 feet and the average length of this population is about 7,182 feet. This value makes the average loop significantly longer than similar value (about 4,250 feet) for the CSA loop environment in the United States. It is to be observed that all the CSA selection rules are not strictly applied to the British environment, since the limiting length of the 28 gauge wire is still to be evaluated. Further, the absence of loading coils in the British loop environment does not curtail the loops in the United Kingdom data base, whereas these coils reduce the number of loops in the American environment from 2,290 to about 1,520.

5.6.2.2 Electrical Characteristics

In Table 5-11, the distributions and a small sampling of the electrical parameters for the five gauges of twisted wire-pairs in these loops are presented. More of the primary and secondary constants are not published, but they are accessed and computed (if necessary) within the computer systems as the simulation progresses. The presence of the 28 gauge wire in mostly 26 gauge loops causes incremental changes in the loop characteristics. However, the effects are no more the effects of the 26 gauge wire in mostly 24 gauge loops. This situation exists in the United States loops around 16,000 ft. and our simulation studies have demonstrated any significant impairment

due to the combined wire sizes in the loops. By the same token, there does
not seem to be any significant impairment due to the presence of the 28
gauge in the United Kingdom loop environments.

TABLE 5-10 Representative Physical Characteristics of the United Kingdom Loop Environment

Loop Length (ft)	Average Loop Length (ft)	Average No. of Sections	% of Loops Represented
0000- 1000	995	1	1
1001- 2000	1662	2	9
2001- 3000	2419	1	6
3001- 4000	3711	10	1
4001- 5000	4340	4	6
5001- 6000	5320	3	10
6001- 7000	6371	6	10
7001- 8000	7425	8	3
8001- 9000	8582	7	14
9001-10000	9684	7	11
10001-11000	10241	9	8
11001-12000	11230	6	7
12001-13000	12250	3	5
13001-14000	13761	12	5
14001-15000	14123	12	2
15001-16000	15850	3	1
16001-17000	-	-	0
17001-18000	-	-	0
18001-19000	18402	15	1

The study was undertaken to determine the electrical characteristics of
the loop plant in the United Kingdom. In order to accomplish this goal,
statistics of the loops in the United Kingdom were gathered to indicate wire
gauges, loop composition and bridged taps. A data base was generated to
contain the loop makeup. A second data base was also generated to include
the characteristics of the 28 (0.32 mm) AWG cable used in British countries.
The objective of the study is to design the interface requirements, such as
extent of equalization and echo cancellation based upon the spread of loop
impedance, image impedance, insertion loss, and loop loss of these high-

speed digital subscriber lines. The general simulation platform (see Part III of the book) also suffices for these loops.

TABLE 5-11 Sample of Electrical Parameters at dc and at 100 kHz in the United Kingdom Loop Environment

Conductor dia. (mm) UK	US equiv. (AWG)	Average proportion of route length (%)	Typical dc loop resistance per km (Ω)	Typical length (km) per kΩ loop resist. (at dc)	Typical length for 40 dB atten. at 100 kHz (km)	Average atten. of 2.6 km loop at 100 kHz (dB)
0.90	19	<.1	86.9	55	10.6	9.8
0.63	22	13.9	174.3	9.1	7.3	14.3
0.50	24	47.5	277.2	5.9	5.6	18.6
0.40	26	33.2	440.7	3.7	3.8	28.2
0.32	28	2.4	692.9	2.3	2.7	38.6

The image impedance at the Central Office and at the subscriber side of the loops at 200 kHz are presented in Figures 5.27 and 5.28. The slight variation between the scatter plots is due to the presence of coarser wire, and thus more gauge discontinuities towards the subscriber*. The loss characteristics are shown in Figure 5.29 at 200 kHz and there are no unexpected results.

The capacity of these loops to carry T1 and E1 (PRISDN) data are evaluated by the same techniques presented in Section 5.5.4 for the CSA loops. On the basis of rather limited data bases in the loop environment, the results of the penetration study are presented in Table 5-12.

The penetration of services are lower than those for the CSA loops in spite of the fact that the loss in these loops (about 3.5 dB/kft. at 200 kHz) is less than that (about 4.2 dB/kft at 200 kHz) in the CSA environment. The reason for the higher loop failure (compare rows in Table 5-12 with rows in Table 5-6) is due to the fact that the CSA loop population displays a uniform distribution of samples versus the loop-length, the United Kingdom samples seem more concentrated around the 10 kft. length, thus forcing more loop failures over the entire 12 kft. band.

---

* The granularity in the location of the individual points in the scatter plot occurs because of the truncation of the Ohmic values of the real and the imaginary components of image impedance.

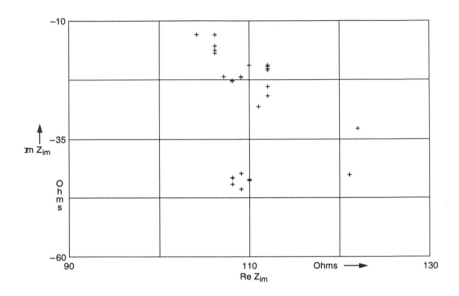

FIGURE 5.27. Central Office Side Image Impedance Scatter Plot for the Loops in the United Kingdom Data Base. Im Z_{im} - Imaginary Part; Re Z_{im} - Real Part

The penetration of services are lower than those for the CSA loops in spite of the fact that the loss in these loops (about 3.5 dB/kft. at 200 kHz) is less than that (about 4.2 dB/kft at 200 kHz) in the CSA environment. The reason for the higher loop failure (compare rows in Table 5-12 with rows in Table 5-6) is due to the fact that the CSA loop population displays a uniform distribution of samples versus the loop-length, the United Kingdom samples seem more concentrated around the 10 kft. length, thus forcing more loop failures over the entire 12 kft. band. When the population is truncated to 6 kft. length, the loop failure in the United Kingdom environment for E1 services falls to 0% even without any coding (i.e., for the uncoded system) whereas the loop failure in the CSA environment is about 5.5%, 1.7% and 0% for the uncoded, trellis, and concatenated (Reed-Solomon and trellis) codes respectively. This is due to differences in the design rules of the two environments practiced during the expansion of the telephone services. The influence of the bridged taps (as they exist in the North American loop environments) is far more detrimental than the effect of loop attenuation alone. The conclusion is substantiated by comparing the loop performance statistics of United States loops with those of the Swiss loops presented in Section 12.4.2.

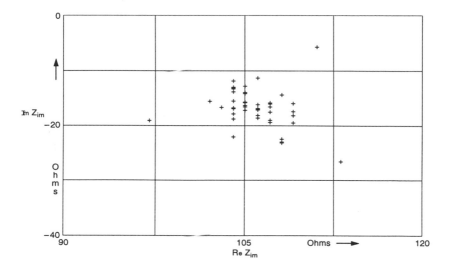

FIGURE 5.28. Subscriber Side Image Impedance Scatter Plot for the Loops in the United Kingdom Data Base. Im Z_{im} - Imaginary Part; Re Z_{im} - Real Part

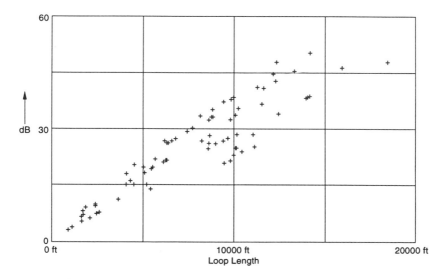

FIGURE 5.29. Loss Characteristics for the Loops at 200 kHz in the United Kingdom Data Base.

5.6.3 HDSL IN THE UNITED KINGDOM ENVIRONMENTS

Adaptive echo cancellation was discarded in favor of time compression mode of data transmission over wire-pairs during 1979 to permit the rapid deployment of the simplistic features of the latter technology. At that time, echo cancelers for HDSL could not be readily designed to meet the 60 dB echo cancellation requirement (i.e., cancellation of all echo signals to within 0.1 percent of their original values) of the non-loaded subscriber loop environment of the American loop plant. With an echo canceler requirement of this magnitude ninety-nine percent of the loops could carry BRISDN data

TABLE 5-12 Percentage of United Kingdom Loops with Negative Margins for Various Transceiver Designs

Data Rate	Signal Constellation	Uncoded	Percentage of UK Loops* Coded	Concatenated
0.772 Mbps	16-CAP	4.3	0.0	0.0
1.024 Mbps	16-CAP	10.1	4.3	0.0
1.024 Mbps	32-CAP	8.6	4.3	0.0
1.024 Mbps	64-CAP	11.6	5.7	0.0
1.544 Mbps	64-CAP	26.1	21.7	8.6
2.048 Mbps	32-CAP	45.0	29.0	21.7
2.048 Mbps	64-CAP	47.8	30.4	21.7
2.048 Mbps	128-CAP	49.2	30.4	21.7

*Most of the United Kingdom loops have length d $\leq$ 12 kft.

at 160 kbps [5.11]. However, through the eighties, the technology had changed enough to provide VLSI echo canceler chips, thus offering ISDN services to most customers in most metropolitan areas. Decision feedback equalizers had also become viable even in the most adverse cases of loop configurations (see Section 5.7). Timing recovery with phase-locked-loop (PLL) technology offers very low timing jitters at the subscriber station. All these factors have contributed to making the HDSL a very realistic way to communicate with almost all subscribers over twisted wire-pairs at the BRISDN rate in the full-duplex mode.

During the transition (early 1980s) to the standard 2B1Q code, for the BRISDN data traffic using the full-duplex echo cancelers, vendors of the VLSI chip sets had introduced the chips for the AMI code. Later, these vendors introduced the VLSI chips for the more widely accepted 2B1Q code for BRISDN. The BRISDN services have been offered by most Regional Bell Operating Companies throughout the United States since 1985. This is now the current Standard for the ISDN platforms (1, and 2) in the country.

In the United States, technical and economic viability of the PRISDN has also been satisfactorily confirmed. The fabrication of the VLSI chips does not pose any serious concerns. The algorithms for convergence of tap weights of an adaptive echo canceler (AEC) have been refined to the extent that existing digital signal processor (DSP) technology can be economically deployed. The interference due the NEXT and impulse noise only limit the range of the PRISDN line, but not its feasibility, especially if dual duplex systems are deployed.

The European primary rate is at 2.048 Mbps. The penetration at this higher rate in the American loop environment is about 76.6%, without any special encoding at the transceivers for 64 point CAP constellation. With trellis codes offering about 3.5 dB NEXT immunity, the penetration increases to about 79.5% and with additional Reed Solomon encoding offering an additional 2.5 dB NEXT immunity, the penetration increases to about 90%.

5.7

SUMMARY OF ANSI AND ETSI LOOPS

Loop environments differ significantly from country to country. For this reason, the two dominant standards organizations, the American National Standards Institute (ANSI) and the European Telecommunications Standards Institute (ETSI), have selected a few topologies representative of the loop environment with the objective that the vendors of DSL (BRISDN services) and HDSL (PRISDN services) may test the digital devices. In addition, different vendors have proposed different methodologies such as discrete multitone (DMT), carrierless amplitude and phase (CAP), etc., for data transmission over the repeaterless digital subscriber line proposed in 1979 [5.11].

5.7.1 ANSI LOOP TOPOLOGIES

Before ANSI had formally announced the loop topologies, Bellcore had announced a typical set of loops to reflect the most frequent and also the very harsh loops from the data bases. These loops were derived from the 1983 Loop Survey and used a de facto standard during the mid eighties. Various alternative coding schemes (e.g., AMI, 2B1Q, 3B2T, and even 4B3T, etc.) were simulated and tested for these loop configurations.

The topologies recommended by ANSI are to exemplify the North American subscriber loop plant. In effect, the harshest loop topologies for BRISDN services become more restrictive for the HDSL and the ADSL services.

Three such sets of loops exist for each environment:

(a) the general ANSI loop-set (15 loops selected to represent loops from the 1983 Loop Survey data) for the DSL-BRISDN service, depicted in Figure 5.30, loops 1 through 15,

(b) the HDSL-CSA type loop-set (11 loop topologies, 0 through 10; 13 loops, 0.1, 0.2, 0.3 versions of loop 0; and loops 1 through 10) derived from CSA type data base for HDSL services, depicted in Figures 5.31a, b, and c. Note that these loops do not exactly meet the CSA requirements,

(c) Finally, the T1E1 committee has specified a set of eight loop topologies (see Figure 5.32) for the ADSL services. The variable length of cable X depends upon the noise environment. The loop insertion loss becomes paramount in these cases because the signal strength has to be maintained. Three such insertion loss values at 300 kHz, are defined for case 1 (i.e., 51.0 dB [all loops]), case 2 (i.e., 49.0 dB [43 dB for loop 8]), and case 3 (i.e., 35 dB [34.0 dB for loop 2]). On this basis, the value of X is tabulated in Table 5-13.

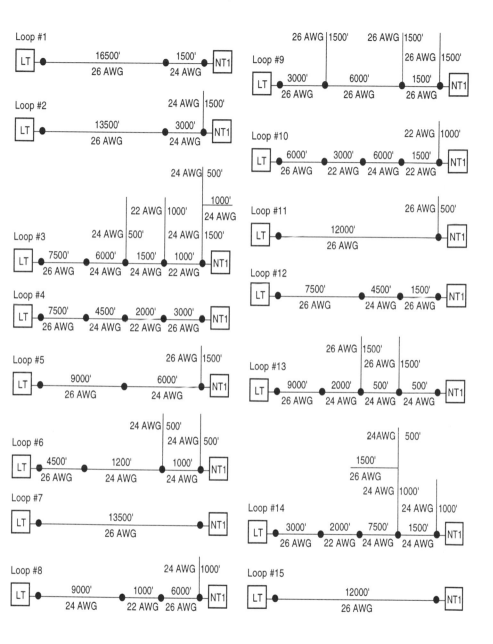

FIGURE 5.30. Bellcore/ANSI Standard Test Loop. ANSI - American National Standards Institute; LT - Line Termination; NTn - Network Termination

TABLE 5-13 Value of Variable Length X in Figure 5.32

CASE 1 51 dB (All Loops)		CASE 2 49dB (43 dB Loop 8)		CASE 3 35 dB (34 dB Loop 2)	
Loop #	Nominal value of adjustable length 'X' km	Loop #	Nominal value of adjustable length 'X' km	Loop #	Nominal value of adjustable length 'X' km
1	3.60	1	3.45	1	2.45
2	4.80	2	4.55	2	3.20
3	2.50	3	2.30	3	1.30
4	2.00	4	1.80	4	0.80
5	2.55	5	2.40	5	1.40
6	2.40	6	2.25	6	1.25
7	1.55	7	1.35	7	0.40
8	2.10	8	1.50	8	1.00

All of these loops are less than 18,000 feet, are free of load coils and there are no non-standard wire gauges (such as AWG 16 or 17) nor any aluminum cable sections. Bridged taps upon bridged taps are limited to two, whereas the actual 1983 Loop Survey data did have an occasional (about 3 or 4 out of 1,520 loops) bridged tap, upon bridged tap, upon bridged tap. The presence of gauge discontinuities, the presence of bridges taps, and of taps on taps, of discontinuous taps is evident in the loop make up. Most of the taps are located closer to the subscriber. The lengths of different sections, the length of taps and the composition are indicative of the most common occurrences in the 1983 Loop Surveys present in this Chapter.

The differences between the loop make up between the North American environment and European environments becomes evident by comparing the loop topologies in Figure 5.31 with those in 5.32 and 5.33. These topologies are chosen to reflect the average statistical characteristics of the loops in that particular environment. More recently the trend is to use the loop simulator boxes in testing the transceivers. The boxes contain different combinations of Rs and Cs to reflect the actual loops selected by the ANSI and ETSI standards committees. They only (and inaccurately) imitate the loop characteristics.

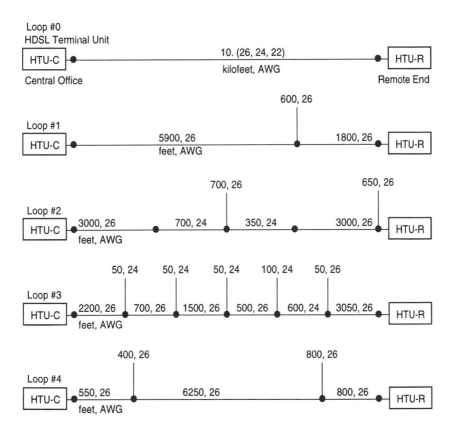

FIGURE 5.31a. HDSL CSA-Type Loops from Bellcore TA-NWT-001210 October 1991. CSA - Carrier Serving Area; HDSL - High-Speed Digital Subscriber Line; HTU-C - HDSL Termination Unit - Central Office; HTU-R HDSL Termination Unit - Remote End

These topologies were proposed during an interval (late eighties and early nineties) of active participation the various committees and cooperation between the numerous North American and European Telecoms. It can be seen that each of the committees has proposed loop topologies consistent with its own network. While this effort can be justified, it does little to make the toplogies representative of the entire world community, especially because the European countries have not have followed any consistent loop design practice. For this reason, it appears that the vendors of the HDSL and ADSL units (CO and Subscriber) will have to contend with a variety of representative loop topologies. This is evident by comparing the standard loops by ANSI and Bellcore with those from ETSI and European Telecoms.

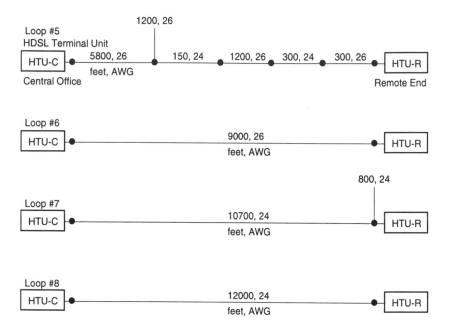

FIGURE 5.31b. HDSL CSA-Type Loops from Bellcore TA-NWT-001210 October 1991 continued. CSA - Carrier Serving Area; HDSL - High-Speed Digital Subscriber Line; HTU-C - HDSL Termination Unit - Central Office; HTU-R HDSL Termination Unit - Remote End

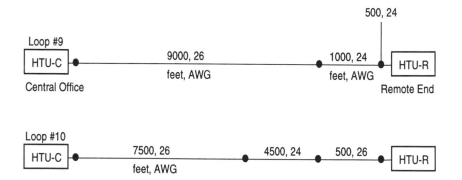

FIGURE 5.31c. HDSL CSA-Type Loops from Bellcore TA-NWT-001210 October 1991 continued. CSA - Carrier Serving Area; HDSL - High-Speed Digital Subscriber Line; HTU-C - HDSL Termination Unit - Central Office; HTU-R HDSL Termination Unit - Remote End

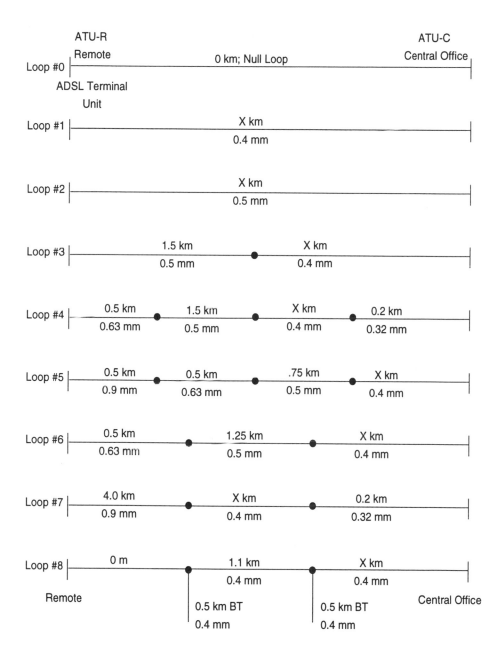

FIGURE 5.32. ADSL Test Loops from T1E1 April 1995 OO7R2. ADSL
- Asymmetric Digital Subscriber Line

5.7.2 ETSI LOOP TOPOLOGIES

Numerous loop environments exist in Europe. The Central Office and other contaminating noise levels also differ.

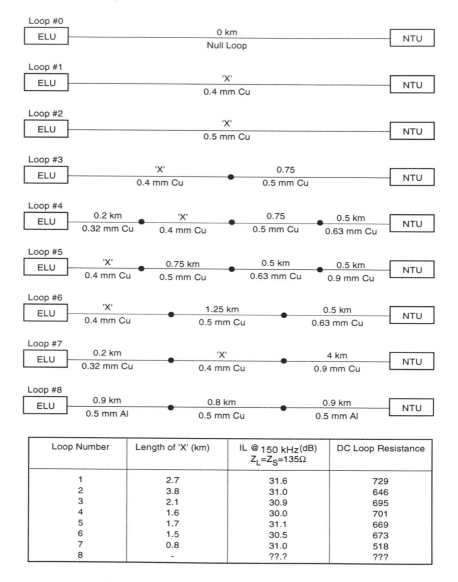

Loop Number	Length of 'X' (km)	IL @ 150 kHz (dB) $Z_L = Z_S = 135\Omega$	DC Loop Resistance
1	2.7	31.6	729
2	3.8	31.0	646
3	2.1	30.9	695
4	1.6	30.0	701
5	1.7	31.1	669
6	1.5	30.5	673
7	0.8	31.0	518
8	-	??.?	???

FIGURE 5.33. British Telecom Test Loops (1993-1994). ELU - End Office Line Unit; NTU - Network Termination Unit

The European loop environments suggest a set of eight loops for the laboratory testing. The first set of eight loops (see Figure 5.33) were suggested by British Telecom and the second set (see Figure 5.34) is an ETSI standard.

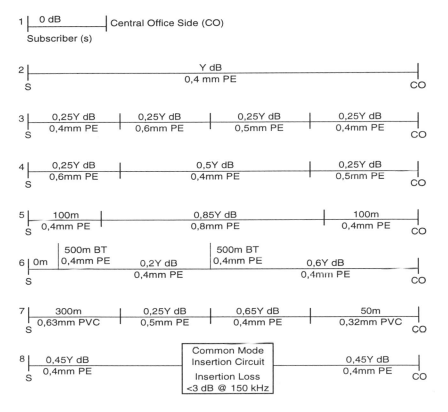

FIGURE 5.34. ETSI Loop Topologies for Digital Loops (for Laboratory Testing). BT - Bridged Taps; CO - Central Office; ETSI - European Telecommunications Standards Institute; PE - Paging Entity in UMTS; PVC - Permanent Virtual Circuit

Note that the customer side for these loops is on the left side for Figure 5.33. When the attenuation values of Y in Figure 5.34 are converted to the lengths of the various gauges, these ETSI loops can be redrawn as the actual loop configurations shown in Figure 5.35. The presence of 0.32 mm (or the equivalent 28 AWG) copper cable, aluminum cable, and only one loop with two taps (loop 6 in Figure 5.34) bridged tap is indicative of the European loop environment for DSL or (BRISDN) services and possibly for the HDSL applications.

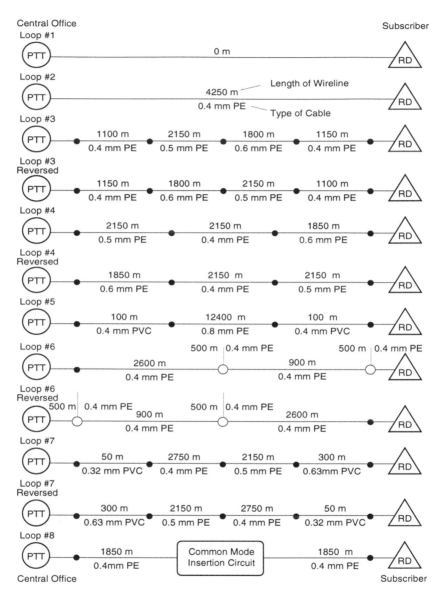

FIGURE 5.35. ETSI Loop Topologies Redrawn to Meet the Loss Requirements. Note that loops 1,2, 5 and 8 are symmetric. ; ETSI - European Telecommunications Standards Institute; PE - Paging Entity in UMTS; PTT - Postal, Telegraph and Telephone (European Government Carriers); PVC - Permanent Virtual Circuit; RD - Remote Destination

Various countries have also suggested loops to typify their own country's loop environments. For example, none of the eight British Telecom loops have no bridged taps, but one loop has a section of 0.9 km (2.95 kft.) of 0.50 mm aluminum cable (Loop 8).

In addition, the seven of the eight non-null loops have an insertion loss between 30 and 31.6 dB at 150 kHz with the source and termination impedance of 135 Ohms. The ETSI loops do not have any aluminum cables, but one loop has two taps. The length X for these loops are tabulated in Figure 5.33.

5.8

RECENT ENHANCEMENT AND COMPETING STRATEGIES

In the more recent telephone environments incorporating the ISDN features, the dc resistance limit (less than 1300 Ohms) does not apply and loop utilization can be thus extended by HDSL, VHDSL and ADSL designs. The more recent architectures of the asymmetric digital subscriber line (ADSL) challenge the older notion that video signals for TV and entertainment can be distributed exclusively by the CATV networks. Frequency division multiplexing techniques separate the two directions of communication for the ADSL.

The CATV subscribers in most CATV networks (see Section 3.5.2), are also connected to a TV signal distribution facility and the numerous houses are served by individual "drop" wires. This modus operandi is feasible in CATV environment because the TV signal flow is mostly unidirectional (down-stream) and there is very little up-stream traffic (see Section 3.5.1). These differences offer some restrictions in the immediate adaptation for one (CATV or telephone) network to perform the functions of the other (telephone or CATV) network. These differences prevail; however they can be used for distributing high-speed data to the customer. The solutions are expensive, but not insurmountable. The regulatory and policy environment is much more restrictive in permitting one network to function as the other. Traditionally, the role of CATV network suits the public entertainment needs and remains isolated from the role of the POTS network.

Fiber to the house (FTTH) and fiber to the curb (FTTC) architectures also bring high quality digitally encoded pictures and video very close to reality for the end user. Within the subscriber loop environment, the current designs of the two wire ADSL aim to permit the downstream data at the primary rate on all loops to 18,000 ft. (no loading coils) and upstream data at 24 kbps with a transmit power at about 13-14 dBm, and a terminating impedance of 100 Ohms. Low rate video at the full range (18,000 ft) becomes feasible. Twice the DS1 rate over short and selected loops also appears feasible. This alternative offers few advantages above and beyond the normal hybrid fiber coax (see HFC architectures in Section 3.6) features from an entertainment perspective. However, if the intelligent network (IN) functions also need to be incorporated, then the HDSL, VHDSL, and ADSL architectures and the FTTH and FTTC architectures offer unique IN capabilities.

Within the subscriber loop environment, the HDSL environment permits bidirectional dual duplex (i.e., two wire-pairs, each pair carrying 784 kbps bidirectional) data at DS1 rate to all the carrier serving area loops (see Section 5.5). The ultimate aim is to provide the full-duplex DS1 rate over one wire-pair in the CSA environment. The proposed two-wire full-duplex HDSL services at (H_0+D) or 456 (i.e., 384+64) kbps is aimed at all customers in the 18,000 ft. (no loading coils) range. In the European environments, the 4-wire dual duplex HDSL systems are proposed to cover a range of 4 to 7 km (13.1 to 22.96 kft.). The transmit power is about 13-14 dBm and the terminating impedance is 135 Ohms. Since all these transceivers were not designed and tested, the viability of such HDSL and ADSL services need to be reexamined from country to country and from time to time.

Cable networks provide TV and entertainment to a large sector of customers in most Western nations (see Chapter 3). Cable trunk networks between Central Offices to carry high density interoffice traffic exist in most nations (see Section 2.5). In LANs, MANs and WANs coaxial cables (Ethernet and IsoEthernet, see Section 4.9) serve very dependably as data pathways. However, their capabilities and costs are continuously challenged by the emerging fiber optic networks (FDDI, FDDI II, and FFOL, see Section 4.9). In the subscriber loop environment, bringing the fiber to the home (FTTH) or to the curb (FTTC) has become a serious concern for many telephone operating companies. Hybrid fiber coaxial (HFC) systems (see Section 3.6) and/or purely fiber carrier systems are also feasible. The economics of the 1990s tends to favor the HFC systems with dual fibers (one for entertainment video and one for telephone services) downstream data. The final distribution to the customer premises is over bidirectional coaxial cable. Other architectures (discussed in Section 3.6) are also feasible.

In the subscriber environment, coaxial cables are sometimes used to distribute high rate data (which can be in the form of PCM encoded speech) between the Central Offices and remote terminals (see Section 2.5). Such remote terminals (RTs) act as miniaturized Central Offices that concentrate and distribute the voice channels to and from a digital carrier system such as DS-1 or DS-3. The projected role of the RTs will include fast packet switching for data services to the customer.

Coaxial cables are primarily used for distributing analog cable TV signals. The potential of these CATV systems for carrying high capacity digital information is enormous, but the demand has been discouragingly low. Equipped with intelligent network components, they can distribute highly customized data to a large number of subscribers bringing in new services generally not available in the cable TV industry. The new era of intelligent CATV networks has still to emerge. The FM hierarchy of the CATV network that permits the transfer of small quantities of upstream data to control large streams of downstream data makes these network potential contenders very personalized and individualized network services.

5.9

PREMISES DISTRIBUTION SYSTEMS

Twisted wire-pairs are generally unshielded. However, during the eighties some of the manufacturers started to introduce shielded wire-pairs to reduce the electromagnetic coupling and thus reduce crosstalk effects from contaminating the signal at high frequencies. The near-end crosstalk (NEXT), which is generally the most serious interference, increases at the rate of 15 dB per decade (see Reference 5.15 and also Chapter 14), and the effect is quite dominant at high frequencies thus limiting the range. The main objective of these new unshielded wire-pairs is also to reduce the electromagnetic coupling, and thus the NEXT interference. However, the goal is achieved by a different fabrication practice. Balancing the pairs is the key to the much reduced electromagnetic (EM) coupling. Typically, this calls for very well matched wires that make up each pair. The balance also depends upon the geometry of the wires. Each wire-pair is also twisted at some twist rate. Both the wire diameter and the insulation thickness become important in maintaining this balance throughout the length of the cable. Typically, sheathing and insulating material do not directly influence the EM coupling. The two conductors of the pair are highly matched, tightly twisted and are precisely placed within the cable geometry, such that the coupling between any one pair and any of the other pairs is almost canceled out. Cable connectors and crossovers also need careful balancing. The UTP cable system design approaches the design of the microwave wave guides. The basic premise is that the differential electromagnetic pickup depends upon the lack of balance and consistency in relationship among wire-pairs, within the structure of the cable. The newer cables (UTP) tend to balance out this electromagnetic pickup in the two conductors constituting the pair.

On the basis of these rather simplistic observations, the category 5 UTP manufacturer calls for random twists and tight twists of the wires in the pair, and a precise geometry within the cable, such that there are no seriously interfering-adjoining pairs. The interference is a random event, but the signal-to-noise ratio (SNR) characteristics of the new cable (category 5) are much higher. The randomness of the tight twist reduces the NEXT interference considerably, and the transposition reduces the unbalanced NEXT still further.

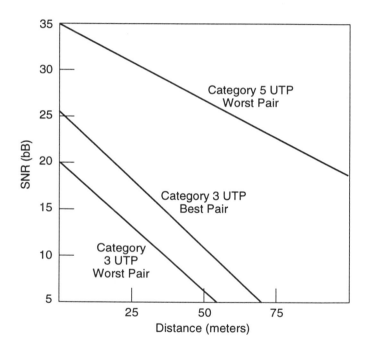

FIGURE 5.36. SNR Requirements Categories 5 and 3 UTP. SNR - Signal-to-Noise Ratio; UTP - Unshielded Twisted Wire-Pair

The UTP wiring standards for the category 3 and category 5 wire-pairs are shown in Figure 5.36. Typically the category 3 UTP is consistent with the normal wire-pairs in the loop plant where little or no care has been taken in the fabrication of cables, i.e., the cables are routinely manufactured. The poor crosstalk coupling characteristics limit the use of category 3 (i.e., the 568 draft standards of EIA/TIA, category 3 or UL Level III) cables to 10 MHz (or 10BaseT) applications. Category 4 (i.e., EIA/TIA category 4 or UL Level IV)

cables are limited to 16 MHz token ring applications and category 5 (i.e.,
EIA/TIA category 5 or UL Level V) cables can range up to 100 Mbps, or at
the FDDI rates and other higher speed applications. The attenuations of these
three categories of unshielded twisted wire-pairs at various frequencies are
tabulated in Table 5-14.

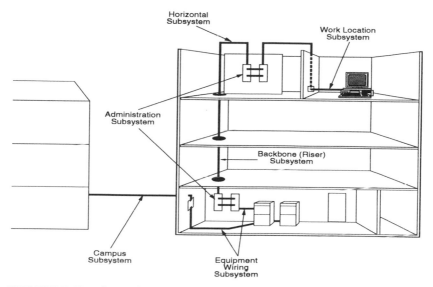

FIGURE 5.37. Overview of Typical Premises Distribution System.

TABLE 5-14 Attenuation of the Various PDS UTP Cables

Attenuation for TIA/EIA-568A UTP   Link @ 20 degrees Celsius
Length of horizontal cable: 90 meters
Length of equipment cords: 4 meters

Frequency (MHz)	Category 3 (dB)	Category 4 (dB)	Category 5 (dB)
1.0	3.4	2.2	2.1
4.0	6.5	4.3	4.0
8.0	9.4	6.0	5.7
10.0	10.7	6.8	6.3
16.0	14.1	8.8	8.2
20.0		9.9	9.2
25.0			10.3
31.25			11.5
62.5			16.7
100.0			21.6

The attenuation of cable segments are subject to a 1.5% per degree Celsius derating for category 3 cable and 0.4% per degree Celsius derating for category 4 and 4 cable when the temperature of the environment is higher than 20 degrees Celsius.

TABLE 5-15 Propagation Loss of the Different UTP Cables at Frequency of 'f' MHz

EIA/TIA-568 Worst Case Propagation Loss at 20°C $L_p(f) = a'\sqrt{f} + b'f$

EIA/TIA cable	Propagation loss dB/kilofoot	Propagation loss dB/100m	Increase per 1° C	Frequency range (MHz)
Category 3	$7.07\sqrt{f}+0.73f$	$2.33\sqrt{f}+0.23f$	1.2%	$0.3 \le f \le 16$
Category 4	$6.4\sqrt{f}+0.08f$	$2.10\sqrt{f}+0.026f$	0.3%	$0.3 \le f \le 20$
Category 5	$6.3\sqrt{f}+0.04f$	$2.07\sqrt{f}+0.013f$	0.3%	$0.3 \le f \le 100$

These values are prepared by TIA/EIA-568A PN-3287 Task Group for UTP at 20° C. Similar, slightly more accurate, worst case propagation loss formulations are presented in Table 5-16.

For the design of the PDS high-speed data distribution, the extent of stray noise pick-up is as important as the attenuation of the cable distributing the data. This is the basis for making sure that the received signal is not overly contaminated by noise. In the PDS environments, the crosstalk pick up is also identified from the earlier NEXT model (introduced in Section 5.5.4 and discussed fully in Chapter 14). These NEXT approximations are tabulated in Table 5-16 for the design of the high-speed premises distribution systems. The basis of these equations is the same as the classic equations for the NEXT crosstalk power presented in Sections 14.4.2 and 14.4.3, based on the Bellcore study of 57 dB at 80 kHz (see Chapter 14 for details). It is interesting to note that for the Category 3 UTP, which corresponds to the normal 24 AWG cable in the subscriber loop plants, the crosstalk loss is indeed 57 dB from Equation (14.2) in Chapter 14 and from the equation

$$\text{NEXT Power Sum Loss} = 40.54 - 15 \log f_{(MHz)} \qquad (5.2)$$

written out in row 1, column 3 of Table 5-16. The coupling coefficient can also be derived from the discussion in Chapter 14 as 7.94E-05 being equal to $10^{-4.1}$ (see Footnote *).

TABLE 5-16 Crosstalk Coefficients and Relations Worst Case NEXT Loss in PDS

EIA/TIA-568 Worst Case NEXT Loss $L_x(f) = K - 15 \log f$ dB

EIA/TIA Cable Range	Coupling Coefficient χ	Pair-to-pair NEXT loss $L_x(f)$ dB	Connecting HW NEXT loss dB	Frequency MHz
Category 3	8.83×10^{-5}	$41^* - 15 \log f$	$58 - 20 \log f$	$f \leq 16$ MHz
Category 4	2.51×10^{-6}	$56 - 15 \log f$	$70 - 20 \log f$	$f \leq 20$ MHz
Category 5	6.31×10^{-7}	$62 - 15 \log f$	$80 - 20 \log f$	$f \leq 100$ MHz

Typical distribution toplogy is shown in Figure 5.37 [5.32] and the length of the untwisted wire-pair (UTP) cables is limited to about 100 m. Fiber and UTP are frequently used in this LAN-type PDS environment, just as fiber and coaxial cables are used in the high-speed data distribution systems in the subscriber loop environment.

It is to be appreciated that the bandwidth efficiency of the code and type of coding algorithm also play a strong role in the actual bit rate that these cables can carry. The preliminary simulation studies have been done for 100 meter length of these three categories (3, 4 and 5) of cables at AT&T Bell Laboratories [5.32]. Such simulations have indicated that the category 3 UTP can carry 136 Mbps carrierless amplitude/phase (CAP) encoded data with a SNR (at average bandwidth) of 10.9 dB margin with one interferer and 79 Mbps with 24 interferers. Similar simulated SNR values are 29.8 dB at 323 Mbps (one interferer) and 22.8 dB at 253 Mbps (24 interferers) for 100 meter of category 4 UTP and the corresponding SNR values are 36.2 dB at 386 Mbps (one interferer) and 29.2 dB at 316 Mbps (24 interferers) for 100 meter of category 5 cable.

---

* The entry of the integer 41, in Table 5-16 (row 1, column 3) is an approximation of the number 40.54. If the precise value of this number is used, the coupling coefficient is 8.83066E-05 rather than 7.94E-05. The number 8.83066E-05 also corresponds to 2.7925E-09 (see Equation (14.3) in Chapter 14) times $1000^{1.5}$. The conversion factor of 1000 arises because f is in kHz in Equation (14.3) of Chapter 14 and the value of the coupling coefficient in Table 5-16 is for f in MHz.

5.10

REFERENCES

5.1   Bell Laboratories 1991, "Siliton Transmission via Optical Gates," *Bell Labs News*, September 3, pages 1 and 5).

5.2   J.D. Dega, G.W.R. Luderer and A.K. Vaidya 1989, "Fast Packet Technology for Future Switches," *AT&T Technical Journal*, March/April 36-50.

5.3   T. Starr, D. Waring and J.J. Werner 1991, "HDSL An Expedient Broadband Access," *Private Communication*.

5.4   Ahamed, S.V. 1982, "Simulation and Design Studies of the Digital Subscriber Lines," *Bell System Technical Journal*, Vol. 61, 1003-1077.

5.5   P.A. Gresh 1969, "Physical and Transmission Characteristics of Customer Loop Plant," *Bell System Technical Journal*, Vol. 48, 3337-3385.

5.6   Manhire, L.M. 1978, "Physical and Transmission Characteristics of the Customer Loop Plant," *Bell System Technical Journal*, Vol. 57.

5.7   S.V. Ahamed and R.P.S. Singh, 1986, "Physical and Transmission Characteristics of Subscriber Loops for ISDN Services," *Proceedings of IEEE ICC-86*, 1211-1215.

5.8   S.V. Ahamed and R.A. McDonald 1985, "Calculation of Insertion Losses in Loops in the 1983 Loop Survey Data Base at Frequencies Relevant for ISDN Basic Access," *T1D1.3 Contribution*, Vancouver, Canada, August 19-23.

5.9   J.W. Lechlieder and R.A. McDonald 1986, "Capability of Telephone Loop Plant for ISDN Basic Access," *Proceedings of ISSLS 86*, September 29-October 3, 156-161.

5.10   R.K. Even, R.A. McDonald and H. Seidel 1979, "Digital Transmission Capability of the Loop Plant," *Conference Record ICC '79*, Paper 2.1.

5.11   S.V. Ahamed, B.S. Bosik, N.G. Long and R.W. Wyndrum, Jr., 1979, "The Provision of High-speed Digital Data Services over Existing Loop Plant," *Proceedings of the National Electronics Conference*, Vol. 33, 265-270.

5.12   M.J. Miller and S.V. Ahamed 1988, *Digital Transmission Systems and Networks, Volume II Applications*, Computer Science Press, Chapter 6, 275-336.

5.13   B.S. Bosik and S.V. Kartalopoulos 1982, "Time Compression Multiplexing for a Circuit Switched Digital Capability," *IEEE Transactions on Communications*, COMM 30, 2046-2052.

5.14   Bell Laboratories 1984, *Special Issue Bell System Technical Journal* dedicated to the SLC-96 Carrier, December, Vol. 63 No. 10, Part 2.

5.15   S.V. Ahamed, P.P. Bohn, N.L. Gottfried 1981, "A Tutorial on Two-Wire Digital Transmission in the Loop Plant," *IEEE Special Issue on Communications*, COM-29, No 11, 1554-64.

5.16 Bell Communications Research 1984, "Circuit Switched Digital Capability Network Access Interface Specification, Switched Network Compatibility and Performance Specifications for 2-Wire Connection to the Digital Public Switched Network," *TR-880-22135-84-01, Issue 1,* July.

5.17 M.J. Miller and S.V. Ahamed 1988, *Digital Transmission Systems and Networks, Volume II Applications,* Computer Science Press, Rockville, MD, Chapter 5, 214-274.

5.18 G.S. Moschytz and S.V. Ahamed 1991, "Transhybrid Loss with RC Balance Circuits for Primary Rate ISDN Transmission Systems," *Special Issue of the 1991 IEEE J-SAC on High-Speed Digital Subscriber Line,* Vol. 9, No. 6, August, 951-959.

5.19 Bell Communications Research 1987, "Characterization of Subscriber Loops for Voice and ISDN Services," *Science and Technology Series, ST-TSY-000041.*

5.20 J.J. Werner 1991, "The HDSL Environment," *Special Issue of the 1991 IEEE J-SAC on High-Speed Digital Subscriber Line,* Vol. 9, No. 6, August, 785-800.

5.21 R.E. Bollen, R.P. Prabhu and J.W. Modestino 1985, "Digital Transmission Performance of GTE's Local Loop," *IEE Third International Conference,* London, England, March 18-21.

5.22 J.W. Modestino, C S. Massey, R.E. Bollen, and R P. Prabhu 1986, "Modeling and Analysis of Error Probability Performance for Digital Transmission Over the Two-Wire Loop Plant," *Journal on Selected Areas in Communication,* J-SAC-4, No. 8, November, 1317-1330.

5.23 R.E. Bollen, "Standard for ISDN Basic Access Interface for Application at Network Side of the NT1," T1D1.3 Contribution 85-003, February 5, 1985.

5.24 K. Mueler, C. Hidelberger 1993, *Private Communication,* Schmid Telecommunication, Zurich, Switzerland, November.

5.25 A. Gillabert 1982, "Bestimmung der eleckrischen Eigenschaften des Ortsnetzes als Grundlage fur seine Digitalisierung," *Record of the PTT, Techniche Mitteilungen,* Berne, Switzerland, 8-16.

5.26 S.V. Ahamed 1994, "Computer Based Optimization Techniques for the Design of HDSL," *International Conference on Communications Record,* May 1-5, 811-820.

5.27 Martin Pollakowski, Hans-Werner Wellhausen 1995, "Eigenschaften Symmetrisher Ortsanschlutz-kabel im Frequenz-bereich bis 30 MHz," *ISSN 0015-010X, 49,* Jahrgang Heft 9/10, September/Oktober.

5.28 K. Maki, K. Yoshida, and R. Komiya, "NTT's Digital Subscriber Loop Transmission System and its Performance," *Private Communication.*

5.29 R. Smith, J. Semple and N. Demytko, 1982, Telecom Australia, Research Laboratories, Victoria, Australia, 3168, *Private Communication,* June.

5.30   Korea Telecom, 1988, "Loop Survey," September.

5.31   Wu-Jhy Chiu, Min-Jung Wu, Wen-King Hwang, and Shyue-Win Wei,
       1991, "Loop Survey in the Taiwan Area and Feasibility Study for
       HDSL," *IEEE Journal on Selected Areas in Communications*, J-SAC-9, No.
       6, August, 801-809.

5.32   Gi-Hong Im and J.J. Werner 1993, "Bandwidth-Efficient Digital
       Transmission up to 155 Mbps over Unshielded Twisted Pair Wiring,"
       ICC Conference Record 93, May, Geneva, Switzerland, 1779-1803.

CHAPTER 6

OPERATIONAL ENVIRONMENT FOR THE HDSL

The DSL and HDSL environments depend upon the physical and electrical constraints of the loops and their ultimate digital capacity. Some of the typical subscriber loop environments are presented in Chapter 5. The BRISDN services are expected to be made available over all telephone lines (without load coils and range extension devices) and their statistics are presented in Chapter 5. The PRISDN services are expected to be available to all CSA type of customers and their loop statistics are also presented in Chapter 5. Even though the simulations indicate that a single wire-pair is not able to carry T1 and E1 rate data over every HDSL CSA type (US) loops in the presence of crosstalk (typical of current Central Offices), there is good reason to believe that the dual duplex systems (i.e., two wire-pairs carrying half rate data in the duplex mode) will be satisfactory even for the most adverse loops. In this Chapter, the total operating environments for the DSL and HDSL including the electrical aspects in the design considerations are presented.

6.1

UNITED STATES OVERVIEW

In the United States, three major Subscriber Loop Surveys (see Section 5.4.1) performed in 1963, 1973, and 1983 yield the major trends and the actual environments for HDSL. The loop environment has been a fairly stable environment for the plain old telephone system for the past three decades. The national average working loop length has increased from 10,613 feet in 1963, to 11,413 in 1973, and then to 11,723 feet in 1983. Whereas the average bridged tap length has come down from 2,478 feet in 1963, to 1,821 feet in 1973, and then to 1,490 feet in 1983 [6.1]. The cable composition has been dominantly #26 AWG, then #24 AWG, and then #22 AWG, with #19 AWG being the least frequently used size. The non-standard cables such as the aluminum and alloy cables have been phased out. Paper and pulp insulated cables are less and less frequently used in the loop plant. Polyethylene-insulated cables (PIC) now dominate the wire-pairs that reach the subscribers.

These statistics can vary considerably from one region to the next, and the 1985 NYNEX survey [6.2] has confirmed such variations. For example, the Boston and Manhattan loops have an average length of about 6,500 feet and 5,500 feet for HDSL applications compared to the national average of 7,535 feet. These statistics can also vary from one operating company to the next with guidelines used in the installation of subscriber loops. For example, the GTE 1982 Loop Survey (see Section 5.7) has an average (non-loaded) loop length of 8,796 feet with an average length of the bridged tap of about 1,300 feet. The corresponding numbers from the 1973 Bell System survey are 7,748 feet and 922 feet; and from the 1983 Loop Survey are 7,535 feet and 1,168 feet.

Loop statistics from the most recent and comprehensive survey are published extensively for the POTS [6.3], for DSL [6.4] and for HDSL [6.5] applications and finally for the very high-speed digital subscriber line (VHDSL) and the ADSL [6.6]. Whereas the performance of the actual CSA loops has been documented [6.7 and 6.8] for the HDSL rates, corresponding results for other countries are not available. These CSA type data bases can be interfaced with the simulation software to evaluate the HDSL, VHDSL and ADSL capabilities of other Loop Plants.

Some of the electrical characteristics of the United States subscriber loops are published both as computer files [6.7] and as technical documents [6.1] to facilitate the vendors and component designers. Such network studies can be thus accurately and efficiently performed by simulation environments described in Chapter 7.

The subscriber loop servicing the individual customers bears the data at enhanced DSL, HDSL, ADSL and VHDSL rates for newer multimedia services. The loop environment is a critical path in the communication between the switching centers and the user of these high-speed services. The major networks around the world (see Section 5.6.1) have attempted to study their loop plants in order to characterize the physical and electrical characteristics. Physical characteristics change very slowly over the decades, because the cable once buried, remains in service until it corrodes, or it is physically disconnected, removed or altered.

6.2

INDIVIDUAL LOOPS FOR HDSL APPLICATIONS

Typical United States subscriber loops are presented in the T1D1.3/86 Ad Hoc Group on Draft Standards, Layer 1 Specification [6.8]. The ad hoc group had addressed the questions for ISDN Basic Access rather than for the higher rates. For this reason 12 of the 15 loops configurations (see Figure 5.30) are unusually harsh for high rate considerations. The remaining three loops (which are also quite severe for the PRISDN) over which the simulation is performed because they have no bridged taps, one tap and two taps, respectively. The loop configurations are slightly altered to satisfy the Carrier Serving Area (CSA) considerations specifying that the maximum loop length be 12,000 feet (9,000 feet of #26 AWG cable), with the composite tap length being limited to 2,000 feet (1,500 feet limit for any single tap). Loop configurations from the CSA environment for the HDSL applications are shown in Figures 5.31a, b and c. These disparity of these loops indicate the range of subscriber environments. The loops in Figures 5.31a and b are typical of United States suburban environments. The European loops are depicted in Figure 5.35.

6.3

IMPAIRMENT FOR DATA TRANSMISSION

The ideal subscriber loop for data transmission in a telephone network would be a uniform gauge wire between the individual subscribers and the Central

Office. The ideal Central Office environment would be completely free from all noise. The lines would carry information and network signals. Each wire-pair would be isolated from other wires, cables, power systems, lighting surges, and other electrical noise pollutants.

This ideal environment is nearly achievable in an all-fiber transmission with all-optical switching facilities. The real network differs from these ideal conditions in a significant number of ways. In a realistic sense, no system (including the all- optical facility) is totally noise free and at some stage the noise power becomes significant enough to cause uncertainty, however small it might be, in the signal recovery. Whereas in extreme cases, (e.g., deep space communications) the noise power may totally obliterate the signal power, the presence of (any) noise degrades the quality of the communication system to that extent. In the DSL environments, the presence of these numerous types of noises (Section 6.3.3) is to ultimately restrict the range and rate for the high-speed data. The reasons for the impairment and their insidious effect on the HDSL data are presented in this Chapter.

6.3.1 IMPAIRMENTS RESULTING FROM DESIGN RULES

Most of the telephone loop plants are conventionally designed, using a maximum loop resistance design rule to provide ringing and voice frequency access of to the customer. This rule sets an outer limit for the dc loop resistance, thus permitting shorter loops to have a finer conductor diameter. The high frequency loss of these lines also prevails, thus attenuating the HDSL signals, and ultimately limiting the range.

In the United States, *three* loop design rules apply. *First*, the value of the dc resistance has been set at 1,300 Ohms to facilitate the flow of supervisory signals, ringing current, and speech signals. Distant subscribers are served with loops that are inductively loaded to compensate for their inherently capacitive nature. The load coils usually appear on loops longer than 18 kft. and spaced at 6 kft. intervals with the first one located at about 3 kft. from the Central Office. *Second*, the unigauge design permits loops as long as about 15 kft. and 24 kft. of the #26 AWG (0.4039 mm) under 48 V dc power and under 72 V dc power from the Central Office. Loops up to 30 kft. would have loading coils located at 15 kft. and 21 kft. from the Central Office. Longer loops up to 52 kft used the 22 AWG cable, but continue to have load coils every 6 kft. after the first one at 15 kft. *Third*, the long range design permits loops up to 210 kft. by progressively relaxing the maximum resistance: (a) from 1,300 Ohms to 1,600 Ohms with a range extender (RE); (b) from 1,600 Ohms to 2,000 Ohms by ranger extender with 4 dB gain (REG) and with 48/96 V dc Central Office supply; (c) from 2,000 Ohms to 2,800 Ohms with 6 dB REG and 96 V supply; and (d) from 2,800 Ohms to 3,600 Ohms, also with 96 V supply and dial long line unit (DLL) with or without a 9 dB voice frequency repeater at about 1,100 Ohms down the line. The actual design depends upon the Central Office being able to send the signaling and supervision information over the long loops. *Four* devices used in conjunction with longer to the longest loops are range extender (RE), dial long line units (DLL), range extenders with gain (REG), and repeaters. These devices are used by themselves or in conjunction for the long loops [6.9]. There are *five* zones for POTS based services. In the *first* zone, the 1,300 Ohm rule applies, and there is no need for any devices in the loop configuration for the other zones. For customers beyond zone 1, the factors that become pivotal in the other *four* zones for providing the service are the appropriate combination of the RE, REGs, DLL units and repeaters. In Table 6-1, the type of loop-electronic device for loops with different resistances are presented.

TABLE 6-1 Resistance Range, Zones and Type of Device Used

Resistance Range	Zone	Type of Electronic Device
To 1300 Ω	1 (#13)	None
1300 to 1600 Ω	2 (#16)	RE (at CO)
1600 to 2000 Ω	3 (#18)	DLL and 4 dB signal repeating (CO) or 3 or 4 dB REG at CO
2000 to 2800 Ω	4 (#28)	DLL and 6 dB signal repeating (CO) or 6 dB REG at CO
2800 to 3600 Ω	5 (#36)	DLL with 9 dB repeater at ≈1200' or 9 dB REG at CO

It is to be appreciated that loops with any load coils (generally prevalent after 18,000 ft.) or frequency-dependent loop-electronics (in the data path) makes them unsuitable for DSL applications. With the 22 AWG, zone 1 has a range of 30,000 ft. With the 26 AWG, zone 1 has a range of about 15,000 ft., and beyond this distance multigauge (26 AWG and 24 AWG) loops are very common. The dc resistance plots of the loops in the 1983 Loop Survey for the residential (R), general (G), and the business (B) loops are shown in Figure 6.1, and the mean value for all the loops is less than 650 Ohms.

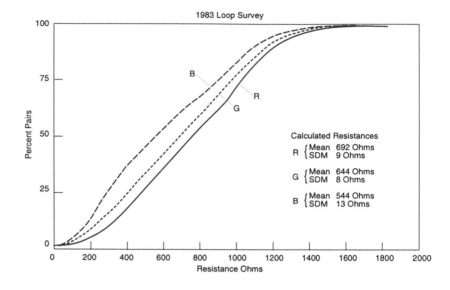

FIGURE 6.1. Calculated Resistance for Residential (R), General (G), and Business (B) Loops. SDM - Standard Deviation of the Mean

In the United States and Canada, four sizes (#26, #24, #22, and #19 AWG) cables are used. The two intermediate sizes (#24 and #22 AWG) correspond to 0.5105 and 0.6426 mm diameter, respectively. The distance limitations vary from one AWG to the next. The gauge distribution in the American loop plant is depicted in Figure 6.2. In Figure 6.3, distribution of the three structures (buried, aerial, and underground) are plotted as a function of the distance from the Central Office.

The European loop plants also have similar design rules, but the wire sizes can be finer, especially in the United Kingdom, because of shorter loop lengths. The metric size generally corresponds to the 0.4 and 0.6 mm of copper conductor diameter in Italy, and 0.32, 0.4, 0.5, 0.63 to 0.9 mm of copper and aluminum in the United Kingdom. The Swedish loop plant has 0.4, 0.5, 0.6 and 0.7 mm wire diameters. The French, German and Finnish (Norway) loop plants have generally used 0.4 and 0.6 mm cables, even though the German loop plant has 0.8 mm cables. The Japanese loop plant uses 0.33, 0.4, 0.5, 0.65 and 0.9 mm cables, with the intermediate sizes dominating the plant. The Australian loop plant has a selection of 0.32, 0.4, 0.51, 0.64, 0.9 and 1.2 mm diameter cables.

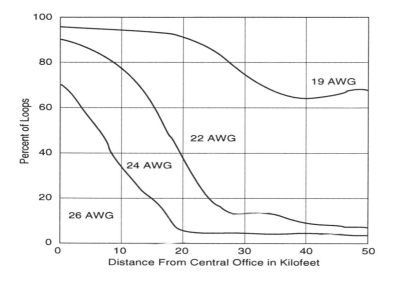

FIGURE 6.2. Cable Gauge Distribution in the 1983 Loop Survey.

However, the three most dominant sizes in the loop plant are 0.4, 0.51 and 0.64 mm cables. The lengths of loops, based upon the loop resistance, varies in these countries, depending upon the cable size and maximum resistance the network can tolerate. Once the various cable characterization parameters and national Loop Survey data bases are coupled to the simulation software, the dc ranges for different gauges can be evaluated, and the penetration based on resistance law can be determined.

The *next* dominant impairment is that which is due to bridged taps primarily caused by the rapid expansion in the number of telephones demanding more loops. One of the ways the operating companies have met this demand is by tapping into an idle loop formerly serving another customer in the vicinity of the new customer. Former multi-party lines also leave bridged taps in the loop plant. Further, when the telephone cables are installed, designers assign a certain number of cables to certain streets. Any unused cable sections in one street get reallocated to other streets, giving rise to more bridged taps. The extension of the loop that served the older customer was left intact in the anticipation of serving him again if the loop returned to its old configuration. The bridged tap statistics for the 1983 Loop Survey samples are presented in Section 5.4.2. The older telephone plants in

various countries (United States, Canada, Australia, Japan, Korea, Italy, etc.)
thus have a history of bridged taps.

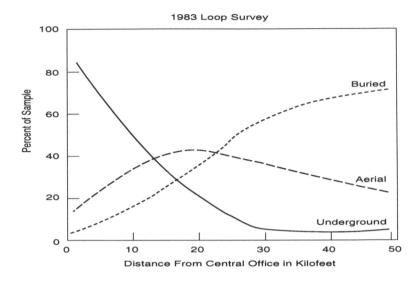

**FIGURE 6.3. Distribution of the Three Cable Structures in the 1983
Loop Survey.**

Bridged taps do offer considerable degradation in the capacity of the loop
plant for HDSL, ADSL and VHDSL types of services and our simulation
studies for the CSA environment (with BTs) and for the Swiss loops (without
BTs) clearly documents this result (see Chapter 12).

6.3.2 IMPAIRMENTS RESULTING FROM ENVIRONMENTAL CONDITIONS

Other physical impairments also exist. Corrosion and moisture in the cables
have been cited as one of the other impairments in the large loop plant of
countries such as Japan, Canada, and some parts of the United States. Open
wires suffer from the climatic effects more severely. Water, frost, sleet, and
ice interfere with the data-carrying capacity. The degradation is gradual.
Corrosion can cause a break in the continuity, especially where splices exist.
Electrical characteristics, especially the line imbalance, can also be altered by
the physical structures. The nearness of power lines causes inductive
coupling. Large signals at power frequency (and its third harmonics
occurring due to switching transient and magnetic saturation) may thus find

their way in the data path. These impairments for the open wire influence the capability to handle higher speed data rates more drastically than for twisted wire-pairs.

Oxidation, water in the cables, intermittent and/or unidirectional contact (i.e., a diode effect) is of serious concern for the transmission of data. This is especially true in the loop plants of some national networks in the United States, Canada, and Japan. Oxidation that occurs at cable splices depends upon the material and exposure to the atmosphere. In corrosive neighborhoods, as well as in very old loop plants, this problem causes severe noise and interruption problems.

Water in the cables influences the capacitive component. Our early studies have shown that the effect of such variation is not too significant up to T1 rates on #22 AWG cables. The different wire sizes and the material changes also cause variations in the primary and secondary constants for transmission. This problem influences the bit error rate in underground cables by closing the eye by an additional 5 to 7%. Short interruptions resulting from oxidation and/or physical conditions make such loops unsuitable for ISDN applications. When the Central Offices become aware of the problem the use of the cable or wire-pairs for data transmission is avoided [6.10].

In the cable environment, copper or aluminum conductors are generally insulated with a plastic material or pulp. A bundle of the conductor pairs are twisted again to form the cable core. Different types of enclosing sheaths are also available. Wet and moist sheathing and insulation cause variation of the conductance and capacitance of the cables. The choice of the cable type and the sheathing become important design parameters to deal with the environmental issues.

6.3.3 IMPAIRMENTS RESULTING FROM ELECTRICAL COUPLING EFFECTS

There are *three* major electrical impairments in the loop environment: *(a)* impulse noise (including lightning and switching surges, fluorescent light effects, and dial pulses from rotary telephone sets, etc.); *(b)* crosstalk; and *(c)* electromagnetic/electrostatic pick-up that arises from imbalance in the cable pairs.

6.3.3.1 Impulse Noise

The subscriber loop environment also is susceptible to spurious impulse noises whose power levels and spectral energy distribution overlaps the signal, especially from distant subscribers. Data bases of these signals exist [6.2] and provide some guidelines to the minimum signal level required for an acceptable error level in data transmission through HDSL.

Residual effects of lightning surges do not have the same statistical characteristics as the Central Office impulsive noise, even though the shape and duration may be similar. Lightning surges can be much more severe and do not follow the distribution approximated by the Equation (6.16) in Section 6.3.4.8. Both lightning surges and impulse noise corrupt the signal received for a finite interval (in the order of 30 to 100 µs) of time. Typically, since the signal level is low at the receiver, the effects of surges and pickup can cause spurious errors. Various surveys of impulse noise in Central Offices have also been reported as far back as the mid 1970s.

The effect of 5XBAR Central Office noise (for the 56 kbps CSDC environment [6.11]) significantly corrupts the low received signal levels at about 25 mV. With the more recent ESS offices, the corresponding received signal level is about 14 mV. A more recent study of the Central Office noise environment is also reported in Reference 6.2. This type of noise varies considerably from Central Office to Central Office. The recent ESS switches (5ESS and 4ESS, and the equivalent European versions) are the quietest, and the very old Central Offices with dc commutator machines (due to brush arcing) to charge the standby batteries for the ringing current and with rotary dial pulse number collection systems offer the loudest impulse noises. The intensity has also been correlated to the magnitude of the dial pulses and the frequency of their occurrence has been tied to the busy hours of the telephone traffic. Power surges due to vicinity of electrical power and distributing stations also cause impulse transients in the electrical communication signals. Automobile ignition and florescent lights are also identified as sources of impulses for the VHDSL rates in premises distribution systems (PDS).

Finally, some of the older Central Offices add short bridged tap lengths on the wire-pair as it is being scanned. When the scanning is based upon the actual sensing of the current flow in the lines, the scanner circuit is seen as an added bridged tap. The sensing is done for a short duration and appears as an impulse. The newer Central Offices and more sophisticated scanners do not cause such transients, nor any impulses associated with them.

6.3.3.2 Crosstalk Interference

Direct crosstalk results from the electromagnetic coupling between physically separated paths. The two direct crosstalk are the near-end crosstalk (NEXT) and far-end crosstalk (FEXT). These crosstalks, represented in Figure 6.4, are a result from the direct coupling between the disturber (primary) and the disturbed (secondary) circuits.

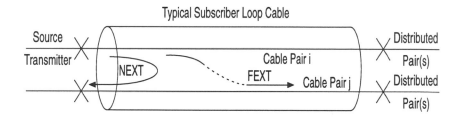

FIGURE 6.4. NEXT and FEXT Coupling in the Typical Subscriber Loop Cables. NEXT - Near-End Crosstalk; FEXT - Far-End Crosstalk

The NEXT crosstalk power travels in a direction opposite to the direction of the disturbing signal. In reality, the capacitive coupling and the mutual coupling between the two circuits generate currents that subtract (but in numerical terms they *add* because the mutually inductive current bears a minus sign) in the disturbed circuit. On the other hand, the FEXT power travels in the same direction as the disturbing signal. For this reason, the FEXT current results as an addition (but in numerical terms they subtract) of the two currents. In the high impedance circuits, (i.e., where the characteristic impedance of the cables is high), the capacitive coupling current dominants the mutual coupling current. The frequency coupling exponent for NEXT is 1.5 (thus giving rise to 15 dB per decade rule) and the frequency coupling exponent for FEXT is 2.0 (thus giving rise to 20 dB per decade rule). For a complete discussion of NEXT in the HDSL and VHDSL environment, see Chapter 14 and for FEXT in the HDSL environment, see Reference 6.13.

In the estimation of the direct crosstalk power, the coupling loss between circuits becomes essential. This loss, known as equal level coupling loss (ELCL), is defined as the ratio of the transmission point signal power in the disturbing circuit to the induced power measured, at an equal transmission level point in the disturbed circuit. Thus by definition, the ELCL becomes independent of the signal amplitude and its transmission level.

Indirect crosstalk results from coupling via an intermediate circuit (called the tertiary circuit) picking the signal from one path and then rebroadcasting it in the path of another, i.e., an additional circuit plays the part of coupling the disturber and the disturbed circuits. *Interaction crosstalk* results from all indirect crosstalk from the tertiary circuit to the disturbed circuit at locations elsewhere along the transmission path from where the disturbed signal entered into the tertiary circuit. *Transverse crosstalk* is the sum of all direct crosstalk, plus indirect crosstalk that involves no transmission along the tertiary circuit.

Statistical estimation for the coupling coefficients of interfering signals through cables is published [6.12, 6.13]. The interference due to two types of direct crosstalks (NEXT, FEXT), and interaction crosstalks (IXT, NE-NE-IXT, NE-FE-IXT, FE-NE-IXT, and FE-FE-IXT) for HDSL can be computed in different spectral bands. The later four of the crosstalks are depicted in Figure 6.5. The nature of coupling paths that give rise to these different types of crosstalks leads to dominance (for the repeaterless systems) of the direct crosstalks over the indirect crosstalks (see Figures 6.4 and 6.5) and of the NEXT over the FEXT type (the addition rather than the difference of the capacitively currents and the mutual induced currents). The FEXT power induced in the disturbed cables suffers line attenuation in addition to the coupling loss, and for this reason, in a majority of the cases, it is not as significant. However, the spectral band, the relative transmit power levels, the characteristic impedance, and the line attenuation do play a part in the FEXT interference.

When evaluating the effects of crosstalks, the relative power levels between the disturbing circuit and the disturbed circuits become important. For example, when repeatered digital lines are considered, the NE-NE-IXT can undergo considerable amplification at the repeater sites and cause a shift in the SNR. Similar effects can exist to a smaller extent for the NE-FE-IXT or the FE-NE-IXT, and to a yet smaller extent in FE-FE-IXT. Such crosstalks exist at apparatus in remote terminals and Central Offices giving rise to apparatus crosstalk or apparatus case crosstalk ACXT.

The composite interference power (integrated in the signal carrying band of the particular code under consideration) computed as a fraction of the signal power leads to the signal-to-noise ratio. The signal-to-noise ratio and the probability of error depend on the type of noise and its distribution. Once again, this ratio is monitored to minimize the error probability during transmission in the HDSL. Typically, in the best of operating conditions and designs, the error probability is as low as 10^{-9} for short T1 lines, and can be as high as 10^{-6} for digital subscriber loop systems carrying encoded speech, such as the very remote SLC-40® carrier systems in the United States.

Crosstalk effects are a severe limitation on the DSL. Defective or inadequate electromagnetic shielding of one wire-pair from the others is the dominant cause of crosstalk. The extent of the actual crosstalk also depends upon the physical placement of the wire-pair in the bundle and the pair number (see Figure 5.36 indicating the extent of difference between the "best wire-pair" and the "worst wire-pair"). Both types of dominant crosstalk (near-end crosstalk [NEXT] and far-end crosstalk [FEXT]) are influenced by the lack of the total shielding within the cables. During the manufacture of the cable, standard techniques to minimize crosstalk effects exist (random number of twists in opposite directions, etc. [see Section 5.1.2]). During installation, interconnecting wire-pairs from different bundles is also practiced in some loop environments.

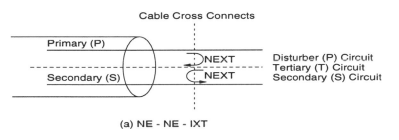

(a) NE - NE - IXT

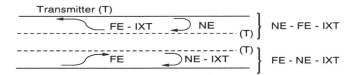

(b) NE - FE - IXT and FE - NE - IXT

(c) FE - FE - IXT

FIGURE 6.5. Interaction Crosstalk (IXT) Coupling in Cables. FE - Far-End; FEXT - Far-End Crosstalk; NE - Near-End; NEXT - Near-End Crosstalk

The effects of crosstalk depend on the physical location of the particular wire-pair with respect to its neighbors and the extent of the longitudinal imbalance along the length, and the overlap of bandwidth with respect to the disturbing signal. During the estimation of the noise level within the cable, various combination of disturbing wire-pairs is assumed (e.g., 49 wire-pairs carrying the same signals as the disturbed wire-pair, giving rise to the "Self-NEXT" limitation, 49 wire-pairs of T1/E1 signals, 9 wire-pairs of SLC-96® signals, etc.). The most important of the NEXT (most dominant form of crosstalk for HDSL) interference is discussed in Chapter 14. An initial review of composite interference from other DSLs, T1, and subscriber loop carriers (SLCs) is provided in Reference 6.14 and again in Reference 6.13.

6.3.3.3 Line Imbalance

Messages in the traditional telephone network are carried by electrical signals between wire-pairs. These wire-pairs are twisted and retain close physical proximity with each other. It is the intent (and desirable objective) that any voltage that one wire develops with respect to the ground be the same as that developed by the other, thus canceling out the voltage between them. To the extent that each of the two wires of the wire-pair maintains exactly the same balanced coupling to the ground as the other, there is no additional voltage in the signal that is being originally carried within the wire-pair. However, small imbalances cause an additional difference in the signal to appear between the wire-pairs. By and large, the imbalance voltage signal frequency is small when the voltage is small with respect to ground, and it is usually blocked by the transformers that provide electrical isolation; and receivers are generally immune to these very low frequency voltages.

However, if the imbalance is due to other voltage sources, such as power transformers, electrical power transmission facilities, power circuit breakers, fluorescent light choking-coils, etc., the additional signal between the wires within the wire-pair can be annoying. By and large, these signals manifest themselves as power frequency hums, clicks and chatter. If the imbalance is with respect to other close wire-pairs, the additional signals produce intelligible crosstalk, distant ringing signals, distant on-hook off-hook clicks, etc. The phenomenon of imbalance is a general one and it exists in numerous cases such as line amplifiers, transformer coils, transistor responses etc. connected to the two wires of the wire-pair.

6.3.3.4 Metallic Noise

In a great majority of cases this is caused by improper electrical contacts and "twisted connections" rather than soldered connections. Oxidation and corrosion of contacts at relays also give rise to metallic noise. Generally insignificant in communication systems with strong signals, they can be a source of problems for HDSL applications because of very low signal level and a wide overlap of the spectral band of the noise and signal. Metallic noise generally appears between the wire-pairs or tip-and-ring. In the voiceband C message filtering is usually done and in most Central Offices about 96% of the loops have this noise under 20 dBrnC. (See footnote in Section 2.4.3.)

6.3.4 BASIC REASONS FOR NOISE WITHIN SYSTEMS

Realization of systems is a noisy and noise-prone process. At the nuclear and molecular level, processes are discrete and every circuit function leading to the recovery of the signal also has a certain amount of noise associated with it. In principle, subfunctions cannot be totally directed towards signal processing without the embedded noise generation and processing.

At the atomic or molecular level, processes occur due to the basic laws of physics that govern the interaction between electrons, holes, photons, and other elementary particles. Most devices achieve a great extent of signal processing but these processes are the laws of physics that are equally applicable to noise. In fact, there is no selective process that distinguishes the elementary particles (and their processing) that arise from noise and from signals. There is no device that performs only signal-related processing and not noise-related processing at an elementary particle level. For this reason, noise is inseparably intertwined with the signal and should be treated as the undeniable contaminant that finally drowns all signals when they reach the noise level. In the rest of this section, the inherent and fundamental reasons for the smallest trace of noise, even in the strongest of the signals, are presented.

6.3.4.1 Thermal Noise

Electrons in conductors move in continuous random trajectories (Brownian movement). Electrons are carriers of electrical charges and their movement gives rise to very, very minute pulses of currents. When the currents are high, the effect of individual electrons is truly insignificant, but at its lowest limits, the noise caused by the Brownian movement of electrons (also known as Johnson noise, thermal agitation, thermal noise, or resistance noise) sets a lowest level for the noise power in a system. The noise depends upon the absolute temperature T, and Boltzmann's constant k (1.3805×10^{-23} joule/K) and the bandwidth over which the measurement is being made. At a room temperature of $17°$ C or 290K, the power in a 1 Hz band is 400.345×10^{-23} watts or -173.979 dBm. This white noise power level is uniform over the entire band (dc to the highest microwave band) can be quantified at the room temperature ($17°$ C) as

$$P_a = kTB_s \quad \text{watts} \tag{6.1}$$

$$P_a = -173.979 + 10 \log B_s \quad \text{dBm} \tag{6.2}$$

where P_a is the power available and B_s is the system bandwidth in Hz. Thermal noise is white and Gaussian and for this reason Gaussian (or normal) density function (see Figure 6.6 a)

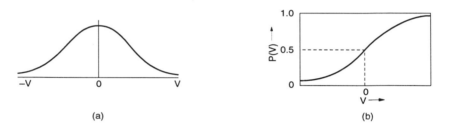

(a)   (b)

FIGURE 6.6. Gaussian Probability Functions: (a) Density Function; (b) Distribution Function.

$$p(V) = \{1/\sigma_n . \text{SQRT}(2\pi)\} \exp(-V^2 / (2.\sigma_n^2)) \tag{6.3}$$

with a corresponding distribution function (see Figure 6.6.b)

$$P(V) = \{1/\sigma_n . \text{SQRT}(2\pi)\} \int_{-\infty}^{V} \exp(-x^2 / (2.\sigma_n^2)) . dx \tag{6.4}$$

where σ_n^2 is the variance and V is the signal level. Standard tables exist for the values of P(V) for different ratios of V to σ_n. The mean square value of V is σ_n^2 and the rms value is σ_n for this distribution. The one-sided folded density function or the average absolute value is determined as

$$E(|V|) = \text{SQRT}(2/\pi) \sigma_n \tag{6.5}$$

by integrating V. p(V) from 0 to ∞. The values of p(V) and P (V) can be used in evaluating the rms open-circuit voltage (e_n) and the rms short-circuit current (i_n) for the thermal noise power source of a noiseless resistor R. If this source is connected to a noiseless load resistance value R (for maximum power into the load) over the bandwidth B_s, then e_n and i_n can thus be written as

$$e_n = \text{SQRT } (4kTB_s.R) \tag{6.6}$$

and

$$i_n = e_n /2R = \text{SQRT}((kTB_s)/R)) \tag{6.7}$$

with the maximum available power $[(e_n)^2 /4R,$ i.e., $e_n/2$ times $i_n]$ into the resistor R, as being p_a defined in Equation (6.1)

6.3.4.2 Shot Noise

Most devices function due to electronic charge transfer by the movement of electron and holes, and this process is discrete. Like thermal noise, shot noise has a Gaussian distribution and a flat power spectrum, but shot noise is not temperature dependent. Shot noise power is directly proportional to the direct current through the device. Over a 1 Hz bandwidth, the mean-square shot current can be derived as

$$i^2 = 2qI \tag{6.8}$$

where q is the charge per electron (i.e. 1.6×10^{-9} coulomb) and I is the current through the device.

6.3.4.3 Low-Frequency (1/f) Noise

This type of noise is due to the contact and surface irregularities of the semiconductor devices and cathodes (that emit electrons). This irregularity causes a fluctuation in the conductivity of the medium closest to such surfaces giving rise to the low frequency noise. This type of noise has a Gaussian distribution and is also called contact noise, flicker noise, or 1/f noise due to its increase towards very low frequencies. The newer device fabrication techniques limit this noise level to negligible values above a few hundred Hz. The power spectral density of this type of noise is approximated as

$$p(f) = K/ f^v \quad \text{watts} \tag{6.9}$$

with the numeric value of v can be between 0.8 and 1.5.

6.3.4.4 Rayleigh Noise

In carrier frequency applications, the noise bandwidth is small compared to the frequency at the middle of the band. Under these circumstances, the Gaussian noise can be viewed as a low frequency modulated sinusoidal carrier (at midband) whose highest frequency component depends upon the noise bandwidth. The low frequency envelope can be physically generated as the output of an smoothing envelope detector that touches the positive peaks of the noise. Now if the noise is Gaussian, the envelope has a Rayleigh distribution specified by Rayleigh distribution function (see Figure 6.7a)

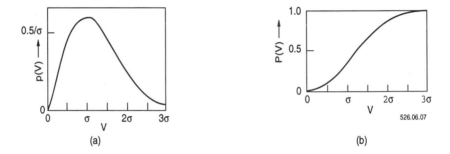

FIGURE 6.7. Rayleigh Density: (a) Distribution and (b) Function

$$p(V) = (V/\sigma^2) \exp(-V^2 / 2. \sigma^2), \quad V > 0 \tag{6.10}$$

with its corresponding distribution function (see Figure 6.7b) for this noise is

$$P(V) = 1 - \exp(-V^2 / 2. \sigma^2), \quad V > 0 \tag{6.11}$$

The average value of this distribution is not zero since there are only positive values for P(V). The average value E[V] and can be obtained by integration of the V.p(V) over the range of 0 to ∞ as

$$E[V] = \int_0^\infty (V^2/\sigma^2). \exp(-V^2 / 2. \sigma^2) dV = \sqrt{(\pi/2)}. \sigma. \tag{6.12}$$

The mean square values $E[V^2]$ and can be obtained by integration of the $V^2.p(V)$ over the range of 0 to ∞.

$$E[V^2] = \int_0^\infty (V^3/\sigma^2) \cdot \exp(-V^2/2.\sigma^2) \, dV = 2.\sigma^2 \qquad (6.13)$$

And finally the value of the mean-square ac component is $0.429\sigma^2$, since 0.429 is $\{2.0 - (\pi/2)\}$.

6.3.4.5 Quantization Noise

In analog-to-digital conversion, every analog value has to be approximated to the nearest digital value. There are only a finite number of digits to represent an infinitely large variation of analog values. Quantization noise arises whenever a selection is made by the A/D converter to select the nearest digital value for every analog value that is sampled. The value of the analog signal less than or equal to half the smallest signal level that can be represented by a 'one' in the digital stream, gets truncated every time a sampling is done by the A/D converter. If V_s is the voltage difference between the steps, the round-off error is bounded by $\pm V_s / 2$. Linear quantizers yield uniform or rectangular density function, (Figure 6.8a), if overload clipping does not occur and the distribution function (Figure 6.8b) becomes a straight line from 0 to 1 at the two bounds .

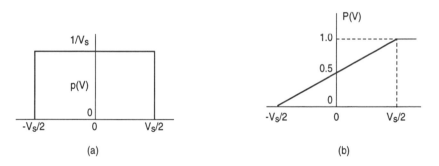

FIGURE 6.8. Probability Functions for Quantizing Noise of Linear Quantizers: (a) Density Function and (b) Distribution Function.

The average value is zero since the plus or minus values of the round-off error are equally likely. The average *absolute* value of the noise is $V_s/4$ and the mean-square value is

$$E[V^2] = V_s^2 / 12 \qquad (6.14)$$

obtained by the integration of (V^2 / V_s) over the limits of $\pm V_s/2$. The rms voltage is $V_s/\sqrt{12}$ with a peak factor of $\sqrt{3}$ and a form factor of $2/\sqrt{3}$ corresponding to 4.8 dB and 1.25 dB, respectively (see Reference 6.15).

Quantizers can be nonuniform, especially for speech signals. Companding permits the system designer to extend the dynamic range of the signal variation with fewer bits at the output of the A to D converters. In essence, these devices permit the step size to increase as the signal amplitude grows. Since the ear responds to the signal power and the relative errors rather than the absolute errors, an appropriate change of step size is adopted. The process of altering the step size with the signal amplitude is called companding. Two laws for companding (μ-law and A-law) exist in most communication systems. The μ-law companding

$$F_\mu (x) = \text{sgn}(x) \cdot \frac{\ln (1+\mu\, |x|)}{\ln (1+\mu)} \qquad \text{for } -1 \le x \le 1 \qquad (6.15)$$

prevails in the digital networks in North America and Japan. The two-segment A-law companding

$$F_A (x) = \text{sgn}(x) \cdot \frac{A\, |x|}{1+ \ln A} \qquad \text{for } 0 \le x \le 1/A \qquad (6.16)$$

for the low level signals and

$$F_A (x) = \text{sgn}(x) \cdot \frac{1+ \ln A\, |x|}{1+ \ln A} \qquad \text{for } 1/A \le x \le 1 \qquad (6.17)$$

for the higher level signals, prevails in the CEPT countries of Europe. The variable x indicates the signal input within the normalized coding range of ± 1. The μ-law companding follows modified logarithmic compression characteristics (to prevent the divergence of the function ln(x) when x approaches zero) and the value of μ determines the extent of compression. This value for μ is mostly chosen as 255 for the speech applications and the function $F_\mu(x)$ approaches a linear function at very low values of x and approaches a logarithmic function as x become large.

The typical value for A is chosen as 87.6 for the speech applications for the function $F_A(x)$. The function $F_A (x)$ is a linear function for small values of x (i.e., for $|x| < 1/A$) and is a logarithmic function at very large values of x (i.e., for $|x| \ge 1/A$). The A-law compander results in a slightly more degraded performance for low level signals, whereas it offers a slightly flatter performance for larger signals ($1/A \leq x \leq 1$) compared to the μ-law companders.

The measure of performance for the codecs (A to D and D to A converters) is computed by evaluating the signal power (S) to quantizing-distortion power (D) for sine waves (in the bandwidth of interest) since the actual speech is dynamic. The error characteristics and thus the S/D ratios can be computed. For example, the S/D ratios for different numeric values of μ are plotted in Figure 6.9 as a function of input sinusoidal power (in dB relative to the full load, see X-axis) for 8-bit codecs. Such performance evaluation can be computed for the two companders.

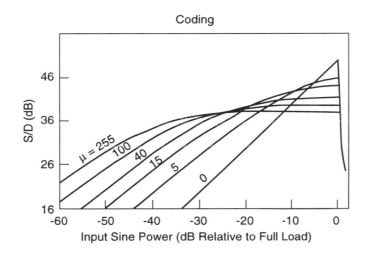

Coding

FIGURE 6.9. S/D Performance of the μ-Law Companding for 8-Bit Codecs. S/D - Signal/Distortion Ratio

Both give better values of S/D for the average talker over the entire band than for very weak or very loud talkers. The codecs in devices do not actually follow the algebraic relations in the equations, instead piecewise linearization is used and for this reason the differences in the performance [6.12] is relatively small.

6.3.4.6 Feed-Through Noise and Misalignment Noise

These two types of noises appear intertwined and to some extent are interdependent during the process of echo cancellation (see Section 6.5.1). Echo cancelers are usually implemented by FIR (finite impulse response)

filters (see Section 6.5.4.2). The device functions by systematically subtracting out the echoes of the prior pulses from the most recent signal received. There are definite limits to the performance of these devices as discussed in the next Section. The noise that is left behind after the imperfect cancellation of the echo, is the sum of the feed-through noise and the misalignment noise. A certain amount of uncorrelated noise is reminiscent at the output of the echo cancelers. When there is no *far-end* transmitted signal, the signal from the echo canceler can be viewed as the difference between the actual echo and the estimate of the echo (that will be used for cancellation). In addition, random signals, such as combinations of the thermal noise of the transmission facility, and/or quantization noise from the digital facilities, and/or the noise generated within the canceler itself, also accrue at the output.

Under these conditions, the convergence time constant for the EC is shown [6.16] to depend directly upon the sample period and inversely upon the variance of the random noise into the echo canceler. In addition, it is also shown [6.16] that the variance of the error signal from the echo canceler has *two* components: *first,* the *feed-through* variance of random signals entering the echo canceller and *second,* an additional variance term which depends of the variances of the first random signal being modified by the imprecise estimation of the tap weights generated within the echo canceler itself. The noise with the first variance is called the *feed-through noise* and the noise with the second variance is called the *misalignment noise.* In practice, they appear (as they should) as closely interdependent. The total echo canceler noise is a composite effect of the system that surrounds it and its own ability to compute the tap-weights.

6.3.4.7 Idle Channel Noise

This is a composite sum or a convolution of numerous random noises in the transmission channel when there is no far-end transmitted signal. There is reason for idle channel to exist. For example, the devices that make up the system have thermal and device noises, however insignificant they may be (see Section 6.3.4). The media itself may exhibit time dependence, non-homogeneity, frequency bumps, and dependence upon random events (such as sun spot activity, meteor trails, galactic events, or even earthly events such as power circuit breakers, transmission line surges, auto ignition systems, lightning, electric arcing, welding, etc.). It is to be appreciated that correlated interference (noise) is predictable and to that extent it can be filtered out. The residual interference due to imperfect devices is generally not considered noise, but it is attributable to the design restrictions of any particular device(s).

Thermal and device noise at the transmitter become the input level idle channel noise sources. In addition the media and the receiver modify and even interject other random noise. The composite signal has the characteristics of the original noise and also of the medium and receiver. The composition (the original and their convolutions) of these random noises gives rise to the idle channel noise. Idle channel noise depends upon the point at which it is measured, the media, and the devices. It can exhibit a variety of statistical properties, but the average and variance offer a clue as to the quality of the digital communication system and the type of noise source.

6.3.4.8 Impulse Noise

Short spikes of power are sometimes encountered in transmission systems in the band of interest in which data is being transmitted. These spikes generally have a flat frequency spectrum. Switching transient, fluorescent light chokes, corona discharges (including lightning), Central Office relays cause impulse noise conditions causing clicks (like dial pulse clicks) in the audio band. In digital systems, these can cause bursts of errors (depending upon the data rate and the impulse noise duration) and even total loss of short-term synchrony. Cables directly affected are in the path of the impulse noise event. Neighboring cable pairs also pick up the event due to the magnetic, capacitive and mutual coupling. The coupling factors (see Chapter 14) used for NEXT and FEXT crosstalks are applicable for the impulse noise between the disturber and disturbed wire-pairs. By and large, the type of Central Office and the time of day (i.e., the traffic intensity) influence the values of a and b in the equation for the probability of experiencing an impulse noise event in Central Offices

$$P(\,|\,X\,|\,>\chi\,) = F(\chi) = \left(\,a/(\chi+a)\,\right)^{\,b}. \tag{6.18}$$

The typical power level of these events in context to the CSDC data transmission at 56 kbps are presented [6.11] and the *four* of the typical characteristics in the American Central Office environment are summarized [6.13] as follows: *first*, the number of events lies between 1 to 5 every minute; *second*, the peak amplitude is in the range of 5 to 20 mV; *third*, most the energy concentration is below 40 kHz and *fourth*, the duration is limited between 30 to 150 µs interval.

Impulse noise exists in almost all systems where extraneous events or the device components can cause an impulsive shift in the performance of the system. Mode hopping and the chirp noise of laser sources, in optical systems also causes transitional noise events. In the FM systems, very short impulses cause a sudden large shift in the carrier phase, causing the devices to slip into nonlinear operation, causing intermodulation noise events in other channels. Most of these conditions generate small amounts of impulse noise events in their own right that is carried into neighboring channels.

6.3.4.9 Intermodulation Noise

In the FM devices, the linearity of the systems is confined to a range of carrier and signal powers. Device nonlinearities cause spurious side lobes that shift energy from one band in another. Typically, these adjoining bands are used by other channels and get affected by the signals from the unrelated bands. In these cases sum and difference frequency effects give rise to intermodulation noise. In most of the digital systems, the effect of intermodulation noise is by the threshold detectors but it can cause errors in the detection process if the sideband energy becomes excessive in relation to the signal power at low amplitudes. The intermodulation noise is of serious concern to the designers of analog signal carrier systems in the design of the repeaters where error is likely to cumulate and grow.

6.3.4.10 Device and Network Noise

Most terminal systems have passive and active electronic components. At very low power levels, these components are not totally quiet. The thermal noise (by agitation of electrons and their collision with molecules), and shot noise (by the uneven flow of electrons within active devices, oscillators, even clocks), can accrue in devices and systems. Hence different interconnections and layouts lead to different noise patterns and levels from these devices. Generally most systems have a large number of components and transistors and an estimate may be made for the noise these systems (such as repeaters and cascaded networks) generate and propagate. Temperature is an important parameter in the evaluation of the device and network noise since the characteristic electron movement and collision probability are temperature dependent. The 2-port network equations for the transfer functions are usually used in determining the noise figure or the noise factor [6.12] of typical cascaded networks.

6.3.4.11 Diode and Transistor Noise

Diode have minority carrier current I_s (i.e. saturation current) and majority carrier current [6.17]

$$I_F = I_s\, e^{\,q V/ kT} \tag{6.19}$$

The noise current (in each of the two currents) are independent and they can be added on a power basis leading to

$$i^2 = 2 q I_s (I + e^{q V/ kT}) B_w \tag{6.20}$$

where B_w is the bandwidth in Hz. In terms of the diode current I (i.e., $= I_F - I_s$) leads to

$$i^2 = 2 q B_w (I + 2 I_s). \tag{6.21}$$

Under ideal conditions of a matched conductance

$$G = dI/dV = q(I + I_s)/k T \tag{6.22}$$

thus leading to the available noise power

$$P_n = i^2 / 4 G = (k T B_w / 2) ((I + 2I_s) / (I + I_s)) \quad \text{watts.} \tag{6.23}$$

Three modes of the diode operation become evident; *(a)* with no diode current (i.e., $I = 0$), the diode acts as a resistor; *(b)* when I is very large in relation to I_s the diode is half as noisy as the resistor (see Equation (6.7)); and *(c)* at reverse bias (i.e., I approaching $-I_s$), the diode becomes a dominant noise source.

For the transistor noise, these equations can be adjusted accordingly by considering the diffusion fluctuations, and recombination fluctuations in the base region. Contributors to the overall transistor noise are *shot noise* due to random passage of carriers through the junction (i.e., diffusion fluctuations), *partition noise* due to random division of carriers between the base and collector (i.e., recombination fluctuations), and finally the *thermal noise* from the base and emitter resistances. These are the three sources of transistor noise.

6.3.5 NOISE CONTAMINATION AND MANAGEMENT

Noise in communication systems is irrevocable.   It exists because the definition of noise encompasses every type of random signal and its contamination. But by design, robust systems tolerate random events and yet function within the limits of tolerance.   In addition, the signal and noise are so closely intertwined that there is no signal without noise from its very source. All the telecommunication systems are less than perfect, thus causing convoluting or to some extent retaining, modifying or enhancing the signal corruption. The management of noise aims at keeping its level to be consistently and significantly less than the signal. This is a crucial issue in the engineering and design of the communication systems. The signal-to-noise ratio is a measure of the signal domination over the noise, and in the digital systems, the bit error rate is also a measure of system robustness. In analog systems, the linearity, net loss variations and envelope delay distortion over the FM band are indicative of the quality of the analog system [6.18].

In the subscriber plant, transmission tests are done to evaluate the general noise level in loops. In the United States environment, the Central Office measurements are done with the 3A type noise measuring equipment with C message weighting for the routine POTS applications. The on-hook and off-hook applications are both taken. The Central Office measurement (done with station set on-hook) indicates the CO noise floor, and the remote tests (done with the station set off-hook) indicates the noise level within the segment of the loop plant. The objective is to maintain the plant and perhaps identify the reason (and correct) for the noise in the loops.

Typical of the noise level in the wire centers of COs is that 95% of the loops should have a noise level of less than 20 dBrnc, and that the number of wire centers having 15 percent or more noisy lines should not exceed 3 percent. When excessive noise is detected, the *three* most common reasons noticed are: *(a)* cable sheath discontinuity make all the wire-pairs in the cable to be noisy, *(b)* pair imbalance arising from party lines and ringer imbalance (in the older type of station sets), and *(c)* maintenance or poor repair service for particular lines which affect the wire-pairs randomly. This type of routine test is made with Automatic Transmission Measuring Systems (ATMS) and are recorded as a part of the OA&M activity. For the DSL and the HDSL applications, these measurements are not meaningful; however, they become indicative of the causes of the noise, and their effects can thus be extrapolated to the spectral band of interest for the newer services.

6.4

MODES OF TRANSMISSION FOR ISDN

The *five* modes for the implementation of digital line are: *(a)* space-division multiplexing; *(b)* frequency-division multiplexing (FDM); *(c)* time-compression multiplexing (TCM); *(d)* adaptive echo cancellation (AEC); and *(e)* adaptive hybrid impedance matching with AEC. These five techniques are discussed in the following sections.

6.4.1 SPACE DIVISION MULTIPLEXING

In this *first* mode, two wire-pairs (i.e. four wires) are used, and the direction (i.e., subscriber to Central Office and Central Office to subscriber) of the signal flow in each wire-pair is different. The electrical isolation between the wire-pairs provides two independent paths for the transmission and reception of data and successfully deployed for repeatered 4-wire T1/E1 carrier systems. In the 1970s, the digital data systems (DDS) were introduced to carry subrate customer data (at 2.4, 4.8, 9.6 and 56 kbps) over the T1 carrier systems. A typical four wire facility consisting of a channel service unit (CSU) at the customer end, interface circuitry, line isolating transformers, two cable pairs, (one for each direction) and an office channel unit (OCU) at the Central Office, is shown in Figure 6.10. Numerous subrate channels could then be multiplexed over the T1 carrier system at 1.544 Mbps. By default, the four wire DDS system at lower rates has assumed the flavor of a spatially divided system. Since the very seminal stages (mid 1970s) of DSL (in contrast to the DDS), this half-duplex method was ruled out for any high-speed digital loop systems.

6.4.2 FREQUENCY DIVISION MULTIPLEXING

In this *second* mode (FDM), the available loop bandwidth is divided into two bands of frequencies, one for each direction of transmission and depicted in Figure 6.11. The directional isolation is achieved by spectral filtering. The major advantage of the FDM approach is that the effect of near-end crosstalk is minimized because the transmitted and the received signal spectra are separated.

However, in most of the practical systems, the filtering, being less than perfect, restricts the band for transmission to less than half the available loop bandwidth, thus effectively reducing the data rate. Frequency shifting may also become a requirement of the FDM systems. These techniques have been effectively used in more cumbersome cable carrier systems for voice signals (e.g., L5E system in Section 2.5) rather than for HDSL applications.

The FDM concept, though not prevalent for the DSL, HDSL or the SDSL (symmetric digital subscriber line at rates higher than the HDSL rates), vaguely reappears in the asymmetric digital subscriber line, or the ADSL. In this instance, the same wire-pair carries the Central Office data subscriber data in a different band from the band for subscriber to Central Office data. There is no justification to label the ADSL concept to be the FDM concept, because there is no carrier signal that is modulated.

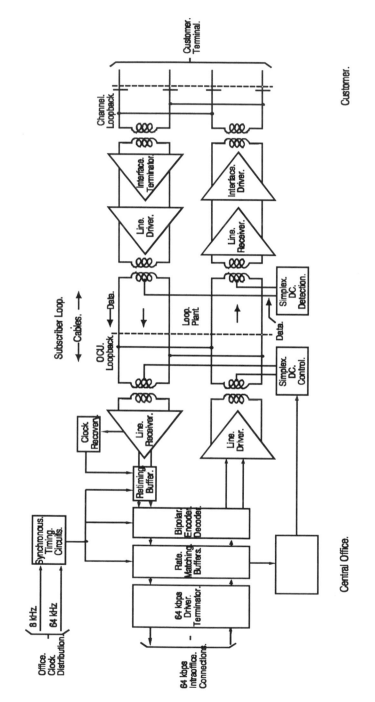

FIGURE 6.10. Typical Layout of a Four Wire System

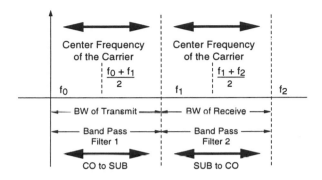

FIGURE 6.11. An FDM Technique for Separation of the Transmit and Receive Signals. CO - Central Office; FDM - Frequency-Division Multiplexing; SUB - Subscriber

However, the two bands for the two directions of the signal flow in the ADSL are different over the same wire-pair. This notion of modulating the data and fitting it in a spectral band without a carrier is called the carrierless modulation; and using a two dimensional cluster code in the modulation of the data (to achieve bandwidth efficiency) is called carrierless amplitude and phase modulation or CAP (see Reference 6.13). Typically, in the ADSL applications, data at the low end of the frequency is for data towards the Central Office and conversely data at the high end of the frequency is for data towards the subscriber. The upper bandwidth depends upon the data rate and the type of CAP coding that is used. The significance of these bandwidth efficient codes in context to the loop plant data accumulated in the 1983 Loop Survey is discussed further in Chapter 14.

6.4.3 TIME DIVISION (COMPRESSION) MULTIPLEXING

In this *third* mode i.e., the time compression multiplexing (TCM) also known as the burst mode [6.19 and 6.11], the intervals of transmission and reception are isolated at either end of the line. The information is collected in the form of a block for a fixed duration 'T' at the transmitter in either end of the line and "burst" out on the transmission medium in less than half that duration. The principle of the TCM is illustrated in Figure 6.12. The process of collecting data at the transmitters and then bursting it in opposite directions is alternated such that the flow of data is essentially unidirectional at any given instant of time.

However, there are certain basic requirements for the TCM system to work well over a sustained period of time. The *first* consideration is the need for two well synchronized clocks at the two transmitters. Typically, the Central Office clock assumes the role of the master clock and the derived subscriber clock becomes the slave clock. The *second* consideration arises from the need for "guard space" or T_g between the transmit/receive functions of the (T/R) switch shown in Figure 6.13.

Typically, this duration consists of *four* components: *(a)* the transit time required for the burst of data to travel down the farthest subscriber or T_t ; *(b)* the duration T_r required for the reflected signal at the receiver to subside enough for the receiver to receive only the distant signal and not be contaminated by the reflection of its own transmitted signal; *(c)* the duration for all the bridged tap reflections to be completely received before the transmitter is activated; and *(d)* the actual duration for the T/R switch to operate or T_s..

The transit time T_t depends upon the range and the propagation velocity of the burst signal. It is derived from the delay term β of the secondary characteristic

$$\gamma = (\alpha + i\,\beta) \text{ per unit length} \tag{6.24}$$

for the cable gauge and its primary constant R, L, G, and C per unit length. In addition, these constants depend upon frequency and temperature.

Hence, the transit, time can vary considerably; however, the actual computation of the delay term β and the propagation velocity shows reasonable constancy over the band of spectral interest for the DSL, HDSL and ADSL band of frequencies and for the four typical gauges (26, 24, 22 and 19). These numbers are presented at $70°$ F in Table 6-2.

From the consideration of bridged tap reflection, the length of the bridged tap becomes more important that its gauge. The gauge influences the extent of attenuation as the incident and then the reflected wave. Hence bridged taps of coarser gauge return more of the reflected signal than the finer gauge taps. From the Loop Survey results, the taps are mostly 26 or 24 AWG with very little 22 or 19 AWG.

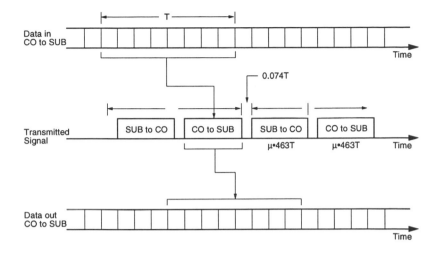

FIGURE 6.12. The TCM Concept. The value of T is 3 ms for the CSDC 56 kbps system with a burst rate of 144 kbps. CO - Central Office; CSDC - Circuit Switched Digital Capability; SUB - Subscriber; TCM - Time Compression Multiplexing

But, by and large, the numerical value is fairly constant and changes typically from 0.63049377 to 0.63689487 for 26 AWG cable at 1 MHz from 70° F to 120° F. In most cases, this number is conservatively taken as 0.5 kft/µs [6.11 and 6.12] of the data pulses/symbols for the various gauges.

TABLE 6-2 Propagation Velocity in *kft / µs* at 70°F

Gauge →	26 AWG	24 AWG	22 AWG	19 AWG
kHz ↓				
100	0.560278	0.580965	0.593203	0.607362
200	0.590154	0.600379	0.609768	0.625932
300	0.600575	0.609875	0.620244	0.636031
500	0.612486	0.622931	0.633790	0.647019
700	0.620936	0.631936	0.641536	0.653257
1000	0.630494	0.640433	0.648943	0.659089
5000	0.662532	0.667799	0.672159	0.677337

The reflected signals at the transmitter do prevail for a finite length of time. The actual value is determined by a simulation study for the particular loop plant environment and its topology. The decay of the envelope of reflections is an exponential process for any loop. When the reflection envelope of an entire population of loop in a national Loop Survey data base is plotted, the duration for the reflected to decay 30 or 40 dB below the received signal can be precisely evaluated. Such results have been reported for the TCM mode for the loops in the 1973 Loop Survey for the 56 kbps CSDC service (in the pre-divestiture United States Loop environment) to be about 7-8 pulse durations for about 40 dB lower reflections, and 12-14 pulse durations for 80 dB lower reflections at 144 kbps [6.14]. The components T_t and T_s are relatively fixed for any given system and only the greater of T_t or T_r is important since they overlap. Only a detailed system design reveals which of the four durations exist in parallel and which will be sequential to make up the actual "guard-duration" for any TCM system.

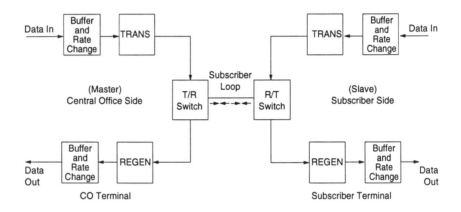

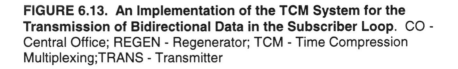

FIGURE 6.13. An Implementation of the TCM System for the Transmission of Bidirectional Data in the Subscriber Loop. CO - Central Office; REGEN - Regenerator; TCM - Time Compression Multiplexing;TRANS - Transmitter

This "guard duration" is a fraction of the total duration cycle time 'T'. However small it might be, it is not zero, thus forcing a "dead-time" over the medium. The actual transmission rate thus becomes faster than twice the data rate. The guard time cannot be too large compared to the duration for transmission, since the transmission rate increases with the guard duration. In this mode, even though the data at the transmitters is collected from the two sources uniformly, it is accumulated in a buffer and then alternately exchanged between the two sides. Typically, with multiple buffers of exponential length (rather than a single buffer) the time to empty the buffer into line can be significantly reduced [6.20] and it has been implemented in the CSDC system. In a sophisticated TCM environment the line rate is about 2.25 times the customer data rate and the delay is one half of the total buffer length times the data rate plus the guard duration [6.21].

The *two* attractive features of this third (TCM) mode for data transmission are: *(a)* there is no need for any echo cancellation since the receiver is deactivated while the transmitter is active; and *(b)* the relatively simple and robust implementation and good stability of the system. However, the system causes an inherent delay in the reception of data at the receiver above and beyond the transit time of the data through the medium due to the buffering requirement. In addition, the range of the TCM system is shorter because of the higher line-rate. Both crosstalk coupling and attenuation increase with the higher line rate. As far as the impulse noise is concerned, the influence is only marginal. Whereas the higher bandwidth is likely to pick up more signal contamination from individual impulse noise events, there is more than 50% probability that the receiver is not receiving during any particular event. The TCM cables also disturb other systems such as the T1 carriers and analog subscriber loop carriers (such as the SLC1™ and SLC8™ systems in the United States [6.22]). These cables also strongly crosstalk with each other via the Self-NEXT. The only way to circumvent this problem is to synchronize all the Central Office bursts of data. However, this would ruin other signals in adjoining wire-pairs.

The implementation of the TCM mode occurred in the United States with the introduction of the circuit-switched digital capability (CSDC) during the mid eighties [6.11]. This service was primarily aimed at the urban business customers needing rates significantly higher than the highest voiceband modem rates (9.6 or 14.4 kbps at that time). Some of these businesses have used the CSDC line to interface distant computers. The service was a precursor to the latter full-fledged BRISDN services in the mid eighties. The CSDC subscriber data rate is at 56 kbps. This value is enhanced to 64 kbps because of the one additional signaling bit per byte and the line rate is at 144 kbps. The guard duration is 16 pulse periods (of the line rate) with a 200 pulse buffer at each end. The cycle time 'T' thus becomes 432 cycles of the master clock comprising of the two-two hundred pulses plus the two-sixteen pulse-guard duration.

Numerous other countries such as Canada [6.23], Italy [6.24], Britain [6.25], France [6.26], Sweden [6.27], Germany [6.28], and Norway [6.29] have documented some trials with the TCM systems. However, the BRISDN system, at a customer rate of 144 kbps with adaptive echo cancellation (AEC, discussed next) and 2B1Q code, has the best transmission performance characteristics, and has survived as the sole winner for almost all digital subscriber lines (DSL) at the lower rates.

6.4.4 ECHO CANCELLATION

In the *fourth* mode or the adaptive echo cancellation (AEC), continuous bidirectional data (i.e., duplex mode) exists over the subscriber line. The canceler provide the cancellation signal to minimize the residual echo of transmitted signal from contaminating the received signal. By and large it succeeds, but in many cases (as is seen later) it is not enough and the limits are explored in this subsection.

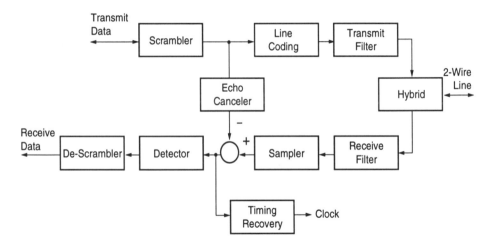

FIGURE 6.14. A Line Termination Unit for the Echo Canceller Hybrid Duplex System.

The principle of operation at either ends of the hybrid duplex system is depicted in Figure 6.14 and the entire echo canceler system with its control of the regenerator and the echo canceler is shown in Figure 6.15.

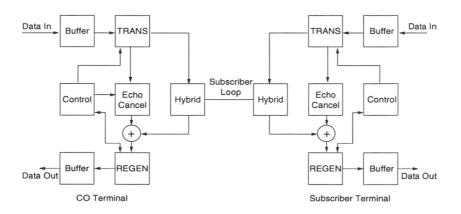

FIGURE 6.15. An Entire Adaptive Hybrid Duplex System for the Transmission of Bidirectional Data into the Subscriber Loop.
REGEN - Regenerator; TRANS - Transmitter

When the signals are not already in a digital format, then A to D and D to A converters become necessary before the scrambler and after the descrambler in Figure 6.15. Some of the isolation of the two signals in the two directions is achieved with hybrid circuits. Both analog and electronic hybrids exist and are discussed further in Section 6.4.5. Generally, a residual echo persists as a residual, yet deterministic, contaminant in the received signal. To the extent that it is deterministic, it is also cancelable. The primary function of the echo canceler is to clean the received signal of the residual echo of the transmitted signal. Perfect echo cancellation is unrealistic, if not impossible. Thus, any trace of the residual echo still contaminates the final signal as it approaches the threshold detection.

All echo cancellation in one device demands an unrealistic dynamic range of its function, i.e., the device has to cancel echoes 60 dB larger than the received signal as well as cancel signals 40 dB lower than the received signal. In this case, the dynamic range from the same device is about 100 dB. Such range of operation is unrealistic, if not impossible, with the current (1990s) device technology, and for this reason, two stage echo cancellation is adopted in HDSL and the VHDSL (if it is in a duplex mode) applications. Traces of the echo, and the noise resulting from the echo cancelers (feed-through and misalignment noise, see Section 6.3.4.6), add to the undeterministic noise in the system causing a bit or a symbol to be in error in rare instances.

With the EC device technology of the late 1980s (that provides up to 60 dB echo cancellation), AEC mode became the ITU-T (formerly CCITT) and ANSI standard for BRISDN in all the loop plants in America and Europe. The advantages of the TCM systems start to lose their appeal as the echo cancelers become cheaper and more dependable. The most rudimentary of bandwidth efficient codes (2B1Q, with one amplitude level [3, 1, -1 or -3] for every two bits) is selected primarily because of its simplicity and ease in implementation.

Echo cancelers are usually tapped delay lines or finite impulse response (FIR) filters. The tap weight adjustment is a continuous process that minimizes the residual echo well within a level 60 to 70 dB below the signal level. In the simpler schemes using amplitude modulation (like the 2B1Q code), the cancelers may be one tapped delay line for modest EC requirement of 50-60 dB cancellation, or two tapped delay lines when there are special requirements, or when additional cancellation is required. The number of delay elements depends upon the persistence of the echo and the sampling rate. Typically, the limit in the EC circuitry is reached by the accuracy of the multipliers and their speed in relation to the data rate. As the data rate starts to increase, the EC circuitry should have enough sophistication to generate the cancellation signal within 70 to 80 dB of its real value precisely at the scanning instant of the symbol. To the extent it fails to meet these two requirements, it leaves behind a trace of the residual echo. This is the residue left behind due to the device imperfection.

In the complex modulation schemes such as CAP, at least two cosynchronous cancelers are necessary to cancel the echo in the real and the quadrature axis. In a majority of the HDSL and VHDSL (duplex mode) applications, two stage echo cancellation becomes necessary, if the loop plant environment has numerous bridged taps with long non-uniform wire-pairs.

In premises distribution systems (PDS), the distances are relatively short (about 100-150 meters) and fiber "riser" architecture permit the data from the curb to be brought into the buildings at very high rates (typically a few Gbps). In such instances, the new modems for the symmetric digital subscriber loop (SDSL) deliver data for multimedia services to the desktop. In essence, the modem technology has taken another quantum jump from rates at 28.8 kbps (V.fast/V.bis rates) to rates between 160 kbps (line rate for BRISDN) to 2.048 Mbps (E1 rate). The new ADSL modem technology permits delivery of data at 6.3 (T2 rate) Mbps. Such modems deliver voice and data over copper PDS systems for video-telephony, medical imaging, or even games and entertainment. The typical mode of its functionality is for symmetric digital transmission. For low rate applications(such as BRISDN), these new modems compete with the traditional ISDN line terminating units.

6.4.5 ADAPTIVE HYBRID WITH AEC

In this *fifth* mode, the magnitude of the echo is reduced by having a impedance adaptation at the two transmitters of the DSL. Electronic hybrid matching [6.30] is preferred over the analog fixed termination hybrid matching.

6.4.5.1 Fixed Termination Hybrids

The analog hybrid with fixed termination is depicted in Figure 6.16. In this Figure, a simplistic and an (overly) approximate formulation for the echo cancellation requirement is outlined. A detailed analysis of the role of hybrids is presented in Chapter 12. Typically, the hybrid exhibits *two* types of losses: *one* for the local transmit signal (L1) and *other* for the received signal (L2).

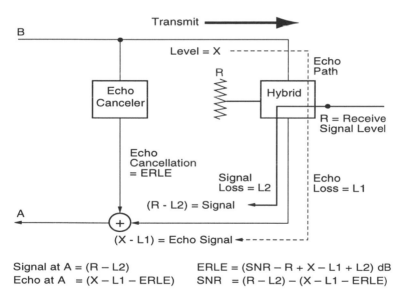

Signal at A = (R − L2)   ERLE = (SNR − R + X − L1 + L2) dB
Echo at A = (X − L1 − ERLE)   SNR = (R − L2) − (X − L1 − ERLE)

FIGURE 6.16. Fixed Termination Hybrid for a Full Duplex Transmission. ERLE - Echo Return Loss Enhancement; SNR - Signal-to-Noise Ratio

If a minimum signal-to-noise ratio level of SNR is required at the center frequency of the received signal (e.g., 40 kHz for the 2B1Q code for 160 kbps,

BRISDN data with a symbol rate of 80 kbaud), then the echo return loss enhancement (ERLE) from the echo canceler can be estimated. as

$$ERLE = SNR + (X\text{-}R) - (L1\text{-}L2)\ dB \qquad\qquad (6.25)$$

where (X-R) is the insertion loss of the line (assuming total symmetry in the system), and (L1-L2) indicates the effectiveness of the hybrid in blocking the transmitted signal (X) from appearing at the output in comparison with effectively conveying the received signal (R).

For the BRISDN applications, the symbol rate is 40 kbuad and the recent EC FIR filter implementations provide an ERLE of about 60 dB relatively easily. In the United States loop environment, if 97% of the loops have a loss less than 40 dB at 40 kHz (see Figure 5.11), then the SNR for the 2B1Q eye-diagrams (due to residual echo only) is $\geq$ 14 dB even with the worst scenarios of a hybrid with +3 dB loss for R (i.e. L2=3) and -3 dB loss for X (i.e., L1= -3). In reality, most well designed analog hybrids are expected to have an L1 - L2 value between +3 to +6 dB. Thus, the SNR for about 97% of the loops (due to residual echo only) will be in the order of 24 dB. Additional noise due to crosstalk and pickup in the wire centers is likely to make the actual SNR less than 20 dB. This method is approximate since other factors such as equalizer tap-weight oscillations, timing jitter in the clock recovery circuit, threshold variations, impulse noise and other devices, such as circuit noise are ignored.

6.4.5.2 Adaptive Termination Hybrids

In the electronic hybrids shown in Figure 6.17, a balancing bridge is deployed. When the line impedance and the matching circuit impedance track well over the band of signal, the echo is greatly (by as much as 30 dB) reduced and thus the dynamic range of the associated echo cancelers may also be reduced by the same extent.

The key element in these hybrids is the presence of a "matching circuit" [6.30] which reduces the echo of the transmitted signal from contaminating the attenuated and dispersed far-end received signal. Perfect match of the impedances of the subscriber line with that of the matching circuit is unrealistic, if not impossible. Generally, the simple configurations depicted in Figures 6.18a, b and c for the matching circuit Z_B (see Figure 6.17) are adequate. For loops without bridged taps circuit 6.18a is sufficient for the HDSL and VHDSL applications, and complex loop structures need circuits 6.18 b and c. Whereas, the circuit depicted in 6.18c provides slightly better performance in blocking the near-end echo for multiple bridged tap loops, its complexity is generally not desirable in the line terminating units for the HDSL applications in the United States environment.

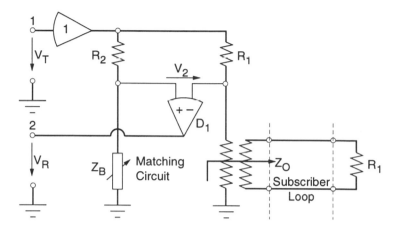

FIGURE 6.17. An Electronic Hybrid Termination with Adaptive Matching Circuit Z_B for Full Duplex Transmission. D_1 - Differential Amplifier; Rn - Resistors for Balancing and Terminating; V_R - Receive Voltage; V_2 - Differential Voltage; V_T - Transmit Voltage; Z_B - Matching Circuit Impedance; Z_O - Line Input Impedance;

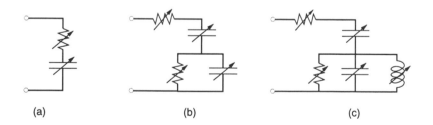

   (a)   (b)   (c)

FIGURE 6.18. Three Configurations of the Matching Impedance of Adaptive Hybrid Termination.

From the simulation studies, it is found that over a 10 kHz to 300 kHz bandwidth for PRIDSN applications (with 5.5 bits per baud), the line impedance matching is most effective (about 30-35 dB) for uniform gauge cable, effective (about 27 dB) for mixed gauge cables, less effective (22-25 dB) for loops with one bridge tap, and least effective (as low as 18 or 19 dB) for loops with two bridged taps. These loop configurations are selected from the ANSI standard loops for CSA environments for 2-wire full-duplex HDSL (i.e., PRISDN) applications.

The matching of the impedance can be done by *two* numerical techniques: *(a)* in frequency domain; or *(b)* in time domain. In the former case, the components for fixed topology impedance (Figure 6.18) network are relaxed iteratively until the highest reflected signal over the entire band, is reduced step by step. Notice that the goal is to minimize the reflections from the loop over the entire band of interest. In the later case, the received cluster is generated by a simulation and then the noise in the cluster is iteratively reduced step by step by relaxing the components of a fixed topology matching circuit. Notice that the goal is to be rid of any noise in the cluster. From our experience, the later approach proved to be more powerful of the two approaches in optimizing the circuit elements of the adaptive hybrid balance network.

Even though the adaptive hybrid circuit matching improves the performance of individual loops to which they are computationally optimized, the technique fails to substantially improve the performance of an entire population of loops, if the loop topology is highly variable. For instance, even though the loop performance of each loop in the ANSI CSA loop data base improved by at least 15-17 dB by individual impedance matching, the overall improvement for all the loops by a single optimized matching circuit varied considerably with a low value of only 3-4 dB for the worst loops to about 17 dB for the simplest loops. On the other hand, when the loop topology is fairly consistent (e. g., the European loops), the improvement was well over 20 dB for all the loops in the data base. These findings are reported in Chapter 12.

6.5

COMPONENT CONSIDERATIONS

6.5.1 EQUALIZERS

ISDN will be realized by deploying the existing telephone lines to reach the customers. Newer networks of communication, such as wireless, fiber nets, local area networks, cable TV networks, etc., to reach subscribers with more elaborate services are being introduced gradually. However, any form of communication over the metallic media is accompanied by *two* distinct effects upon the signal: *first*, attenuation, which leads to the loss of signal level; and *second*, dispersion, which leads to the redistribution of the signal energy from one spectral band to another. When the data is recomposed from the received signal without any form of signal restoration, the recovered signal will be less than satisfactory, making the transmission of the data error-prone. Equalization of the channel-attenuation and dispersion is an attempt to undo the degradation that the signal suffers as it travels the subscriber line. Perfect equalization like perfect echo cancellation is unrealistic, if not impossible. *Two* considerations emerge: *first*, the bandwidth of the original signal; and *second*, the extent of attenuation and dispersion over the band of interest in transmitting the signal in both directions (Central Office to subscriber and subscriber to Central Office).

In the loop plant, an ideal equalizer totally undoes the subscriber line attenuation and dispersion. However, the implementation of this subscriber line inversion device is nontrivial due to the shape of the loss curves for the four (26, 24, 22, and 19) AWG cables in the 10 to 100 kHz band. This transition in the shape of the loss curves is depicted in Figure 6.19 for three wire common wire gauges. In the earlier analog equalizers, designs based on $\sqrt{f}$ loss-inversion algorithm were introduced since the loss curve approximately follows an $\sqrt{f}$ relationship. While this may be the case when the frequency is over 150 kHz, it is not true for the lower band of frequencies of interest for the DSL and HDSL applications. This disparity is illustrated in Figure 6.20. The performance of such equalizers was relatively poor and their use was abandoned in the 1970s.

Subscriber lines can display a wide range of attenuation and dispersion characteristics due to *two* major reasons: (*a*) gauge discontinuities and their associated loss curves; and (*b*) the presence of bridged taps. Both of these are accurately tracked by most of the simulation software. One typical simulation environment is presented in Chapter 9.

Loops offer spectral singularities in the presence of bridged taps. Open-ended (i.e. non-terminated) taps or taps with mismatched terminations return reflection or echoes to any forward signal. Loops also offer temperature dependent loss and dispersion. In the early designs, the effect of increased temperature was approximated by increasing the length proportionately. Except for the design of HDSL (i.e., PRISDN), VHDSL and ADSL where the loop configurations are specifically designed for the American (ANSI) and for the European (ETSI) environments, the simulations need to be carried at the specific temperatures.

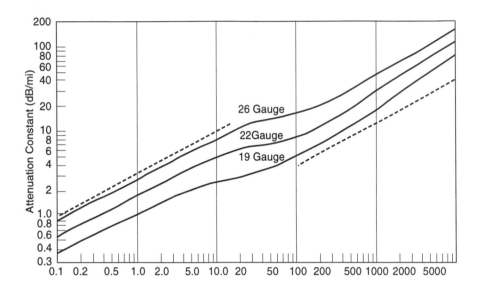

Note: The Dashed lines indicate slopes proportional to √f

FIGURE 6.19. Loss Characteristics of 26, 22 and 19 AWG Cables.

Loops offer spectral singularities in the presence of bridged taps. Open-ended (i.e. non-terminated) taps or taps with mismatched terminations return reflection or echoes to any forward signal. Loops also offer temperature dependent loss and dispersion. In the early designs, the effect of increased temperature was approximated by increasing the length proportionately. Except for the design of HDSL (i.e., PRISDN), VHDSL and ADSL where the loop configurations are specifically designed for the American (ANSI) and for the European (ETSI) environments, the simulations need to be carried at the specific temperatures.

Numerous equalization strategies exist [6.31]. Pass-band, linear and decision-feedback equalizers are grouped as devices having constrained complexity and finite degrees of freedom. Minimum mean-square-error (MSE) solution and MSE gradient algorithm are used to equalize the frequency dependent line loss characteristics with the later two types of equalizers. In the adaptive linear equalizer, stochastic gradient algorithm is commonly used.

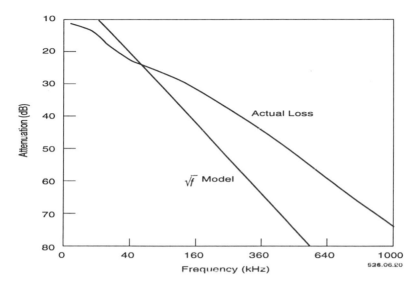

FIGURE 6.20. Actual Loss Curve for a 12 kft 24 AWG Cable and for an Approximation of the Loss.

This method is implemented as the least mean square (LMS) adaptive transversal filter. The MSE is held at it lowest value on a time average basis rather than on an absolute basis. Normalization of the step size and gear shift algorithms are two of the modifications to the original SG approach which may assist in enhancing the MSE.

6.5.1.1 Tapped Delay Line Devices

Decision feedback equalizers (DFE) function by adapting their coefficients to minimize the mean square error.   Finite feed-forward and feedback transversal realizations are used and sometimes necessary depending upon the line to be equalized. Both the   MSE solution and stochastic gradient algorithm are feasible with DFE equalizers.

Fractionally spaced equalizers (FSE) are generally used for the equalization of complex channels for *three* reasons: *first*, matching the transversal or the DFE filter in earlier equalizers may be error prone if not impossible; *second*, the sampling phase become overly crucial; and *third*, a separate matched filter may (and sometimes does) enhance the dynamic range of equalization. FSE on the other hand permit the adaptation of both the matched filter and the equalizer. Drift of the FSE coefficients can cause some instability but the problem can be corrected by saturating the coefficients, or by added white noise at the input to the adaptation algorithms, or by coefficient leakage. However, the later two methods degrade the equalizer performance.

Passband equalizers become important when carrier recovery is involved, as in QAM transmission algorithms. Most of the equalization algorithms used in the baseband applications discussed earlier can also be used with passband equalization. When there is no correlation between the transmitted signal, the convergence properties of the adaptive passband equalizer are the same as those of the adaptive baseband equalizer. In this case, the adaptation is derived from rotated (QAM) symbols and the equalizer indeed inverts the passband channel response rather than the baseband response as it attempted to do in the earlier equalizers. These devices are of no immediate concern in the DSL and HDSL applications except when spectral allocation (e.g., for ADSL) may be of importance.

6.5.1.2 Practical Design Considerations

In order to make the dependable data recovery in the ISDN environment, the line effects are tackled by *two* distinct strategies. *First*, by equalization, the signal degradation in line for the forward transmission of the signal is restored. *Second*, by echo cancellation, the effect of systemic reflections of the transmitted signal that adds to the received signal is accurately subtracted out.

There are two aspects to the line equalization problem. Numerous loop plants (especially in the United States, Canada, Italy, and Australia) have bridged taps. These open-circuited line stubs of variable length offer a branch point in the forward transmission of signals. One component is received at the receiver. The other component travels up the open tap, gets reflected and arrives a little more attenuated, distorted and more importantly, delayed at the receiver. During continuous transmission of data, the delay causes serious concern since the reflected and delayed pulse can occupy the duration in which the succeeding pulses were to arrive. Hence, the line equalizers address the *two* major aspects of signal restoration. *First*, they overcome the line effects. *Second*, they undo the effects of bridged tap reflections. Both these functions of the equalizers have to be adaptive since the ISDN subscriber loop length and configuration are extremely variable.

Equalization can be accomplished in the analog or digital domain. Analog equalizers have been successfully built and deployed especially for the T1, T1C [6.32] and the circuit-switched digital capability (CSDC) systems. However, adaptive analog equalization (using distributed poles and zeros) can be limited at best since the digital signal processing features are absent. Decision feedback equalizers (which are coarse digital type of devices) can offer reasonable adaptability against bridged taps even if their topology is complex. These devices use tapped delay lines with appropriate tap weights to inject a signal equal and opposite to the delayed signal. During the training sequence, these weights are adjusted so that the delays from the taps and/or discontinuities are completely canceled. Any residual effects are continuously monitored such that the average reflections are as close to zero as possible.

Effective equalization for all loops requires generic strategy. In a loop plant where bridged taps prevail, reflections from these open-ended wire-pairs is inevitable. However, the reflection from an average bridged tap can be statistically computed in that particular loop environment; then the basic equalizer can be designed to truncate the tail of the reflected pulse. If this truncation strategy is successfully accomplished at the next scanning instant, the effect of the preceding symbol may be totally eliminated from the symbol recovery of the current symbol. Thus, the extent to which any particular loop deviates from the average loop with an average bridged tap causes an error in the residual tail of the received pulse. The process of total equalization can thus be reduced from that of total tail cancellation to that of the cancellation of the residual tail in the residual pulse. If an elegant algorithm in the adaptation of the residual equalizer is included, then startup and equalization procedures can be significantly simplified. Such a strategy is called for in the detailed simulation of the loop environment.

6.5.1.3 Numerical Design Optimization

Numerical techniques also exist for the design of equalizers when the subscriber loop environment is adequately sampled. *Two* approaches are feasible: *(a)* frequency domain optimization; and *(b)* time domain optimization. A computer aided design facility is deployed for this procedure.

In the design of the analog equalizers with frequency domain optimization, the poles and zeros are distributed along the $j\omega$ axis to create a loss inverse transfer function for a median loop at finite frequencies in the bandwidth of interest for the code deployed. The circuit is then realized by circuit elements (Rs and Cs) with strategic and robust operational amplifiers.

In the design of analog equalizers with time domain optimization, the poles and zeros are distributed along the $j\omega$ axis such that the equalized pulse from a median loop exactly corresponds to the transmitted pulse or an ideal pulse for the detection circuits. The signal is computationally processed to create a time domain response at finite points in the time domain and the optimization process is continued until the least mean square error is achieved to be less than a prescribed number.

In the design of finite impulse response filters (with one overall gain control only, rather than individual tap weight adjustment), the process is repeated to find one set of fixed tap weights for the median loop. These tap weights would then be cast in silicon or ROMs for all the loops in that environment.

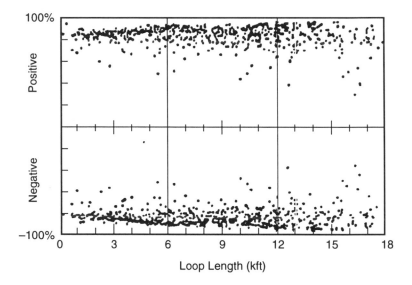

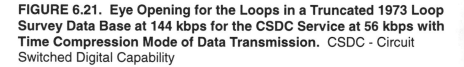

FIGURE 6.21. Eye Opening for the Loops in a Truncated 1973 Loop Survey Data Base at 144 kbps for the CSDC Service at 56 kbps with Time Compression Mode of Data Transmission. CSDC - Circuit Switched Digital Capability

It is to be appreciated that these techniques need an additional final optimization procedure when the initial equalizer design is computationally coupled to the actual loop topologies in the data base to evaluate the performance of the entire population. Such equalizers are adaptive in a limited sense, and can be used where the loop disparity in the plant is not extreme. In a sense, these equalizers perform adequately well for 98% to 99% of the loops, with the ideal performance centered around the "median" loop. When the design calls for the least expensive rather than most adaptive equalizers, such computationally optimized designs can suffice if the loop environment is not extremely diversified.

6.5.1.4 A Case Study of the Equalizer Design

 The results of such an optimization procedure is presented in *two* stages for the loops in the 1973 Loop Survey data base for an analog equalizer for the 56 kbps CSDC TCM system with a line rate of 144 kbps. In the *first* round of optimization, the positive and the negative eyes of the bipolar code are almost one hundred percent open for loops about 7,750 ft. long with an a average bridged tap of about 922 ft. with the initial design. However, in the *second* round of optimization, the goal was to reduce the loop failure percentage to as low as possible (less than 2%) at the slight expense of the 100% eye opening for the median loop. Further, one additional practical consideration was interjected in the CAD process. The optimization was shifted towards the eye diagrams of longer loops, since the signal is more severely attenuated. The result of the second optimization is depicted in Figure 6.21. The Y axis designates the eye opening and the X-axis designates the loop length. As is evident, most of the loops do indeed carry the CSDC data with satisfactory eye openings. Additional results are provided in Reference 6.21.

6.5.2 ECHO CANCELERS

Echoes are a part of the transmitted signal being returned to the sender and thus corrupting the received signal. The effect of any discontinuity in the propagation path results in a reflection. Two such discontinuities presented in Section 6.5.1 are the gauge discontinuities and bridged taps, and both of these are physical. Electrical discontinuities may also prevail when the same AWG wire has pulp insulation rather PVC insulation or when the 24 AWG wire transits from its loop environment to premises distribution systems (Category 5 PDS indoor cabling).

6.5.2.1 Types of Echoes

Four types of echoes can be identified in the metallic transmission media during the transmission of data (see Figure 6.22). *First*, echoes arising from the impedance mismatch between the loop impedance and the source impedance can cause considerable transmission echo signals to be generated at the receiver.

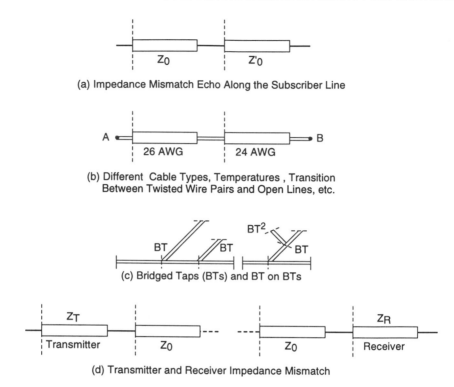

(a) Impedance Mismatch Echo Along the Subscriber Line

(b) Different Cable Types, Temperatures , Transition
Between Twisted Wire Pairs and Open Lines, etc.

(c) Bridged Taps (BTs) and BT on BTs

(d) Transmitter and Receiver Impedance Mismatch

**FIGURE 6.22. Location of Reflecting Points in the Transmission of
Data.**

Second, the reflections caused by the splicing of cables where different
sections of cables are joined together cause echo of signals in one direction to
appear in the reversed direction. *Third,* open bridged tap sections cause
reflections from the distant end of the tap. These echoes are caused by the
forward signal splitting at the junction of the cable and the tap. The signal
down the cable reaches the receiver at the end of the cable. However, the
signal traveling the bridged tap also reaches the receiver as an echo after
being reflected at the open end of the bridged tap. *Fourth,* the mismatch
between the termination impedance and the loop impedance over the
forward signal also causes an echo. It is to be appreciated that the reflection
paths and the corresponding attenuation of the reflected signals will be
different for each of the four type of echoes.

The crucial question in most echo cancellation devices is the duration
over which it has to completely cancel the echo of the previously arrived
sequence of pulses. The bottleneck is generally the arithmetic processing

ability of the digital devices used with the tapped delay line. These FIR filter delay lines generate the signal equal and opposite to the composite echo.

6.5.2.2 EC Design Considerations and FIR Realization

The architecture of an EC chip for the digital subscriber line is shown in Figure 6.23. In the design, there are *four* crucial questions. The *first* question deals with arithmetic processing within the chip.

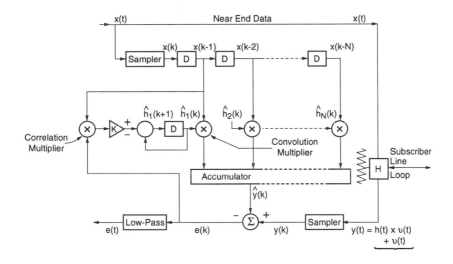

FIGURE 6.23. Internal Structure of Echo Canceler.

In most of the delay line devices, the number of delay periods and the accuracy of the tap weights (i.e., the extent to which the effect of the previous pulse residue will contaminate the present pulse) play a significant role. The number of delay periods indicates the expected length of trail of the farthest pulse that may still linger on to contaminate the signal of the current pulse. The accuracy of the tap weights indicates the extent of signal attenuation in relation to the reflected signal. Both are important in the echo cancellation process. The summing unit of the EC arithmetic unit sums the effect all previous pulse pulses and cancels it out from the signal of the current pulse. In the feed-forward devices, an estimate of the next contamination is also attempted and is also used to enhance the accuracy of the device. The bottleneck is generally the speed and precision of the arithmetic processor unit of the digital devices used with the tapped delay line. This EC device delay line generates the signal equal and opposite to the combined echoes of all the discontinuities in the loop, at finite sampling instants within any pulse period.

Second, the convergence time for the tap weights is also significant in the design of ECs. The convergence time constant is influenced by the sampling steps, the number of delay steps and the variance of the idle channel input noise i.e., the thermal noise in the transmission system, the quantization noise in the A/D converters, etc. *Third* and *fourth,* factors deal with feed-through noise and the misalignment noise, have already been explained in Section 6.3.3.

6.5.2.3 General Considerations for Tapped Delay Line ECs

When the hybrid device is linear and invariant with time, then the time domain representation of the return signal y(t) in Figure 6.23 can be written as

$$y(t) = h(t) * x(t) + v(t) \tag{6.26}$$

where $x(t)$ is the near-end signal, $h(t)$ is the impulse response of the echo path, and $v(t)$ is the far-end signal plus any additive noise in the system. The extent of echo cancellation depends upon the estimation of $h(t)$ by the echo canceler. The sampled and digitized versions of $x(t)$ and $y(t)$ are indeed the sequences $x(k)$ and $y(k)$. Thus, the most recent N samples of $x(t)$ ranging from $x(k)$ to $x(k-N+1)$ are stored in the tapped delay line. They in turn get multiplied by the tap weights $h_1(k)$ to $h_N(k)$ and get added to become the estimate of the echo at a sampling instant k. In terms of the sampled data, Equation (6.26) can be rewritten as

$$y(k) = \sum_{n=0}^{n=N} h_n(k) \cdot x(k-n) + v(k) \tag{6.27}$$

Implicit in Equation (6.27) is the assumption that the impulse response of the echo decays to zero after 'N' sampling times, and thus the finite impulse response (FIR) filter can be constructed out of the 'N' period tapped delay line integrated circuit chip and made an integral part of the EC.

The signal from the accumulator $y'(k)$ can be seen to be the sum of the products of the estimates $\{h'(k)\}$ of ideal tap weights $h(k)$ and the actual samples $x(k)$. This sampled quantity is indeed the same as the first term of Equation (6.27), except that $h'(k)$ are estimates of $k(k)$. The feedback mechanism of the EC circuit adjusts the tap-weights during every sampling period and change it from h'(k) to h'(k+1) by the following constraint

$$h'(k+1) = h(k) + K \cdot x(k-n) \cdot e(k) \qquad (6.28)$$

This Equation (6.28) is essentially a form of relaxation of the tap-weights, which permits them to be modified incrementally by a factor K and the error function e(k). The x(k-n) indicates the transmit level at the $(k-n)^{th}$ sampling instant and is not crucial since the device is a linear device. The term e(k) is the difference between the output from the difference gate Σ (see Figure 6.23) and the ideal value of the signal level at its output. In essence, the device attempts to give rise to ideal levels of the signals at the sampling instants. The factor 'K' also has an influence on the functionality of the EC. Increasing the numerical value of K increases the rate of convergence. On the other hand, if it is too high, it affects the stability of the tap weights, making them noisy and thus increasing the noise at the output of difference gate Σ.

At the implementation stage, the echo canceler circuitry thus consists of a tapped delay line with N delay periods, N tap-weight multipliers and an accumulator, and generates a sampled output y'(k), which is in fact an estimate of the echo at any sampling instant of time k to cancel out the real echo from the subscriber loop. The sampled values of y(k) and y'(k) are subtracted out at the difference gate Σ and if the tap-weight adjustment algorithm is robust, then the error samples e(k) will indeed contain the sampled form of the far-end signal and any additive noise. This signal passes through a low pass filter and can be the input to the first stage analog equalizer. In Figure 6.24, the details of a typical transceiver for the digital subscriber line are presented. The two stages of echo cancellation and two stages of equalization are evident.

A slightly different rendering of the echo canceler can also be implemented [6.37] from the standard cononical form shown in Figure 6.23. In Figure 6.25, a 32-tap filter implementation is shown. In reality, four such filters are cascaded for 128 tap filter for 400 kHz sampling at baud rate for 800 kbps the full duplex H1 rate. A 16-bit accuracy is maintained for coefficients and a 36 bit accuracy (rounded to 24 bits) is maintained for the partial sums. When the tap weights have converged, and for all practical purposes held as fixed coefficients, then the sum at the Output (Figure 6.25) is equivalent to the sum from the Accumulator in Figure 6.23. In Figure 6.25, the coefficients are stored in the device (see Figure 6.26) with a time delay such that the coefficient is at the tap n and time m will be used to generate the partial sum at the n-th position and m-th instant.

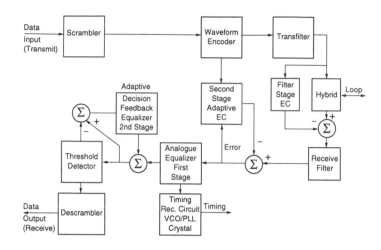

FIGURE 6.24. Typical Transceiver of a Digital Subscriber Line. EC-Echo Canceller; PLL - Phase Locked Loop; VCO - Voltage Controlled Oscillator

The coefficient adaptation board uses a slightly modified LMS algorithm (because of the difference in implementation) and its convergence is demonstrated on the nine CSA loops presented in Section 5.7.1.

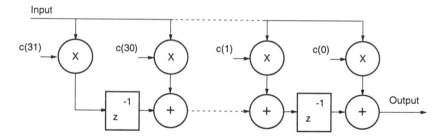

FIGURE 6.25. A 32 Tap Adaptation of the Filter for Echo Cancellation.

Echo cancellation of about 55-60 dB is reported in [6.37] after about 2,800 iterations for Loop 5 from the CSA loops. It is estimated that 13 bit A/D converters and 90 tap baud-rate FIR filters will provide adequate HDSL performance. The word length for the adaptation algorithm is estimated at 23 bits. Fixed point 32 bit accumulator DSP hardware for tap-weight calculation has provided sufficient accuracy.

6.5.2.4 Performance Requirements of EC

The performance of the echo canceler is generally the most crucial requirement that makes or breaks the data transmission in full-duplex systems. The extent of adaptation on the part of this device can be over an immense variety of metallic lines. As indicated in Chapter 5, the copper media especially exhibits wide variation of signal loss and echoes over the frequency band of interest for most of the high-speed digital lines. A wide variety of reflections also exist due to the presence of the bridged taps, multiple bridged taps and bridged taps on bridged taps.

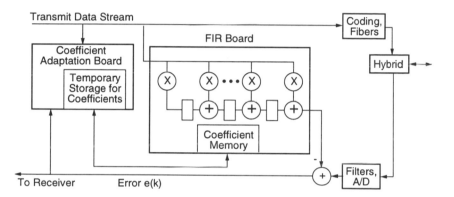

FIGURE 6.26. Implementation of a Prototype Echo Canceler at H₁ Rate for the Dual Duplex PRISDN Systems. A/D - Analog to Digital Converter; FIR - Finite Impulse Response; PRISDN - Primary Rate Integrated Services Digital Network

Such reflections are also prevalent in the wireless environments as the roamer traverses areas surrounded by tall buildings or hilly terrain. The convergence times and the average SNR of the physical layer become crucial in order to satisfy the BER requirements in the network.

The basic rate ISDN channel units at the U interface (see Section 5.2) incorporate the echo cancelers. Numerous vendors (such as Siemens, Ericsson, Northern Telecom, AT&T, etc.) compete closely to mass produce these interface cards and devices with suitable adaptive echo cancelers to meet the requirements of a vast majority of the loops in the loop plant. In most of these devices, the echo cancelers constitute the critical component because of the stringent demands on its capacity to subtract the echo signal to within 0.1 percent or 60 dB of the initial value.

To prevent these reflections from obliterating the received signal, the precise functioning of the echo canceler for a variety of the loops is essential. Its accurate function becomes critical and cancellations in the region of 60 dB become mandatory (in the United States environment) to meet the BRISDN bit error rate (BER) requirement. Longer loops need more cancellation since the echo signal strength remains relatively steady as the received signal becomes more attenuated (and dispersed). In other loop environments, where the loop plant is relatively free from bridged taps (e.g., the Swiss environment), the EC requirements can be relaxed. The limitations of analog signal processing, prevents these devices from being serious contenders for echo cancelers. The digital devices incorporated in tapped delay line filters hold enormous promise provided the data transmission rates are consistently lower than the digital multipliers and adders in the tapped delay lines.

6.5.3 TIMING RECOVERY

The subscriber clock is most frequently derived from the received signal. A typical subscriber does not have access to a local source of accurate and synchronized timing. The recovered clock is then used for transmitting the subscriber data, to shift data in the digital filters used in other devices (such as equalizers and echo cancelers), and also to scan for the received data during an appropriate window of time.

In practice, the subscriber transceiver channel unit relies upon the transition characteristics of the signal that is received. The equalizer (Section 6.5.1) generally precedes the timing recovery circuit (TRC). If the equalizer is functioning correctly, then it would have restored the signal both in its amplitude and against the channel delay-distortion. If the echo canceler is functioning correctly, then it would have removed a very large proportion of the echoes. Thus, the signal at the timing recovery circuits is relatively clean and the transitions are reasonably well placed. The functional requirement of the TRC is to reestablish the clock at which the data was originally encoded at the Central Office and supply it to the local subscriber circuits to be used for data scanning and for transmitting the subscriber data. Some of the other circuits that use the recovered clock are the decision feedback equalizers (DFEs), the finite impulse response filters (FIRs) (see Reference 6.31), the tapped delay lines, and the clocks that are used for signal processing (if any) at the receiver.

It is to be appreciated that DFEs and FIRs for echo cancelers (ECs) depend upon the clock, and the TRC depends upon the accurate functioning of the DFEs and ECs. To break the potential deadlock that leads to a no-win data communication situation, the subscriber circuits need to undergo a finite training sequence before they can function at all. Most ISDN channels resort to this procedure for a finite duration lasting from a fraction of a millisecond to many seconds, depending upon the system and training sequences. These training procedures first ascertain clocking functions under ideal conditions (such as transmitting an alternate binary sequence with a maximal zero-one transitions). They then send single and/or well-defined pulse sequences for equalizer to remove any residual tails from the recovered signals. They then permit the subscriber to be activated such that the echo cancelers may remove most of the reflected signals from the received signals and so on. The training algorithms are generally encoded in the PROMs/EPROMs in the transceiver. Field programmable gate arrays (FPGAs) adapt themselves to the gross environment of the subscriber loops and then the tap weight adjustment algorithms adapt to the dynamic conditions of the line, the temperature variations, etc. These devices that hold the programs to start up the line become an integral part of the transceivers.

The basic methodology in the recovery of time at the subscriber is simple. The partially processed received signal has a certain minimum number of amplitude transitions. Such transitions carry a definite timing stamp at the transmitter. This timing information may get attenuated, distorted and dispersed as it travels down the digital subscriber lines. However, the subscriber transceiver has enough device capability to compensate for the line effects. Hence, the recovery of a subscriber clock becomes that of using the reasonably accurate timing stamps to generate a periodic and well-defined clock. The performance of this circuit is measured as the amount of jitter of the clock. The output of the TRC is generally in degrees or minutes by considering the clock frequency to span 360 degrees. The 144 kHz clock jitter is depicted in Figure 6.27, appearing in the recovered clock during the transmit period of the TCM cycle (see Section 6.4.3) for the CSDC application discussed in this Chapter. There are two important aspects depicted in this Figure. The amplitude starts to decay (Y-axis) and the zero crossing (X-axis) start to wander because there is no received signal during the transmit period. Ideally, the shape of the wave in this Figure should have been a sine wave.

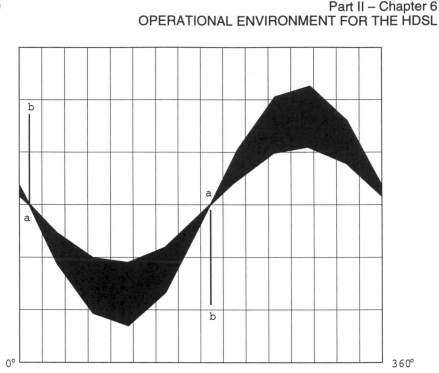

FIGURE 6.27. The Output of the Timing Recovery Circuit for the Loop (from the 1973 Loop Survey) in the TCM, CSDC 56 kbps Data Transmission System. The width of either of the vertical line (a b) indicates the extent of the phase jitter for this system. CSDC - Circuit Switched Digital Capability; TCM - Time Compression Multiplexing

Numerous established filtering techniques are available. The most common class of circuits, known as phase locked loops, exist especially for this purpose. Their function depends upon reinforcing the harmonic of the signal to be recovered in the input of a high gain operational amplifier. These high quality, high gain, operational amplifiers perform to extract the harmonic of interest (i.e., the transmitted clock at the Central Office) and effectively block everything else. Crystal oscillators are often used by making their natural frequency of oscillation correspond to the appropriate ISDN rate. Crystal oscillators are preferred because they are highly stable. System designers have successfully designed every type of TRCs ranging from circuits that use RC (recovery circuits) filters, analog filters (active and passive), digital filters and high quality crystals. Such filters can be conventional, feedback or feed-forward.

6.5.4 OVERALL SYSTEM DESIGN CONSIDERATIONS

There are *four* important parameters in the design of digital transmission systems. *First,* the maximum transmit power, which increases the distance range of the new system, adversely influences the crosstalk in neighboring systems. *Second,* the minimum receive power becomes critical to maintain the essential noise immunity to crosstalk from other systems and the immunity to impulse noise in the Central Offices. *Third,* the loss in signal level depends upon the transmission media characteristics (over the bandwidth of interest) for the particular code chosen to communicate the data. *Finally,* the self-interfering and alien NEXT (near-end crosstalk) and FEXT (far-end crosstalk) becomes crucial in limiting the distance system range.

6.6

SYSTEM COMPONENTS AND THEIR LIMITS

Component limitations are generally explored by simulations. An experimental approach is futile, if not impossible, in view of the topological variations of the loops and component architectures and their design parameters. Most of the HDSL, ADSL and VHDSL designs tend to exploit the components to their limit in order to maximize the bit rate and/or to reduce the cost of the transceivers.

The subscriber line is only one of the several components in high-speed communication systems. One of the approaches to handling many telephone environments is to have an individualized data base and its access for each environment. This approach can also be used for other components since all the system components play a role in the overall performance of HDSL. To provide enough flexibility in the design and study of high-speed data lines, it is necessary to have several models and implementations of each component. For accuracy, it is also desirable to have a specific model for each implementation. Generally, this flexibility is built in the simulation software by organizing each of the major components in a data base of its own. Variation of the individual components is accommodated by a model in the library of devices such as encoders, echo cancelers, equalizers, hybrid matching circuits, timing recovery circuits, decoders, thresh hold detectors, etc. Typically, all echo canceler (EC) models are retained in an EC library. A transversal EC model is implanted in a computational environment by invoking a TEC program module from the EC library. Adequate attention to data and linkage is essential during the overall simulation. Such data bases are discussed further in Chapter 8.

Consider alternative approaches to handle the echo problems in HDSL. The traditional approach of enhancing the echo cancellation requirements leads to diminished marginal return of the echo canceler performance. The alternate approach of matching the line input impedance to the source impedance (at the transmitter) and of matching the line output impedance to the terminating impedance offers substantial promise. The conventional CAD approach calls for a new simulation environment to handle the echo reduction, computation of the residual echo, and to calculate the residual signal-to-noise ratio at the receiver. The problem of optimizing these matching circuits to cover a highly variable topology of the subscriber loops also demands a flexible software organization within the CAD facility. The problem of component optimization is usually coupled to the cost and availability of the VLSI chip sets used in the components. Adaptation of general purpose DSP chips or customized chips also influence how well the components and, in turn the transceivers, will perform.

As a result of such studies, few approaches have emerged and accepted as being suitable to the DSL technology. Only detailed simulation study for each environment can lead to the optimal component designs. Some of the important findings for the American environment have been summarized. *First*, tapped delay lines (FIR filters) rather than adaptive analog devices have dominated most equalizer and echo canceler designs. *Second*, two stage echo-cancellation for the higher rate, long loops is preferred. *Third*, sampling at twice (or higher) the baud-rate is more accepted than baud-rate sampling. *Fourth*, echo reduction (by adaptive impedance matching at the transmitter) and then echo cancellation (by adaptive FIR filtering) is preferred over plain echo cancellation. Phase locked loops for timing recovery are generally more preferred, but are dependent upon the other circuit components and operating environments.

6.7

CODING ALGORITHMS

6.7.1 BIT BY BIT CODES

A survey of codes for the digital subscriber lines has been compiled [6.33]. The simplest of the codes that use bit-by-bit coding with one bit per block is the *alternate mark inversion* (AMI) or the *bipolar* code where every 'one' of any binary sequence of data is coded alternately as a +1 or a -1. The zeros of the binary stream are retained as zeros. Thus, there are three possible levels (+1,

0 and -1) for each bit period. In the *Manchester* code, the transition at the center of the bit period indicates a zero or a one. A transition from -1 to +1 at the center indicates a zero and a transition from +1 to -1 at the center indicates a one, thus the code guarantees a transition (for clock recovery) every bit. The *differential diphase* code retains a transition at the center of the bit period but also in addition, a zero is encoded with an additional transition at the beginning of the bit whereas a one has no such initial transition.

The *dicode* utilizes overlapping symbols. The data stream is passed through a digital filter whose transfer function is 1-D before transmission. It can be seen that this code is very similar to the AMI code because adding a modulo-two adder before the digital filter generates an AMI code. The *modified duobinary* code also uses overlapping symbols. Every information bit leads to a transmitted pulse, but it is also followed by an identical pulse of opposite polarity in a time slot one time interval removed. Consequently, bits for transmission arriving one time slot removed from each other cause overlapping transmitted symbols. Modified duobinary is a ternary code with redundancy and with bit by bit coding without any reduction in the symbol rate. The *biphase* code, like the dicode, has +1 and -1 pulses every bit, but the pulse width is half that of the dicode. For this reason, it does not extend beyond a single bit interval. The biphase is dc balanced and is binary rather than ternary. The bandwidth requirement of the code is high and offers no significant advantage.

6.7.2 BLOCK CODES

The block codes that convert a fixed number of bits into fewer number of symbols according to a prespecified algorithm, offer a reduction in the bandwidth requirement. The amplitude and/or phase of the signal is also encoded. The simplest of the block codes is the 2B1Q code adapted for BRISDN. Two bits of information (00, 01, 10, or 11) are assigned one (3,1,-1, or -3) level of a symbol. The symbol rate is thus half the bit rate and it is considered as a saturated code since all the four possible symbol levels are assigned to the four binary possibilities of the bits. The *3B2T code* converts three binary bits (i.e., 2^3 or 8 possibilities) to one of the two ternary symbols (i.e., 3^2 or 9 symbol pairs). The 00 ternary symbol combination is not used to permit the possibility of all zero transmission. The symbol rate is two-thirds the bit rate, but the symbols have only three (-1, 0, and +1) possible levels. The *4B3T code* casts four binary bits (i.e., 2^4 or 16 possibilities) into 16 of the 27 (i.e., 3^3) possible symbol arrangements. In essence, three, three level symbols are used instead of four binary bits and the symbol rate is thus three quarters of the binary rate. Both the 3B2T and the 4B3T codes were rejected as possible contenders for the BRISDN digital subscriber line by the ANSI and ITU (formerly CCITT) subcommittees.

6.7.3 BANDWIDTH EFFICIENT CODES

As the data rate starts to increase to the T1 and perhaps the T3 rates, additional codes that enhance the bandwidth utilization become attractive. The study of bandwidth efficient Quadrature Amplitude Modulation (QAM) codes indicates that these codes (traditionally used in microwave carrier systems and voice-band modems), have an application in high-speed data transmission. The QAM concept is modified from its traditional context of having the information (symbols) be encoded by the amplitude and phase of a traditional microwave carrier. Being better known as carrierless amplitude phase (CAP) modulation, it qualifies as one of the bandwidth efficient codes viable for HDSL, ADSL and VHDSL applications. The chief advantage of these codes is the lower bandwidth per bit of communicated data. Typically, the bandwidth corresponds to baud-rate and in these codes, a selected number of binary bits are encoded as a symbol.

When the amplitude is held constant at 1.414 (i.e., $\sqrt{2}$), the symbol may have a real (in-phase), and imaginary (quadrature) components, thus leading to the simplest four-point cluster (1, 1; -1, 1; -1, -1; 1, -1) with two binary bits being encoded as one of the four cluster points. If the amplitude can also be used as an information-carrying parameter, then a family of cluster points may be used. Generally, the X (in-phase) and Y (quadrature) components can be encoded independently. This freedom permits three binary bits to be encoded as an 8-point cluster, four bits as a 16-point cluster, or n bits as 2^n point cluster. In practice, the balanced clusters (e.g., 4, 16, 64, etc.) are favored since the transmission medium, i.e., the twisted wire-pairs, can treat the in-phase and quadrature components, as well as positive and negative signal levels, alike. In the application envisaged in this study, 16- and 64-point clusters are considered, which in turn limits the required symbol rate of the transmission medium to about 386 and 257 kilobaud (kbaud), respectively, corresponding to T1 transmission rate (1.544 Mbps), or about 512 and 340 kbauds, respectively, corresponding to the E1 transmission rate (2.048 Mbps) used in Europe.

6.7.4 2B1Q AND PSK CODE COMPARISON FOR TYPICAL LOOPS

Impairments to the transmission of data in the loop plant are presented in Sections 6.5.1, 6.5.2, and 6.5.3. In this Section, the investigations are limited to the effects of line codes and echoes. The major reason for this concentrated area of investigation is to quantify the echo canceler requirements for the data rates exceeding the line rate for BRISDN service. The channel unit echo canceler requirements have been delineated [6.34] and ninety-nine percent penetration of this service capability at 18,000 feet range is established at about 60 dB. No such study has appeared for other enhanced rates with the two competing codes (2B1Q and QAM) with different echo cancellation capabilities.

In this Section, the performance of two major codes to carry HDSL data in three typical subscriber loops is shown in Figures 6.28a, b, and c. The two port analysis techniques (see Chapter 7 or Reference 6.35 for its actual usage in the simulation of digital subscriber loops) provide the individual signals (received signal, and the individual echoes) accurately. The simulation environment used in generating the results presented in this paper incorporates the calculation of each of these echo signals and systematically reduces the amplitude by the extent of echo cancellation specified as the input data.

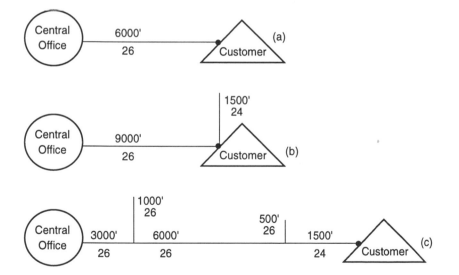

FIGURE 6.28. Three Loop Configurations for the Comparison of 4 PSK and 2B1Q Codes.

In Table 6-3, the S/N ratios for the three loops for the four point constellations at the subscriber and the Central Office are tabulated. In Table 6-4, the S/N ratios for the 2B1Q code are presented. The loop configurations and the symbol rates are identical for simulations for both codes.

TABLE 6-3 S/N Ratios for 4 point Clusters (800 kbps)
(Values are represented in dB)

EC→	40dB		50dB		60dB	
LOOP ↓	S	CO	S	CO	S	CO
Fig.6.28a	19	19	24	24	>30	>30
Fig.6.28b	Fail	8	8	17	17	36
Fig.6.28c	Fail	Fail	4	11	14	20

The reflections tend to affect the energy distribution in the real and imaginary axes equally. The combined effect has become more severe thus decreasing the S/N ratio in the QAM clusters.

TABLE 6-4 S/N Ratios for 2B1Q Code (800 kbps)
(Values are represented in dB)

EC→	40dB		50dB		60dB	
LOOP ↓	S	CO	S	CO	S	CO
Fig.6.28a	22	22	29	29	>30	>30
Fig.6.28b	Fail	10	4	21	12	29
Fig.6.28c	Fail	6	9	14	16	18

Our initial study has indicated that the echo cancellation requirement for the H_0 rate is of the order of 60 dB. This is consistent with the echo cancellation requirement for the basic rate ISDN facility. The penetration is expected to be much lower for the H_0 rate and the number of loops that are likely to carry the H_0 data is also significantly smaller. Since the signal strength is lower at the receiver, crosstalk and impulse noise can become more restrictive than bridged tap echoes. Both QAM-4 and 2B1Q codes suffer severely as bridged taps are encountered. In certain cases, bridged taps affect the QAM codes more drastically than the block codes.

In cluster codes, the Euclidean distance between consecutive symbols plays a crucial role in identifying the symbols. If no form of protection against transmission errors is used, then random noise in the system (which in effect causes a random perturbation in the Euclidean distance between symbols) is likely to cause performance degradation. Hence, in general, some form of trellis encoding is also used. In effect, trellis codes offer protection against random noise, which are likely to shift the position of individual symbols, by placing consecutive symbols in discrete segments of the cluster rather than randomly placing the symbols. In effect, this enhances the Euclidean distance at a small expense of enhanced symbol rate. Typically, a 64 (i.e., 2^6) point cluster reduces the symbol rate to one-sixth the bit rate. Trellis encoding adds one bit to a block of 11 data bits, thus reducing the bits/baud from 6 to 5.5. There are many numerous versions of these trellis codes and are discussed in great detail in Reference 6.31. Newer variations also have been documented in the literature [6.38]. Further simulation studies of the performance of these codes in the CSA loop environment at HDSL rates are presented in Chapter 13.

The impairments to the transmission of high-speed data in the loop environment are presented in this Chapter. The devices and components that are designed to facilitate the higher rates are also presented. By and large, modules for simulation studies need to be assembled before any broad conclusions about algorithms, devices, system performance, or services can be drawn about any particular loop environment. The techniques for analysis and organization of modular software will be discussed further in Chapter 7. The system is configured by selecting the appropriate analysis modules to simulate the components in the HDSL system. The subscriber loop environments are invoked by accessing the appropriate data base for the whole telephone company environment of a major organization, or for any particular country. The self-learning and retention techniques can be invoked by the AI tools embedded in the intelligent CAD that monitors the broad goals of the designer (such as selecting the best design for maximum penetration or maximum range, or highest speed for CSA customers, etc.,). The optimization techniques are invoked by selecting the parameters around which the optimization is based (e.g., maximum echo return loss for a class of

European loops, maximum S/N ratio in an broadly defined loop environment, all loops to have an SNR > 25 dB, etc.).

6.8

REFERENCES

6.1   Bellcore 1987, "Characterization of Subscriber Loops for Voice and ISDN Services: 1983 Subscriber Loop Survey Results," *Science and Technology Series, ST-TSY-000041.*

6.2   Bellcore 1986, "NYNEX Loop Performance Survey: Report of Results," *Loop Survey Report.*

6.3   R.P.S. Singh 1987, "The LAST Bell System Subscriber Loop Survey," *Telephony,* October 5.

6.4   S.V. Ahamed and R.A. McDonald 1985, "Calculation of Insertion Losses in Loops in the 1983 Loop Survey Data Base at Frequencies Relevant for ISDN Basic Access," *T1D1.3 Contribution,* Vancouver, British Columbia, August 19-23.

6.5   S.V. Ahamed and R.P.S. Singh, 1986, "Physical and Transmission Characteristics of Subscriber Loops for ISDN Services," *Record of the ICC '86,* June 22-25, 1211-1215.

6.6   S.V. Ahamed, P.L. Gruber, and J.J. Werner 1995, "Digital Subscriber Line (HDSL and ADSL) Capacity of the Outside Loop Plant," *Special Issue of the 1995 IEEE J-SAC on High-Speed Digital Subscriber Line,* Vol. 13,. 1540-1549.

6.7   Bellcore 1985, "I-Match.1 Loop Characterization Database," *SR TSY-000231,* September 9.

6.8   Ad Hoc Group on Draft Standards 1986,: Composite Draft Standard for ISDN Basic Access for Application at the Network Side of NT1, Layer 1 Specification, *T1D1.3/86-145R1,* October 13.

6.9   AT&T 1985 *Telecommunications Transmission Engineering, Volume 2, Facilities,* New York, NY.

6.10   AT&T, 1985 *Telecommunications Systems Engineering, Volume 3, Networks and Services,* Chapter 19, New York, NY.

6.11   B.S. Bosik and S.V. Kartalopoulos 1982, "Time Compression Multiplexing for a Circuit Switched Digital Capability," *IEEE Transactions on Communications,* Vol. 30, 2046-2052.

6.12   Bell Laboratories 1977, *Engineering and Operations in the Bell System,* Chapter 9, Western Electric Company, New York, NY.

6.13   J.J. Werner 1991, "The HDSL Environment," *Special Issue IEEE J-SAC on High-Speed Digital Subscriber Line,* Vol. 9, No. 6, August, 785-800.

6.14 S.V. Ahamed, P.P. Bohn, N.L. Gottfried 1981, "A Tutorial on Two-Wire Digital Transmission in the Loop Plant," *Special Issue IEEE on Communications*, COM-29, No 11, 1554-64.

6.15 D.A.S. Fraser 1958, *Statistics: An Introduction*, John Wiley and Sons, New York, NY.

6.16 M.M. Sondhi 1967, "An Adaptive Echo Canceler," *Bell System Technical Journal*, 497-511.

6.17 E.G. Nielsen 1957, "Behavior of Noise Figure in Junction Transistors," *Proceedings of IRE*, Vol. 45, 957-993.

6.18 AT&T, 1985 *Telecommunications Systems Engineering, Volume 3, Networks and Services*, Chapter 16, New York, NY.

6.19 B.S. Bosik 1980, "The Case in Favor of Burst Mode Transmission for Digital Subscriber Loops," *Proceedings of ISSLS*, 26-30.

6.20 S.V. Ahamed 1979, "Minimal Delay Rate-Change Circuits," *U.S. Patent 4,316,061*, November, assigned to AT&T.

6.21 M.J. Miller and S.V. Ahamed 1988, *Digital Transmission Systems and Networks, Volume II Applications*, Computer Science Press, Chapter 5, 214-274.

6.22 D.H. Morgan 1975, "Expected Crosstalk Performance of Analog Multichannel Subscriber Carrier Systems," *IEEE Transactions on Communications*, COM-23, February, 240-245.

6.23 E. Aron, E.A. Munter, S.C. Patel, P.A. Roddick and P.W. Willcock 1982, "Customer Access System Design," *IEEE Transactions on Communications*, COM-30, 2143-2149,.

6.24 A. Borosio, V. De. Julio, V. Lazzari, R. Ravaglia and A. Tofanelli 1981, "A Comparison of Digital Subscriber Line Transmission Systems Employing Different Line Codes," *IEEE Transactions on Communications*, COM-29, 1581-1588. Also see J.W. Lechleider 1989, "Line Codes for Digital Subscriber Lines," *IEEE Communications Magazine*, September, 25-32, for line codes. See R. Fossati, S. Galio, V. Lazzari, R. Ravagalia 1980, "A Remote Powered Digital Telephone Set: Problems, Performance and Prospects," *Proceedings of ISSLS*, 60-63, for initial design considerations of the DSL.

6.25 E.C. Vogel and C.G. Taylor 1982, "British Telecom's Experience of Digital Transmission on Local Networks," *IEEE Catalog No., 1686-5/82*, 126-130.

6.26 J-A Le Giillou, F. Marcel and A.J. Schwartz 1982, "PRANA at the Age of Four: Multiservice Loops Reach Out," *IEEE Transactions on Communications*, Vol. COM-30, 2185-2210.

6.27 J-O Anderson, B. Carlquist, and G. Nilsson, "A Field Trial with Three Methods of Digital Two-Wire Transmission 1980," *Private Communication*.

6.28   D. Becher, L. Gasser and F. Kaderali 1981, "Digital Subscriber Loops; Concepts, Realization and Field Experience of Digital Customer Access," *Proceedings of the ISS,* Montreal, Canada. Also see B. Aschrafi, P. Meschkat and K. Szechenyl 1982,   "Field Trial Results of a Comparison of Time Separation, Echo Compensation and Four-Wire Transmission on Digital Subscriber Loops," *IEEE Catalog No., 1686-5/82,* 181-185.

6.29   J. Meyer, T. Roste and R. Torbergsen 1979, "A Digital Subscriber Set," *IEEE Transactions on Communications,* COM-27, 1096-1103,.

6.30   G.S. Moschytz and S.V. Ahamed   1991, "Transhybrid Loss with RC Balance Circuits for Primary Rate ISDN Transmission Systems," *Special Issue of the 1991 IEEE, J-SAC on High-Speed Digital Subscriber Line,* Vol. 9, No. 6, August, 951-959.

6.31   E.A. Lee and D.G. Messerschmitt 1994, *Digital Communication,* Second Edition, Kluwer Academic Publishers, Boston, MA.

6.32   P.C. Davis, J.F. Graczyk, and W.A. Griffen 1979, "Design of an Integrated Circuit for the T1C Low Power Line Repeater," *IEEE Journal of Solid-State Circuits,* SC-14, 109-120.

6.33   J.W. Lechlieder and R.A. McDonald 1986, "Capability of Telephone Loop Plant for ISDN Basic Access," *Proceedings of ISSLS 86,* 156-161.

6.34   S.V. Ahamed and V.B. Lawrence 1987, "An Intelligent CAD Environment for Integrated Services Digital Network (ISDN) Components," *Proceedings of the IEEE International Workshop on Industrial Application of Machine Vision and Machine Intelligence,* February 2-4, Roppongi, Tokyo, Japan, Paper No. 02.

6.35   S.V. Ahamed 1982, "Simulation and Design Studies of the Digital Subscriber Lines," *Bell System Technical Journal,* Vol. 61, 1003-1077

6.36   S.V. Ahamed and V.B. Lawrence 1988, "PC Based Image Processing System for the Two Binary to One Quaternary (2B1Q) Code," *Proceedings of the Southeastern Simulation Conference,* Vol. SESC 88, Orlando, FL, 92-97.

6.37   W.Y. Chen, J.L. Dixon, and D.L. Waring 1991, "High Bit Rate Digital Subscriber Line Echo Canceller," *Special Issue of the 1991 IEEE J-SAC on High-speed Digital Subscriber Line,* August, 848-860.

6.38   H.Y. Chung and S.G. Wilson 1993, "Multi-mode Modulation and Coding of QAM," *IEEE Transactions on Communications,* January.

PART III

THE METALLIC MEDIA: THE COMPUTATIONAL ENVIRONMENT

The third part of the book deals with all the analysis, the computer aided design (CAD) and the data base aspects for the design and analysis of the metallic medium in the data networks. Graphic display of results is essential to facilitate the comprehension of the various design choice. This part answers the design question on broad generic topics such the choice of codes, the line rate, adaptive hybrid techniques, etc. and also at a detailed level, such as the effect a particular bridged tap on a particular loop in the 1983 Loop Survey with 2B1Q (two binary 1 quartenary) code versus the AMI (alternate mark inversion) code. The use of computational facilities for numerical processing, data base management and graphic display is completely exploited to streamline the engineering and design of the metallic media.

There are four Chapters (7 through 10) in this part. Chapter 7 deals with the analytic techniques, the line characterization, the data reduction and display techniques and the use of these techniques in the design of the high-speed digital subscriber line (HDSL). Data bases for the design and engineering of high-speed data transmission systems are discussed in Chapter 8. Data base classification, organization, access and management is also presented from a CAD perspective. The simulation and design of HDSL with more bandwidth efficient codes such as the 2D carrierless amplitude and phase modulation (CAP) techniques is explored in Chapter 9. The options in the North American and European environments are explored for the 1.544 Mbps and 2.048 Mbps services to the subscribers. Finally, the numerical optimization techniques and their effect on the performance of HDSL are explored in Chapter 10.

CHAPTER 7

SIMULATION AND CAD ASPECTS

In this Chapter, an overview of the analysis, numerical and simulation aspects of the high-speed digital subscriber line (HDSL) is presented. The analytical techniques for generation and processing of the *ABCD* matrices are included. Computational data and file handling aspects are addressed. For the physical environment of the HDSL, the typical subscriber loop makeup, the overall topology and its physical distances over which the HDSL data transmission occurs are presented. The important features of the subscriber loops around the world are summarized. This Chapter covers *two* significant aspects of the organization of HDSL: *(a)* computational environment and *(b)* actual physical HDSL environments. The typical features necessary for a flexible approach to the yet evolving HDSL systems are retained in the software design. The evolving integrated network services to the subscribers open a variety of fundamental questions regarding their digital capabilities so far considered unimportant for the loop environments around the world. This software organization permits the designers to probe different national telecommunication environments regarding their status and compatibility to carry ISDN data.

7.1

INTRODUCTION

The detailed analysis of uniform transmission lines to carry telephone voice signals dates back to the fifties [7.1]. The computational approach was streamlined to handle regular transmission lines problems in the voice network at Bell Laboratories during the seventies [7.2]. The mainframe environments dominated most analyses and studies during this time. In the mid seventies, when minicomputers were becoming increasingly popular, a simulation effort was undertaken to study [7.3] the digital capability of the loop plant in the United States. The basic analysis of the uniform transmission line was enhanced to include all the inconsistencies in the topology of the wire-pairs in the actual loop environments.

The primary aim in establishing the simulation software for any network facility is to evaluate the system performance with the various transmission techniques and with the design of associated components. The software designer has to contend with (a) the analysis of electrical circuits (e.g., filters, equalizers, echo cancelers, etc.), that surround the transmitter and/or receiver, (b) the systematic simplification of the transmission medium (the digital subscriber loop in this case) by suitable network reduction techniques, (c) the manipulation of data bases to retrieve the pertinent loop and component characteristics and store away the computed results for further analysis for building a knowledge base, and (d) the organization of the graphic display files to present the performance characteristics in the appropriate format as it is sought by the designer.

The seventh ITU-T (formerly CCITT) Plenary Assembly has defined ISDN as an evolution from the existing telephone network that provides end to end digital connectivity. A wide range of services is anticipated. Both voice and non-voice services will be offered. The network users will have access to these services via the standard multipurpose customer interfaces. ISDN network simulation plays a significant role in the services and the capabilities of the future networks.

The digital revolution, the modern computing environments, and the competition in the telecommunication industry have contributed to the development, use and application of sophisticated CAD techniques for HDSL. From a network consideration, the HDSL is the last and most variable link between the evolving digital networks and the eventual user of the data. The primary consideration in the investigation revolves around evaluating the highest digital rate between the user and the Central Offices in the subscriber loop plant. More recently, this question has reoccurred in the premises distribution systems in order to evaluate the maximum rate (up to 155 Mbps) of data transmission over wire-pairs within large industrial buildings and office complexes. It is noteworthy that this question was asked (and is still being asked) to maximize the trunk capacities between Central

Offices, thus maximizing the digital throughput within the communications network.

In light of emerging multimedia services and intelligent network functions, it is necessary for the vendors of network components to ascertain that the components can perform under widely varying copper data distribution systems. In addition, the vendors need to ascertain that the sensitivity of the components is not likely to cause excessively high bit-error rates. Experimental verification under laboratory conditions with every possible loop impairment is impractical, if not impossible. For this reason, computer simulation studies provide a significant amount of confidence regarding component performance. This approach is highly desirable, because installation of poorly designed components can seriously mar the network performance. Such components can cause bottlenecks in the complete utilization of the network over a long period of time. The simulation and design approach for networks are practical and feasible, as has been documented in the VLSI industry. The network industry stands to benefit greatly by having access to standard design algorithms and sophisticated data bases.

The major validation of vendor components is based upon studying the accuracy of data transmission in the subscriber network. The systems are verified by experimental test setups and the system performance is evaluated by counting the bit-error rates in the (Central Office) noise environments, and measuring the eye openings. Specialized (mostly computer based) test equipment exists for such evaluations. In this Chapter, the study is confined to the computational approach in evaluating eye diagrams or signal constellations and the role they play in the design and optimization of systems.

7.2

SIMULATION ENVIRONMENTS

In the typical computer aided design (CAD) facilities, there are *three* phases of interaction with the designer. The *first* phase consists of defining the system configuration. The transmission components are assembled by a set of interactive graphic routines. These routines spell out to the designer all types of components available in the component library. The system configuration is then specified as the user selects the components (such as equalizers, echo cancelers, timing recovery circuits, etc.) from the library according to the network function. The system then presents the user with vendors in the data bases for each of the components selected. The vendor choice for each component is user defined input. This procedure is also carried out in an interactive graphics mode. The system permits the user to backtrack at any stage and change the components or the vendors selected. Once the system is configured and vendor selection is complete, the system offers the user the choice to generate a hard copy of the final HDSL transmission system before simulation starts.

During the *second* phase dealing with the simulation of the entire system, the user is actively informed of the status of the simulation procedures and any errors resulting from user inputs or posted data base access. Recoverable errors are tackled by the system and non-recoverable errors terminate in the HDSL CAD facility. A large variety of analysis programs are activated in the second reentrant series of computer programs. The organization of the programs that constitute the simulation of the actual data transmission process in the physical medium is elaborate and hierarchical. Interfacing of the various software modules also need considerable care.

The *final* phase permits display (and generation of the hard copy) of the system performance. This phase also operates in an interactive mode. The system offers the designer to see the time domain results as wave shapes at any of the nodes in the HDSL system. It can also display the eye diagrams as repetitive cyclic excitation occurs at the transmit end of the network. Finally, the designer can track signals and their strengths at different nodes in the system (such as line output, the equalized signal with and without crosstalk, etc.). This feature makes the simulation system extremely useful as different vendors of equalizers, echo cancelers, or terminations offer slightly different characteristics from one batch to the next or offer improved system components. Likewise, the different cable manufacturers offer different grades of cables with slightly varying primary characteristics and crosstalk coupling. The user is capable of determining precise optimal combination of components to build any HDSL system. The system performance can be studied and ascertained in any amount of detail. With respect to price and inventory and available data bases, the system yields cost and timing information for any typical HDSL installation in a given geographical area.

7.2.1 ANALYSIS ASPECTS

Twisted wire-pairs generally serve as the physical media for HDSL and ADSL application. From transmission line theory dating back to 1900, the effect of uniform cable in view of voltages and currents have been evaluated. These signals suffer attenuation (or loss in power) and dispersion (or signal degradation), losing disproportionate amounts of energy (from one band of frequencies to the next). Quantification of these two effects is crucial to evaluate the information content in the received signal at the end of a long transmission line. The basic concept behind the design of these data transmission lines is to retain the integrity of data as it traverses long cable sections with an acceptable probability of error (typically 1 in a 100 million). The attenuation and the dispersion of twisted wire-pairs are accurately characterized by the real and imaginary components in the *ABCD* matrix representation.

7.2.1.1 *ABCD* Representation

A uniform cable section is a series of infinitely small elements. Each of these elements has *four* basic electrical parameters, i.e., resistance (R), inductance (L), capacitance (C), and leakage conductance (G). Typically, these four parameters change from cable to cable and from one frequency to the next. However, they are supposedly uniform (within 1%) over the entire length of the uniform cable and are designated as R, L, G, and C per unit length, respectively. These four parameters are the primary cable characteristics. The first, second and fourth of the primary characteristics exhibit a strong frequency dependence over the band of spectral interest for HDSL and ADSL applications. In the computational environment one of the widely accepted ways to consider the frequency dependence is to assign a data base for each of the cables in a typical national subscriber loop environment and direct the simulation process to find the appropriate data as the computations progress. The other basis of representing the frequency dependence is by using published algebraic relations among the four primary constants and frequency. In the earlier applications, where the spectral band of interest is limited (for example in the circuit-switched digital capability at 56 kbps in the United States, or BRISDN throughout the world) the algebraic approximation have provided accurate enough results to build robust transmission systems. However, data bases for primary cable characteristics provide an alternative for PR ISDN at T1/E1 rates, premises distribution systems (up to 155 Mbps), and premises LANs.

In cable manufacturing through the mid-eighties, uniformity of the primary characteristics was within tight tolerances to ensure consistent cable characteristics. More recently, special cables (AT&T grade 5, and others) have been fabricated to distribute twisting of the twisted wire-pairs in a reasonable and methodical pattern that overcomes random crosstalk coupling prevalent in these cables. Considerable reduction of crosstalk is claimed and verified by laboratory measurements in building LAN environments. Since near-end crosstalk (NEXT) coupling sets the ultimate range of very high-speed data. Designers are faced with a certain random distribution of the primary constants in order to reduce NEXT coupling in cables. The compromise effects (i.e., gains for data rate due to reduced NEXT versus the uncertainty in exact characterization) are not completely documented. In these cases, the *ABCD* values would have a statistical variation and the transmission line theory for such cables accommodates a standard set of values and a fuzzy perturbation around these values.

7.2.1.2 Derivation of the Basic *ABCD* Matrix

This matrix is derived for a single uniform section of any twisted wire-pair. The voltage and current relationships for a homogeneous metallic line for data transmission are solved as follows.

7.2.1.2.1 Voltage and Current Relationship at Point x

These classic transmission equations presented in this subsection have been in the literature since the transmission of electrical signals and power was initiated. They do not play as vital role a in the power transmission arena, but dominate the signal flow in the copper media, especially since these lines can be open and the reflected signals are a continuos source of concern over the period of bidirectional communication.

Consider a uniform transmission line (Figure 7.1) carrying data signals from a known point at the transmitter to any location at a distance x from the transmit end.

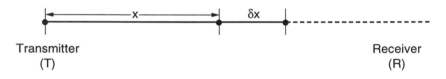

FIGURE 7.1. An Infinitely Long Transmission Line.

If its primary constants are held constant for its entire length, at R Ohms per unit length, L Henries per unit length, G mhos per unit length, and C Farads per unit length, then the voltage and current relationship for the elemental length δx can be derived. *Two* cases are considered. *First*, when the transmit voltages and currents are known, then voltage and current at any point at a distance x can be derived. This is done in this Section 7.2.1.2.1. *Alternatively*, when the voltage and current at any point x are known, then the transmit voltages and current can be derived. This is done in Section 7.2.1.2.2.

Voltage and current at any given point x along a uniform homogeneous data transmission line can be written from Kirchoff's Current Law[*] (see Figure 7.1)

$$I - \left[I + \frac{\partial I}{\partial x} \cdot \delta x \right] = VG \cdot \delta x + C \frac{\partial V}{\partial t} \delta x$$

(7.1)

$$-\frac{\partial I}{\partial x} = VG + C \frac{\partial V}{\partial t}$$

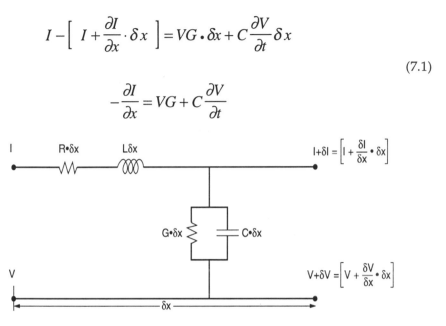

FIGURE 7.2. Voltage and Current Relationship at the Boundaries of an Infinitesimally Small Unit of Length δx.

Similarly, Kirchoff's Voltage Law (Figure 7.2) leads to

$$V - \left[V + \frac{\partial V}{\partial x} \cdot \delta x \right] = \left[IR \cdot \delta x + L \cdot \delta x \frac{\partial I}{\partial t} \right]$$

(7.2)

$$-\frac{\partial V}{\partial x} = IR + L \frac{\partial I}{\partial t}$$

Differentiating Equation (7.2) with respect to x

---

[*] Note that the differential with respect to time is taken for current and voltage relations only. The differential with respect to distance x is for evaluating the potential difference and gradient along the transmission line.

$$-\frac{\partial^2 V}{\partial x^2} = \left[R + L\frac{\partial}{\partial t} \right] \cdot \frac{\partial I}{\partial x}$$

Substituting Equation (7.1)

$$\frac{\partial^2 V}{\partial x^2} = \left[R + L\frac{\partial}{\partial t} \right] \cdot \left[G + C\frac{\partial}{\partial t} \right] V \qquad (7.3)$$

The $\dfrac{\partial}{\partial t}$ operator can be replaced by p or $j\omega$ at any given frequency f with

$$\omega = 2\pi f.$$

$$\frac{\partial^2 V}{\partial x^2} = [R + j\omega L][G + j\omega C] \cdot V$$

$$\qquad (7.4)$$

$$= \gamma^2 V$$

The general solution for Equation (7.4) is of the form

$$V = Pe^{\gamma x} + Qe^{-\gamma x} \qquad (7.5)$$

where P and Q are constants that need to be evaluated and γ is defined as the propagation constant of the line. Differentiating Equation (7.5) we have,

$$\frac{\partial V}{\partial x} = \gamma \left[Pe^{\gamma x} - Qe^{-\gamma x} \right].$$

From Equation (7.2)

$$\frac{\partial V}{\partial x} = -I[R + j\omega L].$$

Thus leading to the value of the current as

$$I = \frac{\gamma}{(R+j\omega L)} \cdot \left[-Pe^{\gamma x} + Qe^{-\gamma x} \right]$$

from Equation (7.4).

Now

$$I = \frac{\sqrt{(R+j\omega L)(G+j\omega C)}}{R+j\omega L} \left[-Pe^{\gamma x} + Qe^{-\gamma x} \right]$$

and we have

$$\frac{1}{Z_0} = \sqrt{\frac{G+j\omega C}{R+j\omega L}}$$

where $Z_0 = \sqrt{\frac{R+j\omega L}{G+j\omega C}}$ and defined as the characteristic impedance of the line.

Thus,

$$Z_0 I = \left[-Pe^{\gamma x} + Qe^{-\gamma x} \right]$$

$$V_x = Pe^{\gamma x} + Qe^{-\gamma x} \qquad\qquad (7.6a)$$

$$I_x = \frac{-Pe^{\gamma x} + Qe^{-\gamma x}}{Z_0} \qquad\qquad (7.6b)$$

Applying the boundary conditions for the two Equations (7.6a and 7.6b) leads to the input voltage and current. At the transmit end of the line we have[*]

$x=0$, voltage $V_{x=0} = V_1 = P+Q$ and current $I_{x=0} = I_1 = \dfrac{-P+Q}{Z_0}$;

$P = (Q - I_1 Z_0)$. P and Q can be thus rewritten as

---

[*] Note that we can solve for V_x and I_x in terms of V_1 and I_1 when x is positive.

$$P = V_1 - Q = V_1 - (Z_0 I_1 + P)$$

$$(7.7a)$$

$$P = \frac{V_1 - I_1 Z_0}{2}$$

$$Q = V_1 - P = V_1 + Z_1 I_1 - Q$$

$$(7.7b)$$

$$Q = \frac{(V_1 + I_1 Z_0)}{2}$$

Substituting for P and Q from Equations (7.7a) and (7.7b) in (7.6a) and (7.6b) we have

$$V_x = \left\{ \frac{V_1 - I_1 Z_0}{2} \right\} e^{\gamma x} + \left\{ \frac{V_1 + I_1 Z_0}{2} \right\} e^{-\gamma x} \qquad (7.8a)$$

and

$$I_x = -\left\{ \frac{V_1 - I_1 Z_0}{2 Z_0} \right\} e^{\gamma x} + \left\{ \frac{V_1 + I_1 Z_0}{2 Z_0} \right\} e^{-\gamma x} \qquad (7.8b)$$

From Equations (7.8a) and (7.8b)

$$V_1 \left\{ \frac{e^{\gamma x} + e^{-\gamma x}}{2} \right\} = V_x + I_1 \frac{\left[e^{\gamma x} - e^{-\gamma x} \right]}{2} Z_0$$

$$I_1 \left[\frac{e^{\gamma x} + e^{-\gamma x}}{2} \right] = I_x + \frac{V_1}{Z_0} \left\{ \frac{e^{\gamma x} - e^{-\gamma x}}{2} \right\}$$

or the voltage and current at any point x can be written in terms of input voltage, current and the characteristic impedances as

$$V_x = \left[\ V_1 \left[\ \frac{e^{\gamma x} + e^{-\gamma x}}{2} \ \right] - I_1 Z_0 \left\{ \frac{e^{\gamma x} - e^{-\gamma x}}{2} \right\} \ \right]$$

$$I_x = \left[\ -\frac{V_1}{Z_0} \left\{ \frac{e^{\gamma x} - e^{-\gamma x}}{2} \right\} + I_1 \left[\ \frac{e^{\gamma x} + e^{-\gamma x}}{2} \ \right] \right]$$

Rewriting we have V_x and I_x in terms of V_1 and I_1,

$$V_x = V_1 \ Cos \ h\gamma x - I_1 Z_0 \ Sin \ h\gamma x \tag{7.9a}$$

and

$$I_x = I_1 \ Cos \ h\gamma x - V_1 / Z_0 \ Sin \ h\gamma x \tag{7.9b}$$

7.2.1.2.2 Voltage and Current Relationships at the Transmit End

To obtain the values of V_1 and I_1 in terms of V_x and I_x, we make $x=-x$ and $V_x = V_1$, $I_x = I_1$ and $P = P'$; $Q = Q'$ in Equations (7.6a) and (7.6b).

From Equation (7.6a) if $x = - x$

$$V_1 = P'e^{-\gamma x} + Q'e^{\gamma x}$$

$$= V_x \left[\ \frac{e^{-\gamma x}}{2} + \frac{e^{\gamma x}}{2} \ \right] + I_x Z_0 \left[\ \frac{e^{\gamma x} - e^{-\gamma x}}{2} \ \right] \tag{7.10a}$$

$$= V_x \ Cos \ h\gamma x + I_x Z_0 \ Sin \ h\gamma x$$

From Equation (7.6b) $x = - x$

$$I_1 = \frac{1}{Z_0}\{-P'e^{-\gamma x} + Q'e^{\gamma x}\}$$

$$= \frac{1}{Z_0}\left\{V_x\left[\frac{-e^{-\gamma x} + e^{\gamma x}}{2}\right] + I_x Z_0\left[\frac{e^{-\gamma x} + e^{\gamma x}}{2}\right]\right\}$$

(7.10b)

$$= \left\{I_x \, Cos \, h\gamma x + \frac{V_x}{Z_0} \, Sin \, h\gamma x\right\}$$

$$= \frac{V_x}{Z_0} \, Sin \, h\gamma x + I_x \, Cos \, h\gamma x$$

The general two port network representation of the transmission line leads to

$$\begin{matrix} V_1 \\ I_1 \end{matrix} = \begin{vmatrix} A & B \\ C & D \end{vmatrix} \begin{matrix} V_x \\ I_x \end{matrix}$$

or

$$V_1 = A \cdot V_x + B \cdot I_x \tag{7.11a}$$

$$I_1 = C \cdot V_x + D \cdot I_x \tag{7.11b}$$

hence,

$$\begin{aligned} A &= Cos \, h\gamma x \\ D &= Cos \, h\gamma x \\ B &= Z_0 \, Sin \, h\gamma x \\ C &= (Sin \, h\gamma x)/Z_0 \end{aligned} \tag{7.12}$$

These equations satisfy the law of interchangeability of the input and output ports and $AD - BC = 1$

$$Cos \, h^2 \gamma x - Sin \, h^2 \gamma x = 1 \tag{7.13}$$

The *ABCD* matrix assumes two versions for the two directions of transmission as depicted in Figure 7.3. The values of A and D are identical only if the loop is symmetric and to the extent to which the characteristics viewed from the two ends of the loop differ, A differs from D.

$$
\begin{vmatrix} A & B \\ & \rightarrow \\ C & D \end{vmatrix} \quad \text{and} \quad \begin{vmatrix} D & B \\ & \leftarrow \\ C & A \end{vmatrix}
$$

FIGURE 7.3. Variation of the ABCD Matrix for the Two Directions of Transmission of Signals.

7.2.1.2.3 The Multisection Composite Matrix

The presence of gauge discontinuities in the loop (see Figure 7.4a) requires the cascading of individual *ABCD* matrices for each loop section that is itself homogeneous.

When only gauge discontinuities exist, the composite matrix for the loop is the product of each individual *ABCD* matrix for the corresponding uniform-gauge section. Cascading then implies generating the overall *ABCD* matrix as the product matrix from the Central Office to the subscriber in the order in which the corresponding cable sections are encountered. The order of matrix reduction is important since the cable sections are in the order in which they occur in the loop. The effect is retained in the equations, since the matrix multiplication is also ordered and noncommutative.

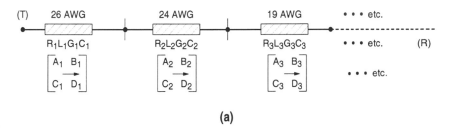

(a)

FIGURE 7.4a. Multisection Loops. R - Receiver; T - Transmitter

7.2.1.2.4 Bridged Taps

When bridged taps are encountered in a loop (see Figure 7.4b), it becomes necessary to derive the ABCD matrix for that bridged tap (taking into account any discontinuities that it may have). The composite bridged tap ABCD matrix is then referred to as $A'B'C'D'$. For inclusion of the bridged tap, which is in parallel with the loop sections, it becomes necessary to compute its input impedance Z_{in} from the composite $A'B'C'D'$ matrix and its load impedance Z_x.

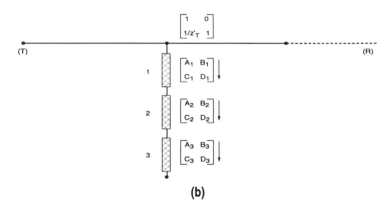

(b)

FIGURE 7.4b. Bridged Tap Cable Sections in Loops. R - Receiver; T - Transmitter

In analogy to Equation (7.1), we obtain $Z_{in} = (A'Z_x + B')/(C'Z_x + D')$.. However, the terminating impedance of the bridged tap is infinite, since the bridged taps are generally left open. Hence, with Equation (7.3) we obtain:

$$\underset{Z_x \to \infty}{Lim} \{Z_{in}\} \to (A'/C') = Z_0' \; cotanh \; \gamma l. \tag{7.14}$$

Thus, the ABCD matrix of the bridged tap, represented by its input impedance Z_{in} across the loop, has the form:

$$[ABCD]_{BT} = \begin{vmatrix} 1 & 0 \\ \dfrac{1}{Z_{in}} & 1 \end{vmatrix} \tag{7.15}$$

where, $1/Z_{in} = Y_0'$ tanh γl, and

$$Y_0' = (1.0 + j0.0) / Z_0' = \sqrt{(C'D')/(A'B')} \tag{7.16}$$

and

$$\gamma l = \Gamma = \text{Complex } [\log \{0.5 * (A' + D'$$
$$+ B' / Z_0' + C'Z_0')\}]. \tag{7.17}$$

In Equation (7.16) Z_0' is the characteristic impedance, and in Equation (7.17), Γ the propagation constant of the entire bridged tap. Similarly, the values of A', B', C', and D' are the values of the entire bridged tap. In this way, the matrix reduction continues in the presence of loop multiple discontinuities and/or multiple bridged taps with multiple discontinuities. The composite $ABCD$ matrix is obtained by normal matrix multiplication techniques for the entire subscriber loop with one or more bridged taps. (See Figure 7.4c.) Recursive computer programming for the matrix reduction saves on the program code. Sometimes this becomes an issue for computer systems with limited amount of memory, but by and large it is not a serious issue.

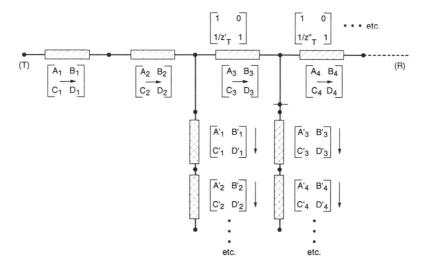

FIGURE 7.4c. Loops with Cascaded Bridged Taps and Multisections.

7.2.1.2.5 The Multiple Bridge Tap Matrix

When bridged taps upon bridged taps are encountered (see Figure 7.4d), the matrix reduction proceeds in a similar way. However, considerations concerning computational efficacy begin to dominate the derivation of the matrix reduction algorithm. In principle, the $ABCD$ matrix of the parallel impedance of the outermost bridged tap is derived from its individual $ABCD$ matrix. This, in turn, is cascaded with the inner bridged tap matrix.

Next, the new matrix of the bridged tap upon bridged tap structure is generated and cascaded with the rest of the loop. This process is repeated throughout the entire loop configuration. An algorithm for generating the general $ABCD$ matrix for any loop whose topology can be represented as a tree structure has been derived from these basic principles, and was used for the simulation of attainable transhybrid loss using this proposed bridge termination.

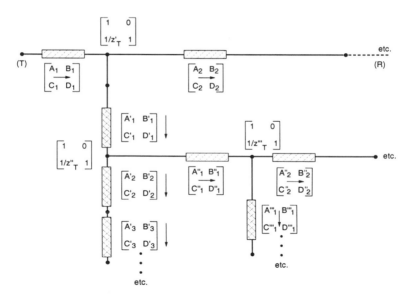

FIGURE 7.4d. Multiple Bridged Tap Matrix Reduction.

7.2.1.2.6 The Composite $ABCD$ Matrix

The entire loop matrix is the product of appropriate matrices in the order in which individual sections or bridged taps occurred in the loop.

Composite cables are represented by the product matrix derived as the segments of individual sections of uniform cable that the signal encounters as it traverses down the composite cable. When the final and composite $ABCD$ is thus derived, the other electrical properties, such as the loop loss, the transmission delay, the input impedance, etc., can be derived from the following relationships.

7.3

LINE CHARACTERIZATIONS

Line Impedance: The line impedance Z_1 depends on the line termination impedance generally designated Z_T. Typically, it is held constant at 110 or 135 Ω as shown in Figure 7.5a.

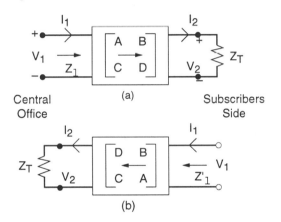

FIGURE 7.5. Evaluation of the Line Impedances Z_1 and Z_1' for the Central Office and Subscriber Side.

Accordingly, the line input (or line) impedance Z_1 from the Central Office to the subscriber may be written as

$$Z_1 = (AZ_T + B) / (CZ_T + D) \qquad (7.18)$$

and Z_1', the line impedance from the subscriber to the Central Office, as

$$Z_1' = (DZ_T + B) / (CZ_T + A) \qquad (7.19)$$

where A, B, C, and D represent the four (complex) elements of the composite $ABCD$ loop matrix. The elements A, B, C, and D for a single-gauge homogenous line with no discontinuity have been derived (Equation (7.12)) as

$$[ABCD] = \begin{vmatrix} Cosh\ \gamma l & Z_0\ Sinh\ \gamma l \\ \dfrac{1}{Z_0}\ Sinh\ \gamma l & Cosh\ \gamma l \end{vmatrix} \qquad (7.20)$$

where γ and Z_0 are the propagation constant* and characteristic impedance, respectively (also called the secondary constants), derived from the four primary constants R, L, G, and C of the homogeneous line of length l units.

7.4

TRANSFER FUNCTION OF THE LOOP

The transfer function of a loop loss of a loop is calculated from the $ABCD$ matrix of the loop. Since the $ABCD$ matrix varies from the Central Office and the subscriber side, the loop loss also varies for the two sides. If the subscriber loop is represented in a two port network shown in Figure 7.5a with V_1 as the input voltage, V_2 as the output voltage, I_1 as the input current, and I_2 as the output current, then

$$\begin{bmatrix} V_1 \\ I_1 \end{bmatrix} = \begin{bmatrix} AB \\ CD \end{bmatrix} \begin{bmatrix} V_2 \\ I_2 \end{bmatrix} \qquad (7.21)$$

or

---

* If R is in Ω/unit length, L in Henries/unit length, G in mhos/unit length, and C in Farads/unit length, then

$$\gamma = \sqrt{(R + j\omega L)(G + j\omega C)}, \text{ and}$$

$$Z_0 = \sqrt{(R + j\omega L)/(G + j\omega C)}\ \Omega$$

at an angular frequency of $\omega (= 2\pi \cdot f)$ radians/second or at f Hz. It should be noted that the real part of γ gives the attenuation coefficient in nepers/unit length and the imaginary part yields the phase angle in radians/unit length.

$$V_1 = AV_2 + BI_2 \tag{7.22}$$

$$I_1 = CV_2 + DI_2 \tag{7.23}$$

Also,

$$V_2 = I_2 Z_T \tag{7.24}$$

$$V_1 = AI_2 Z_T + BI_2 \tag{7.25}$$

$$= I_2 \left(AZ_T + B \right) \tag{7.26}$$

$$\frac{V_2}{V_1} = Loop\ Transfer\ Function\ = \frac{Z_T}{AZ_T + B} \tag{7.27}$$

The loop loss can be represented in decibels as $10 \log \left[\dfrac{V_2}{V_1} \right]^2$ or 20 log

$\dfrac{V_2}{V_1}$. The loop loss increases as the signal level decreases; voltage V_2 starts to decrease in such a case.

In most cases, the source voltage and source impedance are known. Hence, Equations (7.21) to (7.26) can be modified to source impedance and its voltage as shown in Figure 7.6a. Hence, the relations can be written as

$$V_1 = (A Z_T + B) . I_2 \tag{7.28}$$

and

$$I_1 = (C Z_T + D) . I_2 \tag{7.29}$$

The source voltage V_s is

$$V_s = I_1 . Z_s + V_1 \tag{7.30}$$

Substituting for V_1 and I_1. we have,

$$I_2 = V_s \Big/ [(A Z_T + B) + (C Z_T + D). Z_s]. \tag{7.31}$$

Since $V_2 = I_2\ Z_T$, the new transfer function from the Central Office to subscriber (see Figure 7.6a) can be written as

$$V_2 / V_s = Z_T / [(A Z_T + B) + (C Z_T + D). Z_s] \qquad (7.32)$$

and from the subscriber to Central Office (see Figure 7.6b), the transfer function can be written as

$$V'_2 / V'_s = Z'_T / [(D Z'_T + B) + (C Z'_T + A). Z'_s]. \qquad (7.33)$$

Generally, Z'_T and Z'_s of Figure 7.6b is Z_s and Z_T respectively of Figure 7.6a and the subscriber side transfer function becomes

$$V'_2 / V'_s = Z_s / [(D Z_s + B) + (C Z_s + A). Z_T]. \qquad (7.34)$$

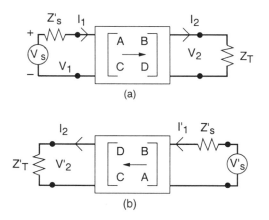

(a)

(b)

FIGURE 7.6. Equation for Duplex Signal Transmission in the Subscriber Loop.

Loss can be computed at specific frequency. Insertion loss of the loop (by definition) is

$$IL = 10.0* \log [\{\text{Real Power without the loop}\}/\{\text{Real Power with the loop inserted}\}].$$

The real power without the loop is

$$P_0 = [V_s / (Z_s+Z_T)]^2 . \text{ Re} \{Z_T\} \qquad (7.35)$$

and the real power with the loop is

$$P_1 = \{V_s / [(A Z_T + B) + (C Z_T + D). Z_s]\}^2 . \text{ Re} \{Z_T\} \qquad (7.36)$$

The insertion loss thus becomes

$$IL = 20 \log [(A Z_T + B) + (C Z_T + D). Z_s] \Big/ [(Z_s + Z_T)] \, dB. \qquad (7.37)$$

In this equation, the *ABCD* constants can be evaluated for the loops in the data base and the transfer functions and the losses can be determined for different values of source and terminating impedances for the characterization of different loop plants.

7.4.1 THE NUMERICAL APPROXIMATION

In the simulation process, these *ABCD* matrices have to be evaluated at fixed frequencies within the band of spectral interest. With the study, analysis or optimization is performed in the spectral domain, these discrete frequencies are specified by the user. However, when the optimization is being carried in the time domain, the discrete frequencies are specified by the number of points on the exciting signal and the data rate. Such a signal can be very, very long or very, very short depending upon the specific study. Typically, the user may specify 8 or 16 points per pulse and there may be 1,024 pulses to 4,096 pulses. When the product of the number of points per pulse and the number of pulses in the exciting signal duration is retained as a binary number, the execution time for performing the time to frequency domain (the forward Fourier transform), and then from the frequency domain to time domain (next Section), can be reduced considerably.

Further, while the discrete frequencies at which the *ABCD* matrices are either specified or computed harmonics, the exact primary constants of the cable may not be stored in the permanent data bases. Hence, the primary or the secondary cable constants at the required frequencies are interpolated from those stored in the data base. Linear interpolation is generally adequate. If an algebraic approximation is used, then these constants are readily obtained from the equation.

7.4.2 FROM FREQUENCY DOMAIN TO TIME DOMAIN

Fast Fourier transform (FFT) algorithm was presented by R. C. Singleton in 1968 to compute a real and complex Fourier transforms [7.4]. The algorithm is accurate and computationally efficient. For these reasons, fast Fourier transforms are used in simulation programs to convert signals from time to the frequency domain. The following Section provides a detailed background of the algorithm in context to the simulations. For the simpler codes involving one real parameter (such as the real amplitude of an excitation signal and its variation with respect to time), the FFT of a real variable (as opposed to the FFT of a complex variable) will suffice. These simpler FFTs are invariably used for simpler codes such as AMI, 2B1Q, Manchester, block codes, etc. The computation is considerably faster than that for the complex FFTs and significantly faster than that for plain Fourier Transforms. In this Section, the details of the complex FFT are presented and the real FFT is a simplification of the complex FFT presented here.

The fast Fourier transform (FFT) is an efficient method for computing the transformation

$$\alpha_k = \sum_{j=0}^{n-1} x_j \exp(i2\pi jk / n) \tag{7.38}$$

for $k = 0, 1, \ldots, n-1$, where $\{x_j\}$ and $\{\alpha_k\}$ are both complex-valued and i is defined $\sqrt{-1}$. The complex Fourier transform can be expressed as a matrix multiplication

$$\alpha = Tx \tag{7.39}$$

where T is an $n \times n$ matrix of complex exponential

$$t_{jk} = \exp(i2\pi jk / n) \tag{7.40}$$

The matrix T can be decomposed as

$$T = PF_m F_{m-1} \ldots F_2 F_1 \tag{7.41}$$

where F_i is the transform step corresponding to the factor n_i of n and P is a permutation matrix. The matrix F_i has only n_i non-zero elements in each row and column and is partitioned to n/n_i square submatrices of dimension n_i.

The matrices F_i are further factored to yield

$$F_i = R_i T_i \tag{7.42}$$

where α_j is a diagonal matrix of rotation factors and T_i can be further partitioned into n/n_i identical square submatrices (each is a matrix of complex Fourier transform of dimension n_i).

The permutation P is required because the transformed result is in digit-reversed order, i.e. the Fourier coefficient α_j with

$$j = j_m n_{m-2} \dots n_1 + \dots + j_2 n_1 + j_1 \tag{7.43}$$

is found in location

$$j' = j_1 n_2 n_3 \dots n_m + j_2 n_3 n_4 \dots n_m + \dots j_m \tag{7.44}$$

The permutation can be performed in place by pair interchanges if n is factored so that

$$n_i = n_{m-i} \tag{7.45}$$

for $i < n\text{-}i$. In this case, one can count j in natural order and j' in digit-reversed order, then exchange α_j and α_j if $j < j'$.

7.4.2.1 The Rotation Factor

The rotation factor R_i following the transform step T_i has diagonal elements

$$r = \exp\left\{ i2 \frac{\pi}{kk} (j \bmod k)[j \bmod \frac{kk}{k}] \right\} \tag{7.46}$$

for $j = 0, 1, \dots, n-1$ where

$$k = n / n_1 n_2 \dots n_i \quad \text{and} \quad kk = n_i k \tag{7.47}$$

and the square brackets [ ] denote the greatest integer function. The rotation factors multiplying each transform of dimension n_i within T_i have angles

$$0, \theta, 2\theta, \dots (n_i - 1)\theta$$

where θ may differ from one transform to another.

7.4.2.2 Decomposition of a Complex Fourier Transform

The complex number $(a_k + ib_k)$ can be decomposed as

$$a_k + ib_k = \sum_{j=0}^{p-1} (x_j + iy_j)[\cos(2\pi j \frac{k}{p}) + i\sin(2\pi j \frac{k}{p})] \tag{7.48}$$

$$= x_0 + \sum_{j=1}^{p-1} x_j \cos(2\pi j \frac{k}{p})$$

$$+i\left\{ y_0 + \sum_{j=1}^{p-1} y_i \cos(2\pi j \frac{k}{p}) + \sum_{j=1}^{p-1} x_j \sin(2\pi j \frac{k}{p}) \right\} \tag{7.49}$$

$$= x_0 + \left\{ y_0 + \sum_{j=1}^{(p-1)/2} (x_j + x_{p-j})\cos(2\pi j \frac{k}{p}) + \sum_{j=1}^{(p-1)/2} (y_j - y_{p-j})\sin(2\pi j \frac{k}{p}) \right\}$$

$$+i\left\{ y_0 + \sum_{j=1}^{(p-1)/2} (y_j + y_{p-j})\cos(2\pi j \frac{k}{p}) + \sum_{j=1}^{(p-1)/2} (x_j - x_{p-j})\sin(2\pi j \frac{k}{p}) \right\}$$

$$\tag{7.50}$$

for k = 0, 1, . . ., p-1.

The remaining Fourier coefficients can be expressed as

$$a_k = a_k^+ - a_k^- \tag{7.51}$$

$$a_{p-k} = a_k^+ + a_k^- \tag{7.52}$$

$$b_k = b_k^+ + b_k^- \tag{7.53}$$

$$b_{p-k} = b_k^+ - b_k^- \tag{7.54}$$

for k = 1, 2, . ., (p - 1)/2, where

$$a_k^+ = x_0 + \sum_{j=1}^{(p-1)/2} (x_j + x_{p-j})\cos(2\pi j \frac{k}{p}) \tag{7.55}$$

$$a_k^- = \sum_{j=1}^{(p-1)/2} (y_j - y_{p-j}) \sin(2\pi j \frac{k}{p}) \tag{7.56}$$

$$b_k^+ = y_0 + \sum_{j=1}^{(p-1)/2} (y_i + y_{p-j}) \cos(2\pi j \frac{k}{p}) \tag{7.57}$$

$$b_k^- = \sum_{j=1}^{(p-1)/2} (x_j - x_{p-j}) \sin(2\pi j \frac{k}{p}) \tag{7.58}$$

Altogether there are $2(p - 1)$ terms in the series to sum, each with $(p - 1)/2$ multiplications, for a total of $(p-1)^2$ real multiplications.

Typically, these equations are embedded in subroutines invoked during the computational process. It is necessary for the system specialist to pay particular attention to the linkage and the exchanges of data/files that accompany the computations.

7.5

CONVENTIONAL FACILITIES

Computer simulation techniques to verify the experimental results from the subscriber loop tests were introduced in the late seventies [7.2]. The hardware and software techniques were quite rudimentary and data bases were simply files generated for use during the simulation procedure. Now the optimization parameters are increasingly complex. For dealing with enhanced data rates, bidirectionality, adaptive echo cancellation, equalization and timing recovery, etc., the design and management of data bases become crucial. It is essential to study the overall system performance to make sure that the results and comparison are accurate, dependable and consistent with other simulations.

Time domain simulations that lead to eye diagrams and eye statistics are essential in studying system performance. *Two* techniques exist. The *first* technique is based on calculating the system performance over a short finite duration of time. Initial conditions are defined, and a predefined sequence of inputs generates the system response. When steady state response is desired, the input sequence (of repetitive pattern) is made arbitrarily long and the output is obtained after transient conditions have settled down.

In the *second* technique, the entire system is excited by the repetitive input sequence of an arbitrarily long duration. The boundary conditions at the start and finish (of this input sequence) are matched and the system is trapped in a cycle of repetitive performance. The input sequence and its duration being user defined, permits the simulationist to study the system performance under a wide variety of input conditions. The accuracy of this technique depends upon the number of Fourier components and the sampling interval of the input sequence. These user parameters are generally programmed and offer the simulationist ample flexibility. In the results presented in this Chapter, the later technique has been used, with the number of excitation pulses ranging from a minimum of 48 pulses (with a sampling frequency of five points per pulse) to 16,384 pulses with one sample per pulse, or 2,048 pulses with 8 samples per pulse.

For the 2B1Q code the line rate is 192 kbps for BRISDN services. Two bits are mapped into one symbol, thus leading to a symbol rate of 96 kilosymbols per second. The symbol rate simply becomes half the bit rate for this code.

The simulation employs (typically 96 and up to 2048) 'n' random symbols or '$2n$' (typically 192 and up to 4096) binary bits for each run. Fast Fourier transforms did not prove to be particularly time saving for very short sequences of symbols. In these simulations, a major fraction of the time is spent in the reduction of the *ABCD* matrices for the entire loop at each of the harmonics.

The individual harmonics of the excitation function are computed and stored away for simulations on numerous loops in the CSA data base. To conserve computer memory, the computation uses *three* overlays. The *first* overlay generates the random symbols and computes the harmonics. The *second* overlay simulates the system performance, and generates the received signals and reflections. The *third* overlay depicts the results as an eye diagram or eye statistic.

The development, deployment and effective utilization of routine CAD for the simulation and design of DSL has been documented [7.5,7.6]. Effective as it has been, it still leaves a number of basic questions (impulse noise conditions, new IC technology, component aging, etc., with respect to their effects), unanswered. It is our contention that HDSL CAD environments will become unified to handle all transmission media, all digital signal processing algorithms and IC chip design techniques. Eventually, the flow of data in the network will be as dependable and secure as the flow of bits within an IC chip.

The ultimate ceiling on the rate and distance for data transmission is the superfluous noise in the network that will obliterate the signal. In addition, the HDSL components are imperfect. For optimal system design, it becomes crucial to investigate the margin for these components, such that they are neither prohibitively sophisticated (and expensive) nor are they under-designed. Such frontiers can be effectively investigated with the recent CAD facilities.

7.6

PROCESSING OF EYE DIAGRAMS

The algorithms for processing the eye diagrams of the alternate mark inversion (AMI) code exist [7.2]. This code is prevalent in the circuit-switched digital capability (CSDC) offered for 56 kbps and also in the T1 carrier system. In Japan, almost all the Nippon Telephone and Telegraph loops are satisfactorily covered by the time compression mode, basic rate ISDN data transmission over the subscriber loops. However, the algorithms are no longer applicable for the 2B1Q code selected for the basic rate ISDN by the Bell Operating Companies. Hence, it becomes necessary to modify the AMI eye-processing algorithms (see Section 7.6.2.1.1) and discuss their limitations. The compromise between disk storage space requirements, displayed information, and the number of scatter plots necessary to convey all the significant information to the systems' designer are highlighted.

The line code selected for BRISDN (2B+D) has been standardized as the 2B1Q block code. In addition, the local characteristics of the loop environment can also vary significantly from one country to the next, and from one geographical area to the next. Further, the local switching environment is not consistently uniform to use the 'D' channel signaling entirely. For this reason, the line rates for BRISDN can differ significantly until the circuit-switched environment is totally unified to handle the CCS 7 signaling protocol. Other universally accepted rates also exist for the subscriber loop environments.

The transmission of data for the emerging Integrated Services Digital Network calls for extensive simulations. The ISDN rates have been standardized at 144, (2B+D); 384, (5B+D); 1,544, (23B+D); and 2,048, (30B+D) kbps with signaling on delta or 'D' channel. Various subrates of 2.4, 4.8, 9.6, 14.4, 19.6 kbps also exist for the local loop plant. In addition, the V.fast and V.bis modems carry data at 28.8 and 32 kbps. More recent modems carry data at 2.048 Mbps in the HDSL mode, 6.3 Mbps for the ADSL mode and even 51.6 Mbps (dual duplex) mode for PDS applications.

The local characteristics of the loop environment can also vary significantly from one environment to the next. Experimental verification of the system performance becomes tedious and time consuming in view of the extremely diversified loop conditions documented in Section 5.6.1 and component characteristics discussed in Section 6.5. Computer simulation studies prove to be viable and the system performance may be ascertained in an intelligent computer aided design environment rather than in the laboratory.

7.6.1 AMI AND 2B1Q CODES IN THE LOOP PLANT

In this Section, the image processing of the eye diagrams obtained with AMI for the circuit switched digital capability (CSDC) and 2B1Q code for the (2B+D) or basic rate ISDN (BRISDN) data is presented. From a simulation point of view, it becomes necessary to manage the information that is displayed. At one extreme, when every detail is depicted, the designer is overwhelmed by the quantity of the data displayed. At the other extreme when the information is too sparse, the system performance is not completely depicted. Hence, the number and types of display should be matched to indicate the major trends in the performance of the system when any particular design step is implemented.

In addition, an enormous amount of computing is required to optimize the system components. Components such as the equalizers, echo cancelers, matching circuits, etc., exert crucial influence on the system performance. These components have to be designed by vendors of channel units at the U-interface of the evolving ISDN. It becomes necessary to validate network performance with the particular component and channel unit that any proposed system deploys. A proven and inexpensive methodology that permits the system designer to automatically process each loop simulation is essential. These techniques are applicable to any simulation environment where large amounts of data have to be systematically processed.

Time domain simulations lead to wave shapes and eye diagrams for the received data. For the AMI code an ideal eye diagram is depicted in Figure 7.7. There are three levels at +1, 0 and -1 and two eye openings between +1 and 0 and between 0 and -1. Since the ideal eye is not realized in the practical loop environment, a degenerate eye diagram shown in Figure 7.8 is generally obtained. In this case, the three levels deviate from their idealized values and form three bands centered at +1, 0 and -1. Superior systems retain integrity of these three levels at the receiver, such that the receiver data is clearly discernible. Hence, the system performance may be measured by the thickness of each band of data.

The image processing algorithms for the AMI code scans the eye diagrams for *seven* eye statistics: *top eye* thickness, *top eye* opening, *central eye* thickness, *lower eye* opening, and *lower eye* thickness and the *two (+ and -) eye heights*. These statistics are normalized to the (numeric rather than algebraic) sum of the two eye heights and stored away in a separate data base. The main simulation program can thus scan many thousands of loop configurations effectively in a single simulation run. The eye statistics are then displayed to indicate broad and significant effects of changing system components like the echo canceler, or the equalizer, etc.

7.6.2 EYE DIAGRAMS AND EYE STATISTICS

Time domain simulation of data transmission through the subscriber loop yields signal pulse shapes at selected points. This signal undergoes a certain amount of signal processing before the data can be recovered. At the subscriber end, the two major aspects of signal processing include equalization and timing recovery. Eye diagrams are generated when the recovered data is superimposed in the same time slot. However, the loop plant consists of a widely diverse collection of loops. If a given component has to perform with a very large percentage of these loops, then significant information is extracted from the eye diagrams of these loops. The eye statistics contain the summary of these significant results.

7.6.2.1 Eye Diagrams and Line Code

The line code influences the shape of the eye diagram. Two major line codes for transmission are the alternate mark inversion (AMI) and the two binary to one quaternary (2B1Q) code. The standard adapted by the Bell Operating Companies is the later 2B1Q code for basic rate ISDN in the United States. For this reason, the techniques discussed are limited to these two frequently used line codes.

7.6.2.1.1 The Generation of Eye Statistics for the AMI Code

The number of pulses and points per pulse both influence the control parameters necessary for the generation of the eye statistics. These two variables are defined and held fixed for the simulation of each loop in any particular system configuration. For the purpose of illustration, we derive the eye statistics for a repetitive pulse duration with 'P' pulses with 'n' points per pulse. There are three signal levels at +1, 0, and -1 in this code and two eye openings are generated.

In the computational environment, a long sequence of AMI data with of 'P' pulses serve as an excitation at the two sides (the Central Office side and the subscriber side) of the loop. Each of these pulses are approximated as a sequence of 'n' straight lines (in the time domain). Out of the 'nP' instants of time for scanning 'P' ones, zeros or minus ones, a subset 3P or 5P instants are selected for the scanning of the 'P' data bits. The first set of 'P' instants contain the instant at which the absolute maximum value of the incoming magnitude is located. The second set of 'P' instants precede the first 'P' instants by $(1/n)$th of the pulse duration and the third set follows the first 'P' instants by $(1/n)$th the pulse duration and so on.

It is also desirable to select the number of pulses and the scanning instants carefully to conserve the computational efficacy of the fast Fourier transforms (FFT). Non-binary products of the number of pulses and the scanning instants per pulse are likely to reduce the algorithmic advantages of the FFT and enhance the computational time. However, it is our experience that the image processing of the generated eyes and scatter plots that is one of the most time consuming activity of any computational systems. Pixel based hard copy generation on color printers is the worst offender for the display of key results. In a sense, the post processing of results (discussed in Section 8.4) is a major concern for the design of the computer based facility for digital subscriber lines. Very long sequence of pulses yields only marginal improvement in accuracy of the numerical results. The computational environment needs to be tailored to the direction of investigation and it needs to be matched to the measuring devices that will be ultimately used in the performance of the DSL systems. In overall systems study involving loop populations, the sampling error is severe limit on the accuracy of the penetration studies.

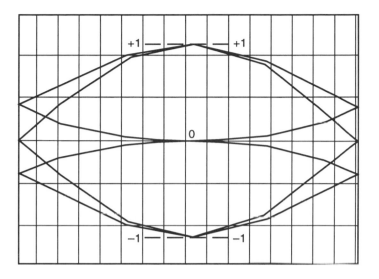

FIGURE 7.7. An Ideal Eye Diagram for the AMI Code. AMI - Alternate
Mark Inversion

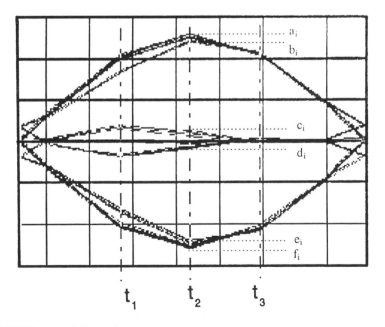

**FIGURE 7.8. A Practical Version of the Eye Diagram for the AMI
Code.** The granularity in the picture is due to the EGA (640X350) display
resolution of the display device.

Next, for each of the three or five sets of 'P' instants for scanning, data is arranged in descending order of magnitude by any one of the numerical sorting algorithms. *Three* clusters of data are generated: (*a*) at the upper level (plus ones) corresponding to the plus ones received in the data; (*b*) at the intermediate level corresponding to the zeros received; and (*c*) at the lower (minus ones) level corresponding to the minus ones received. From these clusters seven eye statistics can be generated. In Figure 7.8, the three scanning instants t_1, t_2, and t_2, and the three data clusters representing the distances a b c d e and f (i = 1, 2, or 3) are indicated. Next, a b expressed as a proportion of the average eye height constitutes the top eye thickness. The distances b c, c d, d e and e f, also expressed as proportions of the average eye height constitute the top eye opening, the central thickness, the lower eye opening, and the lower eye thickness, respectively. Finally, the positive average eye height and negative average eye height constitute the sixth and seventh pertinent eye statistics from the eye diagram at each of three scanning instants t (i= 1, 2, or 3).

However, the scanning instant t is selected to have the maximum average top and bottom eye opening leading to an unique scanning instant and unique set of seven statistics stored away for each eye. Two such (that is, the positive and negative openings) of the seven parameter eye statistical sets are for each of the two eyes. These results are generated for data from the Central Office to the subscriber and for data from the subscriber to the Central Office, and are stored for each loop for generating the eye statistics.

Scanning the eye at three or five rather one instant gives *two* relevant insights into the functioning of the system. *First*, the time differential of the opening (i.e., from t_1 to t_2 as compared with t_2 to t_3) indicates the horizontal eye opening. *Second*, the difference in the positive vs. negative eye indicates the asymmetry of the eye. Generally, any tails in the single pulse response make the eye opening greater at instant t_1 compared to instant t_2. These additional numerical quantities govern the timing recovery circuit on the subscriber side. The display of these quantities depends purely upon the system user.

7.6.2.1.2 Generation of Eye Statistics for the 2B1Q Code

The ideal 2B1Q code has *four* levels at +3, +1, -1, and -3. These levels correspond to three equal eye openings between +3 and +1, between +1 and -1, and between -1 and -3. The idealized eye diagram of this code is shown in Figure 7.9. A realistic version of the eye (Figure 7.10) has finite bands and associated thicknesses at each of the four levels.

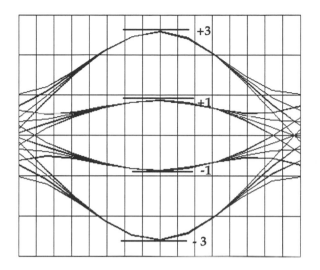

FIGURE 7.9. An Ideal 2B1Q Eye Diagram. The granularity in the picture
is due to the VGA (800X600) display resolution of the display device.

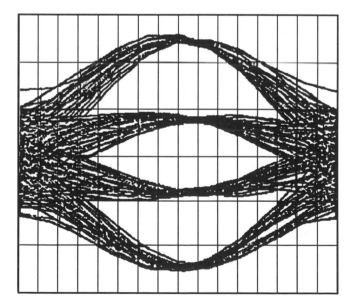

FIGURE 7.10. A Practical Version of the 2B1Q Code Eye Diagram.

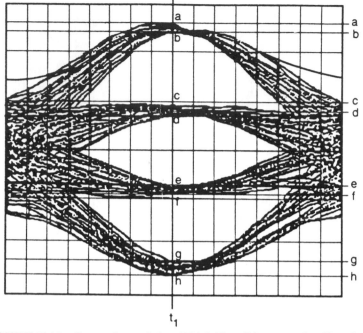

t_1

FIGURE 7.11. Scanning of the 2B1Q Eye Diagram for Eye Statistics.

The number of pulses for the 2B1Q code is typically 200 (with a maximum number of symbols to 4096) and the number of points per pulse is 8 to 16. These two variables are predefined and held fixed for the simulation of each loop in any particular system configuration. However, the 2B1Q code has four signal levels at +3, +1, -1, and -3 and three eye openings are generated. The number of eye statistics increases and the depiction of the composite results needs reappropriation for the entire loop population.

In Figure 7.11, an eye diagram for the 2B1Q code is portrayed. At the maximum eye height at instant t, there are *seven* actual distances: ab, bc, cd, de, ef, fg, and gh that completely characterize the thickness and opening. However, the average heights at the +3, +1, -1, and -3 levels add four additional statistics. There is no elegant way of displaying these eleven statistics even after proportioning the distances to the average values. For this reason, an underlying simplification is made. All the eye diagrams are rescaled to a maximum eye height from +3 to -3, even if the system equalizer cannot perform to yield these ideal values. The seven distances (ab through gh) are then proportioned to generate a composite plot for all the loops.

7.6.3 SCATTER PLOTS FROM EYE STATISTICS

Scatter plots are merely the display of all or a few of the eye statistics for each loop plotted against its length. Shorter loops display the more open eye diagrams and thus the eye statistics for idealized eye openings. On an intuitive basis, the designer would expect to see the degradation of these statistics from short loops to long loops. By and large, a great majority of the simulation results confirm this expectation. However, in some designs (CSDC case study presented in Chapter 4 and also Figure 6.21) where the strategy calls for better performance of the longer loops at the expense of the shorter loops, the scatter plots exhibit a convexity in the display and depict the best performance at the central regions of the loop length. Very long loops suffer a degradation anyway, in spite of the optimal design strategy.

7.6.3.1 AMI Code

A representative eye diagram is displayed (Figure 7.8) that leads to one of the scatter plots (Figure 7.9) obtained by the image processing programs. The scatter plot is generated by displaying the eye opening of all the loops in a data base in one figure. The horizontal axis can depict the loop length, the loop loss, or any parameter that uniquely identifies the loop. Only two (i.e., the positive and negative openings) of the seven statistics are displayed in this Figure. In Figure 7.10, the top thickness, the central thickness, and the lower eye thickness are depicted. Thus two sets of scatter plots display all the necessary information. Figure 7.10 becomes useful in depicting any effects that may help to differentiate eye characteristics of the AMI code. Some of these effects can be viewed, such as the top eye with respect to the lower eye, or the top thickness with respect to central or negative thickness of the eye.

The eye statistics of loops in the data base are generated and stored under a predefined set of terminal conditions. The eye statistics can now be displayed as a collection of scatter plots, each dot denoting a particular eye statistic of a particular loop. The top and bottom eye opening statistics are assembled in one scatter plot (Figure 7.9). Along the X-axis, the physical length of the loop or the equivalent length of 22 AWG cable (or equivalently the loop attenuation) may be plotted. The top central and bottom eye thickness may also be plotted on one scatter plot (Figure 7.10). With two independent directions of transmission a set of four scatter plots are obtained.

7.6.3.2 Composite Plots for the 2B1Q Code

Two types of composite plots (Sections 7.6.3.2.1 and 7.6.3.2.2) are feasible. In the latter types, the system performance can be very tightly depicted, since the effects of echo cancellation are crucial to the success of the BRISDN in any given loop plant. Hence, the study of the 2B1Q code is performed at a much more detailed level than that of the CSDC or the T1 lines. In these latter cases, there is ample margin and the system performance is assured without such detailed studies.

7.6.3.2.1 Linear Scatter Plots

Figure 7.12 depicts a composite 2B1Q plot for the Central Office to the subscriber data transmission for all loops in the data base. Here an ideal 2B1Q eye with zero eye thickness would be represented as four points. The eye thickness in ab, cd, ef, and gh in Figure 7.11 would be four single points at +3, +1, - 1, and -3 levels. However, due to the imperfection of the system vertical lines of finite length appear at the levels. The eye openings start to close for eye thickness in bc, de, and fg in Figure 7.11. A completely closed eye would be a vertical line between +3 and -3.

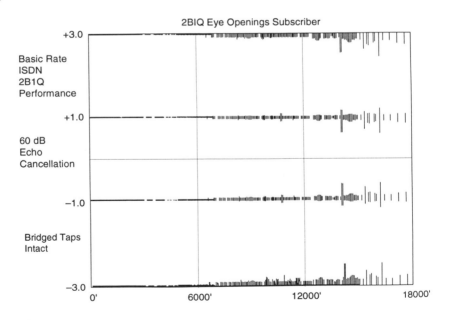

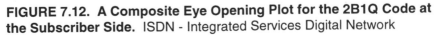

FIGURE 7.12. A Composite Eye Opening Plot for the 2B1Q Code at the Subscriber Side. ISDN - Integrated Services Digital Network

In Figure 7.12, the horizontal axis depicts the loop length. Longer loops display eye closure since the composite plot becomes dense with longer vertical lines in the right side of the display. Two plots are necessary to depict the bidirectional mode of the ISDN data transmission. In Figure 7.13, the composite plot for the subscriber to Central Office data transmission is depicted. Duplex data transmission with adaptive echo cancellation suffers more at the subscriber side because of the increased number of bridged taps. This effect is clear by comparing Figure 7.13 with Figure 7.12. Systematic representation becomes desirable because of the enormous amounts of data generated from each simulation. Typically, the optimization procedure needs as many as a hundred iterations before the components can be mass produced and for this reason, both the representation and management of the results (also see Chapter 8) becomes crucial. Similar computational and data base management techniques are deployed in the VLSI CAD environment. In both situations, the designer interaction is confined to a higher conceptual level and the organization, computation and updating is programmed in the CAD and the graphics routines.

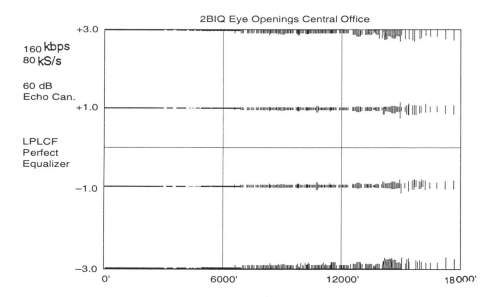

FIGURE 7.13. A Composite Eye Opening Plot for the 2B1Q Code at the Central Office. LFLC - Loop Data Base

7.6.3.2.2 Circular Scatter Plots

ISDN simulations and the component optimization process provide large amounts of significant results. In Figure 7.10, the shape and deformity of the eye diagram with 2B1Q code is displayed as data is transmitted over a typical loop. There are three eye openings in this composite diagram between +3 and +1 levels, +1 and -1 levels and -1 and -3 levels. In 7.14, two sets of three scatter plots of the eye openings with this particular code and 60 dB of echo cancellation for an entire loop population are depicted in a condensed format. As it can be seen from this section, circular scatter plots offer greater representational flexibility than the linear scatter plots and the implications of the echo canceler performance can be more accurately tracked.

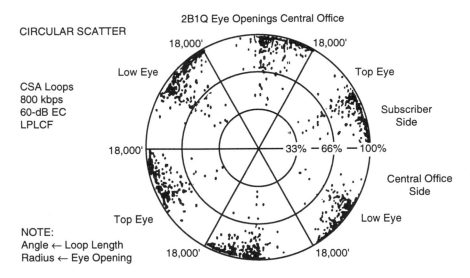

FIGURE 7.14. A Composite Circular Eye Scatter Plot for the 2B1Q Code at Basic Access Rate. CSA - Carrier Serving Area; LFLC - Loop Data Base

The six (60 degree) segments are each depicted on a composite scatter plot. The top half of the plot (Figure 7.14) condenses the information for the Central Office to subscriber transmission. The lower half depicts the subscriber to Central Office transmission. In each of the 60 degree segments, the loop length is represented as the angle and the radial distance represents the eye opening. Ideally the points in the scatter plot should lie on the perimeter of the outside circle for 100% open eye diagrams.

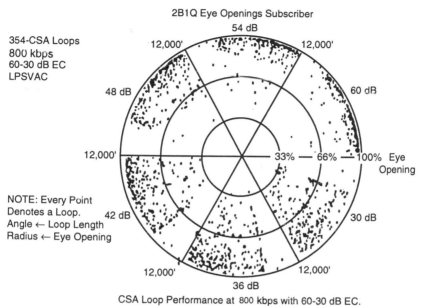

CSA Loop Performance at 800 kbps with 60-30 dB EC.

FIGURE 7.15. The Effect of Echo Cancellation on the Central Eye of the 2B1Q Code at the Subscriber Side. CSA - Carrier Serving Area.

In Figure 7.15, the effect of changing the echo cancellation is depicted. Each of the six scatter plots depict the central eye opening of the 2B1Q code. The entire loop population is equipped with 60 dB, 54 dB, 48 dB, 42 dB, 36 dB, and 30 dB of echo cancellation circuitry. The variety of results and designs becomes enormous. Figure 7.16 depicts the effect of changing the echo cancellation for the data transmission in the opposite direction.

7.6.3.2.3 Signal Degradation and SNR Scatter Plots

The echo canceler for the 2B1Q code has to perform dependably over a very wide range of loops and meet the 60-65 dB cancellation requirement.

Performance variations at such fine levels cannot be depicted by eye diagrams and eye openings, even though they depict gross effects and trends. For this reason, when the components have to be fine tuned, then the degradation of the received and equalized signal becomes crucial. The echo that corrupts the signal depends upon the near-end loop configuration.

2B1Q EYE OPENING AT CENTRAL OFFICE

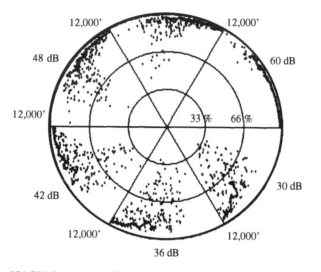

354 CSA loops at 800 kbps. Every point denotes a loop and the (r, θ) coordinates denote the eye- opening and the loop length respectively.

FIGURE 7.16. The Effect of Echo Cancellation on the Central Eye of the 2B1Q at the Central Office Side. CSA - Carrier Service Area.

This signal suffers a steady attenuation and also a degradation, depending upon the entire loop configuration. Sometimes it becomes necessary to isolate the component of the noise and correlate it to that particular component or function in the circuit responsible for the added noise. In these circumstances, the signal-to-noise (S/N) plots become more graphic in their depiction of the cause/effect relationship. In Figure 7.17, the SNR scatter plot for the eye diagram generated at the Central Office side is depicted. A similar plot is generated for the subscriber side. The S/N ratio at the receivers for the 2B1Q code are depicted with a perfect equalizer and an imperfect echo canceler. The X-axis (or the angle θ) depicts the length, and Y-axis (or the radial length r) depicts the S/N ratio in each of the three eyes. It is possible to vary the parameters and components during the actual design process and thus study their effects precisely. When the digital capacity of the loop plant is not exploited to its full capacity, then the component optimization does not have to be optimal. Suboptimally designed components do perform adequately enough in a great majority of the loops. It is the system designer's responsibility to explore the loop penetration in detail or to explore competing technologies such as wireless access, hybrid fiber coax, or fiber to the curb, etc.

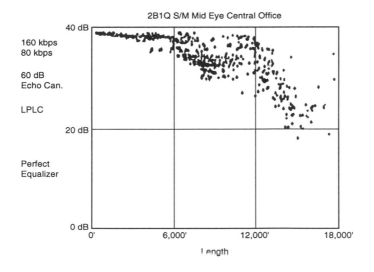

FIGURE 7.17. Signal-to-Noise Ratio (for the Central Eye) Scatter Plot at the Subscriber Side. LFLC - Loop Data Base

7.6.4 DISCUSSION OF THE ALGORITHMS AND THEIR USE

An automatic scanning of the eye evaluates system performance under the highly variable system, and the loop configurations facilitate the comparison of a large cross section of results [7.5]. The algorithms presented in this Chapter generate diagrams and plots that can be easily scanned by the system designer to get a qualitative picture of the acceptance or rejection of a certain system. With this initial inference, the system designer may pursue promising system configurations. When a more detailed evaluation is necessary, additional scatter plots with S/N ratios or circular plots are obtained by supplementary computing and programming efforts.

The time required to execute the algorithms presented in this Chapter represents about 15 to 20% of the total execution time. A considerable part of this time is consumed by disk access and the overall simulation time for the loop. It can be reduced by allocating additional memory in the mainframe environment, or by executing the program in a virtual disk in the PC environment.

The simulation studies that have been attempted incorporate this eye scanning post-processors. This can be for generating eye statistics or for deriving S/N ratio (20 times logarithm to base 10, of the ratio of eye opening to eye closure) for large cross sections of subscriber loops.

7.7

DESIGN CONSIDERATIONS FOR HDSL

The computer aided design of HDSL demands both sophistication and flexibility. Computing environments provide different mixtures of sophisticated algorithmic approaches in the solution of engineering problems, and data base functions for the vendor and component flexibility. Coupled with user-friendly software for the interaction and display of results, the CAD facility becomes a necessary tool for the design of most modern digital communication systems. The recent simulation environments do offer new tools in the design of crucial components for the HDSL, such as echo cancelers, equalizers, timing recovery circuits, detectors, etc. The CAD facility used for the results presented here has an optimal mix of computerized AI functions and designer insight. Computational approach is divided into three categories. Human creativity and insight resist any such categorization.

The mathematical techniques (especially fast Fourier transformations, matrix multiplications, complex algebra, and interpolation) necessary for obtaining the results are well documented [7.1,7.2]. Techniques for systematic reduction of loops with complex topology are well established [7.3]. Overall simulation of the system with inherently non-linear elements especially encountered in the timing recovery circuits is feasible [7.7, 7.8]. In this Section, techniques germane to the manipulation of data bases for computation and graphic display are presented. Special attention is focused upon the incorporation of the artificial intelligence (AI) and expert system (ES) techniques for the intelligent computer aided design (INCAD) environment.

Algorithms follow a hierarchical order. Typical of such a hierarchy is spectral domain computations, time domain calculations, eye opening calculations, eye statistics generation, SNR calculation, etc. A second level hierarchy of these algorithms consists of routines/functions that perform FFT operations, *ABCD* matrix computation, multiple bridged tap matrix reduction, echo signal calculation, equalization, phase jitter calculation, etc.

The realization of any DSL, HDSL, ADSL, or the VHDSL system entails the scientific CAD process for the Physical Layer design. The tools and techniques presented in this Chapter form the core of the component design process for the copper environment. From the OSI viewpoint, such numerous component layers exist. These components may be software modules, specifically designed chips with their specific ROMs embedded within them, or even DSP chips with field programmable gate arrays (FPLAs). The DSP and modular software approach follow their own specific procedures in the system realization. Once the implementational strategy is streamlined for the VLSI approach, *five* dominant levels in the system realization can be visualized.

The *first* level consists of the conceptual domain in which major methodologies for system design and their implementation can be queried, altered, and recast to evaluate feasibility and categorical studies about the rates, modes and access of the DSL (or even wireless) techniques into the backbone network. The result of the first level study is the tentative blueprint of an DSL (or even a wireless) system.

The *second* level consists of compiling and building "an electronic bread-board" of the components that make up the particular system. All component designs, their input/output analyses and their component limitations are accessed from data bases from previous or similar designs. Any recent technological changes are incorporated to reflect the design experience and component performance.

The *third* level consists of evaluating the performance of the "electronic bread-board". This phase is viewed as a *vertical integration* of the component to make up the system. The "time-process" charts for each component and the entire system are developed at this level to ascertain the timing and processing germane to the components selected that make up the system. No clock-cycle and no process is left unaccounted. It is crucial to the overall functioning of the system under the most adverse input-output and/or environmental conditions. The ultimate limit of each component is exercised fully to verify its functionality in context to the system requirements. Signal level study is also performed here to determine the extent of accuracy in the devices and components. Noise problems, if any, are envisioned and controlled at this stage in view of component and transistor noise.

The *fourth* level consists of implementing the components in silicon and designing (or recompiling) the system in a "binary arithmetic" language. Every component (or its binary version from previous designs) is accessed from the appropriate data bases and "copied" into an electronic bread-board of the system. This phase is viewed as a *horizontal* integration of the system. The "time-signal-space (or gate)" relationships on the VLSI chip are completed at this stage. The performance of the chip (which can be the entire system or a subsystem) is validated at a binary level to detect and correct any potential problems or conflicting requirements from the hardware at a gate level.

The *final* level consists of realizing the chip (i.e., fabricating the chip and packaging it) and validating its performance in the actual subscriber loop environments. The DSL, HDSL, ADSL and perhaps the VHDSL is mature enough to realize the silicon version of the system without the need of expensive reiteration of design steps.

AI (Artificial Intelligence) techniques constantly access the previous designs. This activity ascertains previous CAD [7.6] designs can facilitate the current design. This aspect is based upon machine learning from its own interaction with the user, and inference engine rules become applicable. Expert system knowledge base is not compiled for the HDSL components because the HDSL design is too new for "real" human experts from whom design knowledge or experience can be distilled.

7.8

REFERENCES

7.1   F.B. Llewellyn 1952, "Some Fundamental Properties of Transmission Systems," *Proceedings of IRE*, 271-83.

7.2   S.V. Ahamed, B.S. Bosik, N.G. Long and R.W. Wyndrum, Jr. 1979, "The Provision of High-Speed Digital Data Services over Existing Loop Plant," *Proceedings of the National Electronics Conference*, Vol. 33, 265-70.

7.3   H. Seidel and S.V. Ahamed 1978, "Simulation of the Bidirectional Line on Minicomputers," *Private Communication*, June 15, Bell Telephone Laboratories, Murray Hill, NJ.

7.4   R.C. Singleton 1969, "An Algorithm for Computing the Mixed Radix Fast Fourier Transform," *IEEE Transactions on Audio and Electroacoustics*, Vol. 17, 93-103.

7.5   S.V. Ahamed and V.B. Lawrence 1988, "An Image Processing System for Eye Statistics from Eye Diagrams," *Proceedings of the International Association for Pattern Recognition Workshop on Computer Vision*, October 12-14, Nihon Daigaku Kaikan, Tokyo, Japan.

7.6   S.V. Ahamed and V.B. Lawrence 1987, "An Intelligent CAD Environment for Integrated Services Digital Network (ISDN) Components," *Proceedings of IEEE International Workshop on Industrial Applications of Machine Vision and Machine Intelligence*, February 2-4, Roppongi, Tokyo, Japan.

7.7   D.G. Messerschmitt 1984, "A Transmission Line Modeling Program Written in C," *Journal on Selected Areas in Communications*, Vol. J-SAC 2, January, 148-153.

7.8   M.J. Miller, and S.V. Ahamed 1988, *Digital Transmission Systems and Networks, Vol. II: Applications*, Computer Science Press, Rockville, MD.

CHAPTER 8

DATA BASES AND THEIR MANAGEMENT FOR DSL DESIGN STUDIES

The performance evaluation of the high-speed digital subscriber line (HDSL) depends upon the design, linkages, and cooperative role of the extensive data bases during simulation studies. Such data bases are necessary during component design, performance evaluation and the overall feasibility studies of the HDSL at the basic data rate (144 kbps) or at the primary data rate (1.544 Mbps in the United States and 2.048 Mbps in Europe). In this Chapter, the design and procedures to build these interoperable data bases are presented. The need, consistency and interdependence of the data bases that actually supply the simulation programs are presented. The input data to the simulation programs is derived from large national Loop Surveys, the cable characteristics and surveys of noise in the loop plant. The output of the simulation programs generates interdependent and consistent data bases that are essential to evaluate the overall performance of the entire subscriber Loop Survey population at various rates (such as 144, 384, 768 kbps, 1.544 and 2.048 Mbps) in different telecommunication networks of various countries, regions, or even in premises distribution systems. In this Chapter, the success and frustrations in organizing and working with such interdependent and interoperable data bases are also presented.

In addition, the organization of the numerous data bases for effective ISDN simulations and for the design of components is presented in this Chapter. In contrast to the passive computer aided design methodology, an interactive and intelligent CAD environment is discussed. The Chapter focuses upon the use of three categories of data bases. Major loop survey data, crosstalk and impulse noise data, cable characteristics, and relatively static data constitute the permanent data bases. The intermediate computational results, excitation data, Fourier components, and loop responses to impulse noise conditions, etc., which contain information of transitory nature, constitute the interim data bases. Finally, the important simulation results in the form of graphic files that constitute the pictorial data base. Intelligent techniques for managing and accessing these data bases are addressed here.

8.1

INTRODUCTION

Simulation procedures introduced in the late seventies [8.1] assume the role of elaborate computer aided design (CAD) when the components' values used in HDSL design procedures need optimization [8.2]. The consistency and interoperability of the simulation programs play a critical role in the CAD procedure. This aspect has been studied in light of the recommendations of the standards committees, since the study of the digital subscriber line started during the late seventies. The design aspects, data rates, subscriber loop environments, encoding algorithms and actual devices used on the HDSL applications have changed dramatically.

Multiple data bases are essential [8.3] in the optimal design of high-speed digital subscriber lines (HDSL). Typically, on the input side, three or four data bases need to be accessed, and on the output side, three to six matrices of results are generated. The HDSL itself serves as a crucial link in the deployment of integrated services digital networks (ISDN). Most of the Regional Bell Operating Companies and European telecommunication networks are committed to bring the higher data rates to the customer from the Central Offices. There are as many as 5,120 data bases (based upon our designs over the last 10 years) in CAD for HDSL.

An overview of the general purpose CAD software is depicted in Figure 8.1. *Four* major data bases (permanent, component, intermediate and graphical) are activated during the CAD function. The role and usage of these data bases is discussed further in this Section.

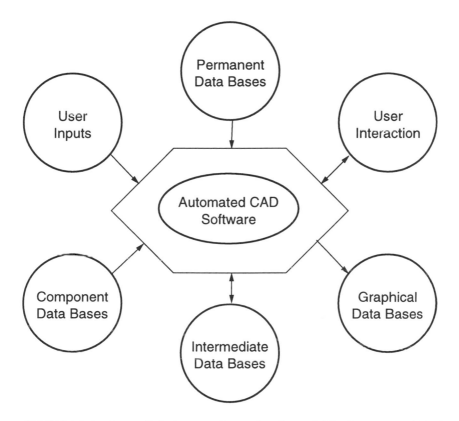

FIGURE 8.1. Overall Software Organization. CAD -Computer Aided Design

User inputs and interaction facilitate the fine tuning of software to the specific application being processed. In particular, this model has been used in most of the software in both copper lines and fiber-optic transmission links (the physical layer of the open system interconnect or the OSI model) used in communication networks. Since the CAD procedure involves the verification of the proposed transmission system for an entire country or countries ranging over numerous subscriber loop environments, the inputs to the central analysis module of the software are derived from national Loop Survey data bases for the copper HDSL environment. In the fiber optic CAD facility, other factors such as multiple vendor fiber characteristics, range of optical characteristics of the devices used, range of fiber splice characteristics, etc., supply the inputs to the central analysis modules of the fiber design capability.

8.2

DATA BASES AND FILE MANAGEMENT

Data bases holding the details of the current environment are essential to ascertain the compatibility of new high-speed lines and new components. The *five* major types of data bases necessary for the design of high-speed facilities are (a) permanent data bases, (b) program data bases, (c) post processing program data bases (d) pictorial data bases, and (f) intermediate data bases. These are discussed next.

8.2.1 PERMANENT DATA BASES

Some of the system components retain specific characteristics for the transmission of data, even though they can be different from country to country or even from one region to another. However, the environmental data that is fairly stationary can be stored in permanent data bases, and intermediate data leading to an optimization or a specific design of a version of the DSL, HDSL, ADSL, VDSL, SDSL, etc. can be stored in over-writable/ transitory data bases. Typical permanent data bases are discussed in Sections 8.2.2 through 8.2.6.

8.2.2 LOOP AND COMPONENT DATA BASES

The variety of these permanent data bases depends upon the extent and nature of the simulations. Truly global ISDN simulation needs the digital subscriber loop characteristics of different countries. The physical and electrical properties of the overall transmission media, the networking and carrier serving features and so on are assembled in these data bases. A generally localized or a national loop data base suffices to initiate a limited study of the ISDN environment. But by and large, the topologies of the loop plant, primary and/or secondary electrical characteristics of the twisted wire-pairs, crosstalk and impulse noise data bases generally constitute the primary data bases for ISDN simulations. In many circumstances, it is necessary to design ISDN components to be functional with those of other vendors. In these cases, it becomes essential to retain the interface characteristics of the vendor products. At a more detailed design level, the vendor data bases have to become more exhaustive requiring the physical, electrical, spatial power, thermal, etc., characteristics of these components. Specialized data bases need to be developed under such conditions when a viable system is designed by using a large number of existing subsystems.

8.2.3 LOOP SURVEY DATA BASES

National Loop Surveys in the United States were undertaken in 1964, 1973, and 1983 for the then existing Bell System. The results pertaining to the design have been published in the sixties [8.4], seventies [8.5,8.6], and eighties [8.7,8.8,8.9]. It is noteworthy that the published results over the three decades are not consistent. The earlier surveys fail to report the features as they would impact the DSL and HDSL capabilities of the loops surveyed. In addition, the Loop Survey data base of the sixties (in its original form) is not consistent with the information in the later data bases. However, it has all the topological information about the loops to be able provide results consistent with the results from the 1973 and 1983 loop data bases. Other telephone companies in the United States (GTE [8.10]) have published their survey results. Countries like Canada, Germany, Australia, and Japan, etc., also have extensive loop survey data bases and similar linkages so that the central analysis module can be established.

8.2.4 CARRIER SERVING AREA LOOP DATA BASE

Three major surveys of the American subscriber loops have been done. The 1964 Survey loop statistics do not significantly differ from the statistics published for the Surveys in 1973 and 1983. The average physical and electrical characteristics for these surveys have been also published. As the DSL technology started to evolve, additional data bases have been generated (especially from the 1973 and 1984 data bases) that exclude loops with loading coils, over 18,000 ft. and otherwise unsuitable for DSL applications. In addition, it is possible to impose the carrier serving area (CSA) constraints on the original data bases and generate CSA loop data bases. The loops in the derived CSA data base are used for the simulation studies presented in this Chapter. Performance of the loops in the derived data base are presented as scatter plots.

8.2.5 DATA BASES WITH CABLE CHARACTERISTICS

The copper medium used in the subscriber loop plant has also significantly changed over a period of time and over the loop plant itself. The characteristics of the medium and their documentation are highly variable. There are four basic primary characteristics (R, L, G, and C measured per unit length) for the copper medium. The central analysis module of the CAD facility also accesses these data bases that are generally not consistent with each other.

8.2.6 IMPULSE NOISE/CROSSTALK DATA BASES

Extraneous noise [8.11] in the subscriber loop plant places an ultimate ceiling on the digital capability of the HDSL. The impulse noise events differ considerably from one Central Office to another, type of peripheral equipment used, time of the day, etc. Crosstalk noise is also a statistically bound phenomenon in the HDSL environment. Impulse noise data bases are accessed by the central analysis modules, then statistical characteristics of the crosstalk noise can be estimated. These data bases are accumulated over a wide range of Central Offices and over wide operating usage conditions, such as: nearness to power line, generating stations, power distribution plants, etc.

In essence, there are *three* major data bases dealing with the environment of the HDSL: *first*, the Loop Survey data bases that are generally not consistent from decade to decade, from country to country, and from one operating company to another; *second*, the copper medium data bases that are generally not consistent from one vendor to another or from one gauge to another, and inconsistent from one nation to another; and *third*, the impulse noise and crosstalk models or data bases that may or may not contain the crosstalk coupling coefficients at different frequencies [8.10]. This last data base is generally inconsistent with similar data bases from other countries, from one telephone company to another, and from one PTT to another.

8.2.7 PREPROCESSING OF PERMANENT DATA BASES

Certain preprocessing of these data bases becomes essential before they can be used interchangeably in the CAD environment for the HDSL. This preprocessing program is individually designed and encoded for the particular application. For the studies in the United States, this proved to be the most expeditious path to regain consistency, since our CAD facility did not generate this reference data base in the first place. In contrast, the data bases generated and used within the facility could be streamlined for the flow of data from one phase to another (e.g., simulation phase to display phase, performance evaluation phase to vendor selection phase, or display phase to documentation phase, etc.). Proper design of these linking data bases made them interoperable over extended simulation/design runs and over numerous design of components such as equalizers, echo cancelers, or hybrid matching circuits in the HDSL transceivers and over the two and half decades of their optimization runs. This learning in the data base environments for copper based facilities has prompted the design of data base environments for the fiber optic facilities to be more flexible and streamlined.

8.3

PROGRAM DATA BASES

In streamlined CAD facilities, the main analysis programs are standardized to the extent that they perform as glorified calculators. However, in the R & D environments, these programs allow such dramatic variations in the inputs, components, and loop environments, that a data base approach to managing their functionality, format, interoperability and use become desirable. For this reason, managing these programs and supporting programs are outlined in the following approaches.

8.3.1 FUNCTIONAL DESIGNATION IN PROGRAM NAME

The name of the program in the program data base is synthesized by its major functions. Each successive alphabet is a code for the major functional module within the program. This type of coding is akin to the Library of Congress classification of subject matters. Each successive letter/number indicates the subspecialty of the subject matter.

For this task, the software designer needs to broadly categorize the major tools and techniques used in this simulation software. The assigning of each alphabet or letter is like a MIS (Management Information Systems) specialist assigning a designation to each human group within the company. Once the tools have been identified, any program is a string of computations, and correspondingly, its name can be synthesized. Alphabetizing these program names by the sorting programs on the computer creates a list of names. For the simulation programs, the input and output become attributes. The input and output lists are also similarly categorized by type, number, and format of the input/output. The program names, and their supporting I/O attributes form the basis of the program data base.

The major constraint in this approach is that some computer systems restrict the length of the files to eight characters. When the program chain becomes too long, the least frequently used characters (such as T, U, X, Y, Z) replaces the most frequently used string of 3, 4, or 5 characters. In this application, the approach has worked out satisfactorily.

8.3.2 CHRONOLOGICAL CLASSIFICATION

This approach uses the computer clock to rubber stamp the original encoding date, the modification date, etc. Being rudimentary and lacking methodology in its organization, this approach is unsatisfactory and does not provide enough information to be able to use significant software over decades of development.

8.3.3 PHYSICAL/LOGICAL CLASSIFICATION OF MASS MEDIA STORAGE

The approach of storing major functional modules of programs in different disks, tapes, or even computer systems appear to provide some relief temporarily. However, it becomes a major stumbling block due to changes in technology and age of the storage media. Logical partitioning is feasible by appropriate directories within the storage medium. Hierarchical combination of the two classifications is also feasible. This approach though effective at the early stages of CAD development proves inadequate for any sophisticated user of the data bases. A more elegant solution in the long run becomes attractive.

8.3.4 DOCUMENTS AND DATA BASES

In order to ensure the interoperability of the multi-data base within the CAD environment, the use of diagrams depicting the various program packages, inputs, outputs, use of permanent data bases, manual interaction/inputs (if any), for the specialist can serve as the strongest documentation. Essentially, encoded in a graphical form, the information of the three prior types of classifications (see Sections 8.2.2.1, 8.2.2.2, and 8.2.2.3), is contained in a one page document for the systems user. In a sense, this diagram becomes a flow-graph for the program. All the key input and output data bases, and the contiguous chain of programs and physical/logical addresses are documented here. The reactivation of any CAD environment, even decades old, can be completed in about half an hour with similar or compatible hardware.

In Figure 8.2, a typical one page documentation of a CAD program "**PNC2B2H0**" (**P**rogram series **N**, for **C**able simulation for **2B1Q** code, Sequence **2** Program, at H_0 rate) is depicted. The subscriber loop simulation program was prepared in the eighties (using the 1983 Loop Survey data base) to determine the performance of the 2B1Q code at 384 kbps or H_0 rate. The location of the programs is on 20-MB mass media "ZZS". This program has one input file "IP2B2" (**I**n**P**ut file for **2B1Q** code, Sequence **2** Program, etc.) and four output files "OP2B21", "OP2B22" (**O**ut**P**ut files **1** & **2** in the last character, from the 2B1Q, Sequence 2 Program. etc.), "OP2B2LG", "OP2B2RG" (two **O**ut**P**ut files and **L** and **R** in the sixth character indicating

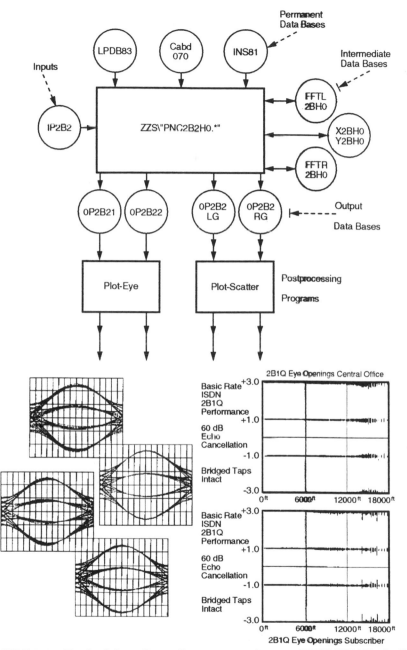

FIGURE 8.2. Typical One Page Documentation on "PNC2B2H0. for" Program File within the Program Data Base. ISDN - Integrated Services Digital Network; LPD - Loop Data Base(s); INS - Impulse Noise Data Base(s) ; FFTL -Fast Fourier Transform, Left Side; FFTR - Fast Fourier Transform, Right Side; OP - Output Files.

transmission from right to Left side and transmission from left to Right side, and finally **G** indicating the Global loop performance) is documented. The use of the permanent data bases "LPDB83" (Loop Survey 1983), "CABD070" (cable characteristics data base 0 for all gauges at 70 degrees F) and "INS81" (Impulse Noise Survey of 1981) is also indicated in the linkages of the program. This figure also depicts the use of two of the four intermediate files "FFTL2BH0" (Fast Fourier Transforms of the Left side excitation function) and a corresponding "FFTR2BH0" for the Right side. The third and fourth intermediate bases "ABCD2B2H0", abbreviated as "X2B2H0" and "DBCA2B2H0", abbreviated as "Y2B2H0" are also shown. These files contain the forward and reverse *ABCD* [8.14] complex matrices of each loop in the loop data base LPDB83. It is noteworthy that these files become full-fledged members of the data bases associated with the simulation studies (for the 2B1Q code at the H_0 rate) in all the subscriber loops from the 1983 Loop Survey data base.

8.3.5 SCANS OF DOCUMENTS ON DATA BASES

This last step is generating page scans of the specialist's first page of each document that completes the hierarchy of data organization. Once the scans are archived, the retention of data (probably the highest hierarchical and conceptual level) is complete. It becomes very simple to return to complex simulation environments by flipping through pages of "scans" after years of abandonment.

8.4

POST PROCESSING PROGRAM DATA BASES

Generally, the output from long simulation runs needs to be graphically displayed. Such graphic capability can be customized or standard display packages may be used. In this case, the display capability has been developed (see Figure 8.3) to customize its performance to the need of the systems designers. The data base of the display or **PL**otting programs (PL-series; note that series PL programs for simulation do not exist), is also organized in a similar hierarchical format. In the example the plotting program for the results of "PNC2B2H0" is simply "PLC2B2H0". Looking for "PLF2B2H0" is futile since the 2B1Q code is not used at H_0 rate in the Fiber environment. This type of syntactic and semantic checks for the proposed human activity for any particular data base can be made by the name of the data base on which the human activity is to be carried out. Experience has shown that this aspect of "sanity check" (dealing with hundreds of intertwined data bases) proved to be the most valuable asset of data base hierarchical organization proposed in this Chapter.

8.5

GRAPHICAL OR PICTORIAL DATA BASES

HDSL simulations provide large amounts of significant results. Most of the linear and circular scatter plots shown in Section 7.6.3.2 (see Chapter 7) are generated by the post processing and display programs. The advantage of separating this set of programs, to be distinct in their own right, is that incremental changes and fine tuning of the software can be most easily accomplished.

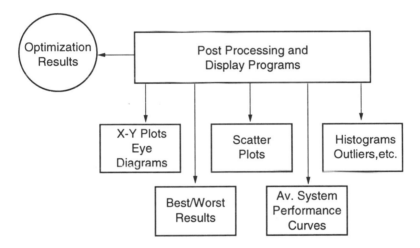

FIGURE 8.3. Organization of the Post Processing Software.

In addition, very long sequences of simulation runs can be executed by "shell" commands by using batch programs. The computer system does indeed keep a running account of the execution sequence and a detailed documentation of the simulation/optimizations when the machine is left unattended.

Unless a systematic strategy for their display and tracking is adopted, the user tends to become overwhelmed in effectively using the results. For this reason, the management of pictorial files for display and interaction fall under the realm of data base design. Considerable flexibility is necessary since the user is generally unaware of the intermediate or the eventual expectations of the CAD facility. As the project progresses, the user needs change. Typically, during the early stages, a feasibility study may be the primary goal. Next, system configuration and component optimization become important. During the third phase, compatibility of the proposed data transmission system with other systems already existing in the loop plant dominates the study. Finally, design rules and installation criteria need to be evaluated. Each of these phases, as the project reaches its goal, needs its own formats for display and evaluation of the results. If existing graphic routines are used, then the amount of new codes can be minimized by categorizing the displays as recurring groups. For example, two eye diagrams (Central Office to subscriber and subscriber to Central Office) are generated for each loop simulation. Likewise, two scatter plots depicting the eye statistics are obtained by one performance evaluation over all the loops in the data base. Hence, a hierarchical design of the files and displays facilitates the data base management strategy.

Tabulated data is sometimes required to scan the marginal loops. Histograms depict the statistical nature of the loop characteristics. Image impedance is most effectively displayed as scatter plots. Flexibility in the presentation of the results generally hastens the next phase of investigation. User requirements and component specification also play an important role in the software design.

8.6

INTERMEDIATE DATA BASES

Numerous intermediate files are necessary to conserve core space. Some examples of these write, read-back files are: (a) files containing spectral components of excitation; (b) files containing eye opening statistics [8.12, 8.13] for each loop in either direction; (c) single pulse reflections for echo cancellation, etc. The management of these files can become the system user's responsibility. With an extremely large number of files and updates, the system designer is faced with this additional task. Ideally, when the generated files are to be reprocessed for graphical display in the form of eye diagrams, scatter plots, signal-to-noise ratios, etc., then the CAD system should provide a built in safeguard to ascertain the accuracy of the displayed results. Effective file management is essential if the system has limited disk space. In addition, long and recurring data files should generally be refreshed from the disk rather than by recomputation to reduce the execution time.

Considerable precaution is necessary while using these data bases. Lack of care in systematically deleting or overwriting the data in these files leads to overpopulation of under utilized files. This is a serious situation in CAD environment due to limited disk space. On the other hand, a policy of successive overwrites may lead to loss of important data or sequence of steps. Some of the proven operating system methodologies, discussed in Part III, prove to be useful. This is especially the case with micro based multiuser CAD facility. Deletion of least recently used (LRU) files can be easily adapted by verifying the directory data for the sequence of any intermediate files created by the same user. Similarly, the least frequently used files in the sequence may be deleted by inspecting the read-access on the particular files.

8.7

INTERMEDIATE OPTIMIZATION

Component sensitivity is usually discovered during the optimization of system performance. The objectives of the component design generally become evident after the completion of the feasibility study. Hence, the system user tends to achieve a maximum benefit-minimum cost system during the later part of the simulation study. The engineering details and the component design start to play a significant role here. Most components, when properly simulated, display their particular influence on the system performance. When maximized, the ability to interact with the simulation system becomes crucial. In the digital subscriber loops for ISDN, equalizers, filters, and the echo cancelers need iterative design procedures. When the display programs are effectively used, a high degree of optimization generally can be achieved at very little additional cost for some of the components. One example of this is an optimal equalizer for line delay and distortion. Echo cancelers, on the other hand, display an opposite effect. When digital penetration of the digital subscriber systems is high in any given loop plant, large efforts to improve the echo canceler performance shows only marginal effect upon the penetration. This is specially the case when the penetration reaches 96 to 98% with 54 dB of echo cancellation (with 2B1Q code) selected by Bell Operating Companies in the American loop environment.

8.8

AI AND ES TECHNIQUES

In order to streamline the CAD function, the uncontrolled parameters for HDSL are stored away in permanent data bases. The CAD facility (see Figure

8.1) accesses user inputs. Permanent data bases generate output files automatically sequenced in the output data bases. If intermediate files are to be generated (to conserve memory space), the machine systematically generates and sequences them in the intermediate data bases for use, until they are either retrieved or overwritten by the user. The output files are processed by the display programs or other post processors to generate the statistical results, histograms, etc. Finally, the machine retains the logical linkage between inputs, permanent data bases accessed, and output files. Thus, any particular simulation run can be tagged for future reference by processing in the AI (Artificial Intelligence) mode of function.

The AI approach permits the CAD environment to learn optimal values of individual parameters and the conditions that lead to desirable system performance. Input parameters (and their combinations) that are responsible for the desired results are automatically sensed and saved. Sequence of user operations can also be retained by the system. When these AI techniques are used, the INCAD user may retract the exact conditions saved by the system as files or data bases. This approach simplifies the CAD steps to the computationally standardized techniques of retrieval (based upon indexing, sorting, and/or hashing) rather than repeated and prolonged computation, inference or system query.

Performance measure of AI techniques includes the number of sequences saved, the amount of effort to learn these sequences, and the length (i.e., the number of primitives) of solutions generated. The likelihood that the sequence will be needed in the future becomes a necessary factor to consider in the design of the INCAD environment. When the process is handled by the system, the machine may be programmed to selectively forget sequences based upon the least frequently used (LFU), least recently used (LRU), etc., classes of algorithms readily employed in most operating system environments for resource management.

Such techniques have been effectively used in other disciplines. Samuel's checkerboard game (a.k.a. checkers) program uses the minmax search. This algorithm looks ahead a few moves (the moves are evaluated into maximum or minimum levels of the gametree), updates the given static evaluation function for a checkerboard layout, and then stores the learned value. Another example where these techniques are used is the STRIPS robot-planning program in which macro operators (i.e., a useful sequence of operators) are stored in a triangle table. Their parameter values are generalized. Adequate care is necessary to make sure that the preconditions of each generalized operator in the sequence still hold in the new application. In the AI field such examples are abundant. With reasonable amount of sophistication, the INCAD designer can emulate proven procedural steps for ISDN system design and facilitate the ISDN implementation to be as efficient as possible. In this particular application, it becomes desirable to store a partial system configurations learned from prior simulations and/or experience of the experts.

Another major class of such expert system applications is the Digital Electronic Corporation's XCON system for the configuration of the VAX computers. Given a customer's requirement, the system generates diagrams displaying spatial relations among its components to fill the customer's need. The complexity of the configuration task is managed by partitioning the task into ordered subtasks, each with their own set of constraints rules. Data bases containing properties of several hundreds of components are used to configure the VAX computer.

Plausible reasoning during the early stages of the ISDN overall system design may become necessary. One of the techniques in expert systems is to follow the generate and test approach. Here, partial solutions are guessed and tested with no commitment to the solution if it fails. Another approach uses top-down refinement. In this case, the refinement occurs as the solution becomes more and more specific. Optimization is postponed to a later stage after global issues are resolved.

The design steps can be partitioned to satisfy a number of electrical, mechanical, size, power, etc., requirements. Linking of appropriate segments at critical stages in the design procedure becomes the function of the INCAD environment for ISDN system and components. Unfortunately, a body of distilled knowledge for the ISDN type of network does not exist as it does for the conventional plain old telephone network. Hence, the refinement of the initial guesses has to come from sophisticated modeling. The feedback of the INCAD environment can be made invaluable by the embedded AI and ES concepts. This approach is far more economical, sophisticated and efficient than the painstaking training of an inexperienced human being in the man-machine feedback loop.

8.9

CONCLUSIONS

A sophisticated human designer can compensate for an archaic CAD environment. The steps that are taken by an INCAD environment can be tediously, but surely taken by an experienced system designer. However, this approach is highly undesirable for ISDN. An INCAD environment can be duplicated. It is portable, flexible and systematic, though not creative. An inexperienced designer does not have either of the two desirable characteristics. The current ISDN growth environment needs sophisticated components as the network evolves and grows over a long period of time. For this reason, an enormous amount of intelligent designing of ISDN subcomponents is necessary. Given these conditions for the ISDN evolution and growth, numerous INCAD environments with a large number of designers are more desirable than a few very good designers and passive CAD environments. Data base management and access based upon the recent AI techniques become a necessary feature in designing these intelligent CAD environments.

The purely humanistic approach is undesirable since the results and the performance of HDSL have to be very dependable. Any probable cause of error in dealing with large data bases needs scrutiny and reexamination, calling for sophisticated data base capability.

The approach proposed in this Chapter uses data base techniques particularly suited to the HDSL CAD environment. This data base environment can be duplicated for other designs, such as the fiber optic system (see Chapter 18). It is portable, flexible and systematic. The HDSL needs sophisticated components as high-speed digital networks evolve and grow over a period of time. For this reason, it becomes essential to cross compare results under various conditions, components and design strategies. Both refined data base management strategy and AI techniques become necessary for the design process.

8.10

REFERENCES

8.1.   S.V. Ahamed, B.S. Bosik, N.G. Long and R.W. Wyndrum, Jr. 1979, "The Provision of High Speed Digital Data Services over Existing Loop Plant," *Proceedings of the National Electronics Conference*, Vol. 33, 265-270.

8.2.   S.V. Ahamed, P.P. Bohn and N.L. Gottfried 1981, "A Tutorial on Two-Wire Digital Transmission in the Loop Plant," *IEEE Transactions on Communications*, Vol. COM 11, 1554-64.

8.3.   S.V. Ahamed and V.B. Lawrence 1989, "Database Organization and Access for ISDN," IEEE International Workshop on Microelectronics and Photonics in Communications, June 6-9, New Seabury, MA.

8.4.   P.A. Gresh 1969, "Physical and Transmission Characteristics of Customer Loop Plant," *Bell System Technical Journal*, Vol. 48, 3337-85.

8.5. L.M. Manhire 1978, "Physical and Transmission Characteristics of Customer Loop Plant," *Bell System Technical Journal*, Vol. 57, 35-39.

8.6. S.V. Ahamed 1982, "Simulation and Design Studies of the Digital Subscriber Lines", *Bell System Technical Journal*, Vol. 61, 1003-77.

8.7. Bell Communications Research 1987, "Characterization of Subscriber Loops for Voice and ISDN Services," *Report Number ST-TSY-000041*, Science and Technology Series, June.

8.8. S.V. Ahamed and R.P.S. Singh 1986, "Physical and Transmission Characteristics of Subscriber Loops for ISDN Services," *Conference Record of ICC '86*, June 22-25, 1211-15.

8.9. J.W. Lechlieder and R.A. McDonald 1986, "Capability of Telephone Loop Plant for ISDN Basic Access," *Proceedings of ISSLS 86*, September 29-October 3, 156-61.

8.10. J.W. Modestino, et al 1986, "Modeling and Analysis of Error Probability Performance for Digital Transmission over the Two-Wire Loop Plant," *IEEE Journal of Selected Areas in Communication*, Vol. J-SAC-4, 1317-30.

8.11. J.J. Werner 1990, "Impulse Noise in Loop Plant," *Conference Record of ICC '90*, June 16-19, 1734-37.

8.12. M.J. Miller and S.V. Ahamed 1988, *Digital Transmission Systems and Networks Volume II, Applications*, Computer Science Press, Rockville, MD.

8.13. S.V. Ahamed and V.B. Lawrence 1988, "PC Based Image Processing System for the Two Binary to One Quaternary (2B1Q) Code," *Proceedings of the 1988 Southeastern Simulation Conference*, Orlando, FL, Vol. SESC 88, 92-97.

8.14. Bell Laboratories 1982, *Transmission Systems for Communications*, Fifth Edition, Western Electric Company, Winston-Salem, NC.

CHAPTER 9

SIMULATION TECHNIQUES FOR THE QAM (2D) CODE

Microcomputer-based simulation facilities, built to investigate the performance of bandwidth efficient quadrature amplitude modulation (QAM) type 2-dimensional codes, may be used to investigate the capacity of loop plants. Modulating techniques at baseband frequency are the QAM type, 2-dimensional (2-D) carrierless amplitude and phase or CAP modulation. There is no RF carrier (generally associated with QAM) per se in CAP. The CAP techniques are specifically important in the maximum utilization of available bandwidth. Data rates in copper media can thus be maximized to the full media capacity. In particular, these codes are likely to enhance the bidirectional data rates to T1 (1.544 Mbps in the United States) and to E1 (2.048 Mbps in Europe) in the subscriber loop plants. The extent and limitations are discussed in this Chapter. The feasibility of providing Basic Rate Integrated Services Digital Services (BRISDN at 144 kbps) in the various telephone environments is well documented with the 2B1Q code and is used as a benchmark (see Section 9.7.1) for the success or failure of the other more efficient modulating techniques.

Simulation methodology and preliminary results of system performance at the PRISDN (at T1 rate in the United States and E1 rate in Europe) are also presented. The management of large data bases that hold specific details of the subscriber loop plants and electrical characteristics of cables needs special consideration. In addition, the large data bases that hold the simulation results also need systemic post-processing to condense them in order to be able to cross compare the system performance under various options of design. Algorithmic indexing of variables for the complex domain spectral components and impedances is essential to generate the large variety of results. Systematic depiction of significant cluster noises in 2-D signals is also presented. This Chapter reports *two* distinct flavors of investigation: the simulation of 2-D modulating techniques; and system performance evaluation in view of the limited (55 to 60 dB) echo cancellation in most transceivers.

9.1

INTRODUCTION

Quadrature Amplitude Modulation (2-dimensional encoding) techniques are being considered for use with the existing twisted wire-pair to transmit data to subscriber premises at 1.544 Mbps. In the western telephone networks, a metallic connection exists to most of the customers from a Central Office or a remote distribution facility. Enhanced use of this existing copper facility to transmit high-speed data appears to be most economical and viable choice towards primary rate ISDN. The advent of fiber in the distribution plant facilitates the availability of very high rate data, ranging from 51.64 Mbps (or the standard OC-1 rate) through 2.4 Gbps (see Chapter 15) at the network distribution centers. However, the use of fiber to every home is not entirely economical at this stage because the copper wire-pairs already reach the individual customer residences and almost all businesses.

The object of the study is to evaluate the response of loop environment, to specify component design, to determine the PRISDN service (at T1 and E1 rates) penetration in existing twisted wire-pair POTS, the CSA type of subscriber loops in the American loop plant and finally to formulate a loop selection criteria. This study aids in the design interface requirements and specifies the performance requirements for echo cancelers in view of the typical spread of loop impedance, image impedance, insertion loss, and loop loss of these high-speed digital subscriber lines.

9.2

CSA GUIDELINES AND THEIR REVIEW

The CSA concepts (see Section 5.5) became important for DSL application in the United States during early eighties. For the sake of designers and standards organizations, the results as they pertain to the earlier CSA guidelines issued by Bell Communication Research for the component manufacturers for United States markets are presented in Chapter 5.

However, from the perspective of the 1990s, these guidelines, issued during the early eighties, seem almost irrelevant, if not obsolete, in light of the recent technological trends. At the time when CSA guidelines were issued, the investigations pertaining to: bandwidth efficient channel coding (Section 6.6.3); trellis and concatenation (see Chapter 11.4); influence of more refined matching circuits [9.1,9.2]; and customized echo cancelers [9.3] were not complete. For this reason, the traditional CSA guidelines in some of the simulation results presented in Section 9.5 can be relaxed (and have been relaxed) to determine the effects of newer trends.

Two viable sets of codes for the HDSL application are the 1-dimensional codes (typically, the multilevel or 2B1Q) and the 2-D codes (typically, 16, 32, 64, 128, etc. point cluster CAP). The former is a simple form of a block code, in which multiple binary bits are assigned an amplitude level. For the simple 2B1Q code, 2 binary bits (2B) from the data stream are encoded as one of the four possible levels (1Q) of the symbol. In the later, a sequence of bits is encoded as a symbol and each symbol is transmitted as two orthogonal signals in the time dimension. This way, the dual signal structure of CAP permits the conservation of bandwidth. They form clusters similar to those of the quadrature amplitude coding (QAM). The actual symbol rate varies inversely (and logarithmically [9.1,9.2]) with the number of points in the constellation. At higher rates for HDSL and for ADSL applications, this feature becomes particularly attractive. Typically, symbol rates, and thus the bandwidth requirement (including the additional bandwidth for overhead) for various HDSL applications, are shown in Table 9.1.

Hard error correction at the bit level (as it is used in the context of single correcting codes) may also be deployed. But by and large, it is not appropriate for the trellis (2-D) codes. Instead of Hamming (or logical) distance, the Euclidean distance between the symbols is increased for trellis codes. The effect of this increased Euclidean distance between successive symbols is that these trellis codes become more resilient in the presence of random (such as crosstalk) noise. Hence, a given probability of error can be maintained at the required level, while the crosstalk interference (i.e., noise) is relatively higher than the noise had there been no coding. Thus, the required SNR with trellis codes for a given bit error rate requirement (BER) can be about 3 dB lower. Concatenation of codes (trellis and Reed Solomon [see Section 5.5 and Reference 9.4]) offers another 3 dB gain. In practice, about 6 dB gain is realizable. Another significant effect of these (trellis and its concatenation) techniques is to enhance the loop length over which HDSL can function dependably or offer greater immunity against random noise in the loop environment for a given range.

TABLE 9-1 Parameters for Uncoded 2-D Constellation (CAP) Transceivers

Data Rate (Mbps)	Constel- lation (Points)	Symbol Rate (kbaud)	Excess Bandwidth (Percent)	Total Bandwidth (kHz)	Center Frequency (kHz)
0.772	16	193.0	8.8	210	105
1.024	16	256.0	9.4	280	140
1.024	32	204.8	7.4	220	110
1.024	64	170.7	11.3	190	95
1.544	64	257.3	8.8	280	140
2.048	32	409.6	7.4	440	220
2.048	64	341.3	11.3	380	190
2.048	128	292.6	9.4	320	160

Carrierless AM/PM techniques for Europe, the Far East and other developing countries cannot be ignored, since the loss characteristics of the loops and noise characteristics of the loop plant can vary significantly from those in the United States. As the rate is enhanced, the very simple codes do not generally suffice. The ITU-T (formerly CCITT) recommendation of the 2B1Q code for Basic Rate ISDN (BRISDN) has facilitated the development of the U interface chips and this design is now well along the way for BRISDN services in the United States and Europe. However, for the Primary Rate ISDN (PRISDN), the differences between the subscriber loop plants in the United States and Europe can be significant and crucial. For this reason, the detailed investigation of the bandwidth efficient codes in both the United States and European subscriber loop environments are reported in this Chapter. Further, the performance of the subscriber loop plant at E1 rates common in Europe may become necessary in the United States for some multinational companies. The differences existing between American and European subscriber loop plants can critically influence the penetration of the PRISDN services in the United States and Europe at the T1 and E1 rates.

9.3

SIMULATION SOFTWARE

The primary aim in establishing the simulation software for any network facility is to evaluate the performance of the entire loop population from any subscriber loop data base (such as the CSA, the United Kingdom, the Swedish, the Australian, the Japanese, etc. loop data bases) with the various types of 2-D and trellis codes. System performance with various transmission techniques and associated system components such as equalizers, echo cancelers and timing recovery circuits can also be optimized.

9.3.1 THE BASIC BUILDING BLOCKS

The software organization for the entire facility has been grouped into *three* categories. The *first* part of the software deals with defining the system prior to the actual simulation.

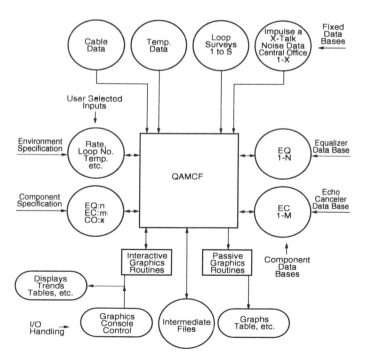

FIGURE 9.1. Block Diagram of Simulation Software for 2-D Codes.

The *second* part deals with the simulation per se and generates numerical results from the input files in conjunction with the permanent data bases that store the electrical characteristics of the system components and of the twisted wire-pairs carrying the QAM data. The intermediate results are stored for visual displays or these files are reprocessed to extract statistically significant result from the simulation. A typical example of such a result will be the fidelity of the transmission and the SNR in each of the 16, 32, or 64-point clusters of the QAM constellation. The *third* part of the software deals with graphics and hard copy generation.

The second part of the simulation software (Figure 9.1) is the most elaborate, since all the numerical and computational algorithms are encoded here. The mathematical techniques (especially fast Fourier transformations, matrix multiplication, complex algebra, and interpolation) necessary for obtaining the results are presented in Chapter 7, along with the techniques for systematic reduction of loops with complex topology. Other simulation techniques for systems with inherently non-linear elements especially encountered in the timing recovery circuits are discussed in Reference 9.5.

For this reason, the structure of the second part of the simulation software is designed to be completely modular. Routines and computational blocks can be imported and exported from other major simulation packages such as the fiber optic simulation systems (Sections 9.6, 9.7, 9.8 and Chapter 18), and 2B1Q BRISDN simulation facility. However, special attention is necessary to handle the complex Fast Fourier Transforms (FFTs) in these 2-D CAP simulations. Another area of specific concern is the two dimensional (real and quadrature) components of cluster noise in the QAM codes. The computational techniques for evaluating the SNR closely parallel the conventional techniques in any other signal processing environment for acoustic signals. An additional extent of uncertainty gets into the evaluation of the SNR, if the actual scanning instant (due to phase jitter of the clock recovery circuit) is not known. If a completely accurate simulation is desirable, then the entire system together with all the transceiver components has to be simulated.

The software is capable of generating both the spectral domain and the time domain results. The spectral domain results capture the flavor of the loop plant *capability* under different coding algorithms, different noise conditions, different ranges, etc.. The time domain results capture the essence of the loop plant *performance* under specific operating conditions, such as a specific code, with a specific echo canceler, a specific trellis, loop failure rates, loop topologies that lead to loop failure, etc. In order to evaluate both the *capability* and the *performance* of the loop plant, both domains of simulation are necessary. Loop data bases are gathered to indicate topology of the loop plant. The cable characteristic are gathered to evaluate the frequency dependent attenuation and delay of the signals.

The simulation facility is partially automated to generate the specific results sought by the designer. Since the data bases can become extensive and the computations can take numerous days to run, effective use of the primary and secondary memories becomes crucial. From a typical time domain simulation where 16,384 symbols are transmitted bidirectionally, only 8, 16, 32 or 64 SNRs are extracted. The actual number of bits depends upon the number of points (i.e., 8, 16, 32, or 64) in the constellation. The SNR at each constellation point is computed by evaluating the deviation of the displaced point from its ideal location. The minimum SNR is thus extracted to meet the BER transmission criterion.

9.4

DATA MANAGEMENT ASPECTS

The manipulation of the primary, the intermediate and the display data needs special care to ensure the accuracy of simulation (and hence the design) of the various DSL systems under different operating conditions.

9.4.1 LOOP TOPOLOGIES AND CABLE CHARACTERISTICS

The detail and variety of this information depends upon the extent and nature of the simulations. Global ISDN simulations need the digital subscriber loop topologies of all the different countries stored away in their respective data bases. The topology and electrical properties (of the overall transmission media), the networking and carrier serving features also need to be stored. By and large, localized or a national loop information is adequate to initiate a typical study of the ISDN environment. The statistical nature of the loop plant topology, the primary and/or secondary electrical characteristics (of the twisted wire-pairs), the crosstalk and impulse noise data bases, generally make up the permanent data bases for DSL simulations.

Statistics of national telephone loop environments are assembled by major surveys. Such surveys have been done in the United States, Canada, Japan, Italy and Australia. The major results in some of the telephone environments have been published. The last major survey of the then-existing Bell System was done in 1983. From this survey, a segment of shorter and less severe loops known as Carrier Serving Area or CSA loops (see Section 5.5.3) has been derived. The feasibility of the HDSL project at T1 and E1 rates can thus be evaluated. Data based upon the European loop surveys is less readily available. To the extent that the data is known, we differentiate between these two (United States and European) loop environments. The loops accessed in this study have been restricted to the CSA loop data bases in the United States and the some of the subscriber loops in the United Kingdom.

Numerous intermediate files are also necessary. Examples of these write-read back files are: (a) files containing Fourier components of excitation or of single pulse; (b) files containing impulse response of a unit/delta function or a single pulse, and (c) reflections of a unit/delta excitation function for echo cancellation, etc. The management of these files can be imposed upon by the system user. With an extremely large number of files and updates, the system designer is faced with this additional task. Ideally, when the generated files are to be reprocessed for graphical display in the form of eye diagrams, scatter plots, SNRs, etc., then the simulation facility provides built in safeguard to ascertain the accuracy of the displayed results. Effective file management is essential if the host computer system has limited disk space. In addition, long and recurring data files should generally be refreshed from the disk rather than by recomputation to reduce the execution time.

9.4.2 GRAPHICAL OR PICTORIAL INFORMATION

Simulations provide large amounts of significant results. These results need to be studied and evaluated by the human designer in some orderly fashion. Ideally, the results generated should track the thought process and logical steps that led to the design procedure. They should provide the human designer the option to fine tune every step of the design at a local level and then at a global level. In the first iteration of the design of any DSL (DSL, HDSL, VHSDL, or ADSL), the insight of a single human designer does not usually suffice and the consultation with the design team becomes essential. One of the effective ways to communicate with the design team is to generate pictorial information after every stage of signal processing in the particular DSL design environment. The common wisdom (innovation) of the design team is thus invoked by "seeing", what is going wrong, where it is going wrong, why it is going wrong, how it can be fixed, and what are the optimal techniques or algorithms to fix it. Typically, this level of detailed design calls for access to the files holding the pictorial data with a accurate file tracking, file recovery and file management system. These pictorial files are accessed/reaccessed by the third part of the software (Sec. 3.1) to display the signal enhancement/contamination before and after every finite stage in the transmission and recovery of 2-D clusters through the DSL and the eventual SNRs at the receiver.

9.5

TYPICAL SIMULATION PROCEDURES AND RESULTS

The simulation employs 8,192 or 16,384 random symbols for each run for the cluster codes. The actual number of information bits to generate any point in the 4, 16, and 64 constellation symbol are 2, 4, and 6, respectively. Complex fast Fourier transforms are necessary to reduce the running time for each loop. Further, the harmonics of the excitation function are computed and stored away for simulations on numerous loops. To conserve computer memory, the computation uses *three* overlays. The *first* overlay generates the random symbols and computes the harmonics. The *second* overlay simulates the system performance and generates the received 2-D clusters as numeric data. The *third* overlay show the results depicting these clusters, the cluster noise, and the SNRs in each cluster. A matrix display is used to identify each cluster as a computed object in its own right.

The two dimensional cluster are generated by the X and Y coordinates of the symbols in the originally transmitted cluster. Cluster codes and PSK codes are handled exactly alike by the simulation software. The four-point cluster (or the 4-PSK) is formed at 1,1; - 1,1; -1,-1; and 1,-1 points where each of the two data bits (00,01,10,11) are placed. This leads to a baud rate of about 200 kilosymbols per sec (kS/s) for H_0 (400 kbps) rate. This rate is the same as that for the 2B1Q code already accepted for BRISDN access. For the 16 and 64-point clusters, the symbol rates become about 100 and 66.6 kS/s. Other symbol rates for T1 and E1 access are tabulated in Section 6.6. The 8-PSK (with $2^3 = 8$) has four symbols at $(\sqrt{2},0)$; $(0,\sqrt{2})$; $(-\sqrt{2},0)$; $(0,-\sqrt{2})$ in addition to those for the 4-PSK, and the symbol rate is at 133.33 (i.e. 400/3) kS/s.

The simulations presented in this Section are performed on typical subscriber loops at T1 rate with commercial grade (about 60 dB) echo cancelers. The loop topology is depicted in Figure 9.2. These loops conform to the CSA guidelines. Loop 1 is very short (without any bridged tap or BT) and conveys the T1 rate data with a high SNR (typically 50 dB) with commercial grade of echo cancelers. Loop 2 with a single BT offers about 30 dB for the SNR and Loop 3 becomes marginal (with its SNR less than 6 dB) especially at the Subscriber end due to the presence of the second BT. Two different clusters (Central Office-end and Subscriber-end) are generated for each loop because of the loop asymmetry. Typical 8-PSK cluster configurations for Loop

No. 3 at T1 rate are depicted in Figure 9.3a (at the Central Office-end) and 9.3b (at the Subscriber-end) with 60 dB echo cancellation.

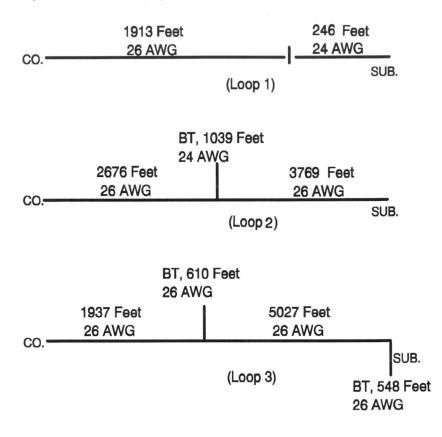

FIGURE 9.2. **Loop Configurations.** AWG - American Wire Gauge; BT - Bridged Taps; CO - Central Office; SUB - Subscriber

Other constellations (sixteen, and sixty-four points) are equally viable. In Figure 9.4, the Subscriber-end 2-D constellation of Loop No. 2 (a representative CSA loop) with 16-points (at 1,1; 3,1;3,3; and 1,3 in the first quadrant and at image locations in the second, third and fourth quadrants) in the constellation are depicted. This loop with one tap is short and does not cause a serious cluster noise at T1 rate with 60 dB echo cancellation.

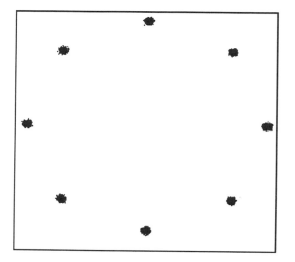

FIGURE 9.3a. 8-PSK Cluster for Loop 3 at 60 dB EC (Central Office).
EC - Echo Cancellation; PSK - Phase Shift-Keying

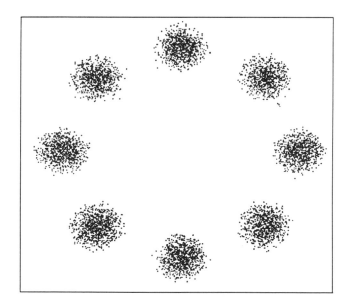

FIGURE 9.3b. 8-PSK Cluster for Loop 3 at 60 dB EC (Subscriber Side). EC - Echo Cancellation; PSK - Phase Shift-Keying

CAP - 16   SUBSCRIBER SIDE PLOT

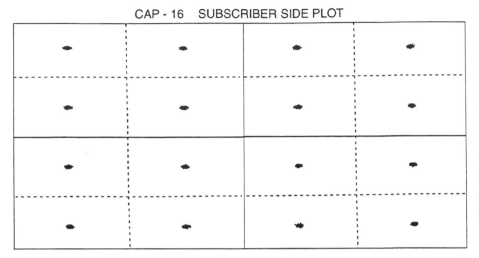

FIGURE 9.4.   A Typical 16 Point CAP Cluster for Loop 2 at 60 dB EC (Subscriber Side). CAP - Carrierless Amplitude and Phase Modulation; EC - Echo Cancellation

CAP - 16 SUBSCRIBER SIDE PLOT

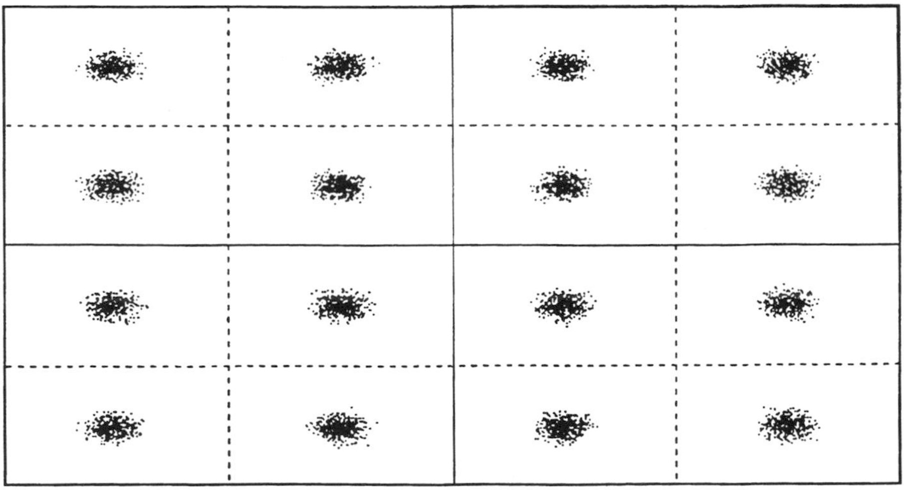

FIGURE 9.5.   A Typical 16 Point CAP Cluster for Loop 3 at 70 dB EC (Subscriber Side). CAP - Carrierless Amplitude and Phase Modulation; EC - Echo Cancellation

However, Loop 3 under the same conditions, displays (Figure 9.5) considerable degradation of the 16-point 2-D constellations even with 70 dB echo cancellation.

In Figure 9.6, the subscriber-end constellation of Loop 2 with 32-points cross is depicted (i.e., the 8 point cluster is located in the first quadrant at 1,1; 1,3; 1,5: 3,1; 3,3; 3,5: 5,1; 5,3: and then at image locations in the second, third and fourth quadrants). Information rate is at T1 rate and the loop can indeed carry the data with about 60 dB of echo cancellation.

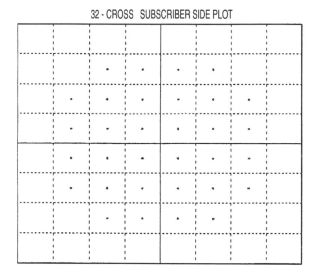

FIGURE 9.6. 32-Cross Cluster for Loop 2 at 60 dB EC (Subscriber Side). EC - Echo Cancellation.

In Figure 9.7, the constellation for Loop No. 3 is depicted under the identical conditions, indicating its failure to carry the data. This particular loop does indeed need more echo cancellation to prevent the residual echo from obliterating the symbols. In these instances, adaptive hybrid impedance matching or two level echo cancellation techniques (see Section 6.5.2) are necessary. The inter symbol distance in the 32 point cluster is shorter than that for the 16 point cluster when the average transmitted power is the same for both constellations.

32 - CROSS SUBSCRIBER SIDE PLOT

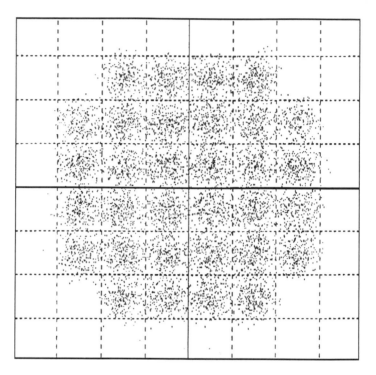

FIGURE 9.7. 32-Cross Cluster for Loop 3 at 60 dB EC (Subscriber Side). EC - Echo Cancellation.

The upper limit on the transmit power is generally a requirement to limit the crosstalk interference into other twisted wire-pair systems, such as the T and E carrier systems, the SLC-1, SLC-96, SLC-2000 systems. For this reason, the SNRs required to maintain a given bit error rate changes from one constellation to another constellation and are tabulated in Table 9-2.

When data is transmitted bidirectionally, two sets of clusters (Central Office end and Subscriber side) are generated for each simulation run. When the loop population and topology are restricted (such as the ANSI loops [Section 5.7.1] or the ETSI loops [Section 5.7.2]), this approach for the time domain simulations is generally adequate. Loop information is stored in these data bases and it is accessed individually to verify its performance. In general, to offer a generalized study, the data bases for the American CSA loop environment (see Section 5.5) and for the sample loops in the United Kingdom (see Section 5.6.3) are used. In most of the European countries there were no bridged taps in the loops surveyed (Section 5.6.3 and Chapter 12). To this extent loops from European environments outperform the loops from the United States, Canadian, Japanese, Australian, etc., environments. For the United Kingdom loops, a data base is generated to contain the loop makeup. A second data base to include the characteristics of the 28 (0.32 mm) American Wire Gauge cable is used in the European countries.

TABLE 9-2 Theoretical SNR Requirements for a BER of 10^{-7} in the Presence of Gaussian Noise

4-point Constellation	16-point Constellation	32-point Constellation	64-point Constellation	128-point Constellation
14.5* dB	21.5* dB	24.5* dB	27.5* dB	30.5* dB
11.0$^+$ dB	18.0$^+$ dB	21.0$^+$ dB	24.0$^+$ dB	27.0$^+$ dB
8.50$^\sim$ dB	15.5$^\sim$ dB	18.5$^\sim$ dB	21.5$^\sim$ dB	24.5$^\sim$ dB

(*) Uncoded Constellations,   (+) Trellis coding,
(~) Concatenated Reed Solomon and trellis coding

The study has *two* distinct directions of investigation. *First*, the performance of cluster codes are investigated in these loop environments to determine the highest margin against the additional crosstalk noise. *Second*, the loop environments are investigated to determine the dynamic range of devices like the equalizers, echo cancelers, timing recovery circuits, hybrid matching circuits, etc. Time domain simulations provide typical results in both these two directions of investigation.

9.6

SUMMARY OF RESULTS

When the performance of the entire loop population needs to be depicted, the representation of the signal-to-noise ratio (SNR) in each cluster of every one of the loops is more effective. Each loop contributes a point to a scatter plot and there are as many scatter plots as there are clusters in the constellation. The collection of 16 scatter plots are depicted (see the Figures in the next Section) as a matrix for the 16 point constellation. However, such display matrices can be (and have been) generated for 4, 32, and 64 for these constellations. Further, at the next level of summarization of results, the SNR data may be averaged over all the clusters for each loop giving rise to only one scatter plot for all the loops in the data base.

In effect, the SNR of the average noise in all the clusters is one point per loop in the final scatter plot. Along the Y-axis, the SNR (in dB) is generally plotted with the X-axis depicting the loop length or the bridged tap length (if any). For the United States, Canadian, Australian and Japanese loop environments, the scatter plot against the bridged tap length carries significant information, since the loop distribution with or without the bridged taps can be immediately identified (by the X coordinate of the points) and secondly, the distribution of quarter-wave-length taps and their effects on the cluster noise can be identified by the location of points with the lowest SNR.

An alternate depiction for the cluster noise is by plotting the SNR as a vertical line. The difference between the maximum SNR (say 75 dB) and the actual cluster SNR for the particular loop is computed and depicted as a vertical line from the maximum SNR value. The best loops have a zero or a minimum line length. The length of vertical lines in the scatter plot indicates the entire loop plant performance. Best performance from the loop plant is a horizontal line at the highest ordinate of the scatter plot. Any deviation of this shape of the scatter plot indicates the imperfection in the loop plant performance.

9.7

DISCUSSION OF RESULTS

The dual duplex mode (with adaptive echo cancellation) of transmission for Basic Rate ISDN with two B (bearer) channels at 64 kbps each, and one D (delta) channel is the standard throughout Canada, Europe, and the United States and the rest of the world. The medium for transmission is consistent with the ISDN objective to deploy the twisted wire-pair copper voice facility for telephone services around the world. These guidelines also suggest a non-repeated facility between the distribution point and end user for the (2B+D) basic rate facility. The viability, design and engineering has already been established by the commercial IC chips that have started to appear for the U interface between the subscriber and ISDN.

The twisted wire-pairs display a dramatic change in the first three of the four primary electrical characteristics (R, L, G and C) as the transmission frequency is swept from about 50 kHz through 300 kHz. Simple line coding techniques cannot facilitate the transmission of data through this dramatic swing (of the primary characteristics) of the existing electrical media. For this reason, at this stage of network evolution, the last link between the distribution centers and the customer appears to be the bottleneck in taking higher data rates to the doorstep. The wireless last mile is a distinct and viable possibility for voice and low data rate services. At very high rates, the cell size has to be very small.

The 2B1Q code is adequate for the BRISDN to reach the telephone customers (at 144 kbps with a line rate of 192 kbps). At higher rates, newer, bandwidth efficient codes hold substantial promise. The transceivers are more expensive and the capacity depends upon the particular loop plant. It is not immediately evident whether the extra complexity and cost for the newer codes is justified in all loop plants. The penetration of the loop plant for the higher rates is also reported in this Section. Optimizing the line termination units for the particular national network is feasible and economic.

9.7.1 2B1Q BENCHMARK PERFORMANCE

For the sake of comparison, the benchmark is the BRISDN performance in the United States CSA environment. The acceptance of the simplistic 2B1Q block code (originally proposed by British Telecom) in the American T1D1.3 committees was after a detailed study of the loop characteristics. Detailed loop surveys and loop characterization do not exist for all the European countries for comprehensive performance studies and for the design of the U interface. However, a study is underway and additional results should facilitate the design of the European U interface at E1 rates based upon the ETSI loops (see Section 5.7.2) rather than the loop surveys.

Simulation results with the 2B1Q code are presented as the transmission rate is enhanced from the BRISDN, H_0 (384 kbps; 5B+D), $2H_0$ (768 kbps, 11B+D) and then to T1 (23B+D) rates for the three loops shown in Figure 6.28. In Table 9-3 the SNR values are presented at these four rates for the three loops. The degradation is evident as the rate is increased or as topology becomes more complex from loop to loop (see Figures 6.28a to 6.28c). The equalizer used is a typical commercial adaptive analog device and the echo cancelers has an adaptive configuration with 60 dB cancellation.

It is also interesting to note that the SNR within the three eye openings of the 2B1Q code can vary by as much as 2 or 4 dB. The reason for this change can be traced back to the number and sequence of symbols, the imperfection of the equalizer, the imperfection of the echo canceler, and the combined (convolutional) effect of the two. For the sake of completeness, the SNR scatter plots for the top-eye and the central-eye of the 2B1Q code for the

United Kingdom loops at 530 kS/s at the Subscriber end are shown in Figures
9.19a and b respectively.

TABLE 9-3 SNR for the 2B1Q Code with 60 dB
Echo Cancellation (Values are represented in dB)

RATE→	Basic Rate		H_0(368kbps)		$2H_0$(768kbps)		T1 Rate	
LOOP ↓	S	CO S	CO S		CO	S	CO	
Fig.6.28a	22	22	29	29	31	31	25	25
Fig.6.28b	17	28	12	29	12	25	Fail	Fail
Fig.6.28c	28	25	16	18	4	15	Fail	Fail

9.7.2 AMERICAN LOOPS AT BRISDN, H_0, $2H_0$, RATES AND PRISDN

The simulation results for the subscriber loops at BRISDN are presented in
Figures 7.12 and 7.13. None of the 1,520 loops in the 1983 Loop Survey fail to
carry the BRISDN data with 60 dB of echo cancellation and high quality of
equalizers. When the rate is enhanced to 768 kbps, the loop failure rate in the
general loop population is very high. When the loop length in the general
population is restricted to 12,000 ft., then the failures are fewer (about 12-15
out of 1200 loops), as depicted in Figure 9.8. Note that any vertical line at a
discrete point along X axis indicates a loop failure to carry the data.
However, for the 354 CSA loops, the failure is limited to two hard failures
(i.e., the data sent is different from the data received) and three loops have an
eye closure greater than 33 percent with 60 dB of echo cancellation. Only the
subscriber end is considered since the BTs are located closer to the subscriber.
The eye closure scatter plot is depicted in Figure 7.14 and also in Figure 9.9 as
a linear plot. When the data is enhanced to the T1 rate, the corresponding
linear plot for the eye closure is depicted in Figure 9.10. The loop failure is
much too high to consider this code for any PRISDN applications.

The loop failure due to crosstalk interference is a serious consideration in
addition to the device limitations considered in this Chapter. We have
quantified the near end crosstalk restrictions in Chapter 14. Three categories
of loop populations are considered. But in overall loop population (United
States populations) studies, the crosstalk restriction claims almost 44 percent
of the loop failures, or about 530 loops from the truncated loop population.
At the present state of device technology, crosstalk effects far outweigh

device limitations for the simple linear (AMI, 2B1Q, etc,) codes. The use of more bandwidth efficient codes becomes more justifiable.

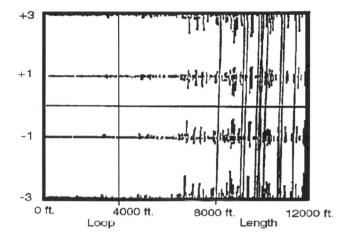

FIGURE 9.8. Composite Scatter Diagram of the 2B1Q Code Eye Diagrams of a Loop Population in a Typical Loop Data Base.

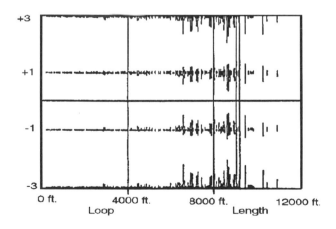

FIGURE 9.9. Scatter Plot of Eye Closure at 768 kbps at the Subscriber Side.

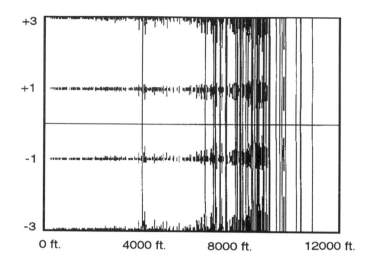

FIGURE 9.10. Scatter Plot of Eye Closure at T1 or at 1.544 Mbps at the Subscriber Side.

9.7.3 QAM 16/64 POINT CLUSTER CODES FOR PRISDN

The effectiveness of the 16 and 64 point, 2-D codes is depicted in Figures 9.11 through 9.17. When the SNR in evaluated in each of the clusters of 16 or 64 point constellations for the CSA loops at PRISDN with 60 dB of echo cancellation, the results may be plotted as scatter plots (see Section 9.6). In Figure 9.11, the simulation results for the 16 point constellation are presented. The display matrix is 4 x 4 having 16 plots and each point depicts a loop with its SNR (Y-coordinate) and its length (X-coordinate). The maximum ordinate value of the SNR is 75 dB and its minimum value is 0 dB as shown in the Figure and the shorter loops have a higher SNR. The plots display a gradual downward pattern as the length of the loop increases. The limitations of the components such as echo cancelers and equalizers is displayed. Loops that do fail have their display points on the X axis of each plot.

The SNR values do not vary drastically for the various clusters. From large screen displays, even the slightest variations are evident and the clusters at the corners generally form slightly different patterns. These effects are observed only when good quality components are used. In practice, noise and misequalization effects generally mask these small differences.

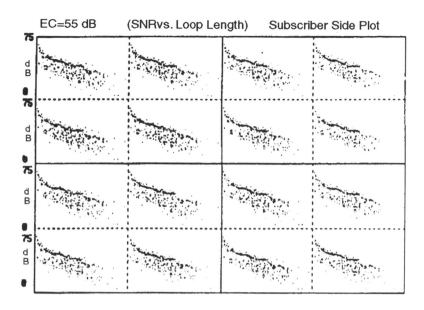

EC=55 dB   (SNRvs. Loop Length)   Subscriber Side Plot

FIGURE 9.11. SNR in Each Cluster of All the Loops in the CSA Loop Data Base. CSA - Carrier Serving Area; SNR - Signal-to-Noise Ratio

In Figure 9.11, the SNR of all the loops in the data base is plotted at each of the 16-points. There are 16 segments in this figure corresponding to the 16 symbol locations. In a noise free environment, all the points will depict a high SNR and the figure will appear blank. The extent to which these points drift down towards the X-axis depicts the inadequacy of the system. The maximum SNR feasible from the computational facility is not infinite due to lower and upper bounds of floating point numbers that can be represented in any CPU. In addition, round-off errors in the numerical processor units leads to very small but finite errors during number crunching, especially while executing the FFT and inverse FFT algorithms. We have noticed these errors yielding SNRs of the order of 85 to 90 dB with a back-to-back QAM simulation over a zero length loop. This residual computational noise is analogous to thermal noise (see Section 6.3.4.1) in analog and digital electrical circuits.

9.7.3.1 Line Scatter Plots for SNR

In Figure 9.12 a vertical line is drawn between the maximum ordinate (60 dB for this Figure) and the SNR of the loop. The darkness of the Figure indicates the poor performance of the components. Three loop failures are evident and most of the CSA loops have a SNR of about 30 or more dB. The echo cancellation for the simulation is also 60 dB.

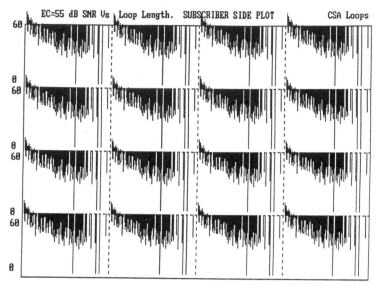

FIGURE 9.12. SNR Plotted as Lines from Maximum SNR Value (60 dB) to the Actual SNR of Each Loop in Every One of the 16 Clusters of the QAM Constellation. CSA - Carrier Serving Area; QAM - Quadrature Amplitude Modulation; SNR - Signal-to-Noise Ratio

In this Figure very short loops have vertical lines over the maximum SNR value of 60 dB. The length of this line segment indicates the SNR in excess of 60 dB. To facilitate the presentation of the SNR in individual loops, the maximum SNR was chosen to be 60 dB rather than the 75 dB depicted in Figure 9.11.

9.7.3.2 Effect of the Bridged Tap Length

In Figure 9.13, the SNR is plotted against BT length. The maximum X-axis value in this Figure is 2,500 ft. The effect of the BT length is evident only at about quarter-wavelength (i.e. 400 ft. since one wavelength is about 1,580 ft.) for the average symbol rate is at about 386 kHz. (i.e. 1.544/4 MHz.) for the 2^4 or 16 point QAM constellation. Quarter wavelength effects do become evident as the low SNR values when the BT length approaches the 400 ft. length. This wavelength computation is based on the propagation velocity of the signals in twisted wire-pairs from their *ABCD* matrices. Other than this well known effect (see Section 5.3.3), there is no other direct correlation between tap length and SNR. Three out of the 354 loops fail to carry the PRISDN data.

EC=55 dB   (SNR Vs BTL QAM 16 - SUBSCRIBER SIDE PLOT   CSA Loops

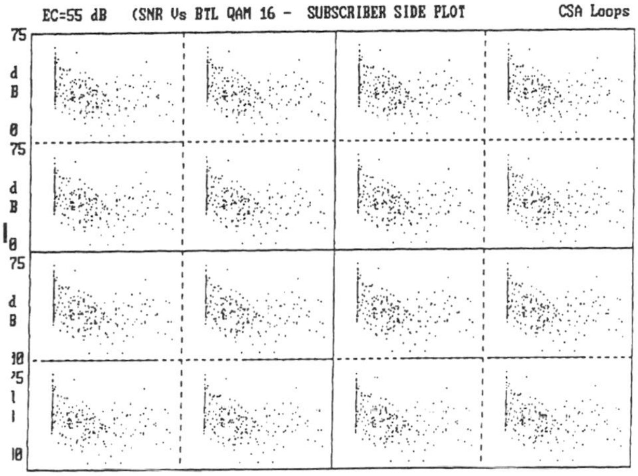

FIGURE 9.13. Cluster SNR Plotted against the Bridged Tap Length for All the Loops in the CSA Data Base. CSA - Carrier Serving Area; SNR - Signal-to-Noise Ratio

9.7.3.3 Effect of Slightly Different Equalization Strategy

The effect of slightly different equalization strategy becomes evident in Figure 9.14. The results are plotted under exactly the same condition as Figure 9.12, and five loops fail to carry the PRISDN data, but the average SNR for the entire population is higher that that for the equalizer for Figure 9.12. Components and their designs may thus be optimized by the simulation package. The average SNR for all the clusters is plotted in Figure 9.15. The 60 to 20 dB segment of the Y-axis is amplified in this Figure to display the details of individual loops. There are six loop failures in this display and their presence may be located by tracking the six full vertical lines that occur in the sixteen sections of Figure 9.14. The length of these loops may also be scaled from the X-axis of Figure 9.15 since the maximum loop length is 12,000 ft.

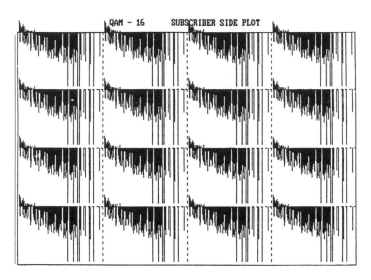

**FIGURE 9.14. SNR Plotted as Lines for Each Loop in Every One of
the 16 Clusters at 60 dB EC (Uncoded, Subscriber Side).** EC - Echo
canceler; QAM - Quadrature Amplitude Modulation; SNR - Signal-to-
Noise Ratio

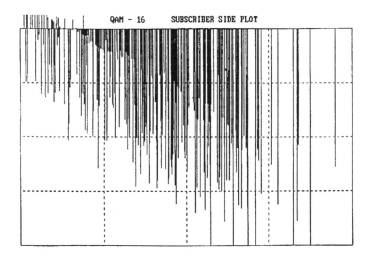

**FIGURE 9.15. Average SNR Plotted as Lines for Each Loop in the 16
Clusters of the CSA Loops at 60 dB EC (Uncoded, Subscriber Side).**
CSA - Carrier Serving Area; EC - Echo Canceler; QAM - Quadrature
Amplitude Modulation; SNR - Signal-to-Noise Ratio

QAM - 64   SUBSCRIBER SIDE PLOT

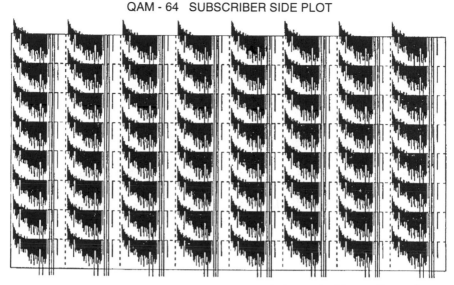

FIGURE 9.16. SNR Plotted as Lines for Each Loop in Every One of the 64 Clusters at 60 dB EC (Uncoded, Subscriber Side). EC - Echo Canceler; QAM - Quadrature Amplitude Modulation; SNR - Signal-to-Noise Ratio

QAM - 64   SUBSCRIBER SIDE PLOT

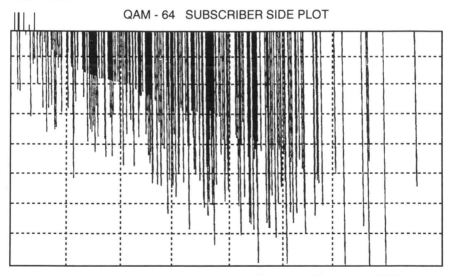

FIGURE 9.17. Average SNR Plotted as Lines for Each the CSA Environmental at 60 dB EC (Uncoded, Subscriber Side). CSA - Carrier Serving Area; EC - Echo Canceler; QAM - Quadrature Amplitude Modulation; SNR - Signal-to-Noise Ratio

Similar results are displayed for the 64 point constellation, and two Figures 9.16 and 9.17 are plotted to indicate the performance of the CSA loops at PRISDN. The symbol rate is only about 260 kHz and there are about 5 loop failures with the 64 point constellation but the average SNR is lower for the longer loops.

9.7.4 NORTH AMERICAN CSA AND EUROPEAN ENVIRONMENTS

The comparison is presented based upon the performance of the 2B1Q code in the CSA and the United Kingdom environments. The CAP simulations in CSA and Swiss environments are presented in Chapter 12. The performance of the loops at various rates is depicted in Figures 9.8 to 9.10. The circular plot for the 768 kbps rate (800 kbps line rate) is depicted in Figure 7.14. The performance of the 2B1Q code for the loops in the United Kingdom data base at 530 kS/s (i.e., half-E1 rate for dual duplex PRISDN systems for data rate of 1,024 kbps) is depicted in Figure 9.18.

2B1Q Code - Subscriber Side (Top Half).

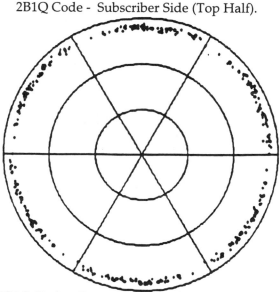

2B1Q Code - Central Office Side (Lower Half.)

FIGURE 9.18. Circular Scatter Plot, 60 dB and 530 Ksps. This shows the circular scatter plot at 60 dB and 530 Ksps to the 100% eye opening mark, which indicates a very good eye opening (or good quality transmission at E1/2 data rate for dual duplex systems).

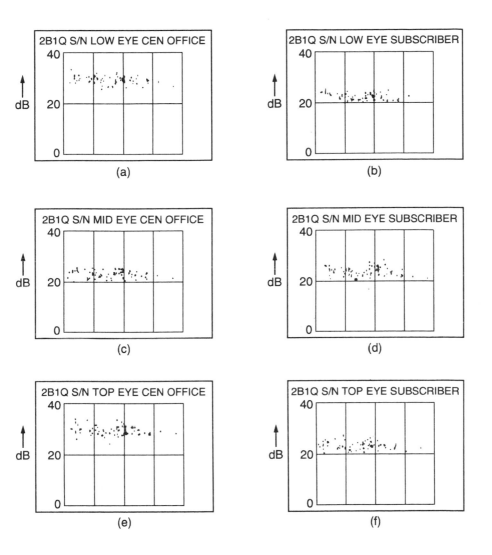

FIGURE 9.19. Signal-to-Noise Ratios in the Three Eyes of the 2B1Q Code for Either of the Two Directions of Transmission for the United Kingdom Loops.

The Central Office side and Subscriber ends both display very similar scatter plots. The eye closure in this Figure for the United Kingdom loops is significantly lower due to the absence of the BTs. In addition, all the loops in the data base up to 20,000 ft. are included, whereas the CSA loops are limited to 12,000 ft. The primary reason for the out-performance of these loops is the reduced reflection due to the absence of BTs, and even a rudimentary (60 dB) echo canceler will perform adequately. This superior performance of the European Loop Plant is proven once again by CAP studies in Chapter 12.

The modems to be designed for Europe need two modifications before they can perform efficiently for the E1 rates. The capability to alter the terminating impedance of the loops may become necessary in the CSA environments. However, the needs to dynamically adjust this impedance does not appear necessary for the European markets. Further, the extent of the echo signal in certain bands of spectral interest is less severe than what the modems are designed to handle. Also, the tail of the echo signal is not as prolonged as it can be in the American loop plant. For this reason, fewer processors are needed to perform both the equalizer and echo canceler functions in European applications. An additional study is necessary for conclusive results to reduce the cost of the modem.

9.8

CODE SELECTION AND NEXT NOISE MARGINS (PERFECT TRANSCEIVERS)

Perfect transceivers perform with infinite echo cancellation, without any phase and timing jitter and with perfect equalization. The current technology does not permit the realization of such devices with digital signal processing (DSP) chips or as VLSI chip embodiments. With perfect devices, the system noise (such as near-end/far-end crosstalk [random in nature], impulse noise [sporadic in nature], stray pick-up, etc. [see Section 6.3.3]) alone provides an upper ceiling to the maximum rate of data transmission. However, an approximation ratio of this maximum rate of data transmission to that realized data rate provides an indication (discussed further in Chapter 11) of the status of the technology. Currently, the ratio is about 2 to 3 depending upon the implementation and the vendor technology. In this Section, the maximum rate is presented based upon the loop make up in the United Kingdom.

In essence, there are *two* restrictions to be considered in the design of the digital systems in the loop plant. The *first* (noise corruption) restriction arises due to the contamination (dominantly the NEXT interference in the HDSL, ADSL and the VDSL systems). Other noises also can cause loop failure under special conditions such as proximity to the power stations, elevator shafts, DC machines, etc.. In Tables 5.6 and 5.12, the capacity of the CSA loops and the United Kingdom loops to carry high-speed data is presented. Perfect devices are used but the balance between line loss (that attenuates the signal) and the crosstalk (that corrupts the signal) also leads to limits of transmission in any loop plant. In Chapter 5, the limitations due to crosstalk are explored.

The *second* (technological) restriction arises due to imperfection of the transceiver devices. In this Chapter, some of the device limitations are explored. In the very early days of the DSL, the technological restriction limited the data rates. More recently, both these limitations overlap in some environments such as the United States, Canadian, and Japanese loop plants for PRISDN in the full-duplex mode. In the European environments, the noise contamination appears to limit the data rate at PRISDN in the full-duplex mode. These findings may be substantiated by Tables 5.6 and 5.12: and then again by studying the scatter plots in Figures 9.8 through 9.17 for the CSA loops and Figure 9.18 for the UK loops. The SNR for each of these loops for the two directions is depicted as dB vs. length in Figure 9.19. The 2-D QAM results are presented in Chapter 12 for the Swiss loops. These results also confirm the findings from Chapter 5 and this Chapter.

9.9

LINE CODE AND NEXT NOISE MARGINS

Line code selection is biased towards bandwidth efficient codes. However, the cost and complexity are the constraints. Trellis and Reed Solomon codes offer additional immunity to random noise in the system. The use of simplest 2B1Q code is justifiably limited to the BRISDN access depending on the loop environment. The degradation of the performance of the code is evident from Table 9-3 and from Figures 9.8 and 9.9 for the CSA environment in the United States. This code is simply unsuitable for higher rates in the United States to achieve a reasonable penetration.

The main contenders are the 2-D QAM type of coding algorithms and the cost and complexity is offset by the much higher rates at T1, E1 in the full-duplex, dual duplex or even in the dual simplex modes. The justification for using the trellis and/or trellis with Reed Solomon codes depends on the particular loop environment, presence of discontinuities, bridged taps, and extent of noise contamination. For these reasons, code selection is an optimization and a numeric process in which the simulation results for the choices of codes and constellations need to be completed. Sometimes this is a contentious process and conformity with the ITU/ANSI standards supersedes any optimization study. This is especially the case because the change in performance is only slight, and the marginal gain of one code over the other does not justify the lack of conformity with the standards.

9.10

CONCLUSIONS

In this Chapter the basic methodology for simulation of QAM codes is presented. The application is directed towards for Primary Rate ISDN data in the North American CSA and the United Kingdom loop environments. The results document the effect of system components, such as echo cancelers and equalizers. The possibility of providing cheap and inexpensive PRISDN services over traditional two twisted wire-pairs becomes more and more real. The techniques for simulations and presentation of the results are well suited to evaluate the response of the entire loop populations of other nations. These results indicate the changes in component design and to effects of the loop selection criterion for PRISDN service in most national networks.

The European subscriber plants differ significantly from those in the United States. Some of the differences are indeed beneficial to the extent that the loops in Europe are shorter, have fewer (if any) bridged taps, and most loops from the Central Office are unigauge (typically 0.32 or 0.4 mm). Some of the differences are indeed objectionable to the extent that the average dc loop resistance is higher than the 900 Ohm limit as it typically exists in the United States, with more number of gauge discontinuities in the typical loop. The study evaluates the net impact of such differences in the loop environments.

The results show that the 28 (0.32 mm) AWG causes some perturbation in the image impedance of the subscriber loops. The network termination impedance of (120+ j0.0) Ohms used in the United States is not likely to provide the best match for the loops of other national networks. In addition, the loop attenuation is slightly higher for this gauge. However the average loop length being shorter, does not unduly stress the equalizer. For this reason, it is desirable to reoptimize the equalizer part of the modem and its special requirements.

In addition, the results indicate the requirements on echo cancelers. If the terminating impedance is not appropriately adjusted, the echo signal can become disruptive since it is not the same as that in the United States loop environment. Because of fewer bridged taps, the exertion of the image impedance over the band of spectral interest is curtailed. Hence, the peaky loop response common in the United States subscriber environment is less frequent in the European loops.

9.11

REFERENCES

9.1   G.S. Moschytz and S.V. Ahamed 1991, "Transhybrid Loss with RC Balance Circuits for Primary Rate ISDN Transmission Systems", *IEEE Journal on Selected Areas in Communications*, Vol. 9, No. 6, 951-959.

9.2   G.S. Moschytz and S.V. Ahamed 1991, "Optimization of the RC Matching Network in Adaptive Active Hybrids for High Speed Data Communications", *Proceedings of the IEEE International Symposium on Circuits and Systems*, Singapore, June, 2741-44.

9.3   J.J. Werner 1991, "The HDSL Environment", *IEEE Journal on Selected Areas in Communications*, Vol. 9, 785-800.

9.4   M. Sorbara, J.J. Werner, and N.A. Zervos 1990, "Carrierless AM/PM", *Contribution T1E1. 4/90-154*, September.

9.5   M.J. Miller and S.V. Ahamed 1988, *Digital Transmission Systems and Networks, Volume II: Applications*, Computer Science Press, Rockville, MD.

9.6   A. Elrefaie, M. Romeiser 1986, "Computer Simulation of Single-mode Fiber Systems", *Conference Record*, Optical Fiber Communications Conference, February, 54-56.

9.7   M. Fashano, A.L. Strodtbeck 1984, "Communication System Simulation and Analysis with SYSTID", *IEEE Journal of Selected Areas in Communications*, Vol. J-SAC-2, January, 8-28.

9.8   K.S. Shanmugan, et.al. 1985, "Simulation of Digital Lightwave Communication Links Using SYSTID", *Record of Globecom 85*, December.

Chapter 10

COMPUTER BASED OPTIMIZATION TECHNIQUES FOR HDSL DESIGN

The design and optimization of the high-speed digital subscriber line (HDSL) need powerful computational strategies. Traditional techniques of distributing poles and zeros on Smith charts generally do not work. In the past, such approaches have lead to suboptimal designs for applications where the data capacity sought is considerably less than the Shannon capacity of the lines. Typical subscriber loops are less than perfect and for the current demands on the HDSL at T1/T2 and E1 rates every possible venue for the HDSL design needs to be investigated, if not exploited. In this Chapter, flexible computational techniques are presented that explore and optimize system components in view of the operating environment of the HDSL and/or the asymmetric digital subscriber line (ADSL) and inherent limitations of system components. The optimization occurs automatically by forcing the computer to track the effects of incremental changes of the subsystem performance (e.g., echo cancelers or equalizers), or the component values (Rs and Cs in the matching circuits) in context to the functional constraints of the (HDSL, ADSL, duplex, dual-duplex, triplex, etc.) line in conjunction with various subscriber loop environments (CSA loops, loops <18 kft., American, Australian, European, ANSI, ETSI, etc.).

10.1

INTRODUCTION

The modern HDSL, VHDSL and ADSL designs lie at the border of exploiting every loop hole in the loop plant. Coding characteristics, matching circuit poles and zeros, the extent of realizable echo cancellation, the nature and extent of equalization all have an impact on the success or failure of the digital subscriber line (DSL) in any particular national loop environment. In order to make the design procedure manageable and also make it evolve out of traditional DSL design procedures [10.1,10.2], *two* initial steps are proposed. These two steps divide the design procedure into *first*, the coarse matching the spectral domain characteristics (wherever it is possible to do so) of the components and *second*, the fine step of evaluating and relaxing the design of the first step to secure the most accurate recovery of the received data encoded at the transmitter. These two steps are discussed in Sections 10.2 and 10.3.

It is also essential to note the realization of the design as circuits where VLSI components are the *third* and final step in the implementation of computer aided design (CAD) tools for DSL. This aspect has been ignored in the earlier low to medium speed DSL, since circuits and systems were not performing at their own limits. However, in the design of HDSL and ADSL, circuit and board noise (however low it may be in laboratory interface boards), can be a source of signal degradation over the spectral band of interest. Thus the laboratory test does not conform to the predicted system performance as the signal attenuation starts to approach 60 dB in some applications. In a true sense, the constraints to build and realize the components need to be factored in the system design to achieve the overall performance. This is usually accomplished at the final stages when the poor performance of certain laboratory components can be compensated by relaxing the optimal design of the system to be able to salvage the HDSL/ADSL transceivers.

10.2

SPECTRAL DOMAIN OPTIMIZATION APPROACH

These techniques have a limited potential in the HDSL/ADSL design. Such techniques are ideally suited to situations where analog circuits dominate the components. The influence of realizable poles and zeros may be visualized in context to the location of other poles and zeros embedded in the system.

Typically in context to the HDSL/ADSL, the line matching circuit for a cross section of subscriber loops can be optimized if the topologies (and thus their own characteristic "frequency bumps") of the individual loops are given. The results of such an optimization strategy are presented in Section 10.4.

Next consider the equalization of low or medium speed DSLs. Here the loop transfer function in frequency domain depends significantly upon the distribution of the bridge taps in the loop plant. If the statistical properties of any particular loop plant are available, then the topology of the Tarbox [10.3, 10.4] or the T1C equalizer [10.5] may be altered to accommodate extra components and optimize their values. Such techniques have worked satisfactorily [10.6,10.7] for the traditional CSDC (56 kbps, TCM subscriber lines) now in use in many countries.

In spite of the limited success of these techniques, *three* severe limitations are evident. *First,* the global optimization of the system is not feasible since this approach matches the properties of one component with those of another (e.g., matching circuit with subscriber line; e.g., "bumpy" line loss with active RC equalizer circuits). The effect of equalization on the performance of other components (such as the clock recovery circuit, or stability of the echo canceler) is not considered. *Second,* the transfer of data over the DSL is essentially a time domain phenomenon, and to this extent a frequency domain optimization can only influence the time domain performance in an indirect way. *Third,* most of the recent components involve digital signal processing and frequency domain optimization does not have any real counterpart in the DSP dominated systems, where sampled error adapts the system parameters.

10.3

TIME DOMAIN OPTIMIZATION APPROACH

From a pragmatic consideration, the frequency domain optimization techniques do not hold significant promise for the HDSL, VHDSL and ADSL designs. However, the computation of the time domain performance is based upon the Fourier components of the time domain signals. To this extent, the frequency domain analysis becomes an integral part of the system simulation dominated by time domain optimization techniques.

At the seminal stage, the coarse separation of signals by the spectral energy distributions facilitates understanding the concept. Through the intermediate stages, frequency domain studies augmented by experience with similar systems identifies major stumbling blocks. However, the final third stage of the HDSL and ADSL projects needs almost all time domain techniques. Here, the entire system is simulated and optimized based upon the two major objectives: maximizing the margin against noise and maximizing the percentile population of loops that satisfy the minimum noise margin. Thus, all the design variables can be optimized in maximizing these objective functions. To some extent the optimization becomes a complex domain operations research problem.

The application of these techniques for the HDSL is presented in Section 10.5. In this case, the combined effects of the loop matching circuit and echo canceler (the equalizer and rest of the system components included) are studied for the CSA [10.8] and ETSI [10.9] loops.

10.4

OPTIMIZATION OF MATCHING CIRCUITS FOR CSA AND ETSI LOOPS

The topology of the matching circuits has been presented in [10.10]. In this Chapter, the optimization of the RC components is directed to matching all 10 CSA loops (see Figure 6.31a, b and c) at a line rate of 784 kbps (i.e., half T1 rate for dual duplex applications) or to match the 20 Swiss loops (see Table 10-2) at a line rate of 1024 kbps (i.e., half E1 rate, also for European dual duplex systems). The objective is to cancel the reflected signal over the band of interest (10-174.0 kHz for the dual duplex T1 HDSL: CSA loops, or 10-226.0 kHz at the E1 rate: Swiss loops). These spectral ranges are for 64-QAM constellation, carrierless amplitude phase [10.11, and also see Table 9-1] modulation with 4D trellis coding. In order to facilitate the computations, the objective function at each discrete frequency is defined as the absolute value of the difference between the reflections on the two sides of the balancing bridge (Figure 10.1). The sign is dropped and the ratio is plotted in dB. This number indicates the value of the uncanceled residual echo.

In Figure 10.2 the real and imaginary loop impedances are plotted over the band of interest. In Figure 10.3, the loop losses of the 10 CSA loops are plotted from 10-170 kHz. In Figure 10.4, the residual echo (voltage at the input of the differential operational amplifier D in Figure 10.1) is plotted.

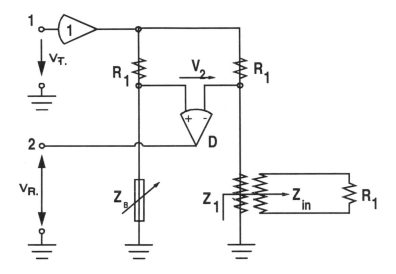

FIGURE 10.1. The Balancing Bridge for the Adaptive Matching Circuit.

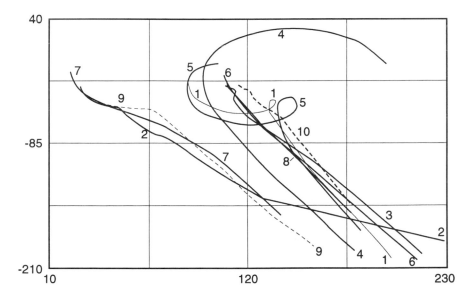

FIGURE 10.2. Real (X-axis) and Imaginary (Y-axis) Components (in Ohms) of the Input Impedance of the 10 CSA Loops from 10 to 170 kHz. The locus of these curves terminates closer to the origin as the frequency approaches 170 kHz.

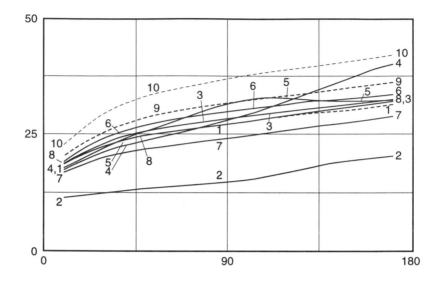

**FIGURE 10.3. Loop Loss (in dB) of the 10 CSA Loops vs. Frequency
(in kHz).** CSA - Carrier Serving Area

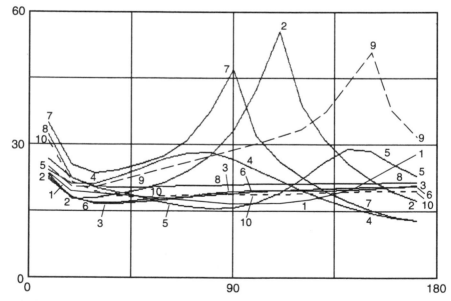

**FIGURE 10.4. Uncanceled Residual Echo (in dB) vs. Frequency (in
kHz) at the Central Office Side.** The optimization is to minimize the
average value of these for all the CSA loops with equal weighting. CSA -
Carrier Serving Area

The residual echo is computed as the difference between the full echo from the loop and the matching echo generated from the left side of the bridged termination shown in Figure 10.1. In addition to minimizing the residual echo over the entire frequency band, one additional objective function is also used. At the receiver, the difference between the residual echo (in dB) and line loss (in dB) is computed. Since the line loss increases along with frequency, the new constraints on the optimization force the selection of components based upon reducing the residual echo at the higher end of the spectral band at the expense of the residual echo at the lower end of the spectral band. Figure 10.5 depicts the corresponding residual echo as the objective function follows the former minimization (i.e., over the entire band). In Figure 10.6, this difference between the transhybrid loss and line loss is plotted, then new optimized RC components in the matching circuit are derived. Both these constraints offered realizable circuits.

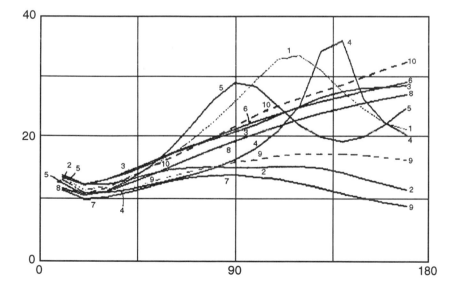

FIGURE 10.5. Uncanceled Residual Echo (in dB) vs. Frequency (in kHz) at the Subscriber Side. The optimization is to maximize the difference between transhybrid loss and the loop loss for all the CSA loops with equal weighting. CSA - Carrier Serving Area

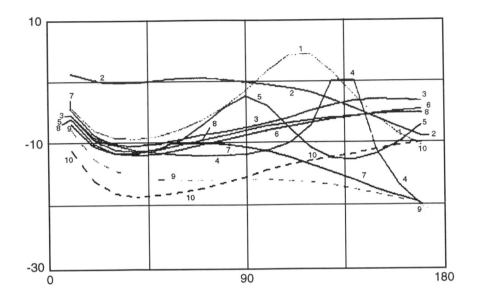

FIGURE 10.6. Uncanceled Echo (Minus Loop Loss) in dB vs. Frequency in kHz. The optimization is to maximize the difference between transhybrid loss and the loop loss for all the CSA loops with equal weighting. CSA - Carrier Serving Area

Convergence of the numerical relaxation of the individual RC component values is achieved by controlling the incremental changes (or the under-relaxation factor) for the critical Rs and Cs. The iterative process (though tedious and time consuming) does converge to realistic values of the individual Rs and Cs.

10.5

MAXIMIZATION OF CLUSTER SNR FOR 2-D (CAP) CONSTELLATIONS

Constellations can be generated by a time domain simulation. In this case, 8,196 symbols at the appropriate baud rate are generated from random bits of bidirectional data. The combined influence of the matching circuit and echo canceler are presented. For the 64-point constellations, the laboratory generated (8x8) point cluster has noise content due to line mismatch and imperfect echo cancellation. Additional influence of the misequalization and imperfect timing recovery also contributes to the cluster SNR. For the purposes of computation, the polar distance between each symbol and its ideal location is computed. Next, the 64-point constellation power is compared against the average noise power in the constellation. This evaluates the transmission quality of the system with one globally optimized matching circuit and the given echo canceler for the system.

Results of frequency domain and time domain optimization are presented in Table 10-1 for the dual duplex T1 rate HDSL. The average cluster SNRs for the 10 CSA loops (subscriber side) after frequency domain optimization of the matching circuit are presented in column 4. As can be seen, loop 10 (which has the maximum line loss) yields the poorest SNR of about 17.4 dB (with 45 dB EC) and the cluster SNRs can deviate significantly over the 10 CSA loops. The corresponding SNR after time domain optimization (column 5) is about 20.0 dB, and is about the same quality for the worst of the CSA loops (4, 7, 9, and 10). The relaxation procedure leads to different R & C values for the matching circuit.

TABLE 10-1 SNR Values for CSA Loops Before and After Time Domain Optimization (768 kbps and EC=45dB)

Loop No.	Loop Length Feet	B.T. Length Feet	SNR after Freq.Domain Optimiz. dB	SNR after Time Domain Optimiz. dB
1	7500	600	25.67	28.80
2	4350	1350	31.66	32.00
3	8550	300	26.77	30.32
4	7600	1200	20.47	20.38
5	7750	1200	23.44	26.64
6	9000	0	25.14	29.08
7	10700	800	20.12	20.00
8	12000	0	28.52	31.14
9	10000	500	18.77	20.44
10	12500	0	17.42	20.02

Results for twenty typical Swiss loops (no bridged taps) at dual duplex E1 rate are presented in Table 10-2. These loops are taken as a representative set of the Swiss loop survey with about 2500 loops in the Swiss Loop data base. Frequency domain and time domain optimization are both performed on these loops. The improvement in the cluster of SNRs is more significant than that in the American CSA loops. The frequency domain optimization of the matching circuits (derived by the scaled T1 optimization results) is presented in column 4. It can be seen that loop 18 (about 6.3 km or 20.67 kft. long) has an average cluster SNR of about 21.6 dB and the SNR for the other loops is not particularly good in spite of an echo canceler with 45 dB of cancellation for the 20 loops in the data base .

TABLE 10-2 SNR Values for Typical Swiss Loops Before and After Time Domain Optimization (1.024 Mbps and EC=45dB)

Loop No.	Loop Length Meters	Loop Length Feet	SNR after Freq. Domain Optimiz. dB	SNR after Time Domain Optimiz. dB
1	197	646	48.26	50.07
2	124	407	48.66	50.12
3	115	377	48.72	50.13
4	224	735	47.54	49.87
5	413	1355	46.48	49.16
6	981	3219	44.38	49.43
7	1747	5732	42.06	50.09
8	2057	6749	38.48	47.58
9	2154	7037	37.73	48.37
10	1724	5656	38.96	48.95
11	2537	8323	35.92	50.03
12	1482	4862	39.62	48.78
13	1901	6237	36.85	46.18
14	1750	5741	41.92	50.04
15	3731	12241	34.50	50.77
16	3435	11270	36.35	50.66
17	4196	13766	32.98	49.71
18	6259	20535	21.64	43.94
19	2245	7365	39.23	50.45
20	5413	17759	34.50	50.76

The corresponding SNRs after time domain optimization are shown in column 5, and increase to 43.9 dB for loop 18. All the 20 loops in the data base also realize significant gain in the cluster of SNRs.

Finally in Figure 10.7, the effect of the optimization techniques is depicted. As an illustration, CSA loop 9 has been selected, and the results are evaluated by computing the magnitude and phase of the residual echo at the receiver.

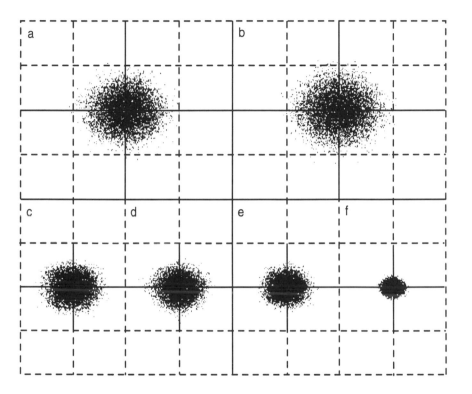

FIGURE 10.7. The Residual Echo in the 8192 Symbol, 64 Point Constellation Depicted as the Noises. The SNR is computed as the ratio (in dB) of the noise power to the power in the constellation. In Figure 10. 7a. the matching circuit is resistive at 110.0 Ohms (SNR=15.7dB). For Figure 10. 7b., the matching circuit of CSA loop 4 is used for loop #9 (SNR=14.9dB). In Figure 10. 7c., the frequency domain match for all the 10 CSA loops is used (SNR=18.8dB). In Figure 10. 7d., the frequency optimization procedure is directed to the CSA loop #9 (SNR=19.4dB). In Figure 10. 7e., the time domain optimization for all 10 CSA loops used (SNR=20.4dB). Finally in Figure 10. 7f., the time domain optimization is directed to the CSA loop #9 (SNR=26.8dB). CSA - Carrier Serving Area; SNR - Signal-to-Noise Ratio

    In Figure 10.7a, the effect of a single resistive matching circuit of 110.0 ohms is plotted as a scatter plot of the echo left over in the 64 point constellation. The average SNR for the constellation is 15.7 dB. In Figure

10.7b, the effect of an unoptimized RC matching circuit (see Z_b in Figure 10.1) is plotted. The average SNR for the constellation is 14.9 dB. In Figure 10.7c, the effect of the optimized RC matching circuit in the frequency domain for all the subscriber sides of the 10 CSA loops is plotted. The average SNR for the constellation is 18.8 dB. In Figure 10.7d, the effect of the optimized RC matching circuit for this particular CSA loop 9 in the frequency domain is plotted. The average SNR for the constellation is 19.4 dB. In Figure 10.7e, the effect of the optimized RC matching circuit in the time domain for all the Subscriber sides of the 10 CSA loops is plotted. The average SNR for the constellation is 20.4 dB. In Figure 10.7f, the effect of the optimized RC matching circuit in the time domain for this particular CSA loop 9 is plotted. The average SNR for the constellation is 26.8 dB. The mode of operation is dual duplex HDSL with 64-point constellation and the symbol rate is about 160 kbaud.

In Figure 10.7a, the matching circuit is resistive at 110.0 Ohms (SNR=15.7dB). For Figure 10.7b, the matching circuit of CSA loop 4 is used for loop 9 (SNR=14.9dB). In Figure 10.7c, the frequency domain match for all the 10 CSA loops is used (SNR=18.8 dB). In Figure 10.7d, the frequency optimization procedure is directed to the CSA loop 9 (SNR=19.4dB). In Figure 10.7e, the time domain optimization for all 10 CSA loops is used (SNR=20.4 dB). Finally in Figure 10.7f, the time domain optimization is directed to the CSA loop 9 (SNR=26.8 dB).

10.6

DISCUSSION OF RESULTS AND CONCLUSIONS

The performance of the Swiss loops at 1.024 Mbps is better than that of the CSA loops at 768 kbps. When the comparison is based on information for loops without bridged taps (CSA loops 6 and 8, Table 10-1), the loop loss of the typical Swiss loops is about 6 to 8 dB less than that of the CSA loops. This accounts for the marginal gain (34.5 dB Swiss loop No. 15 versus 28.52 dB CSA loop No. 8) in the performance levels of the long loops. Only frequency domain optimization results should be considered for comparison, since time domain optimization results involve the Fourier components of the entire band of frequencies. By and large, the lower high frequency loss, no bridged taps, and fewer discontinuities tend to favor the HDSL/ADSL performance in Europe. For providing an good matching circuit for the general loop population, the Swiss/European loops offer a much tighter cluster of scatter points for the image impedance. This observation is also true for the German loops, since there are no bridged taps in the German loop plant.

points for the image impedance. This observation is also true for the German loops, since there are no bridged taps in the German loop plant.

The techniques discussed in this Chapter yield significant improvements in the system performance if the components and subsystems are matched to the local loop conditions. For the CSA loop 9, the SNR for the entire 8192-point constellation can be enhanced from 15.7 dB to 26.8 dB by performing the time domain optimization and customizing it to that particular loop (see Figure 10.7a through 10.7f. The corresponding gain differs substantially from loop to loop and from code to code.

Ideally, an adaptive matching circuit would reduce the residual echo due to line impedance mismatch effects. The effect of standardized matching circuits is to yield a cluster SNR of 20.44 dB, (see Table 10-1 and Figure 10.7e), whereas the customized matching circuit yields a SNR of 26.8 dB (Figure 10.7f) for CSA loop Number 9. Only active hybrid realizations for terminating the line can provide the variable impedance generally needed in the North American loop environment.

Similar results at other rates (H_0, T2, in the simplex, triple duplex, full T1 & E1 rates, with CAP, 2B1Q, 3B2T, 4B3T, etc., codes) have been generated quite simply by changing the call sequence of the modular optimization software. Results presented here demonstrate that when the loops are operating at their limit (e.g., CSA loops at full T1/T2 rate or ADSL applications), then a powerful optimization strategy can lead to improved performance.

10.7

REFERENCES

10.1. Bell Laboratories 1982, *Transmission System for Communications*, Fifth Edition, Western Electric Company, Winston-Salem, NC.
10.2. S.V. Ahamed 1982, "Simulation and Design Studies of the Digital Subscriber Lines", *Bell System Technical Journal*, Vol. 61, 1003-1077.
10.3. R.A. Tarbox 1969, "An Automatic Equalizer for Digital Lines", *Proceedings of the IEEE*, Vol. 57, March, 363-364.
10.4. R.A. Tarbox 1969, "A Regenerative Repeater Utilizing Hybrid Integrated Circuit Technology", *ICC Convention Record*, 46-5 to 10.
10.5. P.C. Davis, J.F. Graczyk, and W.A. Griffen 1979, "Design of an Integrated Circuit for the T1C Low Power Line Repeater", *IEEE Journal of Solid-State Circuits*, Vol. SC-14, 109-120.

10.6. S.V. Ahamed, P.P. Bohn, N.L. Gottfried 1981, "A Tutorial on Two-Wire
Digital Transmission in the Loop Plant", *IEEE Special Issue on
Communications*, COM-29, 1554-64.

10.7. B.S. Bosik and S.V. Kartalopoulos 1982, "Time Compression
Multiplexing System for Circuit Switched Digital Capability," *IEEE
Transactions on Communications*, Vol. COM. 30, 2046-52.

10.8. Bellcore, *TA-NWT-001210*, Issue 1, October 1991.

10.9. CEG Working Group 1993, *Document for DTR/TM 3017 (HDSL)*, ETSI
TM3 WGI RG12, Munich, February 22-26.

10.10. G.S. Moschytz and S.V. Ahamed 1991, "Transhybrid Loss with RC
Balance Circuits for Primary Rate ISDN Transmission Systems", *Special
Issue IEEE J-SAC on High-Speed Digital Subscriber Line*, Vol. 9, August,
951-959.

10.11. M. Sorbara, J.J.Werner, and N.A. Zervos 1990, "Carrierless AM/PM,"
T1D1.4/90-154, September.

PART IV

SYSTEM PERFORMANCE FROM VERY LOW TO VERY HIGH RATES

The benefits of the digital mode for communication started to dawn on the telephone system designers with the advent of digital carriers. This was indeed an innovative concept in its own right. The T1 carrier paved the way to bring digital techniques to the trunk facilities and to some extent to the subscriber plant. Meanwhile, the computer revolution of the 1950s and 1960s was exerting an insidious influence on the telecommunication industry to serve distributed systems. The latter influence shaped the telecommunication industry toward low rate (120 bps, 240 bps, 480 bps, etc.), modem-based digital services from the 1960s to the early 1970s. In those days, the audio frequency based on-off type analog devices served very low speed peripherals, such as electro-mechanical teleprinters. This was indeed a very modest beginning, but it did set an important stage for the *modem* technology of the 1980s and 1990s at rates to 32 kbps (or even higher rates) using the V.bis and V.fast technology. The digital carriers themselves have followed a parallel expansion in their capacity. In this part of the book, the focus is on the copper-based systems in the loop plant and in the premises distribution systems. The impact of fiber is presented in Part V.

Chapter 11 discusses the numerous rates and their viability up to the E1 rates, i.e., 2.048 Mbps for European PRISDN. Chapter 12 presents the loop plant performance at the E1 rate. Chapter 13 presents the simulation results of uncoded or trellis-encoded systems in the American loop plant. In Chapter 14, the ultimate limits of United States loop plants and their Shannon capacity (with only crosstalk setting the upper ceiling for the data rates) are presented.

CHAPTER 11

DATA UP TO PRISDN RATES

The BRISDN, H0 and H1 rates in the loop plant imply the presence of a modern signaling system, if not a full-fledged ITU-T (formerly CCITT) standard SS7 system. Signaling is crucial to the viability of any ISDN service promised by the all-digital networks of the future. The transition can only be gradual since the telephone systems have massive capital investments already buried and sunk as copper wires. Traditional switching systems and the sustained effort make these various components and devices serve as a cogent and coherent communication facility. The vertical integration of these components is a tedious and well planned activity. The major concern of the owners for these systems is the utilization of the capital already committed to its full potential to realize the expected revenue stream. It is here that the higher rates start to become attractive to the network owners such as the European PTTs, AT&T (before the divestiture of 1996), the Bell Northerns and the Telecoms around the world. The challenges from the "wireless last-mile" and the "fiber to the curb" (if not "fiber to the home") are not trivial. The challenge is akin to the challenge that the bubble technology of the 1970s posed to the traditional disk technology for secondary memories. It is this challenge that forced the costs of Winchester disk storage systems to tumble down through the late 1970s and 1980s that led to the demise of the bubble memories.

In this Chapter, the evolutionary path through the H1/E1 rates are presented for the sake of countries where the fiber challenge is not as monumental as it is in the United States. In addition, low rate telephony applications do exist. Other possibilities with copper-based systems that need to be explored are the very low rate (8 to 9.6 kbps) speech oriented CELP techniques. These techniques may be implemented with the CELP modem chips. The cost of these chips is as likely to spiral down as the cost of the ADPCM chips. Such redesigns can reinstate the use of copper using the very techniques developed for wireless to replace the "copper last-mile".

11.1

SIMULATION AND NOT EXPERIMENTATION

While dealing with national environments, the viable approach is to perform feasibility analyses by simulation. The only experimentation is reserved as field trails. While the simulation and CAD (computer aided design) facility routinely generates spectral and time domain characteristics, it can also generate the physical and electrical characteristics of the loops in the loop data base. In most instances, the loop environments have significantly different characteristics. While it is not economical nor necessary to fabricate a customized VLSI chip for every loop environment, it is desirable to know how well each transceiver design will perform in any given environment. As a very first step, the selection can be based upon matching the physical characteristics of the loop plants with the transceiver design. As a next step, the matching can be based upon the electrical characteristics. Both physical and electrical characteristics are statistically evaluated parameters, and thus the penetration of (BRISDN, H_0 or H_1) services is also a statistical estimate.

Loop selection rules and range limitations also influence the physical and electrical statistics of loops for individualized services. Whereas all loops carry voice and voiceband data, selected conditioned loops under 18,000 ft. carry 56 kbps CSDC data (see Section 6.4.3) with a 99.9% probability. The statistics of the population of the general North American loops are harsher than those for the restricted set. For the sake of completeness, the *four* categories of loops in the North American environment are listed.

First, the general telephone loops for voiceband frequency can extend up to 210 kft and have a resistance value up to 3600 Ohms. Numerous architectures are presented in Section 6.3.1 and the type(s) of electronic device(s) necessary are tabulated in Table 6.1. The longer versions of these loops are unfit for any DSL applications. The question about the HDSL (at

PRISDN), SDSL (rate adjusting 160 kbps to 2.048 Mbps) and ADSL (up to 6.34 Mbps) applications does not even arise.

Second, the loops that are less than 18,000 ft. without loading coils have been analyzed for the data rates up to the PRISDN rate. However, the penetration is a foredrawn effect; as the rate increases the penetration decreases. Such loops carry the CSDC data at 56 kbps effectively in the TCM mode discussed in Section 6.4.3. A few loop selection rules may be necessary and some line conditioning also may be necessary because of the bridged tap reflections. In the early 1980s, while the move towards DSL was dramatic, Bellcore had released spectral characteristics of about 100 representative loops from the 1983 Loop Survey in the form of a data base [11.1]. While this data base does have the spectral properties of these 100 loops, the vendor community did not deploy the data base to design the DSL transceivers. These loops are not as widely recognized as the next two groups of loops.

Third, the loops that are under 12,000 ft. and without loading coils have been analyzed up to the H_1 rates. The penetration is less than acceptable with the technology of the 1980s, but starts to become viable with the current technology. In fact, DS3 rates have been achieved with adaptive hybrids and with very good echo cancelers in a subset of these loops.

Fourth, the CSA loops with selection based on loop length and bridged tap length(s) have been analyzed for rates through 2.048 Mbps. The most recent studies offer satisfactory performance with two-stage echo cancellation (see Section 6.5.2). The SDSL (for rates from 160 kbps to 2.048 Mbps) performance for some of the loops is acceptable. In a sense these loops are taxed to their full capacity based upon the transceiver limitation and the acceptable SNR in the cluster of the 2-D codes (see Chapter 9).

Twisted wire-pairs in the subscriber loop plant to carry data at rates significantly higher than the subrates (up to 19.6 kbps) have been investigated since 1976. In a sense, the advent and success of the T1 and E1 carrier systems (during the sixties) at 1.544 Mbps in North America and at 2.048 Mbps in Europe were the precursors to the possibility of high rate data over twisted wire-pairs. However, during the mid seventies, the investigations were directed towards combining the repeaterless features of the modem driven technology with the high-speed capability of the T1 and E1 carrier systems. The era of the digital subscriber line [11.2, 11.3] was thus started.

The success of a particular system depends upon the code, the set of loops selected and the technological sophistication in the transceiver design.

Some of the inferences about the penetration with the most viable codes in different environments at different rates are documented in Sections 11.2, 3 and 4. The decreeing fact is that the improvement in technology is likely to enhance the rates or the penetration to some extent, but not to violate the Shannon limits presented in Chapter 14. These results are in context to the United States loop environment where the loop surveys and documentation are both complete. Similar studies may become necessary for other countries.

11.2

DATA OVER ALL LOOPS SURVEYED

The physical statistics of the entire loop population (from the 1983 Loop Survey) are presented in Figures 11.1a, b, and c. Cumulative distributions are plotted in all three plots indicating, for example, that about 98 percent of the loops are under the 40,000 ft. from Figure 11.1a. Two lengths are plotted here in Figure 11.1a: the working length (i.e., the length of the copper wire between the Central Office and the Subscriber) and the total length, which includes the length of all the bridged taps on that loop.

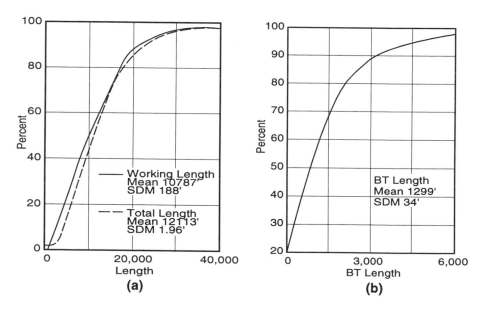

FIGURE 11.1. (a) Cumulative Loop Length Distribution, (b) Bridged Tap Length Distribution (1983 Loop Survey). BT - Bridged Tap; SDM - Standard Deviation of the Mean

The bridged tap distribution shown in Figure 11.1b has a mean length of about 1299 ft., which is slightly less than the difference between the means of the two lengths in Figure 11.1a. Some of the loops do not have bridged taps and this accounts for the lower value of the mean tap length. The drop length is the length of the copper wire (generally 24 AWG cable); (also see Section 14.3.3) between the end of the loop and the point of entry in the customer premises. Similar statistics are generated for residential, business, and special service loops. The means and standard deviation of the mean (SDM) are tabulated in Table 11-1. The loop length distribution in the data base is shown in Figure 11.2.

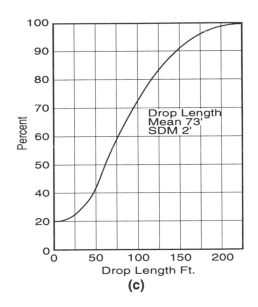

FIGURE 11.1. (c) Drop Length Distribution (1983 Loop Survey).
SDM - Standard Deviation of the Mean

Comparison of the 1983 results [11.4] with results from 1964 indicate a slight increase (about 10 percent in 19 years) in loop length and a significant decrease (about 40 percent in 19 years) in the bridged tap length.

About 24 percent of the loops have loading coils, and a small fraction of the loops have non-standard gauges; neither is included in the computation of the electrical characteristics.

TABLE 11-1 Physical Statistics of the All Loops

	Minimum Ft.	Maximum Ft.	Mean Ft.	SDM
All Loops				
Total Length	250	114,838	12,113	196
Working Length	186	114,103	10,787	188
Bridged Tap	0	18,373	1,299	34
Drop Length	0	1,000	73	2
Residential Loops				
Total Length	495	114,838	13,190	245
Working Length	186	114,103	11,723	236
Bridged Tap	0	18,374	1,490	44
Drop Length	0	1,000	89	2
Business Loops				
Total Length	250	100,613	9,840	302
Working Length	200	99,569	8,816	296
Bridged Tap	0	11,333	894	47
Drop Length	0	800	39	3
Special Services				
Total Length	391	63,348	10,212	658
Working Length	391	63,348	9,059	635
Bridged Tap	0	8,685	736	99
Drop Length	0	300	21	4

Results of a similar nature are summarized for the 1973 and 1963 Loop Surveys in Chapter 5. Comparison between the corresponding results of the various Surveys indicates the national trend and these trends are also summarized in Chapter 5. The results presented here are also indicative of the differences between the general loop population, the truncated loop population, the CSA loop population, and the sample ANSI loops and how they differ from decade to decade.

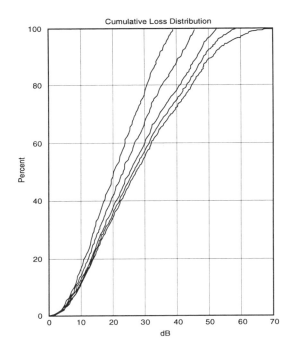

FIGURE 11.2. Cumulative Loss Distribution of All the 1983 Survey Loops (from the left) <10 kft, <12 kft, <14 kft, <16 kft, <18 kft at 200 kHz.

The average length of the remaining subset (excluding those with loading coils and non standard gauges) is 7,535 ft. with a SDM of 116. The scatter plots of the input impedance of all the loops at 1000 Hz (i.e., all loops intended primarily for voice and voiceband data) are shown in Figures 11.3 and 11.4 for the Central Office and the Subscriber side, respectively. Most of the human (male and female) speech has a peak energy level between 800-1000 Hz. The mean value of the loop impedance is about 914 - j 462 Ohms and corresponds to the 900 Ohms resistance with about 0.33 µF capacitance (X_c = - 482.2 Ohms) to terminate the voice lines in the United States. On the Subscriber side, the input impedance is slightly different (747 - j 529 Ohms) due to the concentration of gauge discontinuities and of taps closer to the subscribers. Other input impedances at 1.0, 2.0, 3.0 and 4.0 kHz and the image impedances are presented in Table 11-2. for the sake of completeness. The insertion loss (IL) scatter plot is shown in Figure 11.5 at 1.0 kHz. The mean values of insertion loss at 2.0, 3.0 and 4.0 kHz are 4.89, 6.81, and 12.85 dB, respectively.

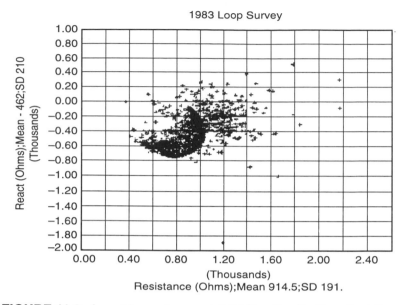

FIGURE 11.3. Input Impedance at 1.0 kHz (Audio Band) at the Central Office Side. SD - Standard Deviation

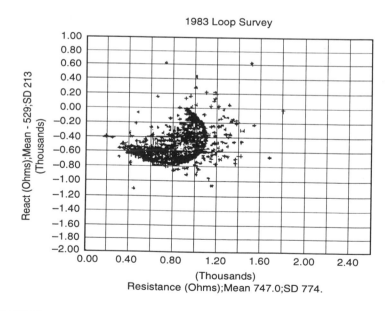

FIGURE 11.4. Input Impedance at 1.0 kHz (Audio Band) at the Subscriber Side. SD - Standard Deviation

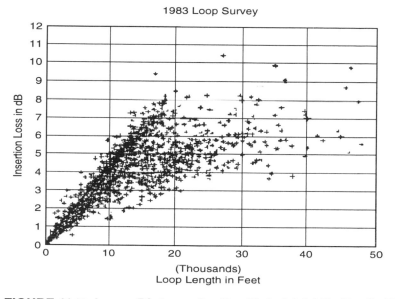

FIGURE 11.5. Loss - Distance Scatter Plot at 1.0 kHz (Audio Band).

TABLE 11-2 Input Impedances of All Non-Loaded Loops in the Speech Frequency Band

Frequency kHz	Real Z_{in} Mean; Ohms	SDM	Imag. Z_{in} Mean; Ohms	SDM
1.0 (CO Side)	914.5	191	$-j\,462.0$	210
1.0 (SUB Side)	747.0	774	$-j\,529.0$	213
2.0 (CO Side)	719.6	431	$-j\,402.5$	378
2.0 (SUB Side)	572.4	277	$-j\,277.3$	192
3.0 (CO Side)	836.4	335	$-j\,388.4$	566
3.0 (SUB Side)	438.8	360	$-j\,438.4$	275
4.0 (CO Side)	323.7	748	$-j\,402.3$	786
4.0 (SUB Side)	367.1	411	$-j\,723.9$	802

These loops can communicate CSDC data and subrate data (see Section 4.6) effectively. The 56 kbps CSDC system (discussed from considerations of line-equalization in Section 6.5.1.4) performs adequately at a burst rate of 144 kbps with the bipolar code (see Section 6.6.1) in the time compression multiplexing or TCM mode. The simulation results (confirmed by field tests in various locations [11.7] of the then existing Bell System of the early 1980s) of the entire system are presented in Figure 6.21.

However, the burst rate may be enhanced to higher rates: 162 (i.e., 2.25 x 72) kbps, 216 (i.e., 2.25 x 96) kbps and finally to 324 (i.e., 2.25 x 144) kbps for BRISDN services. These intermediate rates are at 72 kbps (B+8.0 kbps), and 96 kbps (B+32 kbps). Though not standard rates (126 to 324 kbps), the performance of the loop plant is only slightly worse than that at 56 kbps, and depends on the component optimization with powerful CAD techniques (see Chapter 10). Such systems have not been implemented in any country because of the shift in technology capable of fabricating cost-effective, high-quality echo canceler chips. The standardization of the 2B1Q full-duplex AEC system (Section 6.4.4) was the final blow to the TCM systems [11.5]. The DSL performance of TCM systems under the same conditions at 324 kbps (for BRISDN) is presented in Figure 11.6 to indicate the full potential of the loop plant. In this figure, the percentage eye openings of the bipolar code are plotted. Figures 6.2 and 11.6 are based on the 1973 Survey rather than on the 1983 Survey, but are representative of the performance, since the American loop plant has not significantly changed, except for the shortening of bridged taps, and this is seen as an positive change.

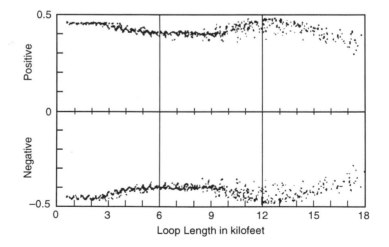

FIGURE 11.6. Eye Opening Scatter Plot for the Loops in the 1973 Loop Survey Data Base at 324 kbps in the Burst Mode of Transmission.

The BRISDN performance is shown as eye statistics scatter plots with the 2B1Q code in Figures 7.12 and 7.13. All the non-loaded loops in the 1983 Loop Survey up to 18,000 ft. carry the data at 144 kbps without any loop failure. The SNR has plenty of margin to satisfy the 10^7 BER criterion. Rates higher than 144 kbps (i.e. H_0 and H_1 rates) are impractical (though not impossible) for all the non-loaded loops with a range of 18,000 ft.. Even though they may work on a subset of such loops, the overall penetration in the range of loops from 12,000 to 18,000 ft. becomes too low, with single stage echo cancellation, for any serious consideration.

11.3

ALL LOOPS TO 12 KFT.

Loops under 12,000 ft. can barely carry H_1 rate bidirectional data with the 2B1Q code and 60 dB of one-stage echo cancellation. The asymmetry of the American loops causes severe problems in the reception of the down stream (CO to Subscriber) data at the Subscriber end. The two eye opening scatter plots for this code are shown in Figures 11.7 and 11.8 for the two directions of data flow (CO to Subscriber and Subscriber to CO).

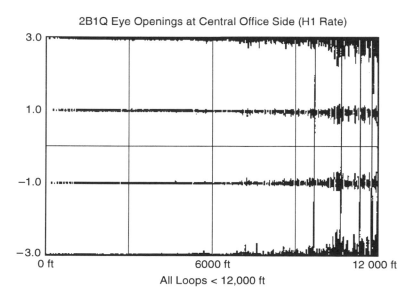

FIGURE 11.7. 2B1Q Central Office Eye Opening Scatter Plot for Loops < 12 kft.

The figures indicate the eye statistics (see Section 7.6) for the 2B1Q eye diagrams. A full vertical line from +3 to -3 level of the code signifies loop failure. Dense and dark figures indicate inferior plant performance. The performance of the code, as depicted in Figure 11.8, essentially terminates the consideration of this code for higher rates in the general subscriber plant with significant bridged taps. However, in loop plants without taps (i.e., Swiss, German, most of the British, Middle Eastern, etc.), the performance of the code can be significantly better than the performance depicted in Figure 11.7 and decidedly better than that depicted in Figure 11.8.

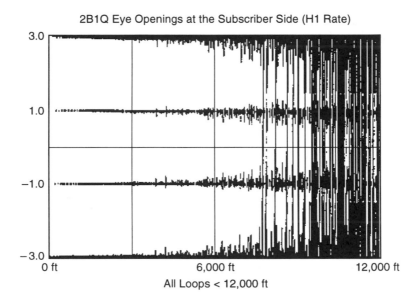

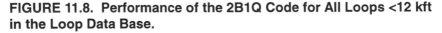

FIGURE 11.8. Performance of the 2B1Q Code for All Loops <12 kft in the Loop Data Base.

Bandwidth efficient cluster codes do perform better in the North American environments (see Sections 11.4 and 12.3). Two-stage adaptive echo cancellation (AEC) technology of the early-mid 1990s makes for the feasibility of the symmetric digital subscriber line, or the SDSL, with an adaptive rate from 160 kbps to 2.048 Mbps. For unconditioned loops these arrangements offer an optimum solution for the high(er)-speed digital subscriber line higher than the BRISDN rates yet consistent with the highest rate that the loop can bear.

11.4

CSA LOOPS

At the H_1 rate, the performance of the CSA loops is appreciably superior, even with the 2B1Q code. All the loops are able to carry the data with 60 dB of echo cancellation to the Central Office. On the Subscriber side, only one failure (out of 354 loops) is identified in Figure 11.9 as the single vertical line from +3.0 to -3.0, compared to the number of such vertical lines in Figure 11.8 for the general loop population. The corresponding plot (see Figure 11.10) at the Central Office side is *totally* open, thus assuring good transmission performance over the CSA loops. At H_1 rates, the cluster codes start to offer better performance.

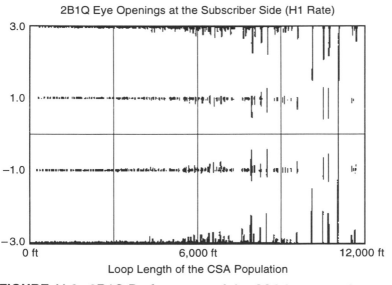

2B1Q Eye Openings at the Subscriber Side (H1 Rate)

Loop Length of the CSA Population

FIGURE 11.9. 2B1Q Performance of the CSA Loops at the Subscriber Side. CSA - Carrier Serving Area

The performance of the cluster codes with the ANSI loops (see Figure 5.30 and note that some of these loops are more severe than the CSA loops) at T_1 and H_1 rates is less certain, especially for loops 9, 13, and 14 with severe line loss and harsh topology. The SNR in these loops does not always meet the noise margin with 60 dB echo cancellation for 10^7 BER criteria. Detailed results, though available, are not presented here. With two-stage echo cancellation, the performance is barely adequate for these loops.

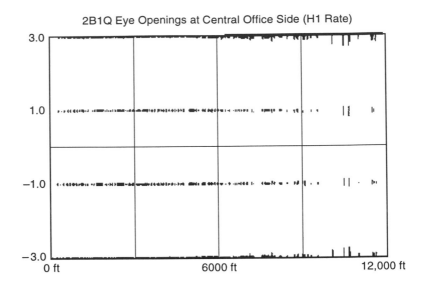

FIGURE 11.10. 2B1Q Performance of the CSA Loops at the Central Office Side. CSA - Carrier Serving Area

11.5

CODING OPTIONS AT LOW BIT RATES

Traditionally, certain codes such as AMI, Walsh, 2B1Q, Manchester, high density bipolar (HDBn), and MBNT (a more general form of 2B1Q) codes or the 1-D family of line codes [11.6], have been used throughout the subscriber plants. In the more recent modem industry, cluster codes or the 2-D codes are in vogue simply because the voiceband frequency (200-3200 Hz) of the telephone system is very restrictive. Whereas the modem data passes through all loops, switching and repeatering systems, the DSL data is for transit to and from Subscriber to Central Office. Whereas the 1-D codes are suited to the low rate DSL (BRISDN, H_0 and H_1, in some cases) applications; the performance of the 2-D codes are for higher rate data, as discussed in Chapter 13.

11.5.1 AMI CODE IN THE SUBSCRIBER PLANT

A widely prevalent code in the trunk and loop plant is the alternate mark inversion (AMI) or the bipolar code [11.6 and/or Section 6.6.1], which is commonly used with the T1 carrier systems (see Section 1.4.2). The AMI code is accepted for easy coding and decoding, for easy timing recovery, and for average zero frequency spectral components. However, it is criticized for inefficient utilization of the bandwidth, causing increased susceptibility to crosstalk and impulse noise. In the loop plant, there is ample opportunity to pick up low power frequency and its harmonics. The average zero frequency signal content of the AMI code provides some immunity at low frequencies. This feature also protects the transceiver against low frequency drifts during signal recovery at the receiver. These features of the AMI code made the code a favored choice in some of the digital carrier applications in the United States and Europe.

The frequent repeating (every 6000 ft.) of the T1 data is not applicable in the digital subscriber environment for ISDN. This consideration has prompted the deployment of more efficient block codes to deal with the line loss of loops to 18,000 ft. In the loop environment, the average frequency loss for the ISDN data is still too large in relation to the crosstalk and impulse noise power, as it prevails in most of the loop plants. In the sophisticated technological environment of the 1990s, the signal contamination, rather than the signal degradation, sets the limit for the rate and the range.

11.5.2 2B1Q CODE IN SUBSCRIBER PLANT

Perhaps the simplest of the block codes, the 2B1Q code, encodes two binary bits on to one of the four quaternary levels. For basic rate access ISDN, the 2B1Q code standardized by the ITU (formerly CCITT), well received by all vendors around the globe and already implemented by the Bell Operating Companies in the United States, dominates the digital subscriber line (DSL) applications. This code offers the desirable feature of the AMI code of having little zero-frequency spectral energy on the average. It also reduces the bandwidth requirement to about half that of the AMI code. However, since there is no zero state in this code as compared to the AMI code, the peak energy content is skewed towards the higher end of the frequency rather than at one-half frequency as occurs in the AMI code. This consideration prevents the 2B1Q code from offering the full 6 dB noise advantage (due to reduced bandwidth) over the AMI code. There are *two* other penalties associated with this code. *First*, the encoders and decoders (or codecs) are slightly more complicated. *Second*, the eye openings are proportionally reduced due to the presence of four levels at +3, +1, -1, and -3 (or three eye openings) as compared to the three levels at +1, 0, and -1 (or two eye openings). This is not seen as a major disadvantage since the signal processing components for data recovery are becoming more and more sophisticated. For the sake of comparison, the AMI and 2B1Q eye diagrams are depicted in Figures 7.7 and 7.9.

The 2B1Q code is only one of the many block codes that appear viable for ISDN. An entire family of the cluster codes or the 2-D coding techniques (thus leading to different types of clusters) is also available. The same reasons that have led to the choice of the 2B1Q code away from the more commonly used AMI code can also be extended to choice of cluster codes away from the 2B1Q code. The bandwidth may be reduced at the expense of component and circuit complexity, or the data rate may be raised to T_1 or E_1 rate without repeaters in the subscriber loops. It is to be emphasized that the actual feasibility of such codes for ISDN applications is not established, even though most of the simulation studies support this rationale.

11.6

DISCUSSION AND CONCLUSIONS

The transmission rates, the coding algorithm, and the particular loop environment play a combined role in the success of the DSL. The 2B1Q code for the BRISDN has been investigated for other rates such as 384 kbps (5B+D rate), and 800 kbps (11B+D, or half-T_1 rate). Serious deficiencies are observed at rates over 384 kbps in loop plants with multiple bridged taps. More recently, echo cancellation has not been taken as a serious limitation for the range over which data is transmitted. One-stage echo cancellation around 60 dB can be achieved and two-stage echo cancellation (incorporating adaptive matching circuits) of up to 70 dB can also be realized for most loop topologies in the CSA environment. Instead, impulse noise and crosstalk are the limiting issues. Other block codes such as 3B2T, or 4B3T offer only limited compression and have been rejected in the standards arena.

In this Chapter, the North American loop plant and it performance limits are presented in view of the imperfect echo cancellation and equalization. The advances in technology still cannot leap frog into the mass production of perfect devices, such as 100 dB echo cancelers or 20 second timing recovery circuits. Hence, the success of the digital subscriber line depends on the collective wisdom in the choice of code, the rate and the system architecture (simplex, duplex, dual duplex, multi-line, etc.) and finally the loop selection rules. If the transceiver is vendor supplied, then an accurate system simulation becomes necessary to evaluate the system performance as it is presented in this Chapter. If the final transceiver is customized, then the design of the devices (equalizer, echo canceler, timing recovery circuits, etc.) need to be negotiated in view of algorithmic, technological and functional limits. Even though these challenges are not serious for the low rate applications, they do take the transceiver design concepts to their limit for the HDSL and VHDSL and ADSL applications.

11.7

REFERENCES

11.1. Bell Communications Research 1985, "I-Match.1 Loop Characterization Data Base", Special Report *SR-TSY-000231*, June.

11.2 S.V. Ahamed 1982, "Simulation and Design Studies of the Digital Subscriber Lines", *Bell System Technical Journal*, Vol. 61, July-August, 1003-1077.

11.3 S.V. Ahamed, P.P. Bohn, N.L. Gottfried 1981, "A Tutorial on Two-Wire Digital Transmission in the Loop Plant", *IEEE Special Issue on Communications*, COM 29, No 11, 1554-64.

11.4 Bell Communications Research 1987, "Characterization of Subscriber Loops for Voice and ISDN Services", Science and Technology Series, Management Information Services Division, *ST-TSY-000041*; Piscataway, NJ.

11.5 B.S. Bosik1980, "The Case in Favor of Burst Mode Transmission for Digital Subscriber Loops," *Proceedings of ISSLS*, 26-30,.

11.6 M.J. Miller and S.V. Ahamed 1987, *Digital Transmission Systems and Networks, Volume I: Principles*, Computer Science Press, Rockville, MD.

11.7 B.S. Bosik and S.V. Kartalopoulos 1982, "Time Compression Multiplexing System for Circuit Switched Digital Capability," *IEEE Transactions on Communications*, Vol. COM 30, 2046-52.

11.8 Bell Laboratories 1984, *Special Issue of the Bell System Technical Journal dedicated to the SLC-96 Carrier*, December Vol. 63, No. 10, Part 2.

CHAPTER **12**

DATA AT PRISDN RATE

Various coding techniques are being considered for higher rates ranging from DS-1 over the existing twisted wire-pairs spanning the ISDN Central Offices to the Subscriber, and up to 155 Mbps from data distribution centers to desk tops. Rates higher than the basic information rate at 144 kbps, i.e., the H0 (384 kbps), 768 kbps, are discussed in Chapter 11. Most telephone networks have at least one metallic connection to the customers from the Central Office or a remote distribution facility. Enhanced use of this existing copper facility to transmit high-speed data appears to be a viable choice towards the ISDN video and imaging services. Simulation results for increased rate with ideal equalizers are presented in this Chapter. The use of fiber in the distribution plant facilitates the availability of very high rate data ranging from 51.64 Mbps (or the standard OC-1 rate) through 2.4 Gbps (OC-48 rate) at the network distribution centers. More recently, the OC-3 rate is gaining most popularity for the fiber coaxial ATM systems. However, the use of fiber to every home is not economical at this stage, because of the existing copper wire-pairs to customer residences and almost all businesses. There is also a lack of justification for such high rates (even the OC-1 rate) to every customer.

12.1

INTRODUCTION

In achieving a satisfactory design, there are *four* important parameters to consider: *first*, the maximum transmit power, which increases the distance range of the new system, but adversely influences the crosstalk into neighboring systems; *second*, the minimum receive power becomes critical to maintain the essential noise immunity to crosstalk and the floor noise from other systems, and the immunity to impulse noise in the Central Offices; *third*, the loss in signal level depends upon transmission media characteristics over the bandwidth of interest (for the code chosen) to communicate the data; *finally*, the self-interfering NEXT and/or FEXT (depending on the rate and code) become crucial in limiting the distance system range.

Other impairments to the transmission of data in the loop plant are presented [see Section 6.3, and/or Reference 12.1]. In this Chapter, the investigations are limited to the effects of echoes. The major reason for this Chapter is to document the effects and perhaps the limitations of single stage echo cancellation for the designers of echo cancelers at T1 the data rate. For the ISDN basic rate channel units, the echo canceler requirements have been delineated [12.3] and over 99% penetration of the basic rate ISDN capability at 18,000 ft. range is established at about 60 dB.

The two-port analysis techniques (see Section 7.2.1.1 and/or see Reference 12.2) provide the individual signals (received signal and individual echoes) accurately. The simulation environment (see Chapter 9) used in generating the results presented in this Chapter, incorporate the calculation of each of the echo signals. This systematically reduces the amplitude by the extent of echo cancellation (EC) specified as input data.

Four types of echoes can be identified in metallic transmission media during the transmission of data. *First*, echoes arising from the impedance mismatch, that is between loop and source impedance, can cause considerable transmission echo signal generated at the receiver. *Second*, the reflections caused by the splicing of cables, where different Sections of cable are joined together, cause echo of signals in one direction to appear in the reversed direction. *Third*, open bridged tap sections cause reflections from the remote end of the tap. These echoes are caused by the forward signal splitting at the junction of the cable and tap. The signal down the cable reaches the receiver at the end of the cable. However, the signal traveling the bridged tap also reaches the receiver as an echo, after being reflected at the open end of the bridged tap. *Fourth*, the mismatch between terminator source impedance and loop impedance (for the forward signal as it enters the loop), also cause an echo. It is to be appreciated that the reflection paths and the corresponding attenuation of the reflected signals will be different for each of the three (aforementioned) type of echoes.

12.2

TRANSMISSION CONSIDERATIONS

12.2.1 LOOP SELECTION GUIDELINES

The selection of the loops based upon the service is perhaps the first step toward assuring the transmission quality. The degradation of the services (from POTS to H1 rates) in the four broad varieties of loop populations (from all telephone loops 18,000 ft. to CSA loops) is documented and the choice of the loops plays a prominent role in the penetration of service. In this Chapter, the degradation of services at and above PRISDN rates in the ANSI, ETSI and the CSA loops and some of the Swiss loops is documented. It also quantifies the slow, but certain, dominance of the residual echo in copper-based systems over the slow and certain decay of the information bearing signal as the loop lengths increases.

12.2.2 ENCODING AND SPECTRAL SHAPING

Data signals have spectral energy. The subscriber loop plant unwittingly offers an excellent electromagnetic environment for exchanges of spectral energies of various correlated and unrelated signals. Correlated signals give rise to echoes and the unrelated signals give rise to noise. In a sense, the signals suffer both degradation and contamination as they are propagated, switched and processed. For this reason it becomes essential for the PRISDN system and component designers to contain the signal energies of any particular data stream within an allocated and well-chosen spectral band. The choice of bandwidth has to meet *two* rather conflicting objectives. *First*, it should be wide enough to contain the signals to communicate the data effectively. *Second*, it has to be narrow enough such that it does not contaminate other signals flowing in the loop plant and also not be contaminated by other signals. The choice of codes, discussed in Chapter 11, plays a role in the compatible coexistence of the many carriers and data signals in the subscriber plant. The study of these objectives and the associated techniques leads to spectrum management in the loop plant.

Spectral shaping is a powerful technique practiced by ISDN systems designers to bound and contain spectral energies of ISDN signals. The basic principle behind this technique lies in evaluating the segment of the spectral band that is most effective in signal transmission, then comparing it with the bands where it may crosstalk, contaminate, or otherwise influence the other signal carriers. By appropriate choice of spectral shaping, certain regions may be emphasized or de-emphasized to serve the duality of requirements presented in the previous paragraph. Signals flow from one circuit to another by stray capacitance, leakage currents through the conductance of copper wires, improper shielding and even by stray electromagnetic coupling. The coupling parameters (i.e., the M term of the $M.di/dt$ component in the pick-up voltages) are generally frequency dependent as are the signals that "crosstalk" into each other. Hence, one set of spectral shaping rules suitable for one carrier system become inapplicable for another environment. The designers generally individually tailor the spectral shaping for a particular application.

At the implementation level, spectral shaping can be accomplished by appropriate filters. This can be done by analog or digital devices. The eventual objective for ISDN is to limit and contain the signal energies in the various subscriber circuits and ISDN Central Offices. It is imperative that the ISDN bearer service meets the bit error rate (BER) error-margin for the data services. It is also essential that the signal-to-noise ratio and delay requirement (as they are both influenced by the filtering) for the voice and video services be within the range of tolerance.

One of the examples of spectral shaping exists in the circuit switched digital capability (CSDC, see Section 6.4, and/or see References 12.3, 12.4) introduced by pre-divestiture Bell System. When this service was introduced, there was a finite probability that the TCM mode (see Section 6.4.3) at 144 kbps burst-rate, used by CSDC in the loop plant would adversely effect the analog carrier systems (SLC-1[®] or SLC-8[®] slightly prevalent [12.5] in the United States) by contaminating the lower end of the spectral band. For the CSDC system with the AMI code (see Section 3.1), there is very little signal energy prevails at the lower end of the spectrum where contamination of the analog subscriber carriers is most likely. For this reason, spectral shaping of the CSDC AMI signals transmitted into the loop plant by high pass filters to block out the low-frequency low-energy signals is incorporated in the system. The crosstalk gain for the coexisting analog carriers is quite large compared to the penalty for the CSDC system.

12.3

CODES FOR PRIMARY RATE ISDN

The spectral energy content of the code is vital to its capacity to convey information over imperfect media. The viable codes are investigated, debated and standardized before the transceivers emerge for any application. The codes selected for the primary rate ISDN (PRISDN) need a high bandwidth efficiency to become viable in the subscriber loop plant. *Three* possibilities are discussed in this Section.

12.3.1 ONE-D CODES

The poor performance of these amplitude modulated codes over any sizable population of the loops may be inferred by the 2B1Q code at the 800 kbps (half-T1 or 11B+D) rate. Results are shown in Figure 11.8 for loops under 12,000 ft.. It is conceivable that the two-stage echo cancellation (providing up to 70-75 dB EC) may improve the performance, but the wider bandwidth requirement of this code still leaves it susceptible to NEXT and Self-NEXT noise contamination. For this reason, there is no serious effort to investigate this code for PRISDN in the United States or European environments.

12.3.2 TWO-D CODES

Numerous possibilities exist. The bandwidth requirement of the 4-PSK is the same as the 2B1Q and the same limitations exist. For this reason, the serious contenders are 16, 32, and 64 constellations to carry 4, 5 or 6 bits as symbols over the twisted wire-pairs. Theoretical limits where the Self-NEXT starts to totally obliterate (i.e. SNR of zero) the signal at the high-band edge are presented in Chapter 14. Before this limit, the residual echoes start to enhance the BER above an acceptable standard of 10^{-7} or 10^{-8}. Hence, the results presented in this Chapter pertain to the SNRs generated by such echoes, and they are used as a primary measure of the quality of transmission.

12.3.3 N-D CODES

N-D (n-dimensional) codes are a super set of 2-D codes. A multiplicity of 2-D codes are used in n/2 blocks constituting a N-D code symbol. For example, a 4-D code using two 2-D constellation points (say 64 points in each constellation) would make a 4-D code symbol. The receiver decodes one 4-D symbol at a time, i.e., it waits until the two neighboring 2-D constellation points are both received and decodes the encoded bits that gave rise to the particular combination of the two points in two constellations.

The desirability of N-D codes is explained as follows. Once again consider the two 64 point constellations. The possibility of any two particular sequential points is one in 64x64 or the probability of the event is 64^2. However, if all the possibilities are disallowed and only certain sequences are permitted by the code, then the distance between the 4-D code symbols is increased and thus leads to better immunity against Gaussian noise in the transmission system. Another advantage is that the total cluster power of the N-D codes can be less than that for the plain 2-D codes where all combinations of the two constellations are allowed. The example can be extended to multi-dimensional N-D codes with increased noise immunity at the cost of complexity of the codec and reduced effective number of bits/baud. Statistically, their performance can be enhanced over that of the plain 2-D codes, and in some cases over that of trellis codes (see Chapter 13). These codes need to be studied individually in context to the loop environments and the media characteristics. The performance of N-D codes for PRISDN in the American Loop environment is not seriously investigated because of the promise of the 2-D and DMT (discrete multi-tone) coding techniques in most of the loop environments.

12.4

THE 64 POINT 2-D CAP RESULTS AT E1 RATE

12.4.1 GENERAL UNITED STATES LOOP ENVIRONMENTS

Perfect transmission of all subscriber loops with 64 point constellation CAP would generate a perfectly spaced 8x8 matrix of points at the receiver. Each cluster becomes a point. The presence of echo degenerates the constellation from its ideal configuration of equally spaced points. If the polar distance between the received symbols and its idealized location is computed in terms of the Euclidean distance between idealized symbol points (i.e., the signal without any contamination, degradation or trellis encoding (see Chapter 13), then it becomes a measure of noise. The SNR can thus be computed as a aggregate of all the noise in all of the clusters for any given loop.

The performance of a subset of CSA loops, as specified by the ANSI Standards Committee for dual duplex T1 system (two wire-pairs) transmission, is presented in Table 10.1. The longer loops with two bridged taps suffer the most (SNR of about 20 dB), but can handle the data rate by enhancing the EC to 60 dB from 45 dB. Not all the CSA loops from the data base fare as well.

The performance of all the CSA loops in the CSA data base for T1 rate under (single wire-pair) full-duplex transmission is presented in Section 9.7.3. The ultimate limits of the performance of both 16 and 64 point constellations are presented as a series of Figures 9.11 to 9.17. However, some of the multinationals have been interested in obtaining the communication capability between the computer networks and the North American and European data centers at the E1 rate. For this reason, the performance limits of the North American loops at the E1 rate is presented in this Chapter.

Two sets of results are presented; the *first* set deals with the CSA loops and the *second* set deals with all loops under 12,000 ft. In the second set, the inappropriate and outdated CSA guidelines are ignored.

For the *first* set of results, the system performance is thus depicted as the plot of the SNR for all the CSA loops. In Figures 12.1 (CO side) and 12.2 (Subscriber side), the average SNR plots are plotted for all the loops in the CSA loop data base for 2-D 64 point constellations at the E1 rate. There are no loop failures and the average SNR in the worst of the loops at the subscriber end is about 10 dB. Longer loops suffer due to greater signal attenuation and the attenuation characteristics for the CSA loop population is depicted in Figure 12.3. The cumulative loss distribution of these loops is plotted in Figure 12.4. The image impedances of the CSA loop populations are plotted at the CO and the Subscriber side in two sets in Figures 12.5 and 12.6.

For the *second* set of results, the corresponding plots of the SNRs for all loops less than 12,000 ft. are shown in Figures 12.7 (CO side) and 12.8 (Subscriber side) at the E1 rate. The loop failure is significant in the population of loops over 7,500 ft, but a vast majority of loops do carry E1 rate data even though they do not comply with the CSA guidelines. Some of the designers have suggested the variable speed HDSL modems that monitor the bit error rate and calibrate the speed depending on the line condition. The highest speed in the range of 384 kbps to 1.544 or even 6.34 Mbps is thus achieved.

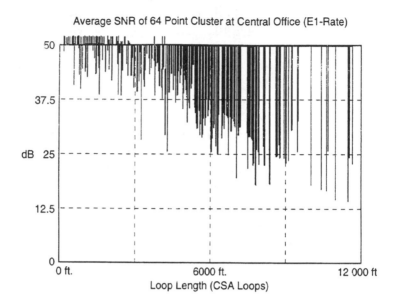

FIGURE 12.1. Signal-to-Noise in the 64 Point QAM Central Office Side. CSA - Carrier Serving Area; SNR - Signal-to-Noise Ratio

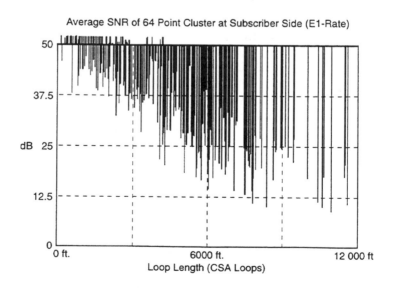

FIGURE 12.2. Signal-to-Noise in a 64 Point QAM Subscriber Side.
CSA - Carrier Serving Area; SNR - Signal-to-Noise Ratio

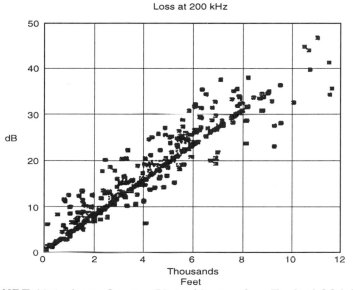

FIGURE 12.3. Loss Scatter Plot of Loops in a Typical CSA Data Base. CSA - Carrier Serving Area

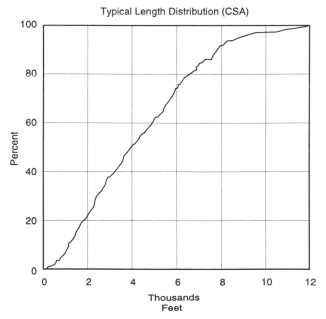

FIGURE 12.4. Length Distribution of Loops in a Typical CSA Loop Data Base. CSA - Carrier Serving Area

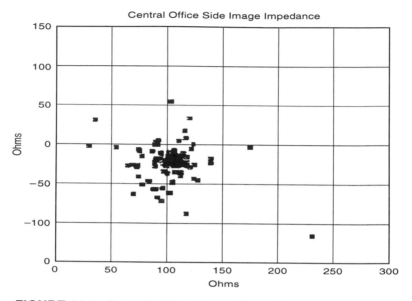

FIGURE 12.5. Central Office Image Impedance Scatter Plot for the CSA Loops. CSA - Carrier Serving Area

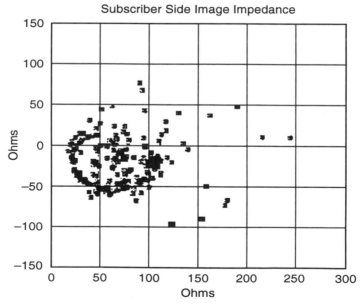

FIGURE 12.6. Subscriber Side Image Impedance Plot for the CSA Loops. CSA - Carrier Serving Area

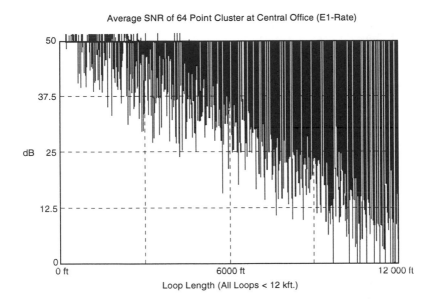

FIGURE 12.7. Average SNR in the 64 Clusters at the Central Office Side Constellations of All CSA Loops <12 kft. SNR - Signal-to-Noise Ratio

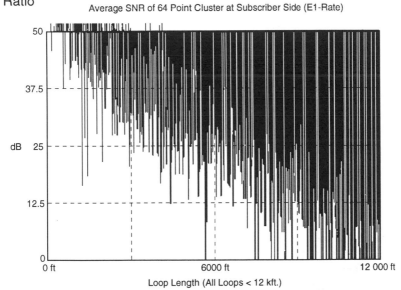

FIGURE 12.8. Average SNR in the 64 Clusters at the Subscriber Side Constellations of All CSA Loops <12 kft. SNR - Signal-to-Noise Ratio

The image impedances of the loop populations under 12,000 ft. are plotted at the CO and the Subscriber side in two sets in Figures 12.9 and 12.10. The signal attenuation for loops less than 12 kft. is the truncated set left of Line *ab* in Figure 12.11. The cumulative loss distribution of loops are plotted in Figure 12.12 by loop lengths in 2 kft. increments.

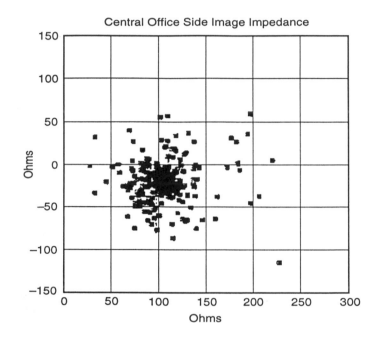

FIGURE 12.9. Central Office Image Impedance Plots for All Loops <12 kft in the Data Base.

Image impedance of loops in discrete populations becomes important in estimating the echo of the transmitted signal, which contaminates the weak received signal. Ideally, if the image impedance and the line termination impedance is the same over the entire band of the transmitted energy, then there would be no echo and no need for echo cancelers. To the extent that the image impedance cluster becomes dispersed, so does the demand on the echo canceler. By and large, the sensitivity of the echo canceler is also increased to the same extent, and the transceivers become more expensive. Some loop populations exhibit a greater diversity (e.g., United States loops vs. European loops; Japanese loops vs. German loops, etc.) and thus need more sensitive devices.

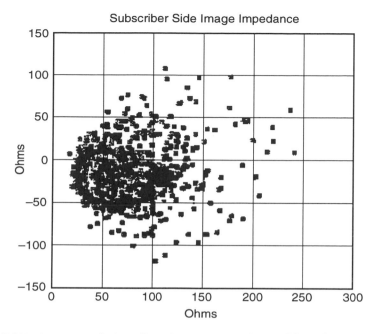

FIGURE 12.10. Subscriber Image Impedance Plots for All Loops <12 kft in the Data Base.

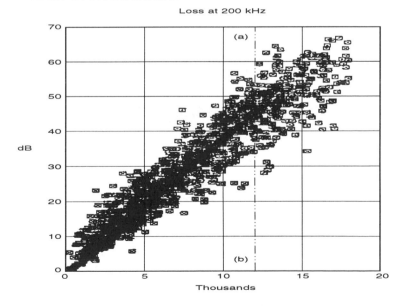

FIGURE 12.11. Loss Scatter Plots of All Loops <18 kft in the Loop Data Base.

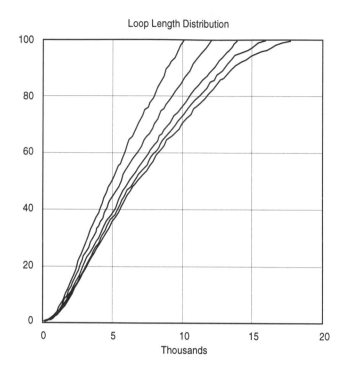

FIGURE 12.12. Cumulative Loop Length Distribution of All Loops in the 1983 Loop Data Base. The five curves (from left) are for loops <10 kft, <12 kft, <14 kft, <16 kft, and <18 kft respectively.

12.4.2 SWISS LOOPS PERFORMANCE AT E1 RATE

European loops individually perform much better, and the improvement in the SNR is evident by comparing Tables 10.1 and 10.2 for the dual duplex systems (i.e., T1/2 rate for the CSA loops and the E1/2 rate for the Swiss loops). Broad comparison of the performances of all the loops in the European data bases has not been attempted. Limited simulations of the selected loops from the United Kingdom and German environments tend to reinforce the better performance at both the dual duplex E1/2 and the full-duplex E1 transmission systems than that in the North American environments. Even though the simulation for all environment are not complete, it is possible to extrapolate the performance of various loop environments by examining the statistical parameters of the loops.   Only a computer based simulation will yield the precise numerical results.

12.5

DISCUSSION OF RESULTS

When the results of scatter plots for CSA loops and the all loops less than 12,000 ft. are compared, it is clear the a large fraction of the loops excluded by the older CSA selections procedure (see Section 5.5) do indeed carry the PRISDN rate quite effectively. This is especially the case up to about 6,000 ft. As a matter of fact, the SNR of the very first loop failure (see the full line from 50 dB to 0 dB in Figure 12.4 at the Subscriber end for the general loop population) occurs when the range is just over 5,500 ft.. This is significant when the average non-loaded loop length of the business customers is only 6,326 ft. (see Chapter 14 and/or Reference 12.7). Data services at this rate are generally solicited by businesses. The average residential customer has a non-loaded average loop length of about 7,865 ft. Noise contamination and timing recovery is not a serious problem when the loop length is less than 9,000 ft. because the loss at 200 kHz is less than 40 dB. This attenuation of 40 dB is the highest loss that the BRISDN 2B1Q signal suffers at its mid band frequency of 40 kHz in serving 99% loops less than 18,000 ft. (see Figure 6.11). In essence, a 40 dB loss limit for 99% loops under 9,000 ft. will permit a 64 point 2-D constellations to carry data at E1 rate. (See Table 9.1 signifying the center frequency of 190 kHz). This inference is also supported by Figure 14. 5(a) and Figure 14.9 in Chapter 14.

The other significant finding is about the adaptive hybrid techniques (see Chapter 10). The adaptation of the matching impedance start to lose its effectiveness as the variance of the image/input impedance start to increase dramatically, as it does in the American loop environments. This conclusion is evident from an analysis of Figures 12.5 and 12.7 in relation to the SNRs in Figures 12.1 and 12.3, in comparison to the relation of Figures 12.6 and 12.8 to the SNRs in Figures 12.2 and 12.4. In essence, the tightly packed impedance clusters (low variance of the European environments) offer higher SNRs and better noise immunity. A perfectly adaptive hybrid would compensate for all variations in the loop impedance and ideally make the CSA loop environment "appear" like a European environment. In Europe, the loops naturally have a much lower variation in the loop impedances (no BTs and fewer discontinuities). The extent of adaptation of the impedance can be judged by the gain in the SNR for the entire population of loops by the techniques presented in this Chapter.

Crosstalk and impulse noise dominate the received signal and range of transmission for the longer loops. Only a limited number of loops in the 9,000 to 12,000 ft. band from the entire loop population can carry the data at E1 rates, even with bandwidth efficient coding in the full-duplex mode. Bridged taps cause severe reflections that may appear at scanning instants, thus causing a maximum damage to the symbol recovery process.

12.6

CONCLUSIONS

The system performance is extended to all loops under 12 kft. rather than the preselected CSA loops. *Three* significant results ensue. *First,* that cluster codes (i.e. CAP techniques) outperform the conventional 2B1Q with a very significant margin. The gain in the penetration of the services can be quantified by comparing the results in Chapter 11 and Chapter 12. *Second,* adaptive hybrid is essential for HDSL rates approaching 2 Mbps. *Third,* the 1984 CSA loop selection guidelines are simply not applicable to the HDSL applications. The studies currently underway include the effects of phase jitter in the timing recovery circuits and the effects of impulse noise in bandwidth efficient coding with and without trellis encoding algorithms.

The loop configurations severely influence higher rate transmission, since the crosstalk and impulse noise become dominant in relation to the signal attenuation. A perfect equalizer and a high quality echo canceler will not significantly enhance the higher rate data penetration in the subscriber loop plant, especially for the longer loops. Intricate loop selection procedures can reduce the device requirements at T1 and E1 rates. But the current technology by itself can bring about faster rates and deeper penetrations than those currently marketed. The 2B1Q code that works well with simple loops, but suffers the effects of bridged taps is not a serious contender at all at higher rates. Shorter loops are indeed capable of carrying higher speed data and scatter plots indicate the eye closure and distance compromise with perfect equalization. Since equalization is not a severe constraint, the reflection constraint on the transmission rate was depicted in this Chapter.

12.7

REFERENCES

12.1 Bell Laboratories 1982, *Transmission System for Communications*, Bell Telephone Laboratories, Inc., Fifth Edition.

12.2 F.B. Llewellyn 1952, "Some Fundamental Properties of Transmission Systems," *Proceedings of IRE*, March, 271-283.

12.3 B.S. Bosik 1980, "The Case in Favor of Burst Mode Transmission for Digital Subscriber Loops," *Proceedings of ISSLS*, 26-30.

12.4 B.S. Bosik and S.V. Kartalopoulos 1982, "Time Compression Multiplexing System for Circuit Switched Digital Capability," *IEEE Transactions on Communications*, Vol. COM. 30, 2046-52.

12.5 Bell Laboratories, 1984, *Special Issue of the Bell System Technical Journal dedicated to the SLC-96 Carrier*, December, Vol. 63 No. 10, Part 2.

12.6 G.S. Moschytz and S.V. Ahamed 1991, "Transhybrid Loss with RC Balance Circuits for Primary Rate ISDN Transmission Systems", *Special Issue of the 1991 IEEE J-SAC on High-Speed Digital Subscriber Line*, Vol. 9, No. 6, August, 951-959.

12.7 S.V. Ahamed, P.L. Gruber, and J.J. Werner 1995, "Digital Subscriber Line (HDSL and ADSL) Capacity of the Outside Loop Plant," *Special Issue of the 1995 IEEE J-SAC on High-Speed Digital Subscriber Line*, Vol. 13, 1540-1549.

CHAPTER 13

PERFORMANCE OF TRELLIS CODING

Trellis codes combine channel coding and modulation techniques for data transmission over band limited channels. Theoretically, it achieves significant coding gains over conventional uncoded multilevel modulation without compromising bandwidth efficiency. In this Chapter, the effect of trellis encoding algorithms are studied, and the effectiveness for Carrier Serving Area (CSA) loop environments in the presence of imperfect echo cancellation at primary rate ISDN applications are evaluated. A basic methodology for the simulation of trellis codes for QAM clusters is utilized. The effect of trellis codes for 16 and 64 QAM is outlined. Results for the coded 16 and 64 QAM constellations for three typical CSA loops with 60 dB of echo cancellations are presented along with the effects of trellis codes on the entire CSA loop population.

13.1

INTRODUCTION

The 2-D Quadrature Amplitude Modulation (QAM) or the Carrierless Amplitude Phase (CAP) offers one of the most viable bandwidth efficient data encoding techniques. Data are coded in two orthogonal components in time dimensions. In these codes, a selected number of binary bits are encoded as a symbol. When the amplitude is held constant at $\hat{0}2$, the symbol may have a real (in-phase) and an imaginary (quadrature) component, thus leading to the simplest four-point cluster (1,1; -1,1; -1,-1; 1,-1), with two binary bits being encoded as one of the four cluster points. If the amplitude also can be used as an information carrying parameter, then the family of cluster points may be used. Generally, the X (in-phase) and Y (quadrature) components can also be directly encoded. This freedom to encode the two components independently permits three binary bits to be encoded as an eight-point cluster, four bits to be encoded as a 16 point cluster, or n bits to be encoded as a 2^n point cluster. In practice, the balanced clusters (e.g., 4, 16, 64, etc., 8-PSK, 32-CROSS, 128-CROSS, [13.1] and so on) are favored since the transmission medium (i.e., the twisted wire-pairs) can treat the in-phase and quadrature components and also the positive and negative signal levels alike [13.2].

Many line codes and variations of line codes were studied by the Exchange Carrier Standard Association (ECSA), that led to the standardization of the 2B1Q line code (every two bits of input signal are represented by a four-level or quaternary line signal) for the *Basic Access* DSLs. The studies assumed that Adaptive Echo Cancelers (AEC) and Decision Feedback Equalizers (DFE) were used in the transceivers [13.3]. Two dimensional (2-D) techniques for use with the existing twisted wire-pairs to transmit data to the subscriber premises for *Primary Access* have been presented in Section 11.9 and in Chapter 12. In this Chapter, the precise performance of trellis codes in American CSA loop environments is presented. Typically, five information bits are encoded into one six bit symbol. Trellis coding needs additional symbols (64 vs. 32) in the cluster due to the decreased information bits per baud. Whereas, five information bits could be mapped into a 32 point-CROSS, the six bits need a 8 x 8 (or 64) point cluster. Hence, the symbol rate is the same; it is just that there are more symbols with trellis encoding algorithms.

Trellis schemes employ redundant non-binary modulation in combination with a finite state encoder, which governs the selection of modulation signals to generate coded alphabet sequences. In the receiver, the noisy signals are decoded by a soft-decision maximum-likelihood sequence decoder. Simple four-state trellis schemes can improve the robustness of digital transmission against additive noise by 3 dB, compared to the conventional uncoded modulation. With more complex trellis schemes, the coding gain can reach a maximum of 6.28 dB [13.4]. The term "trellis" is used because these schemes can be described by a state transition (trellis) diagram similar to the trellis diagrams of binary convolutional codes.

We gratefully acknowledge that the results in this Chapter were generated by one of our students Dr. Ahmed Rashid Khalifa (now teaching at Alhazr University, Egypt) during his doctoral dissertation at the City University of New York. We also acknowledge his contribution in methodically enhancing the basic simulation software to generate the results.

For trellis codes, the bandwidth is no larger than an uncoded system, so the technique is particularly useful for band limited channels. However, since the transmitted power is the same in both the coded and the uncoded systems, the average energy per symbol for the coded system is less than that for the uncoded system. Detailed results of this simulation study under various conditions of the loops and bit rates at various echo cancellation values have been presented in Reference 13.1.

13.2

SOFTWARE ENHANCEMENTS FOR TRELLIS CODES

System performance is evaluated by the use of the generic software simulation platform discussed in Chapter 7. The computational algorithms and methodologies presented in Chapter 7 suffice for the performance evaluation of the uncoded systems. However, the Euclidean distances (EDs) and the free distances (FDs discussed in Section 13.3.1) in the presence of the residual echo (which is akin to colored noise) for the trellis coded system need to be carefully modified from the computed location of the cluster points in view of the reduced power of the coded system. When this additional step is incorporated, the simulation software is depicted in Figure 9.1. The computational modules closely parallel those described in Chapters 7 and 8. Individual loops selected for the evaluations of the trellis gains are depicted in Figure 9.2.

13.3

LOOP GAINS

13.3.1 16 QAM/8-PSK RESULTS

In this Section, the enhancement of the gain in the SNR is compared with the predicted theoretical gains of 5.33 dB [13.1 and 13.4] for the 8-state trellis encoded 16 QAM system over the uncoded 8-PSK system. The derivation of this gain is based upon its basic definition that

$$\text{GAIN} = 10 \text{ LOG}_{10} \{ (d^2_{ED.COD} / P_{AV.COD}) / (d^2_{ED.UNCOD} / P_{AV.UNCOD}) \} \text{ dB} \quad (13.1)$$

where $d^2_{ED.COD}$ is the minimum squared Euclidean distance or ED between the coded signal sequence for the coded systems. This distance is known as the free distance or FD. The denominator $P_{AV.COD}$ is the average signal power for the coded scheme. Likewise, $d^2_{ED.UNCOD}$ is the minimum squared Euclidean distance between the uncoded signal sequence for the coded systems. The denominator $P_{AV.UNCOD}$ is the average signal power for the uncoded scheme. The free distances for different trellis clusters are typically expressed in decibels (dBs) relative to the free distance of the uncoded system. The minimum squared Euclidean distance is evaluated by determining the minimum distance error event from the corresponding trellis diagram. This differs from one trellis encoding scheme to the next and the type of QAM cluster, thus a large number of theoretical gains may be derived. The average power is derived from the total constellation power divided by the number of points in the constellation. The generic details are presented in Reference 13.4 and the specific details for the 16 QAM/8 PSK scheme are presented in Reference 13.1.

Whereas the results presented in this Section include the effect of residual echo in the loops, no such noise (or more precisely no noise at all, other than the white Gaussian noise) is considered in the computation of the theoretical gain [13.4]. For this reason this theoretical gain of 5.33 dB is the same for all loops and independent of the loop topology. The simulated 16 QAM clusters for the typical CSA loops (i.e. loops 2 and 3 in Figure 9.2) are shown in Figures 13.1 and 13.2. The corresponding cluster for loop 1 is not noisy because of its simpler topology, and is not shown.

The corresponding clusters for 64 point QAM are depicted in Figures 13.3 and 13.4. These clusters (Figures 13.1 to 13.4) are in turn used to compute their SNRs. The cluster for loop 1 (Figure 9.2) is not noisy. Similar reasoning for the gains for the 64 QAM clusters over the 32-CROSS is consistently the same for all the loops at 3.77 dB. These gains are achievable under ideal conditions with perfect devices, and the practical device limitations offer less.

Loops that fail to provide coherent transmission are simply ruled out of any consideration in this Chapter since their SNR is zero. Such loops will fail with or without trellis coding and are of no significance except in providing some guidance in selecting the loop selection rules for PRISDN data transmission.

QAM 16   SUBSCRIBER SIDE PLOT

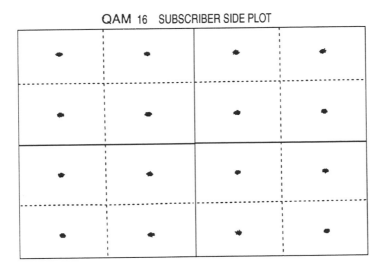

FIGURE 13.1. A 16 Point QAM Cluster Configuration for Loop 2 (Figure 9.2) with 60 dB Echo Cancellation at the Subscriber Side. QAM - Quadrature Amplitude Modulation

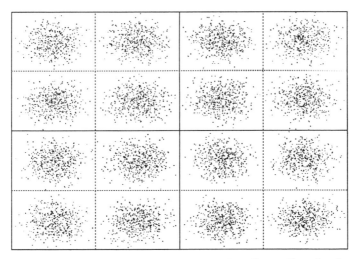

FIGURE 13.2. A 16 Point QAM Cluster Configuration for Loop 3 with 60 dB Echo Cancellation at the Subscriber Side. QAM - Quadrature Amplitude Modulation

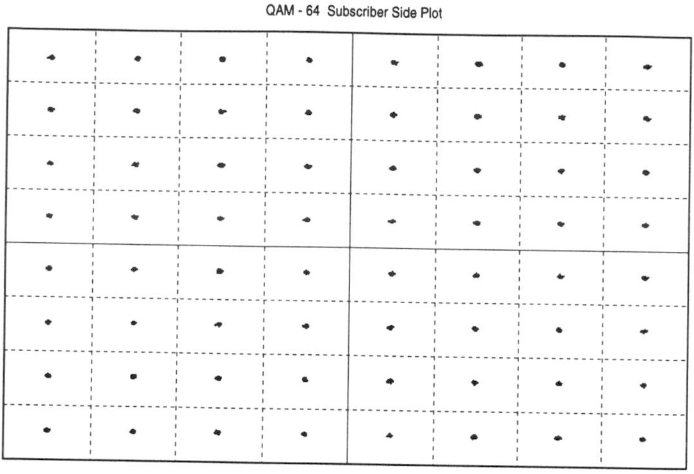

QAM - 64 Subscriber Side Plot

FIGURE 13.3. A 64 Point Cluster for Loop 2 (Figure 9.2) at 60 dB Echo Cancellation. QAM - Quadrature Amplitude Modulation

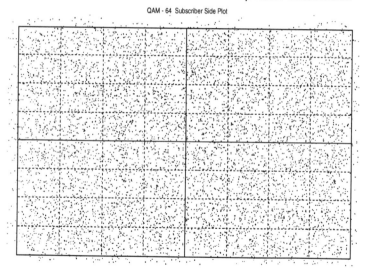

QAM - 64 Subscriber Side Plot

FIGURE 13.4. The 64 Point Cluster for Loop 3 (Figure 9.2) with 60 dB Echo Cancellation. QAM - Quadrature Amplitude Modulation

Only the Subscriber side clusters are shown, since they tend to be noisier due to the closeness of the bridged taps (see the topology of loop 3 in Figure 9.2). Similar cluster configurations are generated for the Central Office side and are used in the calculation of the trellis gains.

The 8-PSK simulations at 60 dB EC are used to calculate the noise in every point in each cluster. The noise is the Euclidean distance by which every point deviates from its ideal position. The eight ideal locations are situated on a circle of unit radius at 45° intervals. After the noise in every point in the cluster is calculated, average noise in the entire cluster is computed. It is used to calculate the gain in the 16 QAM clusters at 60 dB echo cancellation (EC). The trellis gains at both sides for the three loops (Figure 9.2) for various echo cancellations are presented in Table 13-1.

TABLE 13-1 Actual Gain *in dB* for 16-QAM/8PSK

Echo Cancel- lation	Loop # 1 no BT		Loop #2 1 BT		Loop # 3 2 BT	
	SUB	CO	SUB	CO	SUB	CO
55 (dB)	5.2980	5.3082	4.8966	4.8007	Error	2.8894
60 (dB)	5.3133	5.3189	5.0960	5.0456	Error	4.1084
65 (dB)	5.3221	5.3255	5.2026	5.1766	2.2406	4.6861
70 (dB)	5.3263	5.3283	5.2601	5.2440	3.8438	4.9744
75 (dB)	5.3293	5.3299	5.2917	5.2835	4.5655	5.1350

Note: The theoretical maximum for the trellis gain is 5.33 dB. Loop #1 (no BT) and almost no echo (last line) approaches this limit at the CO end.

13.3.2 64 QAM/32-CROSS RESULTS

In this Section, the enhancement of the gain in the SNR is compared with the predicted theoretical gains of 3.77 dB [13.1 and 13.4] for the 8-state trellis encoded 64 QAM system over the uncoded 32-PSK system. The evaluation of this gain follows the procedure identical to those outlined in Section 13.3.1. Whereas the results presented here include the effect of residual echo in the loops, no noise at all, other than the white Gaussian noise is considered in the computation of the theoretical gain [13.4]. For this reason the gain of 3.77 dB is the same for all loops and independent of the loop topology.

The simulated 64 QAM clusters for the three loops shown in Figure 9.2 are used to compute the trellis gains. A typical 64 point cluster for loop 2 (subscriber side) is depicted in Figure 13.3 for consistent and accurate transmission and the cluster for loop 3 (Subscriber side) is shown in Figure 13.4 for faulty and error-ridden transmission. Six such clusters are necessary to compute the trellis gains for the two sides of each of the three loops. The cluster for the coded 64 QAM for loop 1 is not noisy and thus not depicted

The 32-CROSS simulations at 60 dB EC are used to calculate the noise in every point in each cluster. After the noise in every point in the cluster is calculated, the average noise in the entire cluster is computed for the purpose of calculating the gain in the 64 QAM clusters at 60 dB echo cancellation (EC). The trellis gains of the 64 point trellis encoded system over the 32 point-CROSS at both sides for the three loops under various conditions are presented in Table 13-2.

TABLE 13-2 Actual Gain in dB for 64 QAM/32 CROSS

Echo Cancel-lation	Loop # 1 no BT		Loop #2 1 BT		Loop # 3 2 BT	
	SUB	CO	SUB	CO	SUB	CO
55 (dB)	3.6850	3.6953	3.1918	3.1408	Error	Error
60 (dB)	3.7215	3.7273	3.4555	3.4285	Error	1.7648
65 (dB)	3.7417	3.7448	3.5936	3.5817	Error	2.7965
70 (dB)	3.7528	3.7547	3.6704	3.6626	0.4468	3.2503
75 (dB)	3.7592	3.7600	3.7133	3.7089	2.2768	3.4879

Note: The theoretical maximum for the trellis gain is 3.77. Loop #1 (no BT) and almost no echo (last line) approaches this limit at the CO end.

13.3.3 LOOP TOPOLOGY; RESIDUAL ECHO AND TRELLIS GAINS

Two observations are in order. *First*, the highest realization of the trellis gain (close to its theoretical limit) is for loop 1 at 75 dB of echo cancellation and *second*, the poorest value of the trellis gain is for loop 3 at 55 dB of echo cancellation. *Two* inferences are evident. *First*, an undesirable loop topology (i.e., bridged taps) has severe ramifications upon the realizable trellis gain and *second*, low performance devices (an echo canceler in this case) in relation to the loop topology and the nature and location of the impairment (length and nearness of the bridged tap(s), in this case) has an equally detrimental effect on the predicted trellis gains.

13.4

SYSTEM PERFORMANCE AND DESIGN CHOICE

Two options are available for the designer. *First*, the designer may keep the same symbol rate, but enhance the number of symbols from 16 to 32 (four information bits/baud), from 32 to 64 (five information bits per baud), or from 64 to 128 (six information bits per baud), etc. This is the conventional choice by which the comparison is made for the trellis encoding. The performance under this option is delineated in Section 13.3 for the three typical CSA loops.

Second, the designer may keep the same number of symbols, but increase the symbol rate. For the 16 symbol cluster and the T1 rate, this means that the rate would increase from 386 kbaud (no coding and 4 information bits per baud) to 514.66 kbaud (coding and 3 information bits per baud). The results are presented in Section 13.4.1. For the 64 symbol cluster and the T1 rate this means that the rate would increase from 257.33 kbaud (no coding and 6 information bits per baud) to 308.8 kbaud (coding and 5 information bits per baud). The results are presented in Section 13.4.2.

The implication of these two choices are significant. In the first choice, the bandwidth requirement and the noise exposure to crosstalk are the same, but the SNR is increased by the trellis gain, while the average symbol power is lower with the trellis codes. In the second choice, the bandwidth requirement and crosstalk are both increased but the symbol power is the same and the SNR is increased by the trellis gain. It is the second of these options that is investigated in this Section.

13.4.1 COMPARISON OF 16 QAM TRELLIS CODED (520 KBAUD) WITH UNCODED 16 QAM (390 KBAUD)

In the trellis coded 16 QAM scheme, three bits are directed to the convolutional encoder, becoming four bits at the encoder output, which are then applied to the line coder input for symbol transmission. For primary rate data, the transmission rate is 514.67 kbaud and approximated in the study to 520 kbaud. In the uncoded 16 QAM scheme, no convolutional encoder is being used. Therefore, four bits are input directly to the line coder. As a consequence, the transmission rate decreases to 386 kbaud approximated in the present study to 390 kbaud.

Figures 13.5 and 13.6 illustrate the SNR plots for the entire CSA data base (354 loops, see Section 5.5), using the trellis coded and uncoded 16 QAM schemes at 60 dB EC for Subscriber sides, respectively. The two figures are to be compared at the PRISDN (T1) data transmission. There are no errors at the Central Office side; the corresponding SNR plot for the uncoded system is shown in Figure 13.7. The lines in this figure do not reach the floor in any of the 16 clusters, thus assuring error free transmission. The corresponding SNR plot for the coded system also indicates error free transmission.

Trellis codes provide noise immunity and higher gain approaching the ideal gain for some loops. Other loops fail to carry the PRISDN data using trellis codes. The reason is due to the loop topology. The uncoded scheme carries the data for some of the loops that failed the PRISDN transmission using trellis coded scheme. The reason for this result is due to the lower symbol rate. The loop configuration has a tremendous effect on the ability of a loop to carry coded or uncoded data. The length of the loop and of the bridged taps (BTs), the number and total length of BTs, and the location of each BT with respect to the subscriber side, each has its own individual effect. These factors govern the quality of transmission in any loop environment.

In summary, at 60 dB EC with uncoded 16 QAM, six loops (1.7% of CSA loops) failed to carry the PRISDN data, compared to thirteen loops (3.7% of CSA loops) in the coded 16 QAM scheme. Table 13-3 presents the details and topologies of loops unable to carry the PRISDN data for the coded and uncoded 16 QAM schemes.

Note that the severity of the failure (in Table 13-3) is indicated by the number of clusters in error. Loops with most number of sections (or discontinuities) and the two bridged taps are more likely to fail than others. Loops from the 354 CSA Loop data base without any BTs have all carried the PRISDN data with 60 dB echo cancellation. The margin may still be unsatisfactory for any practical system. At other lower rates the systems start to become attractive for the North American unconditioned subscriber loops. It has also been observed the loops longer than 12 kft and without bridged taps (such as those in the Swiss, German and other European countries) carry the PRISDN data effectively. For this reason, we contend that the CSA concept of limiting the loop length to 12 kft be abandoned for the new HSDL and VHDSL applications. In a sense, the SDSL line which monitors its own SNR in the CAP clusters and adjusts the rate accordingly is a compromise for achieving the highest rate without conditioning the subscriber lines.

TABLE 13-3 Loops that Failed to Carry PRISDN with 16 QAM at 60 dB EC

LP#	Sects.	BT	BT Len. ft.	LP Len. ft.	Code 16-QAM	Clus in Err.	Unc 16-QAM	Clus in Err.
3	4	2	1158	6964	Fail	8	Fail	12
18	2	1	234	7390	Fail	8	Pass	--
22	2	1	340	8447	Fail	15	Fail	10
40	4	2	443	7562	Fail	9	Pass	--
53	5	2	557	6022	Fail	13	Pass	--
119	2	1	241	8107	Fail	16	Pass	--
129	2	1	120	8857	Fail	7	Pass	--
131	4	2	988	7800	Fail	16	Fail	16
186	3	1	905	10642	Fail	16	Fail	15
239	2	1	125	10032	Fail	13	Fail	6
301	4	2	1229	7286	Fail	13	Pass	--
331	2	1	323	6587	Fail	14	Pass	--
350	4	2	1746	6862	Fail	7	Fail	2

Note: Pass/Fail simply indicates data-transmission integrity between the transmitter and receiver in the bidirectional mode. The SNR requirements are not addressed in these results. However loops that show fewer clusters failures have the higher SNRs than loops with higher cluster failures.

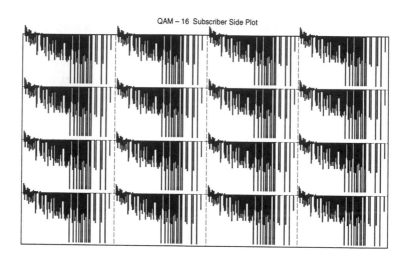

QAM – 16 Subscriber Side Plot

**FIGURE 13.5. Signal-to-Noise Ratios Plotted as Lines for Each of
the Loops in Every One of the 16 Clusters at 60 dB Echo
Cancellation on the Subscriber Side (Coded System).** QAM -
Quadrature Amplitude Modulation

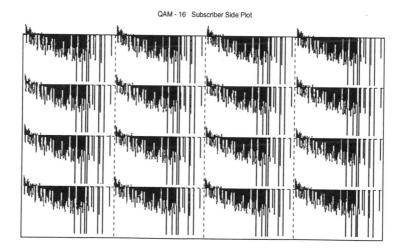

QAM - 16 Subscriber Side Plot

**FIGURE 13.6. Signal-to-Noise Ratios Plotted as Lines for Each of
the Loops in Every One of the 16 Clusters at 60 dB Echo
Cancellation on the Subscriber Side (Uncoded System).** QAM -
Quadrature Amplitude Modulation

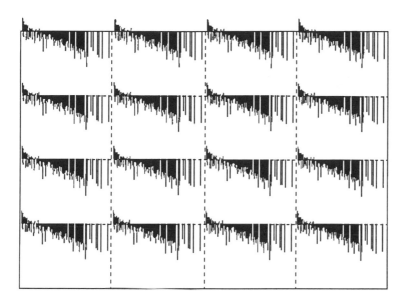

FIGURE 13.7. Signal-to-Noise Ratios Plotted as Lines for Each of the Loops in Every One of the 16 Clusters at 60 dB Echo Cancellation on the Central Office Side (Uncoded System). QAM - Quadrature Amplitude Modulation

13.4.2 COMPARISON OF 64 QAM TRELLIS CODED (310 KBAUD) WITH UNCODED 64 QAM (260 KBAUD)

In the trellis coded 64 QAM scheme, five bits are input to the convolutional encoder, becoming six bits at the encoder output. They are then applied to the line coder input for symbol transmission. The transmission rate is 308.8 kbaud and approximated in the study to 310 kbaud.

In the uncoded 64 QAM scheme with no convolutional encoder, six bits are input to the line coder. The symbol rate decreases to 257.33 kbaud that is approximated in the study to 260 kbaud. Figures 13.8 and 13.9 illustrate the SNR for the entire CSA data base, using the trellis coded and uncoded 64 QAM scheme at 60 dB EC for subscriber sides, respectively. The corresponding SNR plots (coded and uncoded) on the Central Office side, shown in Figures 13.10 and 13.11, respectively, indicate fewer transmission errors. The two figures depicted in 13.8 and 13.9 are to be compared for the PRISDN (T1) data transmission. As presented in Section 13.4.3, some loops offer the ideal coding gain. Other loops fail to carry the PRISDN data.

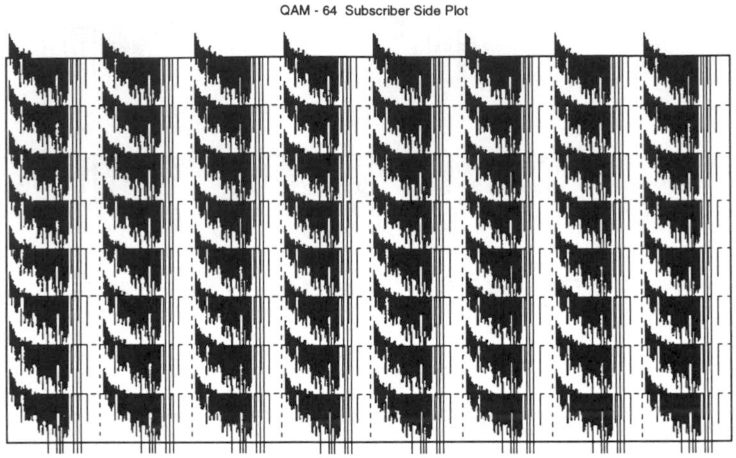

QAM - 64 Subscriber Side Plot

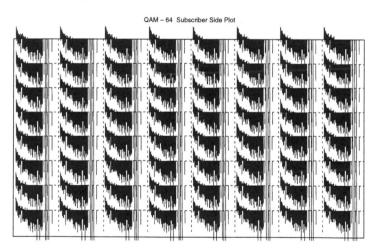

FIGURE 13.8. Signal-to-Noise Ratios Plotted as Lines for Each of the Loops in Every One of the 64 Clusters at 60 dB Echo Cancellation on the Subscriber Side (Coded System). QAM - Quadrature Amplitude Modulation

QAM – 64 Subscriber Side Plot

FIGURE 13.9. Signal-to-Noise Ratios Plotted as Lines for Each of the Loops in Every One of the 64 Clusters at 60 dB Echo Cancellation on the Subscriber Side (Uncoded System). QAM - Quadrature Amplitude Modulation

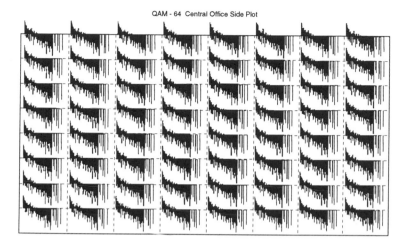

FIGURE 13.10. Signal-to-Noise Ratios Plotted as Lines for Each of the Loops in Every One of the 64 Clusters at 60 dB Echo Cancellation on the Central Office Side (Coded System). QAM - Quadrature Amplitude Modulation

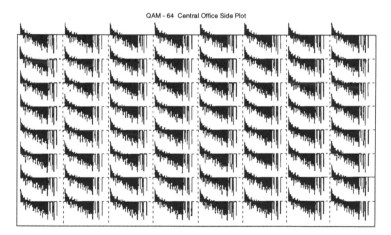

FIGURE 13.11. Signal-to-Noise Ratios Plotted as Lines for Each of the Loops in Every One of the 64 Clusters at 60 dB Echo Cancellation on the Central Office Side (Uncoded System). QAM - Quadrature Amplitude Modulation

For some loops, the uncoded scheme is found to offer more SNR over the trellis coded scheme in spite of the trellis gain. This is due to the lower symbol rate. This results presented here consider the effects of the residual echo as it varies from one rate to the next. The mismatch effects between the input impedance of the loop and its fixed source impedance at the transmitter and also the loop impedance and its fixed termination at the receiver are also both considered. In the traditional analysis, perfect equalizers and echo cancelers are assumed and this assumption is likely to give the trellis coded systems an unrealizable advantage in practical applications. By comparing the performance of the 16 vs. 64 trellis encoding, the trellis coded 64 QAM scheme has fewer (7 versus 13) loop failures than the trellis coded 16 QAM scheme (see Table 13-3). At 60 dB EC with *uncoded* 64 QAM, five loops (1.4% of the CSA loops) failed to carry the PRISDN data, compared to seven loops (2% of CSA loops) in the *coded* 64 QAM scheme. Table 13-4 presents the details and topologies of loops unable to carry the PRISDN data for the coded and uncoded 64 QAM schemes.

TABLE 13-4 Loops that Failed to Carry PR-ISDN with 64 QAM at 60 dB EC.

LP#	Sects.	BT	BT Len. ft.	LP Len. ft.	Code 16-QAM	Clus in Err.	Unc 16-QAM	Clus in Err.
3	4	2	1158	6964	Fail	61	Fail	60
103	4	2	1173	5683	Fail	6	Pass	- -
131	4	2	988	7800	Fail	61	Fail	5
186	3	1	905	10642	Fail	62	Fail	63
209	3	1	800	9464	Fail	4	Fail	2
257	3	1	608	10164	Fail	58	Fail	33
265	4	1	515	7493	Fail	44	Pass	- -

Note that the severity of the failure is indicated by the number of clusters in error. Longer loops with most number of sections (or discontinuities) are more likely to fail than others with the 64 point QAM code.

13.5

DISCUSSION OF RESULTS AND CONCLUSIONS

The effects of the components and the limitations in their realization are combined with the theoretical gains to predict the actual advantages of the well known trellis codes. For each constellation, the average noise in both schemes (i.e., coded and uncoded) is calculated. The coding gains for the 16 and 64 QAM using *trellis codes* are evaluated with respect to the *uncoded* 8-PSK and 32-CROSS schemes, respectively for three loops typical of the CSA loops. In addition, the comparison of performance of the *trellis coded* 16 QAM scheme (520 kbaud or kS/s, i.e., three information bits per symbol) with the *uncoded* 16 QAM scheme (390 kbaud or kS/s, i.e., four information bits per symbol) is also presented.

The 16 and 64 QAM constellations are presented throughout this study. The response of the typical CSA loop environment is evaluated. A comparison between the *coded* and *uncoded* schemes for the two constellations (i.e., 16 QAM and 64 QAM) at 60 dB EC is also presented. It is noticed that the trellis coded schemes work well with some loops and fail with others due to the loop topology and the critical locations of BTs. The uncoded schemes are also influenced by the same factor, providing a lower failure rate with 16 and 64 QAM clusters.

The symbol rate for the coded 16 QAM scheme (520 kbaud) is 25% more than that of the uncoded 16 QAM scheme (390 kbaud). The loop failure rate is thirteen versus six, respectively for the typical loops in the CSA loop data base. The symbol rate for the coded 64 QAM scheme (310 kbaud) is about 17% more than that of the uncoded 64 QAM scheme (260 kbaud). The loop failure rate is seven versus five, respectively for the loops in the CSA loop data base. The reason for the increased loop failure arising in the trellis coded system is also due to the increase of residual reflected signals at higher rates as well as the attenuation of the received signal level. These two reasons together do not compensate for the trellis gain of the coded system. In conclusion, the noise advantage of the trellis codes needs to be critically evaluated for the particular loop environment. The practical limitations of the system components and characteristics of the transmission media can prevent the complete realization of the ideal noise advantage of the trellis coding algorithms.

If a perfect transceiver can be fabricated, and if the media approaches its ideal status (no discontinuities of any sort), then the theoretical gains of the trellis encoding algorithms will be fully realized in practice. Undesirable loop topologies undermine the trellis gain more dramatically than for loops with fewer discontinuities.   In a sense, the results presented in this Chapter quantize the expected degradation in the typical CSA environment, but the inference reaches further. When the media characteristics are better than those in the CSA environment (e.g. in the European loop plants) the expected average trellis gain over the entire loop population can be closer to the theoretical predictions (see Tables 13.1 and 13.2 for the reasoning). When the media characteristics are only slightly better than those in the CSA environments (e.g., Australia, Japan, Korea, etc.), the performance of trellis encoding for worst loops will get more and more marginal.   In loop environments where more and more proportions of the loops have undesirable loop topologies, the overall gain of the trellis encoding algorithm is less and less effective.

13.6

REFERENCES

13.1   A.R. Khalifa 1993, "Trellis Encoding for Quadrature Amplitude Modulation Clusters for Primary Rate Integrated Services Digital Network", *Ph.D. Dissertation*, City University of New York, February.

13.2   G.S. Moschytz, S.V. Ahamed 1991, "Transhybrid Loss with RC Balance Circuits for High-Speed Full Duplex Modems," *IEEE Journal on Selected Areas in Communications*, Vol. J-SAC-9, No. 6, August, 951-959.

13.3   J.W. Lechleider 1989, "Line Codes for Digital Subscriber Lines," *IEEE Communications Magazine*, Vol. 27, No. 9, September.

13.4   G. Ungerboeck 1987, "Trellis-Coded Modulation with Redundant Signal Sets-Part II: State of the Art," *IEEE Communications Magazine*, Vol. 25, No. 2, February.

13.5   S.V. Ahamed and G.S. Moschytz 1991,   "Optimization of the RC Matching Networks in Adaptive Active Hybrids for High-Speed Data Communications," *Proceedings of the International Symposium on Circuits and Systems*, 1701-1704.

13.6   S.V. Ahamed 1993, "Simulation Environments for High-Speed Digital Subscriber Line (HDSL)," in *Conference Record International Communications Conference '93*, Geneva, Switzerland, May 23-26, Vol. 3, 1791- 1796.

CHAPTER 14

DIGITAL SUBSCRIBER LINE (HDSL AND ADSL) CAPACITY

The fundamental limits of the outside loop plant to carry high-speed digital (bidirectional HDSL and ADSL) data are reported in this Chapter. This Chapter also focuses on the copper wire-pairs, drop lengths, and interconnects that convey the data over the "last mile" to almost all businesses and residences in the United States. The two major customers, businesses and households are classified, characterized, and categorized by the distance to the Central Office or the remote terminal. Their spectral capabilities are computed and their ultimate digital capabilities are reported. The crosstalk limitations inherent in the plant are also computed and verified against the loss of signal due to distance and spectral constraints. The signal-to-noise ratio can thus be estimated to ascertain the transmission quality through the plant. Accordingly, this Chapter reports losses, bandwidths, and bit rates in the first half, and major bottlenecks and their implication in the second half.

14.1

INTRODUCTION

Multimedia services using the recent developments in video coding and compression techniques permit real time videophony. The ability to deliver these services depends upon the limitations of the loop plant. The major inputs used in this study are the data bases portraying the make up of business and residential loops. The minimum signal-to-noise ratio is retained to evaluate the percentages of loops that support the new breed of HDSL and ADSL service envisioned in the future.

The trunk facilities in most telephone systems around the world have evolved systematically. The subscriber plant, on the other hand, being more dispersed and driven by the demographic character of a region, tends to become a degraded star network emanating from the Central Office. This is particularly the case in the more established central cities where the growth in the number of residences was haphazard. This almost random growth has resulted in a highly diversified distribution of the physical loop plant and its topology.

Statistically, the attenuation characteristics of the copper cable prevail in limiting the spectral bandwidth in the loop plant. The objective of the investigation is to evaluate the ultimate capacity of providing high data rate services over existing unshielded twisted wire-pairs in the loop plant. Our initial findings show that the bit rates corresponding to the spectral band reported here include and extend beyond the current HDSL and ADSL applications.

At high data rates, *two* major effects are evident. *First*, the loss due to loop plant inconsistency becomes exaggerated, and *second*, the crosstalk coupling increases due to increased bandwidth. These effects are also influenced by the bridged taps, drop wires, outside plant (OP) interconnects, loading coils, distribution cabinets, and finally the customer premises wiring.

This research was presented at the 1994 International Workshop on Community Networking in San Francisco during July 13-14, 1994 and also published in the IEEE Journal on Selected Areas in Communication, Vol. 13, No 9, December 1995 pp. 1540-1549, by S. V. Ahamed, P. L. Gruber and J. J. Werner. We thank P. L. Gruber for her initiation, definition and sponsorship of this investigation.

The effects of these are discussed later in the Chapter. The object of the study is to evaluate the feasibility of providing additional high-speed data services over the unshielded twisted wire-pairs. The platform for the study is a specifically fabricated computer aided analysis and design facility [14.1, 14.2]. The major inputs are the data bases portraying the OP for business loops and the residential loops. Spectral domain characteristics and crosstalk models supplement the study to evaluate the signal degradation on one hand, and the extent of the contaminating noise on the other hand. Between the two lies the estimate of the digital capacity of the OP. Results up to 5 MHz have been generated and reported for a possible array of high speed services. These results are reported and discussed in Sections 14.2 and 14.3.

14.2

THE SUBSCRIBER ENVIRONMENT FOR HDSL AND ADSL APPLICATIONS

The data base for HDSL and ADSL applications is a subset of the 1983 Loop Survey data base. Results from two earlier surveys (1963 and 1973 Loop Surveys) also exist. The 1963 Survey resulted in the characterization [14.3] of the OP for voice grade services. The 1973 Survey was characterized twice. Results germane to the voice grade services were reported [14.4] in 1978. However, results germane to the design and characterization of the digital subscriber line (DSL) were reported [14.5] in 1982. The DSL, as it was perceived then, did not have a firm design and for this reason the spectral domain results were limited to about 250 kHz to cover the basic rate ISDN (BRISDN) services. Finally, the last and most recent of the Bell System Loop Surveys was characterized primarily for the evolving digital services ranging from BRISDN (144 kbps) to possibly the H0 rate (384 kbps) services. Results characterizing the OP to 400 kHz were reported [14.6] in 1986. The current Chapter contains the characterization of loops (amenable to the current HDSL and ADSL applications) for frequencies up to 1 Mhz. Also included are the results up to 5 MHz including most of the future services. Further, characteristics of loops are presented by the type of service: general, business and residential for future designers to specifically target devices/services to certain segments of subscribers. At the next level of detail, the loop penetration data in nine distance bands (of all the 18000 ft. loops) in two kft. increments are also classified.

The general 1983 Loop Survey data base [14.6] with 2,290 loops is reclassified into three major Sections: general loops, business loops and residential loops. Whereas all the loops were designed to carry voice traffic, a smaller fraction of shorter loops can carry digital data (from ISDN rates to multi-megabits per second). The most dominant impairment is the presence of loading coil(s). Almost all loops longer than 18,000 ft. have loading coils, and these coils make them unfit for consideration for any HDSL/ADSL application. Other loops with older aluminum alloys or with nonstandard gauge numbers (such as 14, 16, etc.) also are removed from consideration for HDSL applications. The fraction of loops with the nonstandard gauges is much less than 1 percent and these cables have not been characterized sufficiently to provide an accurate picture of their capabilities at very high rates. Thus, 66.37% of the loops surveyed have the features and topology to carry the HDSL/ADSL rates. The distribution of loop lengths in the general, business, and residential loop data bases are presented in Figure 14.1.

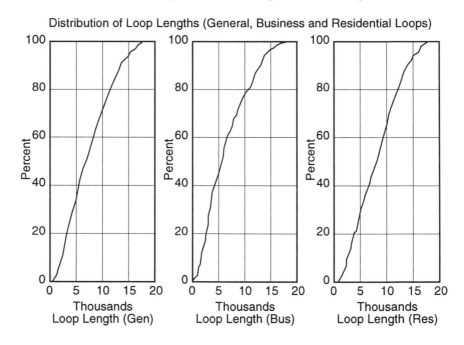

Distribution of Loop Lengths (General, Business and Residential Loops)

FIGURE 14.1. Distribution of All the Loop Lengths in the Data Base.

14.2.1 THE PHYSICAL CHARACTERISTICS

The physical characteristics of 1,520 loops (i.e., 66.3% of all 2,290 loops surveyed) are presented in Table 14-1, starting out with this truncated population and then indicate the penetration based on the number of loops (i.e., 502 or the 1,018 loops) in each of the two (business or residential) categories, respectively.

TABLE 14-1 Physical Characteristics of the OP for HDSL and ADSL Applications

	General Loops	Business Loops	Residential Loops
Number of Loops	1520	502	1018
Percentage	100%	33%	67%
Average Length	7354'	6326'	7865'
Loop Gauge Discontinuities	3.6	3.3	3.7
Bridged Tap (BT) Length	953'	477'	1186'
BT Gauge Discontinuities	1.37	1.03	1.52

About 33% of the 1,520 loops of the OP loops serve businesses. The percentage of loops which serve business in the entire population of 2,290 loops in the 1983 Loop Survey is estimated at 32%. The average length of the business loops is 6,326 ft. compared to 7,865 ft. for the residential loops because of the closer proximity of the businesses to the Central Offices and the centralized location of the Central Offices.

About 67% of the 1,520 loops of the OP serve residences. The percentage of loops that serve residences in the entire population of 2,290 loops in the 1983 Loop Survey is estimated at 68%. These results are consistent with the greater dispersion of the residences. A finer sub-classification for the special services loops was not attempted because of the limited number of loops for these services.

14.2.2 LOSS CHARACTERISTICS

In this Section, the statistical distribution of the loss characteristics of all the loops in the three data bases (general, business, residential) are presented. The results are generated by the computer simulation of each set of loops and then the statistical summary is derived. The spectral band, which is crucial for the HDSL or ADSL applications, depends upon the code used. To make these results as general and useful as possible, the loss over the spectral bands of interest for most HDSL and ADSL applications and for most of the common codes have been investigated. Thus, it becomes necessary for the user of these generic results to know where the spectral band of the particular interest lies. For example, if a 2B1Q code is used with 10% excess bandwidth, then the bandwidth of interest is from 0 to 1+0.10*1,544/2/2 or 424.6 kHz. On the other hand, if CAP 64 is to be used for T1 data using 5.5 customer bits per symbol (plus trellis coding) with 10% excess bandwidth, then the bandwidth of interest is from 0 to 1+0.10*1,544/5.5 or 308.8 kHz. It is customary to start the lower band edge at about 10 kHz, since very little spectral energy is actually present in most of the codes for both the HDSL and the ADSL applications.

14.2.2.1 Cumulative Loss Distribution

Three sets of curves are presented for the three (general, business, and residential) types of loops. However, the range between 100 kHz to 1 MHz is of immediate interest for the HDSL/ADSL application and the range between 1 to 5 MHz may have some interest in the future. Figures 14.2a and 14.2b depict the cumulative loss distribution curves in the lower band (100-500 kHz and 500-1000 kHz, respectively, with 100 kHz steps) for all loops. For higher band (1 to 5 MHz), the cumulative loss distribution is depicted in Figure 14.2c.

Likewise, Figures 14.3a, 14.3b and 14.3c depict the loss curves for business loops and Figures 14.4a, 14.4b and 14.4c depict the curves for residential loops. In the upper band of frequencies ranging from 1 MHz to 5 MHz, the cumulative loss becomes significantly higher and the percentage of loops able to carry data at VHDSL (very high-speed digital subscriber line) rates can be interpolated from the data presented in these figures. If a simple signal loss criterion is used, then the penetration can be directly read from these curves.

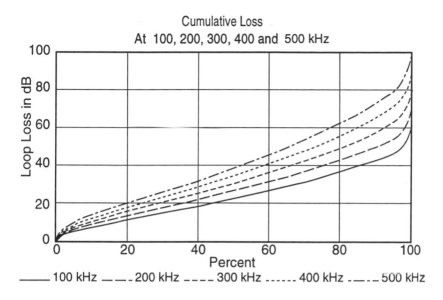

FIGURE 14.2a. Cumulative Loop Loss of All Loops at 100, 200, 300, 400, and 500 kHz.

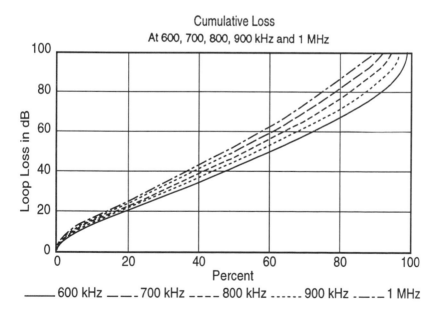

FIGURE 14.2b. Cumulative Loop Loss of All Loops at 600, 700, 800, 900 kHz and 1.0 MHz.

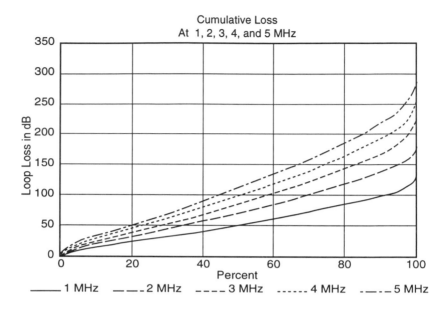

FIGURE 14.2c. Cumulative Loop Loss of All Loops at 1.0, 2.0, 3.0, 4.0 and 5.0 MHz.

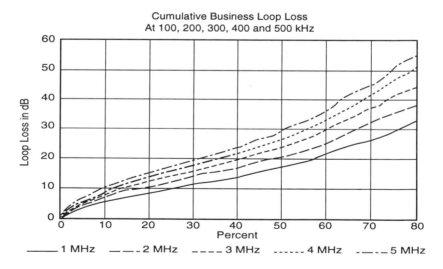

FIGURE 14.3a. Cumulative Loop Loss of Business Loops at 100, 200, 300, 400, and 500 kHz.

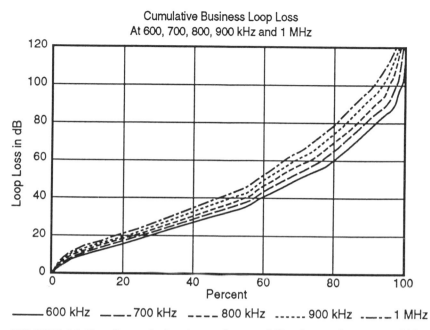

FIGURE 14.3b. Cumulative Loop Loss of Business Loops at 600, 700, 800, 900 kHz and 1.0 MHz.

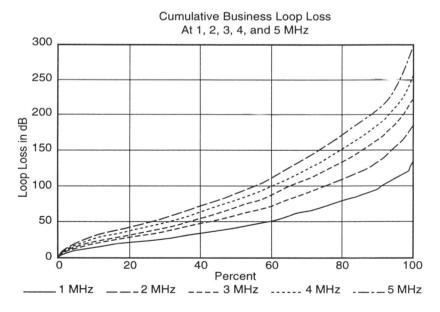

FIGURE 14.3c. Cumulative Loop Loss of Business Loops at 1.0, 2.0, 3.0, 4.0 and 5.0 MHz.

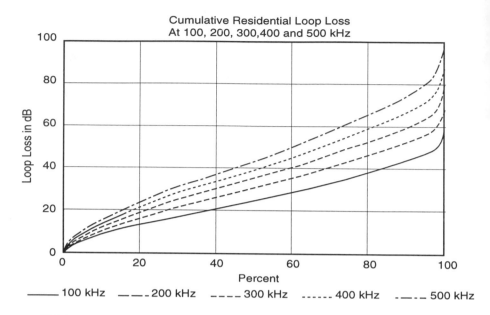

FIGURE 14.4a. Cumulative Loop Loss of Residential Loops at 100, 200, 300, 400, and 500 kHz.

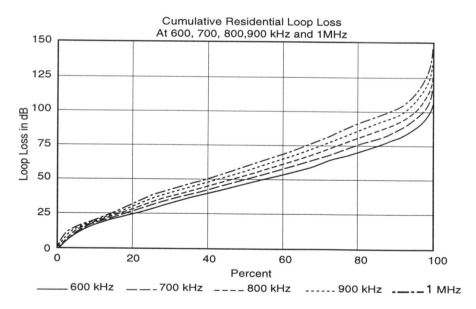

FIGURE 14.4b. Cumulative Loop Loss of Residential Loops at 600, 700, 800, 900 kHz and 1.0 MHz.

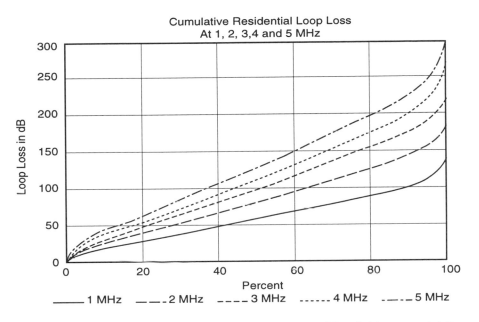

FIGURE 14.4c. Cumulative Loop Loss of Residential Loops at 1.0, 2.0, 3.0, 4.0 and 5.0 MHz.

14.2.2.2 Multiple Interferer Crosstalk

Any individual wire-pair carrying data into the Central Office suffers from both near-end crosstalk (NEXT) and far-end crosstalk (FEXT). Both types of interference depend on the pair number bearing the disturbed signal and the vicinity from the disturbing source. The extent of twist in each of the two wire-pairs also influence the coupling between the two. The estimation of NEXT and FEXT becomes simplified by assuming the same power spectral density for the disturbing and the disturbed signals. For the estimation of the interfering signal, the approach has been partially theoretical, based on the coupling factor, and partially experimental, based on the measurements. The basic shape of the curve is derived from the coupling and signal strength along the cable length.   The constants are then evaluated to force the theoretical curve to overlap the experimentally observed curve.

Two major conclusions are derived from the theoretical considerations. *First*, the single interferer NEXT coupling gain follows the $f^{3/2}$ characteristic [14.7] when the cable length is sufficiently long. *Second*, the FEXT interference signal-to-noise ratio (in dB) decreases by 20 dB per decade with frequency and by 10 dB per decade by cable length. The role of FEXT is only slight in most of the current HDSL/ADSL applications and for this reason, this interference can be ignored, even though it may become important for other VHDSL applications. The findings for NEXT (or self-NEXT in this case, since the disturbing signal has the same power spectral density) can be expressed in terms of loss in dB as

$$\text{NEXT Power Sum Loss (dB)} = \text{Constant} - 15 \log f_{(kHz)}. \tag{14.1}$$

The simple one disturber model (above) is refined by Bellcore [14.8] and extended to a 22 AWG, 18000 ft. cable with 1, 10 and 49 disturbers. Above 20 kHz (the bandwidth of interest for the HDSL and ADSL applications), the loss with 49 disturbers is approximated as

$$\text{NEXT Power Sum Loss (dB)} = 83.6 - 14 \log f_{(kHz)}. \tag{14.2}$$

At 80 kHz, the effect of 49 disturbers on a single wire-pair is computed as 57 dB. This value has been experimentally verified as a satisfactory limit for Central Offices self-NEXT with most gauges. When Equations (14.1) and (14.2) are recombined to yield the Bellcore standard of 57 dB at 80 kHz, and account for the difference between the coefficient of $\log f$, the constant of 83.6 dB in Equation (14.2) can be evaluated as a constant of 85.54 dB for Equation (14.1). This constant (i.e., 85.54 dB) can also be alternatively represented as 2.7925E-09 in the equation for the expected gain (G) of the squared NEXT crosstalk signal

$$G = 2.7925\text{E-09} * f^{3/2} \tag{14.3}$$

since 2.7925E-09 can be verified as $10^{-8.554}$.

Equation (14.1) is used twice. In the discussion of NEXT for the HDSL case, this widely accepted Bellcore value of 57 dB at 80 kHz for interference is used, and the penetration is computed by equating the attenuated signal strength due to line loss to approach this level of interference.

In the ADSL crosstalk computations, the interference for the most reasonable combination of interferers (average white Gaussian noise, plus NEXT from T1, DSL, and HDSL) is taken into consideration in order to determine exactly where this NEXT starts to completely obliterate the signal at the Central Office (a 60 dB NEXT loss limit [i.e., 60=85.54-15*log 50] at 50 kHz) and where 50 kHz is the limit of the bandwidth for upstream data. This computation is based upon the coupling constant for all interferers (at this 50 kHz frequency) as 2.7925E-09, computed to be consistent with the Bellcore standard. The NEXT loss is equated to the attenuation loss and the penetration is then evaluated. Inherent in this assumption is the fact that the upstream ADSL channel has a transmit energy consistent with the transmit energy of other signals in this band. Also inherent in this assumption is the fact that the ADSL downstream transmit level is consistent with the upstream transmit level. When these conditions are satisfied, then the penetration may be evaluated by determining the 60 dB loss (relative to the transmit level for ADSL).

14.2.2.3 Symmetric Frequency Dependent Crosstalk Limits

For the HDSL applications, bidirectionality is essential and in this case, self-NEXT interference is the most important disturber. The crosstalk power [14.7] depends upon the well documented 15 dB per decade rule in Equation (14.1), consistent with the Bellcore BRISDN standard of 57 dB NEXT loss at 80 kHz. When the transmitted signal attenuates down to the crosstalk, there is no signal left to recover. In an extreme sense, this becomes the limit of any signal transmission. In the practical applications, there is some comfortable margin (to be selected by the designer of that particular HDSL environment and the bit error rate requirement) that is retained above this extreme limit. As presented in Section 14.2.4, the practical limit lies around 60 to 70 percent of this limit with well designed equalizers, echo cancelers and possibly adaptive hybrids [14.9] for echo reduction. In this Section, this extreme limit by comparing the loss in the population of loops in the data base with the symmetric frequency dependent NEXT limit is presented.

Since the OP has a wide variety of subscriber loops, the data was classified into nine (0-2, 0-4, 0-6, 0-8, 0-10, 0-12, 0-14, 0-16 and 0-18 kft.) ranges. For the lower band, self-NEXT is computed at 100 to 1000 kHz at 100 kHz steps and the percentages of loops in each distance range that do not obliterate the signal are evaluated. Similar results are also generated for the 1 MHz to 5 MHz band at 1 MHz step.

14.2.2.3.1 Business Loops

Results are depicted in Figure 14.5a for the business loops. There are nine curves and each curve is for a distance range. There are 10 points on each curve in Figure 14.5a to signify the ten discrete frequencies in the lower (100-1000 kHz) band. The corresponding set of curves for the higher frequency band (1 MHz to 5 MHz) is shown in Figure 14.5b. The percentages of loops drop significantly in the 1-5 MHz band.

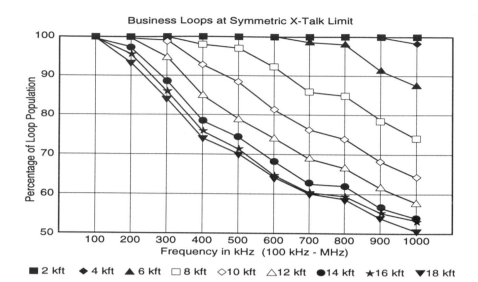

**FIGURE 14.5a. Penetration of Business Loops at Different Limits of
Loop Length from 100 kHz through 1.0 MHz with Symmetric
Crosstalk Limitation.**

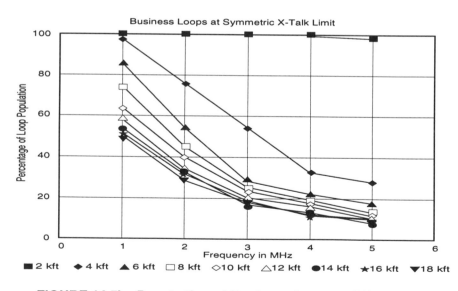

**FIGURE 14.5b. Penetration of Business Loops at Different Limits of
Loop Length from 1.0 through 5.0 MHz with Symmetric Crosstalk
Limitation.**

These curves offer the spectral capacity of the OP for HDSL applications. For example, if 2B1Q needing 424.6 kHz is to be used for business loops with the HDSL mode, then an ordinate at this frequency shows that 72% of 18 kft., 74% of 16 kft., 77% of 14 kft., 84% of 12 kft., 92% of 10 kft., 97% of 8 kft., and almost 100% of loops less than or equal to 6 kft. will carry the data as an extreme case. Performance limitations for other systems, such as the dual duplex HDSL systems needing 212.3 kHz. can be as easily read from Figure 14.5.

14.2.2.3.2 Residential Loops

Curves depicting corresponding results are depicted in Figures 14.6a and 14.6b for residential loops. These two figures (14.5a and 14.6a and then 14.5b and 14.6b) are plotted in complete conformity, and results may be read from either sets of curves in the same way. For example, from Figure 14.6a, the 2B1Q data penetration (at 424.6 kHz) in the residential loops is 60%, 64%, 68%, 75%, 89%, 97% and almost 100% of all loops in the 0-18, 0-16, 0-14, 0-12, 0-10, 0-8, and 0-6 kft. distance ranges.

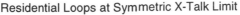
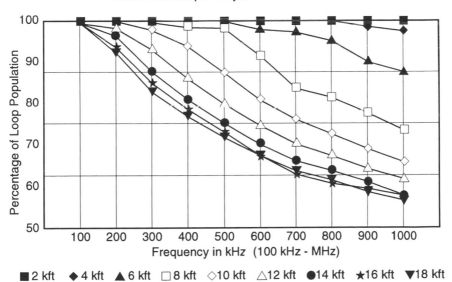

FIGURE 14.6a. Penetration of Residential Loops at Different Limits of Loop length from 100 kHz through 1.0 MHz with Symmetric Crosstalk Limitation.

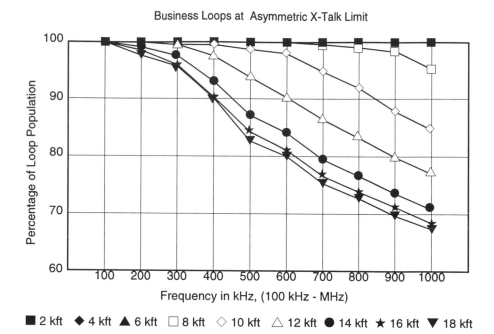

Business Loops at Asymmetric X-Talk Limit

■ 2 kft ◆ 4 kft ▲ 6 kft □ 8 kft ◇ 10 kft △ 12 kft ● 14 kft ★ 16 kft ▼ 18 kft

FIGURE 14.6b. Penetration of Residential Loops at Different Limits of Loop Length from 1.0 through 5.0 MHz with Symmetric Crosstalk Limitation.

The results for the general loop population can be derived as the weighted sum from the data for the 1,018 residential loops and from the data for the 502 business loops. For this reason, there is no dedicated Section for the data for general loops.

14.2.2.4 Asymmetric (Frequency Independent) Crosstalk Limits

For the ADSL application, the composite interference from CO noise and crosstalk from other systems is the limitation. For most of the ADSL applications, the more modern ESS-based COs are quieter than the older crossbar Central Offices and the Central Office noise is not considered as a serious limitation. However, the typical CO crosstalk pickup as estimated in Section 14.2.2.2 still remains at about 60 dB. When the extreme limit for the obliteration of the upstream signal is held at 60 dB, the penetration of the ADSL services may be expected to improve over the HDSL services where the self-NEXT is the dominant limitation. The curves indicating such penetration of the ADSL services are presented in the following Sections.

14.2.2.4.1 Business Loops

The percentages of business loops that offer a loss less than or equal to 60 dB in each of the nine distance (0-2, 0-4, 0-6, 0-8, 0-10, 0-12, 0-14, 0-16, 0-18 kft.) ranges depicted in Figure 14.7a for the spectral band of 100-1000 kHz are scanned at 100 kHz increments. Similar curves for the 1 MHz to 5 MHz band are plotted in Figure 14.7b. These curves offer the spectral capacity of the OP for ADSL applications. For example, if 2B1Q needing 424.6 kHz is to be used for business loops with the ADSL mode, then an ordinate at this frequency, shows that 87% of 18 kft., 88% of 16 kft., 96% of 14 kft., 96 % of 12 kft., 98% of 10 kft., and almost all loops less than or equal to 8 kft. will carry the data as an extreme case. Other performance limitations can be easily read in Figures 14.7a and 14.7b.

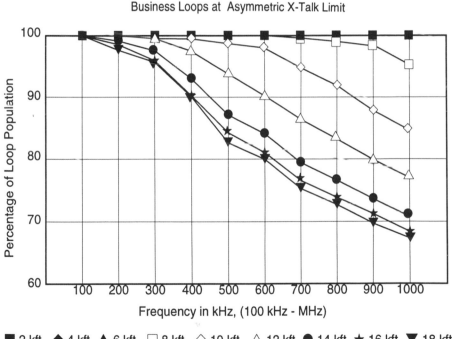

Business Loops at Asymmetric X-Talk Limit

■ 2 kft   ◆ 4 kft   ▲ 6 kft   □ 8 kft   ◇ 10 kft   △ 12 kft   ● 14 kft   ★ 16 kft   ▼ 18 kft

FIGURE 14.7a. Penetration of Business Loops at Different Limits of Loop Length from 100 kHz through 1.0 MHz with Asymmetric Crosstalk Limitation.

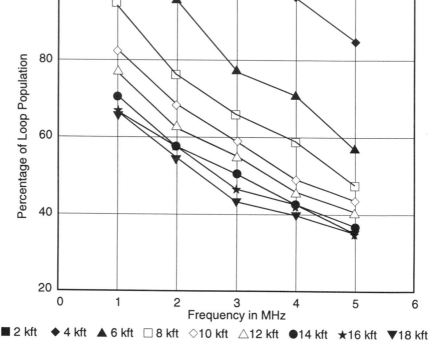

14.7b. Penetration of Business Loops at Different Limits of Loop Length from 1.0 through 5.0 MHz with Asymmetric Crosstalk Limitation.

14.2.2.4.2 Residential Loops

Curves depicting the percentages of residential loops that meet the extreme loss (60 dB; see Section 14.2.2.2) criterion for most of the current ADSL applications are depicted in 14.8a for the lower band 100 kHz to 1 MHz. For the higher bandwidth from 1 MHz to 5 MHz the percentages are shown in Figure 14.8b. It is seen that the percentages of residential loops start to fall more rapidly than for the business loops. Longer loop lengths, more bridged taps and less stringent requirements on the residential loops are seen as the primary reasons for lower performance.

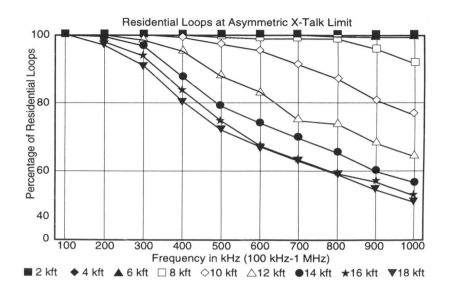

FIGURE 14.8a. Penetration of Residential Loops at Different Limits of Loop Lengths from 100 kHz through 1.0 MHz with Asymmetric Crosstalk Limitation.

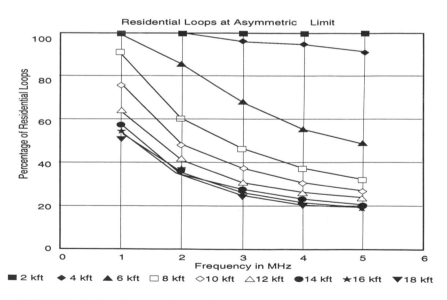

FIGURE 14.8b. Penetration of Residential Loops at Different Limits of Loop Lengths from 1.0 through 5.0 MHz with Asymmetric Crosstalk Limitation.

These curves offer the spectral capacity of the OP for ADSL applications. For example, if 2B1Q needing (1+0.10)*1,544/4 or 424.6 kHz is to be used for residential loops with the ADSL mode (with a downstream rate of 1.544 Mbps), then an ordinate at this frequency in Figure 14.8a shows that 77% of 18 kft., 82% of 16 kft., 86% of 14 kft., 93% of 12 kft., 96% of 10 kft., and almost all (99-100)% of loops less than or equal to 8 kft. will carry the data as an extreme case. On the other hand, if the performance limitations for 6.34 Mbps is to be evaluated, then the data from Figure 14.8b needs to be read. The frequency of interest with 2B1Q becomes 1.698 MHz. An ordinate at this frequency shows that about 42% of loops in the 0 to 14, 0 to 16, and 0 to 18 kft. range, 47% of the 12 kft., 55% of the 10 kft., 68% of 8 kft., 90% of 6 kft. and 100% of the residential loops less the 4 kft. will carry the data as an extreme case.

14.2.3 SPECTRAL LIMITATIONS AND DIGITAL CAPACITY

In Reference 14.10, the Shannon capacity of channels with colored noise is discussed. The relationship between bandwidth and capacity is quite fundamental, and is essentially independent of the gauge. In this Chapter, the theory has been applied to the case of 26, 24, 22, and 19 AWG cables assuming self-NEXT and 0 dB margin. These results are plotted in Figure 14.9. The X axis depicts the available bandwidth. The lower curves give the maximum distance for each of the four gauges. The upper curve gives the Shannon capacity multiplier.

For example consider the top curve alone: a bandwidth of 400 kHz provides a capacity of about 3 (i.e., 0.4*7.2) Mbps. In reality, computer simulations and experiments have shown that about 5 bps/Hz can be transmitted in this 400 kHz bandwidth using state-of-art digital transmission techniques. This leads to a practical achievable speed of about 2 (i.e., 0.4*5) Mbps. Now consider the lower curves. This 2 Mbps rate can be achieved for a range up to 22 kft. of 19 AWG, 15 kft. of 22 AWG, 12 kft. of 24 AWG or 9 kft. of 26 AWG. Simulations and experiments of the achievable bandwidth efficiencies of about 5 bps/Hz have also been found to be close to optimum for other values of bandwidth at the current state of technology. For this reason, the limit of the outside plant can be estimated from the results presented in this Chapter for most of the DSL, HDSL, ADSL and possibly the VHDSL applications.

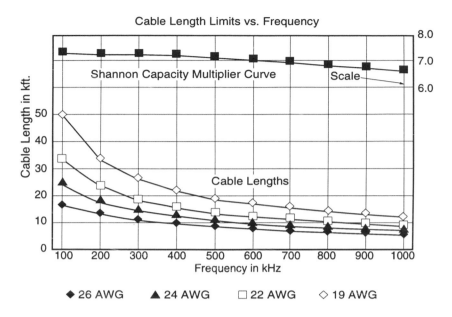

FIGURE 14.9. Bandwidth and Capacity Limits of 26, 24, 22, and 19 AWG Cables.

14.3

EFFECTS OF IMPAIRMENTS

14.3.1 BRIDGED TAPS

Bridged taps are prevalent in many loop plants (United States, Canadian, Australian, Japanese, some Italian). Taps do offer mismatch effects in certain bands of frequencies depending upon their length. The worst length of the tap offers a quarter wave-length resonance at the mid-band of the code carrying the data. The "frequency bumps" that they cause has been documented and in the analog domain compensated by "bump equalizers". The problem is solved effectively by decision feedback equalizers (DFEs) and their tapped delay line implementation. However, the increased loss due to their presence is an inherent limitation to the data traffic on the subscriber lines.

The physical stripping of bridge taps is an expensive and an error prone proposition for the Operating Companies. From the evolution of the DSL in the seventies, the BOCs have resisted the idea of implementing any manual and labor intensive alterations to the loop plant. For this reason, the investigation of the effects of bridge taps is limited to a subset of business loop in the 0-12 kft. range.

To investigate the effects of taps, they were removed from all the loops in the 0-12 kft. range from the business loop data base. The penetration of the loops for the HDSL (symmetric) and ADSL (asymmetric) applications is shown by the two curves in Figure 14.10.

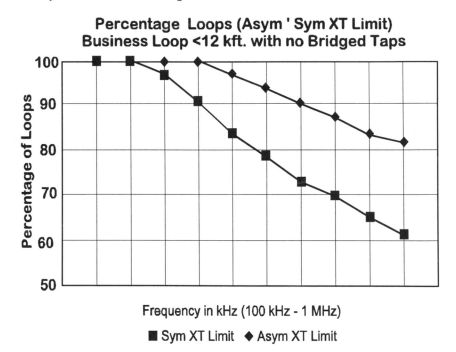

FIGURE 14.10. Penetration of Business Loops with Symmetric and Asymmetric Crosstalk Limitation with Bridged Taps Stripped from Loops.

Corresponding figures (loops with taps intact) for comparison are shown in Figures 14.5 and 14.7. Whereas, 308.8 kbauds data with CAP-64 application, 95% (Figure 14.5) of the loops with taps will carry HDSL data (symmetric case) data, the entire 100% of the loops without taps are expected to carry the same data. If the symbol rate is increased to about 1 Mbaud (for the 6.3 Mbps HDSL), 57% of loops (Figure 14.5) with taps will carry HDSL data, and 62% (Figure 14.10) of loops without the taps will carry the data. In the former case the expense of physically altering the topology of all loops in the 0-12 kft. band yields about 5% increased penetration. Even in the latter case, stripping the bridge taps increases the digital penetration by only five percent. For the ADSL case, the corresponding numbers are a 100% at the lower symbol rate (compare Figures 14.7 and 14.10) and 77% (Figure 14.7) vs. 82% (Figure 14.10) for the higher symbol rate. It is thus concluded that the expense of tap removal from the OP is not justifiable for the HDSL, ADSL and VHDSL applications. A similar argument for the DSL case [14.5, 1982] based on the eye opening and time domain simulations from the 1973 Loop Survey [14.4, 1978] has been made.

14.3.2 DIGITAL LOOP CARRIERS (DLC)

The effect of digital loop carriers (DLC) is transparent for most applications. Interfacing and protocol conversion functions are necessary. However, increased use of the DLCs has the effect of shortening the loop length since the DLC carries the data over the longer distance. The loop topology from the Remote Terminal (RT) to the customer premises is likely to be similar to the loop topology from the CO to the Subscriber. For this reason the increased use of DLC is seen as an impairment in the OP. Most crosstalk studies for DLC systems have indicated that crosstalk noise is far less than encountered in the OP subscriber environment. It is, however, recommended that the interfacing at the RTs be standardized and consistent with requirements of Section 14.3.4.

14.3.3 DROP LENGTHS

The final connection from the curbside to the actual residence (known as the drop length) varies between 50-600 ft. and consists of 24 AWG cable. The median length is about 200 ft. and has an effect of increasing the actual loop length by this distance with a 24 AWG cable. In a general sense, the effect of the drop length may be read in all the preceding figures by simply enhancing the loop length by about 200 ft. This approximation appears justifiable since the characteristics of the 24 AWG cable are close to those of the average loop characteristics and the effects of any impedance mismatches are minimal.

However for the sake of completeness, the effect of drop length from *two* perspectives is presented. *First*, the additional loss of the longest loops (i.e., loops in the 17.5-18.0 kft. band) with the longest drop length (600 ft. of 24 AWG) is reported. There are six loops in the general data base and the losses with and without the drop length are presented in Figures 14.11a and 14.11b.

The loops show an increased loss between 2 to 4.5 dB over the range of interest for the HDSL and the ADSL applications, and its actual value should be estimated for any particular application and the code that is used.

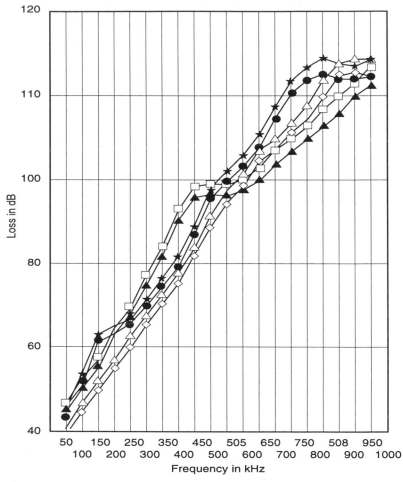

Effect of 600 ft Drop Length on Loop Loss

▲ 17870' lp   □ + drop len   ◇17740' lp   △ + drop len   ●17720'   ★ + drop len

FIGURE 14.11a. Effect of Drop Length on the Loop Loss of Three Loops between 17.5 and 18.0 kft.

Second, the penetration of the business loops under 12 kft. for the HDSL/ADSL applications are reported. Business loops (with and without bridged taps) are added an average drop length of 200 ft of 24 AWG cable.

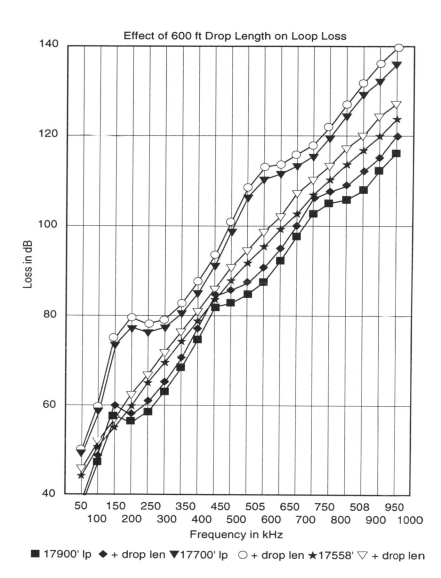

FIGURE 14.11b. Effect of Drop Length on the Loop Loss of Three Additional Loops between 17.5 and 18.0 kft.

The penetration of the loops under the three cases (business loops with taps, business loops with taps stripped, and business loops with taps and drop length added) is reported in Figure 14.12. Two sets of curves are evident. Both set of curves depict the effects of taps, and taps and drop lengths both, on the subscriber loops. Whereas removing the taps may enhance the penetration at any frequency by a maximum of about four percent, the added loss due to the drop length reduces the penetration by about a maximum of two to three percent.

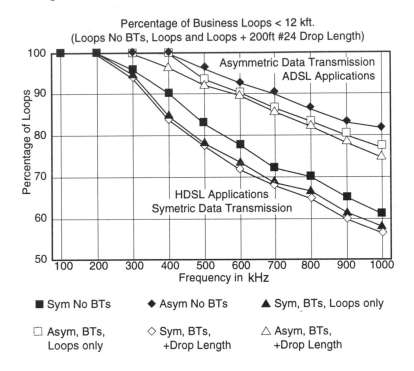

FIGURE 14.12. **Penetration of Business Loops for Symmetric (HDSL) and Asymmetric (ADSL) Applications.** The three curves in each application are for loops with bridged taps removed, loops with bridged taps in tact, and loops with bridged taps intact and with an additional 200 ft. of #24 AWG drop length at the customer premises.

14.3.4 OTHER IMPAIRMENTS

Interconnects within the outside plant can offer serious mismatches in the spectral range of interest for the subscriber line data. When the mismatch is severe, the flow of data suffers immensely. This effect has been documented in many HDSL tests. Typically, in the Swiss HDSL CAP-64 tests [14.11], "twisted wire" connections between subscriber loops and the customer premises equipment were used and resulted in approximately 5-6 dB loss in the measured SNR. The suggested solution to such problems is to require soldered connections. This approach corrected the problem in the Swiss HDSL tests.

Distribution cabinets at RTs and COs can cause mismatch effects and also cause spurious crosstalk and EM pickup. In one of the recent HDSL tests, short lengths of non-standard untwisted wire-pairs were used to connect the channel units to the CO end of the lines. This type of inter-cabinet wiring resulted in 3 dB loss in the crosstalk measurement. The suggested solution to such problems is to require standard wiring and interconnects at the CO and subscriber terminations of the line.

Loading coils cause the most severe impairment for DSL data traffic. Loops with load coils are systematically eliminated from any DSL applications. This is not seen as a severe deterrent since loading coils generally exist in loops longer than 18 kft., where the natural loss limits also occur for any DSL data. The loops that indeed have loading coils, generally have the first loading coil at about 3,000 ft. from the Central Office and one every 6,000 ft. from there on. There is no elegant way for signal processing at either end of the loop to undo the damage sustained to the signal as it passes through a series of these load coils.

Home wiring is also a known cause for the degradation of high-speed data traffic to the desk top. Typically, low-quality non-twisted biquad is used for home wiring. Two balanced wire-pairs exist in the biquad and lack of proper care in choosing the balanced pair causes a serious crosstalk problem in delivering high-speed data for symmetric services such as HDSL. Because of the extensions cords, which act as bridged taps, the existing home wiring may also prove to be a harsh environment for asymmetric services such as ADSL. This issue is presently under intense study.

Within the office environment, unshielded twisted pairs are used. This type of wiring is not seen as a major bottleneck for DSL applications. Even the lowest quality cables, i.e., category 3, have a data transport capability that far exceeds the speeds that can be provided in the OP [14.12].

14.4

CONCLUSIONS AND DISCUSSION OF RESULTS

The Chapter provides an insight into the balance between the signal loss in the loop plant and the noise within it. These two boundaries that define the subscriber digital line capacity at various frequencies have been plotted for frequencies to 5 MHz. The results are generic to the extent that they provide the user with these inherent limitations regardless of the code and the limitations of the line terminating units. Results presented offer the capabilities of the outside plant over the spectral band of interest (100-1000 kHz) to the current and proposed HDSL and ADSL designers. The data is also categorized in nine 2 kft. bands to cover the entire range of DSL, HDSL, ADSL and VHDSL applications. The extent of degradation due the seven bottlenecks, namely bridged taps, digital loop carriers, drop wires, OP interconnects, distribution cabinets, loading coils and customer premises wiring has been discussed and delineated.

14.5

REFERENCES

14.1   S.V. Ahamed 1993, "Simulation Environments for the High-Speed Digital Subscriber Line (HDSL)", *Proceedings of ICC '93*, Geneva, Switzerland, May 23-26, 1791-1796.

14.2   S.V. Ahamed and V.B. Lawrence 1993, "Interoperability of Multiple Data Bases for the Design and Simulation of High-Speed Digital Subscriber Lines", RIDE-IMS '93, *Proceedings of the Third International Workshop on Research Issues in Data Engineering: Interoperability in Multibase Systems*, Vienna, Austria, April 19-22, 262-267.

14.3   P.A. Gresh 1969, "Physical and Transmission Characteristics of Customer Loop Plant," *Bell System Technical Journal*, Vol. 48, December, 3337-3385.

14.4   L.M. Manhire 1978, "Physical and Transmission Characteristics of Customer Loop Plant," *Bell System Technical Journal*, Vol. 57, January, 35-39.

14.5   S.V. Ahamed 1982, "Simulation and Design Studies of the Digital Subscriber Lines", *Bell System Technical Journal*, Vol. 61, 1003-1077.

14.6   S.V. Ahamed and R.P.S. Singh 1986, "Physical and Transmission Characteristics of Subscriber Loops for ISDN Services", *Proceedings of ICC '86*, 1211-1215.

14.7   Bell Laboratories 1982, *Transmission Systems for Communications*, Bell Telephone Laboratories, Inc., Fifth Edition.

14.8   R.A. McDonald and C.F. Valenti 1989, "Assumptions for Bellcore HDSL Studies," *T1E1.4/89-066*, March.

14.9   G.S. Moschytz and S.V. Ahamed 1991, "Transhybrid Loss with RC Balance Circuits for Primary Rate ISDN Transmission Systems", *Special Issue of 1991 IEEE J-SAC on High-Speed Digital Subscriber Line*, Vol. 9, August, 951-959.

14.10 J.J. Werner 1991, "The HDSL Environment", *Special Issue of 1991 IEEE J-SAC on High-Speed Digital Subscriber Line*, Vol. 9, August, 785-800.

14.11 C. Heidelberger and K. Mistelli 1993, "CAP - HDSL on ETSI Loops, 2 Mbit/s Interface Performance Tests", Technical Seminar at Swiss PTT, Berne, Switzerland, Nov. 26.

14.12 Gi-Hong Im and J.J. Werner 1993, "Bandwidth-Efficient Digital Transmission up to 155 Mbps over Unshielded Twisted Pair Wiring," *Proceedings of ICC '93*, May 23-26.

PART V

RECENT HIGH-SPEED NETWORK ENVIRONMENTS

Recent high-speed networks have two major additions: information highways (mostly fiber) that span the regions, the nation and the globe, and distribution systems (fiber or hybrid-fiber coaxial) to carry the information to residences, homes, desk tops, campuses and even the private LANs and MANs within corporations. Supporting switches and signaling networks, components and multiplexers and de-multiplexers are also essential to build viable networks from transmission systems. More recently, knowledge banks and intelligent network components (the SSPs, the STPs, the SCPs, the INAPs, the IPs, the SMSs, etc. listed in companion volume, *"Intelligent Broadband Multimedia Networks"*, see Reference 15.1) have also been introduced to make networks service and society oriented. The older concept of considering the networks being broadly subdivided into trunks and subscriber systems is challenged, if not eradicated by the quantum leap of the transmission rate up from voiceband data to optical rates, and the spiraling down of cost from the expensive electronic switching systems to the (very) cheap ATM switches.

CHAPTER 15

KNOWLEDGE HIGHWAYS

Knowledge highways in an informed society are as generic as lines of commerce in a capitalistic society. The information explosion of the 1990s is not unlike the economic explosion of the 1940s and 1950s. Both these expansions are catalyzed by the growth of the communication and the transportation industries. Whereas, the economic growth of the mid-twentieth century has been the primary result of the evolution of the mass production, transportation and mega corporations, the growth of the knowledge society has been the primary result of the broadband communication and the massive information processing capabilities. In this Chapter, the two of the most important of the catalysts in the communication industry i.e., the SONET frame relay and ATM cell relay methodologies are presented. Intelligent networking and knowledge manipulation, their counterparts in the information processing environments are presented in the companion volume *"Intelligent Broadband Multimedia Networks"*.

In a sense, there can be no recent information network that does not incorporate the far reaching advantages and impacts of

- SONET digital hierarchy,

- ATM architectures that engulf most of the existing circuit, packet, message switched networks including the high-speed DS-1, DS-3, and private LANs, and

- ■ The extended ITU-OSI signaling hierarchy to span the numerous networks

to gracefully integrate the networks from across the street to those across the world. In this part, the overall perspective of the newer networks is presented.

15.1

MODERN NETWORK ENVIRONMENTS

Optical transmission was proposed during the early sixties after the invention of laser in 1960. The multi-layered glass structure was suggested during the mid sixties and a rough specimen of fiber with about 20 dB/km loss was produced by Corning Glass in 1970. Optical systems have been functioning with ever increasing capacity and sophistication. The highest quality single mode fibers offer about 0.16 dB/km loss (see Chapter 16) at 1.55 µM wavelength and less than 0.35 dB/km at 1.30 µm wavelength. The chromatic dispersion (see Chapter 16) at approximately 15 Picosecond/km-nanometer in the low loss window of 1.5 µm (with zero dispersion at 1.3 µm) is also equally attractive. Both of these very attractive attributes makes fiber an essential component of most modern networks.

Two major driving forces behind most broadband digital networks are the availability of high grade fiber for transmission and the profusion of the rather inexpensive ATM switches. The transmission issue has been successfully resolved during the seventies and early eighties. Presently, the data rates and ranges are well beyond the projected needs of even the most digit- intensive society. Where fiber is easily available, the new national and global trunk data communication is mostly optical in networks. In the United States, Canada, Europe, Japan and Australia, fiber has been deployed for the older trunk and shared transmission facilities.

High-speed transmission on copper facilities is an economic alternative to homes and locations where the copper already exists for transmission. With the 2B1Q code BRISDN data is feasible, if not already available to individual homes. With the 2-D trellis, QAM, etc. coding methodology, data to the T1 rate at 1.544 Mbps (in the HDSL mode of operation) is feasible to the individual customers. The same methodologies can take data to 6.34 Mbps in the ADSL mode, and data from variable rates of 384 kbps to 6.34 Mbps in the symmetric digital subscriber line (SDSL) mode. Fiber capacity, even at the lowest SONET rate at 51.64 Mbps can reach individual homes in a star or bus topology from a centralized data distribution point. It is approximately 30 times greater than the T1 rate at 1.544 Mbps. And then again, in the hybrid fiber coaxial environment, fiber serves like a backbone video/multimedia network that takes data to the video hubs (see Chapter 3) at multigigabits per second making digital images appear and sound as clear and crisp as the toll quality PCM encoded speech.

15.2

OPTICAL TRANSMISSION VERSUS OPTICAL NETWORK

At present, optical switching is not yet incorporated within the optical fibers to offer true optical networks. However, fibers carry multigigabit data quite dependably. Hence, public domain networks of the late twentieth century are very likely to deploy semiconductor switching under stored program control and deploy fiber for high capacity transmission. Optical computers must evolve before the networks can assimilate their switching capacity. This scenario is similar to the delayed impact of semiconductor digital computers of the mid fifties on switching systems of the mid sixties. The computer industry was already well into its prime before the first electronic switching system (#1 ESS) was introduced in 1965.

The transmission capacity of the fiber has been deployed by national and international communications networks. On the national scene, fiber have been routinely introduced for interoffice high data capacity channels in the United States, Canada, Europe and Japan. Optical rates (see next Section), OC-3 (at 155.52 Mbps) and OC-12 (at 622.080 Mbps), have been achieved by commercial systems. Universal information service (UIS), deploying data links at these rates, announced by AT&T, is for handling voice, video, and data communication needs of large and small customers. The manifestation of these universal services is via the BRISDN (144 kbps) and PRISDN (T1/E1 rates) bearer capability, progressing into the broadband ISDN (155 to 600 Mbps) services, and then to the ATM based bandwidth-on-demand capability. The running cost for such services appears to be in the maintenance of the transceivers rather than the glass (SiO_2 sand-based) transmission media.

On the international scene the first undersea transatlantic telephone cable (TAT-8) was commissioned in late 1988. The capacity is limited to 40,000 voice calls, which is double the capacity of all other copper transatlantic cables and satellites. The next fiber cable system TAT-9 commissioned in the early nineties handles twice this capacity at 80,000 voice calls.

The TAT-8 high capacity fiber optic transmission system transmits 296 Mbps data over 3,646 miles. This system is equipped with 109 repeaters spaced approximately 70 km (230 kft.) apart using coherent laser sources at 1.3 μm wavelength. Six single mode fibers constitute the cable. The combined capacity of TAT-8 and TAT-9 serves the transatlantic telecommunication needs through the mid nineties. Transpacific fiber optic cable systems link Japan, Guam, Hawaii and the Pacific Rim states.

These fiber optic transmission systems do not constitute any part of the intelligent network environment. In conjunction with intelligent nodes and intelligent data base systems (strategically located within the network), the fiber optic transmission systems offer high performance multimedia digital access and transmission services for computers, knowledge bases, multimedia, video, educational, medical imaging, facsimile, and even plain old telephone services throughout the world.

15.3

SONET: THE SYNCHRONOUS TRANSPORT STANDARD

SONET stands for the Synchronous Optical Network. It is also associated with the well accepted standards for the transport of pre-defined frames through the network. The SONET standards, originally proposed by Bell Communications Research, specifies the standardized formats and interfaces for optical signals. Historically, the work started as far back as 1985 to provide standard optical interface between major carriers (such as NYNEX, MCI, Sprint, etc.) At the outset, the broad goals of the SONET standard are

(a) to provide a broad family of interfaces at optical rates (i.e., OC-1 through OC-192),

(b) to provide easy and simple multiplexing/demultiplexing of the signals (such as headers and play-loads) within the network components,

(c) to exploit the growing trend toward the communication networks becoming synchronous, and

    (d) to provide ample overhead channels and functions (via the header blocks) to perform and support the maintenance of the network facility.

Total synchronism of all networks is unachievable and for this reason the standard committees have accepted the interfacing signals to be plesiochronous. The plesiochronous nature permits a small variation in the significant timing instants of the nominally synchronous signals, i.e., the recovered signals may be permitted to have small, but controlled, variations in their instantaneous rates or their instantaneous frequencies.

These standards, originally approved by ANSI's T1X1 committee for standard fiber optic transmission, were reviewed and accepted by CEPT, ITU (formerly CCITT) and the European Telecoms in the 1987-1988 time span [15.1] after considerable debate (9-row/13-row debate) about the number of rows in each of the frames. The 9-rows x 90 columns frame has been the favored by some of the European Telecoms and the 13-row x 60 columns frame was favored by the United States. In February 1988, the Synchronous Transport Signal - Level 1, shown in Figure 15.1 was finally accepted with 9 rows x 90 columns and at a bit rate of 51.84 Mbps. At this rate, the time duration of the frame is exactly 125 µs and this duration is necessary to get 810 bytes (or octets with 8 bits per byte or octet) at a rate of 51.840 Mbps. Thus, one entire frame gets into the network in 125 µs (at OC-1 rate) and this is the same as the sample time of the PCM encoded speech of one byte (8 bits) at 8 kHz, thus maintaining the 125 µs clock duration for both the traditional voice circuits and the new SONET network.

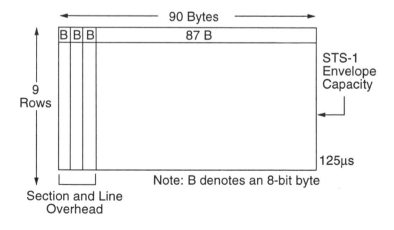

FIGURE 15.1. The Composition of an STS-1 Frame. STS - Synchronous Transport Signal

The approvals by CCITT (now ITU) have been granted with the complete standards issued in 1992. SONET standards are being incorporated for asynchronous transfer mode (ATM) in broadband ISDN applications for voice and data. The standards for fiber optic digital hierarchy range from the OC-1 rate of 51.84 Mbps through OC-3 (widely accepted) rate of 155.52 Mbps to the OC-96 rate at 4.97664 Gbps. More recently the specialized fibers fabricated with self-focusing properties have opened the OC-192 rate for possible applications.

TABLE 15-1 Rate Hierarchy for the SONET Synchronous Data Transport

SONET Transport	Optical Carrier	Data Rate	Signal (STS)
STS-1	OC-1*	51.840 Mbps‡	x 1
STS-3	OC-3*	155.520 Mbps‡	x 3
STS-9	OC-9*	466.560 Mbps	x 9
STS-12	OC-12*	622.080 Mbps‡	x 12
STS-18	OC-18*	633.120 Mbps	x 18
STS-24	OC-24*	1.244160 Gbps	x 24
STS-36	OC-36*	1.866240 Gbps	x 36
STS-48	OC-48*	2.488320 Gbps‡	x 48
STS-96	OC-96	4.976640 Gbps	x 96
STS-192	OC-192	9.953280 Gbps‡	x 192

* Only the OC-1, OC-3, OC-9, OC-12, OC-18, OC-24, OC-36 and OC-48 are allowed by the American National Standard. ‡Currently accepted OC rates. The ITU (formerly CCITT) has the STM-i designation, with i (= $N/3$) for the STS-N designation. The numerical value of i ranges from 1 to 64 corresponding to SONET rates from STS-3 to STS-192 with i limited to 1, 4, 16 or 64 (or with N = 3, 12, 48, or 192) for the currently accepted rates. The play load capacity for each STS rate is exactly 87/90 times the data-rate tabulated (see Equation (15.1).)

The multiplicative relation [15.3] between the hierarchical rates greatly facilitates the framing format for SONET synchronous data transport. The STS-N signal is scrambled and converted to the OC-N signal and this line rate is exactly N times the OC-1 rate. Multiplexing of the STS signals needs intricate byte encoding in the header blocks, and this becomes a crucial concern in the "SONET ready or compatible" switches that link [15.4] various types of networks (such as DS-1, DS-3, all-digital synchronous networks, all optical networks, etc.) via the Digital Cross Connect Systems (DCS). The standard 125 μs time slot is chosen for all the hierarchical OC rates. This leads to simplistic calculations for the number of bytes for the OC-1 through OC-192 rates as follows:

Number of bytes (or octets) in a 125 μs for the OC-1 rate is

$$= 51.84 \times 1.E+6 \times 125 \times 1.E\text{-}06 \, / \, 8$$
$$= 6{,}480 \text{ bits } / \, 8 \text{ or } 810 \text{ bytes}$$
$$= 9 \times 90 \text{ bytes (i.e., 9 rows} \times 90 \text{ columns)}$$
$$= 9 \times (3{+}87) \text{ bytes}$$
$$= (9 \text{ rows}) \times (3 \text{ bytes OH} + 87 \text{ bytes PL})/\text{frame.} \qquad (15.1)$$

Here OH signifies overhead for the SONET header block and PL indicates play load for carrying the customer data. All the 9×3, or 27, bytes are used for overhead and the 9×87, or 783, bytes are used for playload in the data transport function. The rows are transmitted one by one and the byte integrity is retained through the ATM network. It is thus implicit that when the network is operating in its normal synchronous mode, the 8 sequential bits are also the bits of the same byte. The byte (or octet) classification for other rates follows the same pattern. For example, for OC-N rate the byte classification is as follows:

Number of bytes (or octets) in a 125 μs for the OC-N rate is

$$= 51.84N \times 1.E+6 \times 125 \times 1.E\text{-}06 \, / \, 8$$
$$= 6{,}480N \text{ bits } / \, 8 \text{ or } 810N \text{ bytes}$$
$$= 9 \times 90N \text{ bytes}$$
$$= 9 \times (3N + 87N) \text{ bytes.} \qquad (15.2)$$

The breakdown between the overhead and the playload is similarly segmented with $27 \times N$ bytes for overhead and $783 \times N$ bytes for playload. The arrangement of the SONET frame can be simply derived from the considerations listed above. For example, the SONET frame for the commonly used OC-N rate is shown in Figure 15.2.

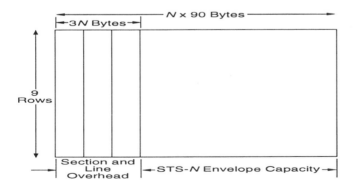

FIGURE 15.2. SONET OC-N Frame. STS - Synchronous Transport Signal

The starter frame STS-1s of an N-th level STS-N frame needs special attention. Byte interleaving and frame alignment are both necessary to form the STS-N frame. Byte integrity at the OC-3 level (to meet the ITU [formerly CCITT] standards) is thus implemented; and it also permits STS-N to carry broadband playload at 150 and 600 Mbps. Multiplexing of various SONET frame and digital cross connects are handled by the SONET Central Offices. A typical configuration of this type of Central Office is shown in Figure 15.3.

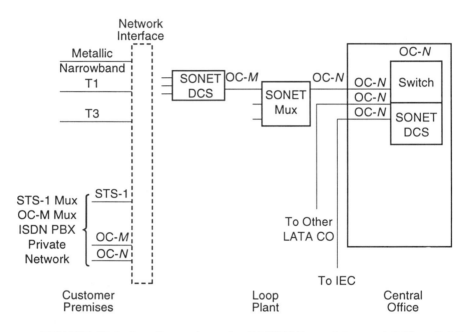

FIGURE 15.3. Configuration of a SONET Based Central Office. DCS - Digital Cross-connect Switch; DLC - Digital Loop Carrier; IEC - Interexchange Carrier; ISDN - Integrated Services Digital Network; LCO - Local Central Office; OC-M - Optical Carrier; OC-N - Optical Carrier; PBX - Private Branch Exchange; STS - Synchronous Transport System

Here, the various services such as T1, T3, HDSL, VHDSL, and other STS-N based services are reduced to the common basis of their synchronism (or the 125 μs), appropriate switch and then the information is then reconstituted in the appropriate signaling format and sent to the customers. The coexistence of broadband ISDN and ATM was proposed as early as 1990 via the ATM cell mapping into the SONET playloads for easy, elegant transport via the optical information highways of the future.   This will be necessary until all the systems assume one standard format, preferably the ATM format (see next Section), and then the ATM switch in conjunction with the bridges and routers will perform all the switching. Meanwhile, a number of the digital cross connect systems, such as AT&T's DDM-2000, FT-2000 SONET multiplexers and many versions of digital access cross connect systems (DACS) and SONET DCS are available to perform high-speed optical-rate switching. Newer services such as broadcast TV facilities, interfacing HDTV quality CAD/CAM terminals, virtual reality devices, medical image transfer, super-computer applications, etc. will be the prime beneficiaries of these DCS and DACS switches.

15.4

SONET ARCHITECTURES

The deployment of SONET occurs in numerous architectures and applications. Within the broadband network, the point-to-point linear deployment for access and interoffice capabilities (Figure 15.4a), integrated access (Figure 15.4b), tree configurations with limited DS-1, DS-3 switching (Figure 15.4c), and hub sites architectures with extended switching (Figure 15.4d) are used.

Rings using Add-Drop Multiplexers (ADMs) can also be constituted for broadband data traffic (Figure 15.5a), and hubs with Digital Cross Connect Switch (DCS) provide a localized broadband switching capability (Figure 15.5b). The DCS are generally built with wideband (DS1 and VT) and broadband (DS3 and STS-1) interfaces. SONET interfacing with both the WDCS and BDCS is already available. The virtual terminals have two modes of operation floating and locked. The floating mode is suitable for bulk transport of channelized or unchannelized DSn and for distributed VT grooming. The locked VT mode is better suited for integral numbers of STS-1s transport and for distributed DS0 grooming between DS0 path terminating equipment.

SONET deployment in phases can enhance the capacity and the switching of the modern information highways quite efficiently and inexpensively since the ADMs and DCSs are relatively inexpensive compared to the traditional ESS type of switches. SONET may also be deployed for DS-1 and DS-3 transport in the trunk environment using the Bellcore developed TIRKS provisioning system [15.5]. SONET is also a contender in the integrated digital loop carrier (IDLC) environments [15.6]. Depicted in Figure 15.6, IDLC applications can provide access to the DS1, baseband, broadband (OC-3) services from any broadband network and local digital switch (LDS), thus facilitating broadband and baseband networks to coexist via ADMs and remote data terminals (RDTs in Figure 15.6).

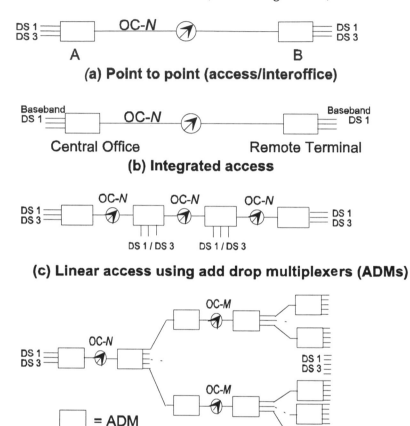

(a) Point to point (access/interoffice)

(b) Integrated access

(c) Linear access using add drop multiplexers (ADMs)

(d) Tree structures using hub sites and ADMs

FIGURE 15.4. Four Stages of Deployment of the SONET in the Existing Broadband Networks. The use of hubs and ADMs is most common in the recent implementations of SONET and finally the ATM environments.

Low power complementary metal oxide semiconductor (CMOS) technology devices at OC-3 and higher rates are makes these architectures easy to implement for generic and programmable data distribution systems to reach homes, businesses, and special services customers. SONET offers direct mapping of the SONET frame to the byte synchronous DS0 (64 kbps) 125 μs time slot (in the floating VT1.5 mode of operation) for encoded voice/speech applications.

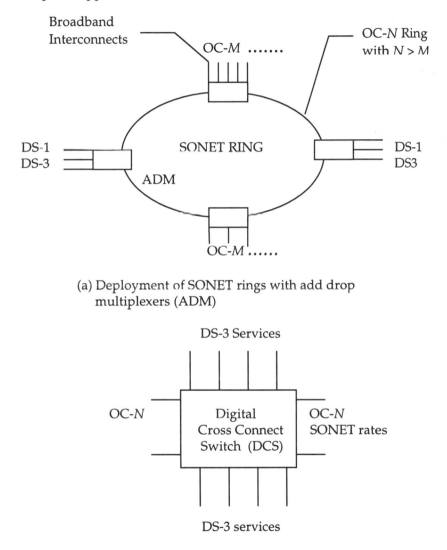

(a) Deployment of SONET rings with add drop multiplexers (ADM)

(b) Digital cross connects for full fledged switching capability at the hub sites.

FIGURE 15.5. The Integration of the Sonic Transport System (STS) Based Optical Networks with the Current Digital Networks. DS - Digital Signal; OC - Optical Carrier

This advantage in the IDLC environment removes the need for DS1 (1.544 Mbps) realignment, thus offering these carrier systems better utilization of the switching facilities. Distributed VT cross connections, bulk transport/switching of channelized and unchannelized, synchronous or asynchronous DS1s (mapped asynchronously), unchannelized bit-synchronous DS1 transport, DS0 circuit-switched traffic and IDLC byte synchronous mapping are feasible in the floating VT modes of the RDTs (see Figure 15.6).

Three general architectures for the SONET transport systems providing point to point DS1, DS1 & DS3 to STS-1 & STS-3, and combinations of OC-N, using add-drop multiplexers, and digital cross connects are shown in Figure 15.7. New networks can thus be fabricated with wide transport and switching capabilities.

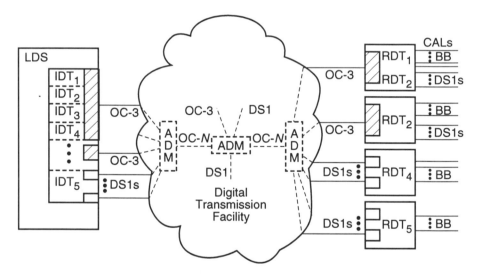

FIGURE 15.6. Typical Configuration for SONET and Integrated Digital Loop Carrier (IDLC) Facilities. ADM - Add Drop Multiplexers; BB - Baseband; CAL - Customer Access Line; DS1 - Digital Signal, level-1; IDT - Integrated Digital Terminal; LDS - Local Digital Switch; OC - Optical Carrier; RDT - Remote Digital Terminal

These architectures indicate the strategies for the incorporation of the evolving SONET and ATM components into the existing framework of telecommunication networks. Over a period of time the older network rates such as DS1 through DS3 will be substituted by the more accepted OC-N rates.

15.5

ATM FUNDAMENTALS

The optical rates currently available (up to 4 to 8 Gbps) for the transport of data are much too high to be directly integrated in the existing telecommunication networks. The SONET frame, lasting for 125 microseconds, can carry a large amount of information as the playload (2,349 bytes) even at the OC-3 rate. While this size may appear perfectly reasonable in the LAN/MAN/WAN data environment, it can be perceived as being too large for other (like switched virtual service, SVC, or permanent virtual service, PVC) types of application in business and industry. In the trunk environment, numerous low rate data channels are multiplexed onto the fiber trunk with its own particular protocol, framing and bit designations for rates ranging from T0 to T3 or E0 to E3. However, recent applications are not always circuit switched, but are becoming more and more packet and message switched. Hence, the idea of having the SONET carry smaller size packets or "cells" becomes more appealing. In a sense, asynchronous transfer mode (ATM) is an extension of the signal transport concept that SONET itself relies upon. Cells provide user level interfacing packets and SONET frames become the photonic network transport packets. ATM standards outline the efficient (and perhaps elegant) techniques to bridge the gap between multiplicity of users and the vast variety of lightwave devices that make up the network. However, until such networks are a reality, ATM can be used to supplement DS-2 and DS-3 networks.

Asynchronous transfer mode (ATM) of data transfer is a worldwide standard. It stems from the availability of a high-speed backbone in the national or global environment. It has started an era of cell relay technologies and falls under a wide umbrella of packet switching. The technological fit of ATM can be seen as a progression of X.25 technologies to frame technologies and then into the cell technologies, as depicted in Figure 15.8 It also depends upon the SONET standard for the transport of data. ATM is a standard for most networks to directly communicate with each other provided the SONET and ATM protocol and header block information is consistently followed.

ATM is based upon packet switched technology. The ATM packets or cells are relayed and routed throughout the network and thus get communicated between the source to the appropriate destination. Since the protocols and functions are unique to ATM, the technology for these ATM based networks is called the "cell relay" technology as opposed to the "frame relay" technology used to carry data in the circuit switched digital capability or CSDC at 56 kbps, fractional T1 rates via T1 carrier, and even DS-1 and DS-3 networks.

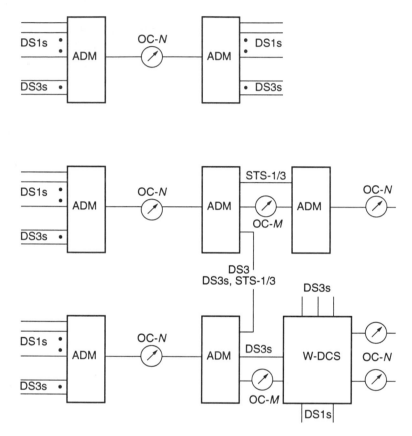

FIGURE 15.7. Three Additional Scenarios (see Figure 15.4) for the Deployment of Wideband Networks. (a) Point-to-point using ADMs, (b) Network extensions using ADMs and (c) Combination of ADMs and Wideband Digital Cross Connect Switches (WDCS). DS-n - Digital Signal; OC - Optical Carrier

ATM functionality has been embedded within a fixed size link-layer (i.e., the OSI layer six, data link-layer) entity. This link-layer entity is the "cell" and the size of this fixed length entity is 53 octets (bytes). There are 5 octets for the header block and 48 octets for the playload (or the data to be transported.) Similar to the header of most packet networks, ATM cell header carries information unique to the nature of data, its functionality, disassemble and reassemble of cells and their playloads. The typical ATM cell format for the user-network interface and network-network interface is shown in Figure 15.9. The information contained in the header is encoded to conform to the functions within the ATM sublayers.

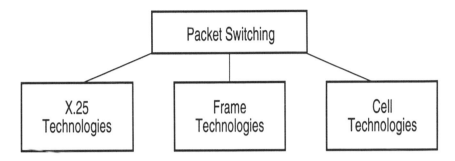

FIGURE 15.8. The Progression of the Packet Switching Capability from X.25 to ATM Based Cell Relay Technologies. The features are listed as follows:

X.25:	Frame Relay:	ATM:
Connection Oriented Service	Connection Oriented Service	Multiple Services/Capabilities
- Dial-up - Dedicated	- PVC and - SVC (future)	- Constant Bit Rate (circuit emulation) - Variable Bit Rate Services - Connection Oriented Data Services - Connectionless Services
Range of Bit Rates	Range of Access Bit Rates	Wide Range Access
- Dial up 300 bps to 19.2 kbps - Dedicated 2400 bps to 56 kbps	- 56 kbps - F T1 (64-1.544 Mbps) - T1	- DS-3 (45 Mbps) - OC3 (155 Mbps) - OC12 (622 Mbps)
Performance	Higher Performance	High Performance
- High reliability & - Wide availability	- Low latency & - High throughput	- Low latency & - High throughput
Network Error - detection and correction	Network Configuration - streamline	ATM Network - scaleable

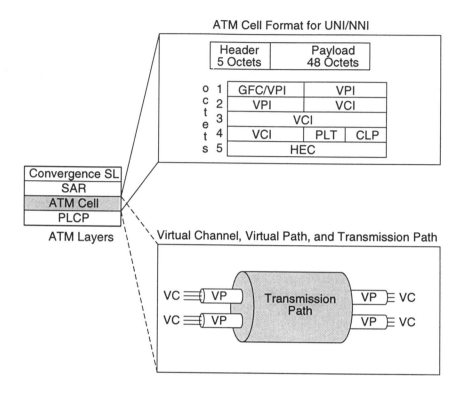

FIGURE 15.9. ATM Cell Format: Structure and its Transport on the Virtual Channel, Virtual Path and Transmission Path. ATM - Asynchronous Transfer Mode; NNI - Network to Network Interface; PLCP Physical Layer Convergence Protocol- ; SAR - Segmentation and Reassembly; SL - Sublayer; UNI - User Network Interface

ATM cells are routed and relayed by virtual circuits. Such circuits may be fabricated depending upon the traffic demands and switched and scaled by the ATM switches. These virtual networks may also be switched in the hardware at the ATM switches and/or nodes.

ATM also offers high-speed data transport capabilities. Since the ATM cells "ride" within the SONET frame, they carry all the advantages offered by transport systems embedded within SONET. The speed of transport is derived from optical rates (and thus low buffering delays), optical speeds and low switching times (because of the simplicity/elegance in the SONET header information).

ATM standards are applicable within LAN/MAN/WANs. ATM standards are applicable to private LANs, campus networks, or any other data network, and thus these networks can be connected to public, regional or national ATM networks. They can also serve the needs of broadband ISDN by functioning via connection oriented ATM switches.

15.6

OSI AND ATM

The OSI (open system interconnect) Reference Model (see Chapter 4, of the companion volume, *"Intelligent Broadband Multimedia Networks"* [15.1]), and its adaptation to the ATM networks is shown in Figure 15.10.

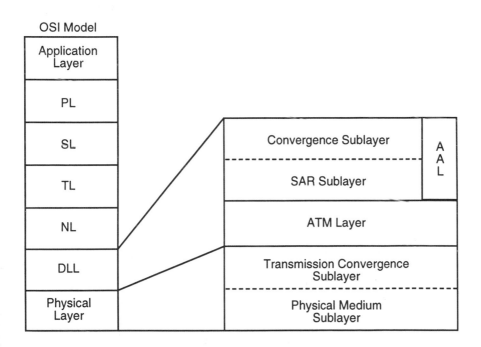

FIGURE 15.10. Encapsulation of the ATM Layers within the Data Link and the Physical Layers of the OSI Reference Model. DLL - Data Link Layer; ISO - International Standards Organization; NL - Network Layer; OSI-RM - Open Systems Interconnection-Reference Model; SL - Session Layer; TL - Transport Layer ; VCI - Virtual Channel Interface; VPI - Virtual Path Interface

The ATM functions occur at the sixth (data link-layer, DLL) and seventh (physical) layer of the OSI model. The top five OSI layers remain intact, and after the fifth layer, the ATM adaptation layer begins. The organization, and conversion to the ATM environment is shown in Figure 15.11.

Layer   Function

ATM Adaptation Layer	CS	Convergence
	SAR	Segmentation
ATM Layer		Generic flow control Cell header generation/extraction Cell VPI/VCI translation Cell multiplex and demultiplex
Physical Layer	TC	Cell rate decoupling HEC header sequence generation/verification Cell delineation Transmission frame adaptation Transmission frame generation/recovery
	PM	Bit timing Physical medium

5 Octets						48 Octets
GFC	VPI	VCI	PLT	CLP	HEC	Playload

ATM Cell Format

FIGURE 15.11. ATM Layer Functions and ATM Cell Format. CS - Convergence Sublayer; CLP - Cell Loss Priority; GFC - Generic Flow Control; HEC - Header Error Check; PLT - Playload Type; PM - Physical Medium; SAR - Segmentation and Reassembly; TC - Transmission Convergence; VCI - Virtual Channel Interface; VPI - Virtual Path Interface

In OSI standards, *seven* functions are identified for the network access protocol. These functions are: framing and transparency, sequencing, error control, flow control, connection setup/shutdown, addressing/naming. These functions (regrouped as *six* functions as shown in Figure 15.12) are handled by the two sublayers: AAL and ATM layer as shown in Figure 15.13.

The traditional DLL now has two distinct layers: the ATM adaptation layer (AAL) and the ATM layer. The ATM network starts from the ATM layer down and this layer may interface with the traditional DS-1, DS-3, and SONET networks. This is where the existing trunk and high-speed networks (DS-1, DS-3) and evolving SONET get integrated via the ATM layer. ATM node-to-node connection may be established at the ATM layer.

	Class A	Class B	Class C	Class D
Timing relation between source and destination	Required		Not Required	
Bit Rate	Constant	Variable		
Connection Mode	Connection-Oriented			Connection-less

Examples of Service Classes
  • Class A: Circuit emulation; constant bit rate
  • Class B: Variable bit rate video and audio
  • Class C: Connection-oriented data transfer (FR)
  • Class D: Connectionless data transfer (SMDS)
Source: ITU 1.362

Current AAL Mappings

Class A	Class B	Class C	Class D
AAL1	AAL2	AAL3/4 or AAL5	AAL3/4

(a) Transmission Error Handling;
(b) Quantization/Segmentation Effect Handling;
(c) Flow Control;
(d) Timing Control;
(e) Lost/Misread Cell Handling;
(f) Service Class (A through D) Handling

FIGURE 15.12. Six Functions (a) through (f) of the AAL to the OSI-RM Higher Layers for the Different (A-D) Classes of Services. AAL - ATM Adaptation Layer; CBR - Constant Bit Rate; FR - Frame Relay; ITU - International Telecommunications Union (formerly CCITT); OSI-RM - Open Systems Interconnection-Reference Model; SMDS - Switched Multimegabit Data Service; VBR - Variable Bit Rate

15.6.1 ATM ADAPTATION LAYER (AAL) FUNCTIONS

The AAL also is subdivided into two sublayers: convergence sublayer (CS), and segmentation and reassemble sublayer (SAR). For the upper layers, AAL handles the transmission errors through the network: single errors are corrected and multiple errors are detected, leading to cell rejection. The effects of segmentation of data (into discrete playloads) among cells are also handled by this sublayer. The flow and time control functions are also enforced within this sublayer. Lost or error ridden cells are also resecured from the network (by requesting retransmission).

In addition the necessary DLL functions of the OSI, the AAL also performs a distinctive ATM function to the extent that it provides *four* (A through D) classes of service (see Figure 15.12). The A and B class service requiring timing relation between source and destination is connection oriented. Only class A service is at a constant bit rate. Class C and D do not require timing relation between source and destination and are both variable data rates. Class C is connection oriented, whereas class D is connectionless.

Class A is exemplified by circuit emulation mode with constant bit rate service, e.g., 56 kbps CSDC lines. Class B is exemplified by variable bit rate video and audio services. Class C is exemplified by connection oriented variable rate data transfer, such as frame relay applications, permanent/switched virtual circuit, ranging from 56 kbps, 64 kbps to 1.544 Mbps etc. Finally, class D service is exemplified by switched multimegabit data service or SMDS. To provide these four classes of services, AAL assumes four formats shown in Figures 15.12 and 15.13.

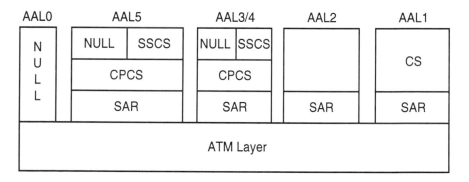

FIGURE 15.13. AAL Structure between ATM Layer and Network Layer of the OSI-RM. CPCS - Common Part Convergence Sublayer; SSCS - Service Specific Convergence Sublayer; CS - Convergence Sublayer; SAR - Segmentation and Reassembly

15.6.2 ATM LAYER FUNCTIONS

The ATM layer functions with the AAL to make the OSI Reference Model's DLL functions complete; whereas, the convergence and segmentation and reassemble parts of the OSI level two functions are handled by AAL. The ATM layer provides for: (a) the multiplexing/demultiplexing of cells; (b) translation the virtual path identifier (VPI) and virtual channel identifier (VCI); and (c) management of the user-network interface (UNI). The third function is generally broken down as two subfunctions: the generic flow control (GFC) and the generation/extraction of the cell header. In a sense, it completes the functions that AAL did not finish, but allows for the node-to-node connection in ATM networks.

15.6.3 ATM PHYSICAL LAYER

The ATM physical layer is also divided into *two* distinct sublayers: *the* transmission convergence sublayer (TCS) and *the* physical medium sublayer. Together they perform all the OSI reference model physical layer (layer one) functions or level one functions.

15.6.4 TCS FUNCTIONS

There are *five* functions at the TCS sublevel: cell rate decoupling; header error check (parity) header sequence generation/verification; cell delineation; transmission frame adaptation; and transmission frame generation/recovery. Each of these functions are tailored to fit the cell structure (5 octet header plus 48 octet playload) and the "frame" that carries these cells. Though tailored to map efficiently into to SONET/SDH frames, these cells may also be mapped into DS-3 or even DS-1 frames.

15.6.5 PHYSICAL MEDIUM SUBLAYER

This last sublayer performs two basic functions: it permits the bits to communicate between any given points in the network by maintaining the bit timing function and it facilitates such bits to actually traverse the physical medium.

15.6.6 ATM HEADER AND ATM FUNCTIONS

The interface becomes important in the format of the header block. At the user-network interface (UNI) and at the network-network interface (NNI), the five header octets convey the information to the ATM layer to service the ATM adaptation layer, or AAL (see Figure 15.11). This five-octet header block of ATM cell permits the ATM switch to perform the GFC (generic flow control), VPI (virtual path identification), VCI (virtual channel identification), playload type (PLT) identification, the cell loss parity (CLP), and the header error check (HEC). The arrangement of the bits and octets at UNI/NNI are shown in Figure 15.9.

It is to be appreciated that the ATM adaptation layer (AAL) function becomes crucial, since the four classes of services (A through D, discussed earlier) have their own specific requirements. For this reason, the AAL has a flexible functional structure (format) depending on the class of service. *Six such structures (AAL0 through AAL5) are identified and shown in Figure 15.13.*

AAL0 is transparent and has a null structure. It does not perform any function and is specifically not used (at present) for any of the service classes. AAL1 structure performs the convergence sublayer (CS) and segmentation and reassemble (SAR) functions. It is suitable for class A, constant bit rate, and connection oriented services needing a timing relation between the source and destination. AAL2 structure performs only SAR functions. It is suitable for class B, variable bit rate, connection oriented services not requiring timing relation, and video/audio services. AAL3, AAL4 and AAL5 structures greatly overlap. All of these perform the SAR function. However, the CS sublayer is subdivided into yet another two sublayers: the service specific convergence sublayer (SSCS) and common part convergence sublayer (CPCS). All three AAL structures handle class C, variable rate, connection oriented, and frame relay type of services. For class D, it is a connectionless variable rate and switched multimegabit data service (SMDS) that is handled by AA3 or AAL4 structures.

15.6.7 DS-1, DS-3 AND ATM TO SONET SIGNALS

SONET can carry data from a variety of existing carriers such as DS-1, DS-3. It is also well suited to carry the ATM cells. However, when the data/cells arrive, the information generic to the data type or the cells has to be made available to the terminal, SONET repeaters and multiplexers. This information is mapped into a new header by three new overhead blocks: the path overhead, the line overhead and the section overhead.

Accordingly, the data/cells flow into the path layer that generates the path overhead, the line layer that generates the line overhead, and then the section layer that generates the section overhead. It is the data/cells that are packed into the SONET frames with an additional header having these three (section, line and path) overhead blocks. This data structure becomes the SONET signal and it is ready to be carried by the photonic layer of the terminal, the SONET repeater or the SONET multiplexer.

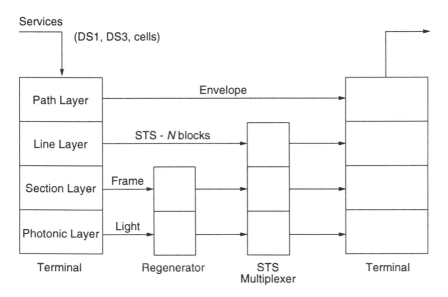

FIGURE 15.14a. Logical Hierarchy of DS-1, DS-3 and ATM Cells Mapped into the STS Signal (Frame).

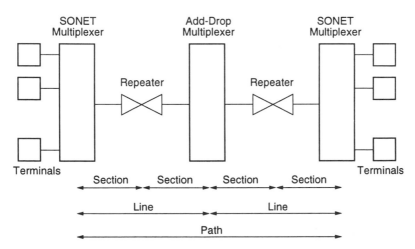

FIGURE 15.14b. Physical Hierarchy of SONET. ATM - Asynchronous Transfer Mode; DS - Digital Signal; STS - Synchronous Transport System

This mapping of the data/cells is represented in Figures 15.14a and b as the logical and physical hierarchy. In Figures 15.15a, b and c, the mapping of the ATM cells (Figures 15.15a) into ATM cell streams (Figure 15.15b) and then into the playload of the SONET STS-3c frame (Figure 15.15c) are depicted.

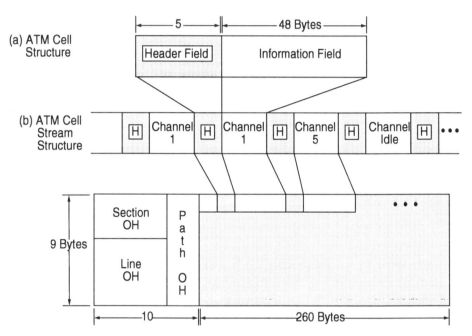

(c) ATM Cell Stream mapped into the playload of the SONET STS-3C signal.

FIGURE 15.15. Mapping of the ATM Cells into an STS-3c Signal.
ATM - Asynchronous Transfer Mode; OH - Overhead; STS - Synchronous Transport Signal

15.7

ATM AND ITS ROLE IN EVOLVING NETWORKS

ATM is a generalized standard that can encompass a large number of existing networks (such as the general T1/E1, T3/E3; optical, OC-1 to OC-48, computer, campus, LANs, MANs, WANs, etc.) with appropriate bridges and routers. Networks have their own idiosyncrasies based on the economic forces that make them viable. For example, the T1/E1 networks are geared towards twisted wire-pairs and used in the trunk/subscriber environments, and CATV networks have been geared towards carrying analog TV information in the homes, etc., but by and large, networks carry information through complex hierarchy of media switches by responding to the network commands via the signaling systems. The photonic network carries pockets (rather than packets of data) of photons through the optically sealed fiber environment. However, at the transceivers (transmitters/receivers), the bits of data get encoded/decoded as pockets of photons and bits of data are recompiled as bytes of information.

The ATM standards particularly facilitate an easy and elegant integration of the existing digital carriers and the 53 byte ATM cells onto to the SONET frames (see Figure 15.15, 1-33) transported by these networks. The integration into the broadband ISDN is equally simple and elegant (see Section 15.6.6) Most general networks may be integrated with the ATM environment by appropriate protocol converters and bridges/routers.

15.7.1 ATM BUILDING BLOCKS

Typically, there are *six* building blocks: the hub, bridge, router, brouter, the LAN/MAN/WAN switch and transport media. Numerous varieties of terminal equipment are connected to the ATM network access node, which interconnects to the ATM hub, and other nodes, etc. The configuration and the interconnection depend upon the type of application (LAN/MAN/WAN) for which the ATM is being deployed.

The interface issue is generally handled by ATM interface/adapter cards. Currently there are *four* interfaces: the high performance parallel interface (HIPPI), the standard ATM, the standard SONET and DS-3. The ATM bridge handles the connection between typically Ethernet and ATM networks, thus facilitating their coexistence in the same campus or local area network. The ATM router interfaces between two ATM networks.

15.7.2 ATM SWITCH

The ATM switch is significantly different from the traditional switch (e.g., ESS) and functions by multiplexing (based upon the virtual channel identifier/virtual path identifier) and path availability, not on time division multiplexing. The cell size is fixed at 53 octets. This does not have variable length frames, nor is it fixed at length time slots. The capacity is determined by the peak rate cell allocation and is not a fixed rate allocation of frames or time slots. Finally, the cell multiplexing is statistical rather than deterministic. Essentially, the ATM switch provides the switching fabric (i.e., interconnection between any input to output ports), the cell buffering, and the contention resolution.

The ATM switch (broadly characterized as campus, LAN/MAN/WAN and hub) has also been operational at numerous (eight; 1993 data) locations providing ATM, frame relay, SMDS, Ethernet, FDDI, broadband ISDN and T3 to OC-3 (to other ATM switches) switching. Such switches are commercially available (from at least seven vendors; 1993 data). As many as 32 ATM ports (1993 data) can be supported. The ATM switch throughput is as high as 12.8 Gigabit/s (1993 data) and is likely to increase by many orders of magnitude over the next decade. The ATM interface is as high as OC-12, even though typical rates range from DS-3 to OC-3 rates (1993 data). Most of the ATM interface is for fiber, even though some ATM switch vendors provide an ATM interface at 155 Mbps over class 5 unshielded and shielded twisted wire-pairs for campus and premises wiring. The LAN interfacing for most of the ATM switches is Ethernet, token ring, FDDI (supervision, gateway, and even node-to-node communication).

ATM hubs are also being manufactured by numerous vendors throughout the world with AAL5 structure and for the older PVC and SVC switching. Typical (1993 data) interfaces are at 45, 100, and at 155 Mbps. Bridge, router and brouter devices are also available to build complete ATM networks.

15.7.3 ATM MEDIA

Three viable media exist. Fiber offers very high bandwidth at a modest cost, some amount of implementation difficulty and high skill maintenance. Coaxial cable offers relatively high bandwidth (up to 400 MHz; the digital capacity depends on coding algorithms) with a high cost and a medium amount of difficulty in implementation and maintenance. The unshielded twisted wire-pairs have the lowest capacity (about 100-155 Mbps at about 30 meters), but cost, implementation and maintenance are quite low.

15.7.4 ATM NETWORK CONFIGURATIONS

For LAN applications, the architecture usually incorporates ATM switches, hub(s), and interfaces. In MAN applications, the architecture typically includes the ATM router, ATM switch, and ATM mux/demux (in addition, to the ATM-LAN building blocks). For wide area public networks, the architecture includes ATM frame relay switch, ATM-SMDS, ATM access node and ATM multiplexing facility. For the wide area private network access, the ATM building blocks that are necessary, may be as few as the switch, concentrator, and gateway.

15.7.5 ATM AND EXISTING DATA NETWORKS

ATM (or a major section of the standard) has been adapted by numerous private networks. Stratacom, Inc. provides for frame relaying at data rates ranging from 56 kbps to the DS-3 rates with all five (AAL1 to AAL5) types of ATM adaptation layer structures. Northern Telecom provides for DS-3 and isochronous T1 data with AAL1, 3 or 4, and five structures. Lightstream Corporation offers frame relay for X.25, HDLC, SNA/SDLC at 56 kbps to E3 rates with the AAL5 structure. Hughes Network systems T1/E1, T3/E3 and OC-3 rates with the AAL5 structure. General Datacom offers frame relay Ethernet, T1/E1 by circuit emulation, also with its own AAL. Cascade Communication Corporation offers SMDS and selected rates between fractional T1 to DS-3 rates with AAL1, 3/4, and 5 structures. Alcatel Data Networks offers frame relay, HDLC (high level data link control), SDLC (synchronous data link control), 64 kbps to E1 rates and SMDS rates ranging from 2 to 16 Mbps.

15.7.6 ATM AND ISDN

The ATM standards have a well defined strategy for the incorporation of broadband ISDN (BISDN) per se. Most of the ATM switches and hubs have provision and facility for ISDN ports. Such ports are connected to the standard BISDN ports of other ISDN networks.

The typical "R" interface for broadband ISDN terminal equipment TE-1 or TE-2 (see Chapter 7, companion volume, *"Intelligent Broadband Multimedia Networks"*, listed in Reference 15.1) reaches the line termination (LT) or exchange termination (ET) via the standard S, T and U interfaces. For broadband applications, these interfaces are prefixed with a B and shown in Figure 15.16 (1-28).

The functions of the terminal adapter (TA), and the network termination NT1 and NT2 are discussed in Reference 15.1. The connectivity at different terminations are also shown in Figure 15.16 (1-28). The protocol reference model for ATM/B-ISDN is depicted in Figure 15.17 and discussed in detail in Reference 15.1.

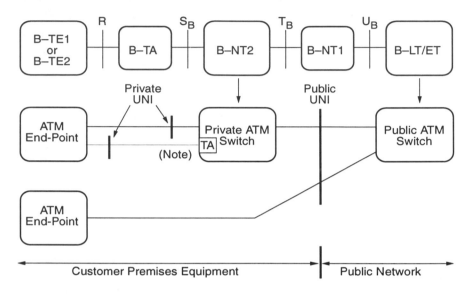

FIGURE 15.16. Broadband ISDN Access from ATM Switch Facilities.
ATM - Asynchronous Transfer Mode; B-NT - Network Terminator for Broadband ISDN; B-TA - Terminal Adapter for Broadband ISDN; B-TE - Terminal Equipment for Broadband ISDN; ET - Exchange Termination; LT - Line Termination; UNI - User Network Interface (For access configurations from conventional facilities see Chapter 7 of Reference 15.1).

The ATM/B-ISDN signaling is accomplished by providing the SAAL (signaling ATM adaptation layer) sublayer between ATM and the Q.2931 layers as depicted in Figure 15.17 (1-43). The standard ISDN user network interface and the ATM/B-ISDN layers are shown in the Figure. A broad global overview of the ATM reference model for BISDN services is shown in Figure 15.18. A graceful integration of the ATM and B-ISDN networks paves the way to beginning and the end of the information highways at the customer premises. The digital revolution that has started in the 1980s would have culminated as information revolution for the customer with the new mode of services that are knowledge related rather than being purely service related. Intelligent knowledge components [15.1] become essential in realizing the futuristic networks.

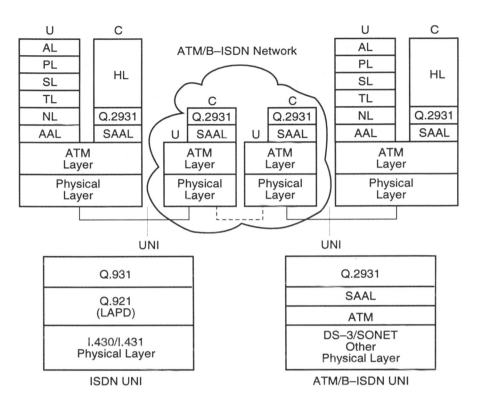

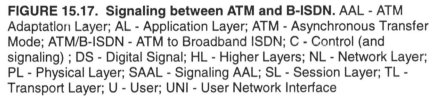

FIGURE 15.17. Signaling between ATM and B-ISDN. AAL - ATM Adaptation Layer; AL - Application Layer; ATM - Asynchronous Transfer Mode; ATM/B-ISDN - ATM to Broadband ISDN; C - Control (and signaling) ; DS - Digital Signal; HL - Higher Layers; NL - Network Layer; PL - Physical Layer; SAAL - Signaling AAL; SL - Session Layer; TL - Transport Layer; U - User; UNI - User Network Interface

15.7.7 ATM AND IN NETWORKS

ATM networks, though sophisticated, are not intelligent at their current state. Network intelligence, as it is applied to network functions, is not a part of the ATM standards. In a sense, ATM has no direct bearing on intelligent network functions (such as those in SSP, STP, SCP, IP or FNs; see Chapters 8-12 of the companion volume, "*Intelligent Broadband Multimedia Networks*," listed in Reference 15.1), even though the future signaling (or backbone) network of the communication systems may adapt ATM standards. It is possible and foreseeable that if the need for intelligent networks evolves and continues to grow, the ATM standards may also evolve and grow to perform these functions. ATM currently has already impacted a variety of services such as

CAD/CAM, teleconferencing, computer communication, multimedia, HDTV, etc. For the most part, the impact stems from broadband capabilities of fiber and easy integration of such high rates by using the ATM standards. The intelligent photonic switching as applied to network customers is still to come. To forge the architecture of intelligent photonic networks appears to be a non-trivial academic pursuit at this stage, even though information society will demand it if it explodes.

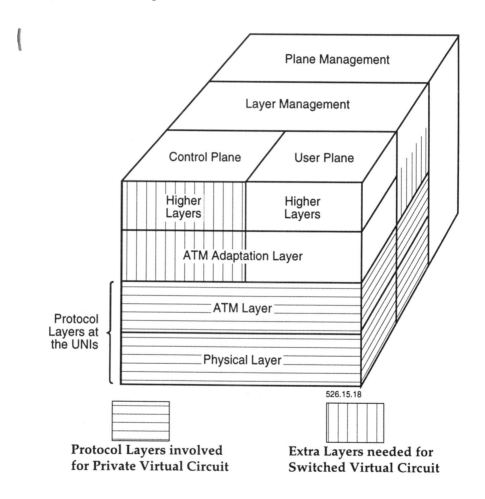

526.15.18

**Protocol Layers involved
for Private Virtual Circuit**

**Extra Layers needed for
Switched Virtual Circuit**

FIGURE 15.18. Overview of the ATM Reference Model for B-ISDN Applications. ATM - Asynchronous Transfer Mode; B-ISDN - Broadband ISDN; UNI - User Network Interface

15.8

CONCLUSIONS

In essence, there is a *threefold* impact of modern networks encompassing SONET and ATM. *First,* the on-going digital revolution that is alive and well in the computer and communication field since the days of von Neumann, is coming closer and closer to the telephone customer. *Second,* the on-going technological revolution in the fiber optic transmission since the fifties is being implemented closer and closer to work and home environments. *Third,* the recent, but strong, social revolution demanding more access, precision, and information/knowledge orientation in daily lives is being made possible. The convergence between the *three, the* digital revolution, *the* information and knowledge processing and *the* transmission technologies, is the apparent source.

ATM plays a focal role in making the current networks exploit the state of the art technologies (VLSI, optical transmission and optoelectronic) to meet the social and social-economic demands. Even though ATM standards are not direct contributors (as yet), to the state of the art intelligent networks or intelligent communication systems, they play a vital role in making intelligent communication systems become economically viable, and able to provide newer services from anywhere to anyone at anytime, network permitting.

15.9

REFERENCES

15.1   S.V. Ahamed and V.B. Lawrence 1996, *Intelligent Broadband Multimedia Networks*, Kluwer Academic Press, Boston, MA.

15.2   T.J. Aprille 1990, "Introducing SONET into the Local Exchange Carrier Network," *IEEE Communications Magazine*, August, 34-38.

15.3   CCITT (now ITU) 1988, "Synchronous Digital Hierarchy Bit Rates," *Recommendation G. 707,* Blue Book.

15.4   N.B. Sandesara, G.R. Ritchie, B. Enggel-Smith 1990, "Plans and Considerations for SONET Deployment", *IEEE Communications Magazine*, August, 26-33.

15.5   Bellcore 1990, "Integrated Digital Loop Carrier System Generic Requirements, Objectives, and Interface," *TR-TSY-000303,* Issue 1, September 1986; Rev. 3, March.

Also see:

Bellcore 1989, "Synchronous Optical Network (SONET) Transport Systems: Common Generic Criteria," *TR-TSY-000253*, September.

Bellcore 1989, "SONET Add-Drop Multiplex Equipment (SONET ADM) Generic Criteria" *TR-TSY-000496*, Issue 2, September.

Bellcore 1989, "SONET Add-Drop Multiplex Equipment (SONET ADM) Generic Criteria for a Self-Healing Ring Implementation" *TA-TSY-000496*, Issue 2, November.

Bellcore 1989, "Wideband and Broadband Digital Cross Connect Systems Generic Requirement and Objectives," *TR-TSY-0002336*, Issue 2, September.

CHAPTER 16

IMPACT OF FIBER OPTIC TECHNOLOGY

Fiber optic technology has catalyzed the information revolution. Its impact is as profound on telecommunications as the impact of the internal combustion engine on transportation. It is just that fiber has not reached every home, even though the automobile has; but then the fiber optic industry is a full generation younger than the automobile industry. The quantum jump in the information rate to the subscribers from that in copper based systems can only make the customers demand and get newer services such as multimedia, video on demand, televideo at telephone prices, high definition TV at CATV prices, instant classrooms, and a very real, accurate and personalized knowledge society just for the asking.

The data rates through the fiber have climbed almost as fast as than the data stored in storage technology, especially the Winchester, the CD ROM, and even the ULSI devices. Together, they move the digital multimedia information to every household over the glassy highways much like the CATV industry has moved the analog entertainment for the last three decades.

16.1

THE ROLE OF FIBER

The impact of fiber optic systems to carry voice and data entering the telephone plant is akin to the impact of digital carrier systems in the early 1960s. The survey of the fiber optic systems over the last decade reveals that the systems are experiencing about two orders of magnitude enhancement in the ratio of component costs to system performance every five years. *Three* typical landmarks of achievements are *(a)* the 45 Mbps (0.82 µm) fiber optic system in 1982, *(b)* the 1.7 Gbps (1.3 or 1.5 µm) fiber optic system in mid 1988, and then *(c)* the 10 Gbps transoceanic fiber optic segment in 1994/5.

There are *two* major areas of the impact of both these digital systems; the trunk systems and the subscriber systems. Earlier digital carriers (T1/E1, T1C, T2/E2, etc.) are now deeply entrenched in most telephone systems around the world and have literally brought the digital revolution closer to the routine telephone users: *first,* via the trunks and long distance PCM (64 kbps) digitally encoded telephone voice service and *second,* via the subscriber loop plants and the PCM (64 kbps), the ADPCM (32 kbps), the ADM (37.6 kbps, via the SLC-40 system) digitally encoded telephone services to the remote areas of countries.

In the wideband and broadband arenas (see Chapter 15), the impact is equally profound with inexpensive, crystal clear bandwidth-intensive multimedia services to every integrated digital remote terminal with optical carrier capacity, also, fiber has already made deep inroads in the trunk systems. Most of the modern *trunks* are digital and fiber based. Fiber-switch interface is almost as prevalent as the D-channel banks (see Sections 4.5 and 4.6) of the 1970s. The rate of data transfer now is three or four orders of magnitude faster and more dependable. In fact, it is uneconomical (if not down right foolish) to think of the older copper based analog carrier systems in the 1990s. In fact, it is unrealistic (if not inconceivable) to think of the *plain old* copper digital carrier systems for the newer networks. Digital copper systems still have a tiny niche in the modernization of services and/or maximization of the utility of the copper already installed in place. Fiber for the *distribution* of data is used via the hybrid subscriber loop plants using the modern subscriber loop carriers (e.g., fiber optic SLC-96 or the more modern SLC-2000). It has the impact of bringing the information revolution to the end users via wideband multimedia services. Multiplexed visual/picture and video information is carried at hundreds of megabits per second (at OC-N rates) in the 1990s rather than the multiplexed voice information that the digital carriers brought at a few megabits per second (at T1/E1, T1C and T2/E2) systems carried in the 1960s and 1970s. Whereas the plain old copper (transmission and D-channel) based carrier systems brought in digital revolution to the telephone plant, the fiber optic (transmission and DCS switches) based facilities is bringing the information revolution to the customers.

Fiber as a viable and efficient photonic carrier over very short to very long distances and the immense capabilities of fiber as an information highway (already proven in society) are explored in this Chapter. Its potential as an processor of data in optical processing systems is not even broached here simply because optical processing is not as yet viable. Its potential in being a significant partner to other systems such as microwave, wireless, and transoceanic facilities is clearly viable, but the protocol and interfacing issues are not addressed in this part. The full utilization of bridges and routers needs to be invoked in making the fiber optic networks work consistently with the existing copper, coaxial, microwave and wireless systems.

16.2

OPTICAL COMMUNICATION SYSTEMS

Most of the modern communication systems are partially optical and some of the possible architectures are presented in Chapter 15 (see Figures 15.5 to 15.7). Transmission is mostly optical, and processing is mostly electronic. The optical communication systems generally have an optimal blend of both technologies. Though the trend has been the steady progress towards building all optical networks, the optical device technology is not mature enough to compete and win over the electronic switching and device technology.

Switching and information processing calls for functions from the computational environment; Stored Program Switching systems have been discussed in detail in the companion volume, *"Intelligent Broadband Multimedia Networks"* [16.1], and the more recent ADMs, DCS, DACSs, IDTs, and ATM switches are presented in Chapter 15. For this reason, only the fiber optic *devices* that make up optical systems, their basic principles, their fabrications and their limitations are discussed here. The devices that perform the basic transceiver functions, and also systems which integrate them into viable networks, are discussed next.

16.2.1 GENERAL OPTICAL DEVICE CONCEPTS

Optical devices are routinely used in the transmission environment of telecommunication systems. These devices (fibers and their transceivers) are now in a mature phase of fabrication and manufacture for optical rates ranging from a few megabits per second to ten or more gigabits per second. These devices are discussed in some detail in this Chapter. Optical transmission systems generally interface electronic systems,. and the transceivers have their footing in both the electronic and optical technologies. However, in earlier composite receiver devices, the optical signal was converted into the electronic signal and most of the processing was done by the electronic devices. A similar approach was practiced at the transmitters when the electronic signal was converted to the optical devices at the last stage. The more recent compound semiconductor materials have distinct properties that permit devices to generate optical signals at the transmitter, that are specially suited to longer and longer fiber lengths and its optical properties. Conversely, these materials are specially designed to recover weaker optical signals at the receiver. They become specially suitable for transoceanic regenerators and make optical systems more and more robust. These devices and the related technologies are also discussed in this Chapter.

In a sense, the optical systems are being designed as composite systems rather than a simple aggregation of the individual components. To this extent they reflect the maturity of the fiberonics (integrated fiber and electronics) technology. To a large extent, the microwave and satellite technology also use the same care in the design of systems. The source characteristics (of the traveling wave tubes, varactors, and the semiconductor power devices) are accurately matched to those of the waveguides that carry the microwave energy [16.2]. In the VHDSL technology (see Chapter 10), the source and terminating impedances are matched as closely as possible to the line impedances over the spectral band of interest to maximize the coupling between the transceivers and non-uniform copper wire. The differences in the technologies offer new boundaries and options for optimality in the design of these different systems.

In addition, optical devices have also become possible contenders for *three* of the other basic functions; switching, processing and storage. Compound semiconductors offer properties attractive to their use in photonic devices with new functions thus far not possible with conventional semiconductor materials. The dominant property of these compound semiconductor materials lies in their ability to emit or to absorb photons efficiently as a result of recombination or formation of electron/hole pairs. The direct bandgap of the compound semiconductors offers this desirable property. The photon/carrier transitions thus facilitate the fabrication of the photonic devices. Further, matched heterojunctions or lattice structures can be formed to cascade the photon or the carrier characteristics to offer new photonic devices capable of more complex functions right at the photon/carrier level. The transition of optical (photonic environment) to the electronic (electron or carrier environment) technologies is thus elegantly made with higher efficiency and higher speeds.

The spectrum of visible light ranges from 0.3 μm (blue) to 0.7 μm (red end). *Three* regions of wavelengths (outside the region of visible light) have been considered for the fiber optic systems. The band gaps in the *0.8 μm* region, the *1.3 μm* region and *1.55 μm* region have been investigated over the last 30 years. These numbers arise from the locations of the three hydra-oxyle or (OH) peaks for the glass as it is drawn into the fiber strand (see Section 16.2.2.3). Earlier devices and fibers had shown some applications (the 1982 FT3 45 Mbps 0.82 μm system) tailored for the 0.8 μm. However, most of the devices manufactured currently are for the longer wavelength systems typically in the 1.3 to 1.55 μm range. The compound semiconductor material frequently used is indium gallium arsenide phosphide/indium phosphide (InGaAsP/InP). For the reduced wavelength typically around 0.8 μm range, the material used is aluminum gallium arsenide/gallium arsenide (AlGaAs/GaAs). Relatively few devices are fabricated since they operate over shorter distances and at lower bit rates around the 0.8 μm wavelength systems. For most communication devices, the compound semiconductor materials used are InP, InGaAsP, and indium gallium arsenide (InGaAs). Lattice matched, epitaxial layers of these materials are grown on a frequently used n-type InP substrate to fabricate the devices.

For the semiconductor devices, GaAs technology offers higher carrier mobility and lower junction capacitance. The resulting circuits operate at a much higher rate (typically five times faster) than conventional silicon based circuits. For photonic applications, the devices built with InP (indium phosphide) technology outperform the devices built with GaAs (gallium arsenide) technology. Devices particularly useful for photonic applications with ultra high-speed functionality are the heterostructure field effect transistor (HFET) and the heterojunction bipolar transistor (HBT). These devices function because of their precisely created microstructure. which uniquely combines the optical and electronic functions. The properties of the materials are thus engineered to suit the functionality. This approach to creating the microstructures with optimal combination of their properties in the photonic and electronic domains is called bandgap engineering, and the materials capable of being thus engineered are called III-V compound semiconductors.

16.2.2 OPTICAL FIBER

The possibility of using lightwaves for communication was proposed as far back as 1880 by Alexander Graham Bell; 26 years after John Tyndell performed the light transmission experiments over a dielectric conductor before the Royal Society in England in 1854. Optical transmission was seriously considered during the early sixties after the invention of lasers at Bell Laboratories, in 1960. The multilayered glass structure was suggested during the mid sixties and a rough specimen of fiber with about 20 dB/km loss was produced by Corning Glass in 1970. Optical systems have been functioning with ever increasing capacity and sophistication. Large capacity, immunity from stray electromagnetic interference, low weight and cost are the chief attractions of the optical fiber waveguide (simply stated, optical fiber) medium. The highest quality single mode fiber drawn during the late eighties offers 0.16 dB/km loss at 1.55 μm wavelength and less than 0.35 dB/km at 1.30 μm wavelength. These lower loss limits are due to intrinsic scattering (see Section 16.2.2.3) within the silica glass rather than manufacturing tolerances in drawing the fiber glass. The chromatic dispersion is also equally attractive at approximately 15 pm/km-nm (picosecond/km-nanometer); see Section 16.2.2.4, for the explanation of the units of measurements for chromatic dispersion) in the low loss window of 1.5 μm (with zero dispersion at 1.3 μm). Specially designed fibers further reduce the overall dispersion to about 17 ps/km-nm (picosecond/km-nanometer) at 1.5 μm band, and prove satisfactory for laser sources without chirp or drifting wavelength of the photonic energy emitted. These two effects, light loss (or attenuation) and chromatic smearing (or delay distortion effects) eventually limit the product of bit rate and distance through the fiber carrying information. Since there is only a single mode in the single mode fibers, the light through these fibers does not suffer modal dispersion effects, but is susceptible to chromatic dispersion effects.

Fibers are dielectric waveguides. A subset of the classic Maxwell's equations [16.3] lead to the fields that coexist in the fiber. Since the material is (by and large) linear, and isotropic without free circulating currents and devoid of free charges, the basic equations can be written as:

$$\nabla \times \mathbf{E} = \frac{\partial \mathbf{B}}{\partial t} \qquad \nabla \times \mathbf{H} = \frac{\partial \mathbf{D}}{\partial t} \qquad (16.1,2)$$

$$\nabla \cdot \mathbf{B} = 0 \qquad \nabla \cdot \mathbf{D} = 0 \qquad (16.3,4)$$

with the traditional relations between **D** and **E** defining ε, and also **B** and **H** defining μ as follows:

$$\mathbf{D} = \varepsilon\mathbf{E} \qquad\qquad \mathbf{B} = \mu\mathbf{H} \qquad\qquad (16.5,6)$$

The solution of the two leads to the two scalar wave equations

$$\nabla^2\mathbf{E} = \mu\varepsilon\frac{\partial^2\mathbf{E}}{\partial t^2} \qquad\qquad \nabla^2\mathbf{H} = \mu\varepsilon\frac{\partial^2\mathbf{H}}{\partial t^2}. \qquad (16.7,8)$$

From the point of view of applications, the complexity of the detailed analysis (discussed in Reference 16.4) for such field distributions in general environments is unwarranted since the structure is cylindrical. Considerable simplification can be gained by the approximate, but adequate, model known as geometrical optics, due to the cylindrical structures of the core and the cladding. To explain the concept, consider a fiber with distinctly different optical densities for the core and cladding. At the interface, there are two effects: *refraction and reflection*. Both effect the propagation of light. When the optical density is graded as a function of its diameter, a combination of the refraction and reflection along the radius effects the recapture and the propagation of light through the fiber core.

Three crucial layers of the fiber are: the core or the inner most layer, the cladding and the outer protective jacket [16.5]. The core and cladding together provide the optical environment for the light wave(s) through the fiber. The cladding is tailored to be an integral part of the fiber geometry to contain and propagate the light. The core has a *slightly higher* optical density (index of refraction) than that of the cladding. The optical density of the core material or silica glass is higher (1.46 to 1.5) than that of the air from where the light is incident. This effects the propagation *two* ways: *first*, the velocity in the denser medium is reduced; *second*, the direction is changed by the conventional rules of refraction. The net effect of the two is a realignment of the rays of light towards the axis of reference or the axis of the fiber.

At the core-cladding boundary, *two* types of rays exist. *First*, consider a ray of light at the core-cladding boundary that is *refracted* in the cladding and is bent towards the axis of reference. This ray travels some distance in the cladding until it approaches the cladding-jacket boundary where the absorption loss is high, due to the lossy portion of the outermost cladding. and due to the scattering of the ray at the rough cladding-jacket interface.

This ray, which enters the cladding and moves toward the jacket, is lost and will not appear to any measurable extent at the end of the fiber.

However, the *second* type of ray of light, which is incident at the core cladding boundary at a low angle of incidence (i.e., with respect to the axis of the fiber), does not enter the cladding at all, but is *reflected* internally and guided towards the axis of the fiber. Hence, the extent of capture of the light in the core depends upon the angle of incidence at the core-cladding boundary. The laws of reflection dominate the propagation in the recapture of the light within the core, and this is the carrier of information to the far end of the fiber. The limiting angle for the cutoff for reflection to prevail over refraction depends upon the relative difference between the optical densities of the core and cladding.

The problem can also be completely stated by considering *three* propagating electromagnetic waves: *incident wave* at the core-cladding boundary; *refracted wave* propagating away from the boundary; and *reflected wave* that re-enters the core. The boundary conditions from both sides of the interface may be enforced, and simply enough, such an algebraic exercise leads to Snell's law, which asserts that the ratio of the optical densities equals the ratios of the cosines of the angle of incident ray and the angle of trajectory ray. Stated simply with respect to Figure 16.1, the relation becomes

$$n_{core} \cdot \cos(\theta_{core}) = n_{cladding} \cdot \cos(\theta_{cladding}) \qquad (16.9)$$

or the reciprocal relation

$$n_2/n_1 = \text{cosine } \theta_1 / \text{cosine } \theta_2 \qquad (16.10)$$

subscript 1 refers to the core and subscript 2 refers to the cladding. The critical angle is simply inverse cosine (n_2/n_1) these ratios (n_2/n_1) can be made very close to unity (typically the value is 0.99), thus most of the light gets trapped or internally reflected totally in the core, and is received at the far end of the fiber. The critical angle in this case is

$$\theta_c = 8.109°, \text{ at the core-cladding boundary} \qquad (16.11)$$

and it signifies that rays of light incident from the core with an angle < 8.109° (with the longitudinal axis of the fiber) will be totally internally reflected. Extending this computation of the critical angle, the rays that can possibly subtend an angle of < 8.109° at the core-cladding boundary can enter the core (from the air) only if they subtend an angle $<$ (n_{core} . 8.109°)   Stated alternatively, the critical angle θ_c, in this case is approximated as

$$\theta_c = 11.84° \text{ to } 12.16°, \text{ at the air-core boundary,} \tag{16.12}$$

with the longitudinal axis of the fiber for typical fiber waveguides, since the typical values of n_{core} lie between 1.46 and 1.5.

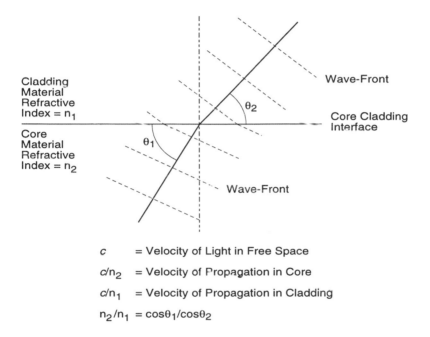

Cladding Material Refractive Index = n_1

Core Material Refractive Index = n_2

Wave-Front

Core Cladding Interface

Wave-Front

c = Velocity of Light in Free Space
c/n_2 = Velocity of Propagation in Core
c/n_1 = Velocity of Propagation in Cladding
$n_2/n_1 = \cos\theta_1/\cos\theta_2$

FIGURE 16.1. Refraction at Core-Cladding Interface.

16.2.2.1 Single Mode Fiber

Physical Features: The fiber core made of silica glass (Figure 16.2) has a diameter of about 5 μm. This diameter corresponds to 0.19685 mils (1 mil = thousandths of an inch) for single mode fibers, which sustain a single mode of propagation through it. The wavelength of the light through the fiber is generally chosen to be a fraction of the core diameter. For the fine core-diameter or single mode fibers, the diameter is generally a small multiple of the wavelength of the light through it.

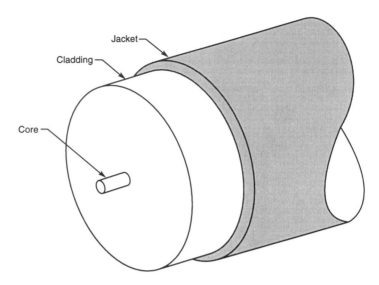

FIGURE 16.2. Physical Configuration of a Typical Single Mode (SM) Fiber. Any deviation from a perfectly cylindrical structure leads to a degenerate single mode. Both the true SM and the degenerate modes generally exist and give rise to slight time lapse between the signals carried by the two modes.

The cladding has a refractive index or optical density slightly lower than that of the core. The difference of the optical densities can range between a fraction of a percent to a few percentage points. However, the discussion to follow asserts the need for proper proportioning of the optical densities in relation to the ratio of the fiber diameter to the wavelength of the light through it. The cladding diameter is of the order of about 125 µm or more, corresponding to 4.92 mils (about the diameter of a human hair) or somewhat larger. The cladding is generally surrounded by a protective jacket for handling. The source wavelength thus plays a part in the choice of the fiber. Other special purpose fibers have been designed and constructed for particular applications.

The perfect single mode through a single mode fiber is more a myth than a reality. The single mode is achieved only if the fiber is perfectly cylindrical and the manufacturing tolerances (however small) force the tiniest modal degradation or polarization. Both modes traverse the length of the fiber at very slightly different velocities and this effect, though very small, does linger on as the polarization mode dispersion effect.

16.2.2.2 Multimode Fiber

Physical Features: The core diameter (Figure 16.3) of the multimode fiber is much large ranging from ten times the wavelength to a hundred times the diameter of the wavelength of the light through it. This has a profound effect upon the attenuation and dispersion of light through it because rays of light trapped in different modes travel different distances through the fiber, and thus the extent of attenuation and dispersion will not be the same. The cladding is thinner and has about the same external diameter; and it is enclosed in a protective jacket like the single mode fiber.

Physical Makeup: Typically, numerous optical fibers are bundled to make up the inner core of a installation cable. Circular and ribbon structures are used. For the circular cables, the strength members are at the center, with the fibers around the periphery, with a sheath to hold the entire structure. For the ribbon structure, encapsulant is used to fill the space between a bundle of fibers and the outer most jacket, which also holds the strength members. As many as twelve ribbons with twelve fibers in each have been fabricated and enclosed in circular copper-clad steel sheath helical wiring for strength with an outer diameter of 1.2 cm. This ribbon structure permits easy gang splicing of the different ribbons.

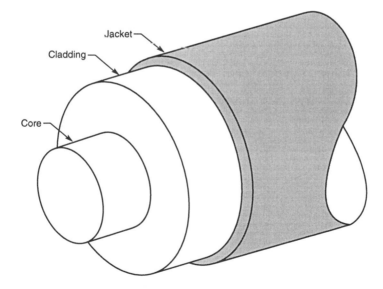

FIGURE 16.3. Physical Configuration of a Typical Multimode Fiber.

The outside diameter of the fiber cables is as little as 2.5 mm (single fiber) through 6.73 mm (5 fibers). Composite fiber and copper cables are also built with six inner fibers surrounded by two electrically isolated copper serve wires (for repeatering under water for transoceanic applications), and with galvanized steel wire armor for strength, all at a nominal diameter of 1.766 cm. The tensile strength of these oceanic cables is typically 147 Newtons and the copper wires can carry about 10 A and sustain a voltage of 3000V. There are large varieties of optical cables with different geometrical arrangements that are being currently manufactured. The tolerance for the attenuation is < 0.1 dB/km over a range of 55° C to 80° C.

16.2.2.3 Optical Properties

Attenuation: The loss of optical energy that a light pulse experiences as it travels down the fiber is called attenuation. *Two* causes for the loss of energy are *absorption* and *scattering*. In the early fabrication of fibers (mid-sixties to early seventies), impurities by absorption was the major reason for high attenuation of the optical signal. An extremely high level of purity against some elements is necessary. For example, two parts per billion (ppb) of cobalt can induce a 10 dB/km loss; twenty ppb of nickel, iron or chromium, or fifty ppb of copper, or even a hundred ppb of manganese or vanadium can each induce 10 dB/km loss through the fiber. In 1970, the quality control of the fiber manufacture process was poor enough to induce a 20 dB/km fiber. In 1972, the loss was reduced to about 4 dB/km and fiber for communication became scientifically and economically feasible for longer distance trunk applications.

Effects of Absorption: Absorption is caused by the photon-electron interaction, which results as the propagating light prompts the electrons to undergo state transitions. Impurities and the silica glass material both absorb energy. However, the impurities absorb substantial amounts of energy from the pulse in the wavelength, which carries the pulse, and thus obliterates its amplitude and its shape. Silica glass, on the other hand, does absorb the energy, but in a waveband generally beyond the region of interest where the pulse energy is concentrated. The energy absorbed by the electrons is eventually released as light of other wavelengths or heat due to mechanical vibration within the material.

Effects of Scattering: Scattering, as the term implies, is caused by the energy in the rays the leave the fiber due to the imperfection of the geometry of the fiber. The measure of the imperfection is its relation to the wavelength of the light through the fiber. Thus, at the imperfection, a certain amount of light leaves the fiber at the same wavelength as it reaches the imperfection.

Silicon dioxide (SiO$_2$ or silica glass) is a non-crystalline material. The atoms are arranged in a somewhat random fashion where any incremental volume of the material does not hold the same number of atoms. The light through the glass does interact with the electrons in the material however small the interaction may be. Rayleigh scattering caused by the weak interaction (absorption and reradiation at the same wavelength, but delayed in phase) of the light with the electrons in the glass structure represents a theoretical lower limit on the attenuation of the particular type of glass for a given wavelength. A typical loss curve at various wavelengths is depicted in Figure 16.4. The Rayleigh scattering limit for 0.85 µm wavelength (from GaAlAs light sources) is about 1.6 dB/km, and varies inversely with the fourth power of the wavelength. However, the lower limit of the loss is about 0.5 dB/km at 1.3 µm and less than 0.2 dB/km at 1.55 µm. The newer sources of light and matching optical detectors have been investigated and successfully fabricated and deployed for high quality fibers fabricated with 0.16 dB/km loss at 1.55 µm light.

The presence of the OH radical (water) offers strong absorbing resonance at about 1.4 µm wavelength. This "water peak" effect is due to the lingering presence of some of the OH radical in the fiber core material. Total eradication of this peak has not been successful and the fiber system designers avoid the 1.4 µm wavelength for this reason for silica glass fiber core material. At wavelengths in excess of 1.6 µm, the absorption loss in silica glass in and by itself starts to increase rapidly, and viability of optical systems using this material does not exist.

This does not mean the other materials have the same restrictions (OH and Rayleigh limits) and cannot be used for other types of fibers. Some materials ("dark fibers") offer a theoretical loss limit of 0.01 dB/km, but the practical use within a scientific and economic system have yet to be demonstrated.

16.2.2.4 Optical Properties, Delay Distortion or Dispersion

Pulse spreading caused by delay dispersion in optical systems arise for *two* major reasons; *first*, the differential lengths of the paths traversed by different modes of propagation and *second*, the lack of a perfectly coherent source of optical power, which gives light at a fixed wavelength. The effects of each is discussed next.

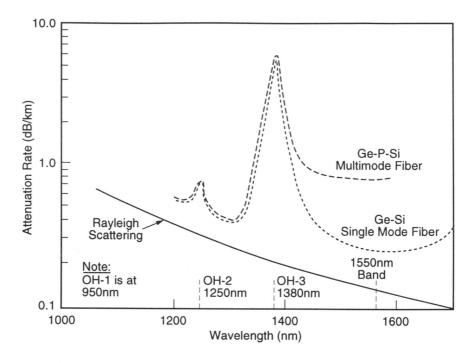

FIGURE 16.4. Typical Loss Curves for Silica Glass Fibers. The lowest theoretical loss (about 0.16 dB/km) occurs at about 1550 nm where the scattering loss and the absorptive loss are about the same.

Incremental Delays: The presence of numerous paths affects the actual distances that rays of light traverse in the core, and this depends upon the angle of incidence at the core-cladding boundary. The fastest ray travels the shortest distance, and thus the least angle of incidence. Zero angle ray (see 1 in Figure 16.5) would travel the axial length of the fiber along its axis, and thus reaches the end of the fiber first. On the other hand, the ray (see 2 in Figure 16.5) at the critical or the maximum angle gets reflected numerous times and arrive delayed by an increment of time

$$\delta_t = n_1 /c \cdot \ell \cdot [(n_1 - n_2)/n_2]. \tag{16.13}$$

where ℓ is the axial length of the fiber in meters, n_1 is the optical density of the core and n_2 is the optical density of the cladding and c is the velocity of light in meters per second.

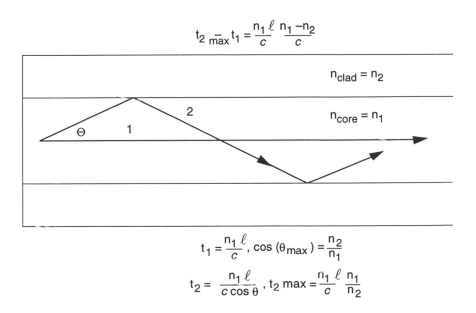

$$t_2 \overline{\max} t_1 = \frac{n_1 \ell}{c} \frac{n_1 - n_2}{c}$$

$n_{clad} = n_2$

$n_{core} = n_1$

$$t_1 = \frac{n_1 \ell}{c}, \quad \cos(\theta_{max}) = \frac{n_2}{n_1}$$

$$t_2 = \frac{n_1 \ell}{c \cos \theta}, \quad t_2 \max = \frac{n_1 \ell}{c} \frac{n_1}{n_2}$$

FIGURE 16.5. Delay in Arrival Times of the Two Rays 1 and 2 Traversing the Shortest and the Longest Paths.

The minimum duration to travel the axial length is

$$t_{min} = \ell . n_1 / c \qquad (16.14)$$

and the maximum duration to travel the length of the fiber with multiple reflections is

$$t_{max} = \ell . n_1 / c . 1/\cos \theta_{max} \qquad (16.15)$$

with $\cos \theta_{max} = n_2 / n_1$ from Snell's law. For silica fibers, the value of t_{min} is 5 μs/km. This number arises from a refractive index for glass being 1.5 and the velocity of light in free space being 3×10^8 M/sec. The typical value of $\cos \theta_{max}$ is 0.99. Stated alternatively, the fractional difference between the refractive indices of the core and the cladding is only 0.01. Thus, the results of the delay difference is about 5 μs x .01 or 50 ns/km, since the longest path through 1 km of the fiber is 1.010 km. One of the standard techniques to reduce the modal delay spreading is to use the graded index fiber shown in Figure 16.6. This fiber has a parabolic distribution of the optical density being highest at the center and tapering off towards the cladding. Rays do travel different distances in the core, but the longest distance is in the region of the lowest optical density. Thus, an equalization effect takes place and all rays tend to arrive at the same time, thus reducing the modal delay spreading that occurs in the core. In practice, the pulse spread of 50 ns/km with step index fiber can be reduced to about 0.5 to 2.5 ns/km or about 100 to 20 times reduced pulse spread.

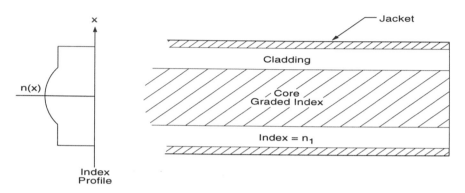

FIGURE 16.6. Typical Configuration of Graded Index Fiber. The refractive index is plotted on the left.

Theoretical calculations have predicted a 1000 to 1 gain with optimal profile of the grading for the optical index. Such gains have not been realized due to manufacturing tolerances.

16.2.2.5 One Single Mode through the Fiber

Consider very small diameter single mode fiber with a core diameter D of about 6.6 µm, for a light source with a wavelength of 1 µm free-space [or λ = 0.66 µm silica glass (refractive index = 1.5)]. The ratio of the wavelength to the core diameter is 1/10, thus the spacing between the rays is 1/10 radian or about 5.7 degrees. This inference is based upon the solution of Maxwell's equations or the laws of physical optics, which assert that only certain discrete angles (corresponding to the discrete set of guided modes, typically the lowest four the HE_{11}, the TE_{01}, the TM_{01}, and the HE_{21} for the step index fiber, which is indeed a dielectric waveguide) can propagate in the fiber. The spacing (measured in radians) between the angles of propagation of the individual rays in the set corresponds to the ratio of optical wavelength of the light λ through the fiber to the fiber diameter D.

Next, consider a fiber consisting of the core and cladding with fractional difference between the refractive indices of the core and the cladding of 0.5

percent, with the first allowed ray away from the axis above the maximum angle θ_{max} (i.e., arccosine 0.995 or 5.73 degrees). Thus, by accurate control on the wavelength of the light that is incident at the entry into the fiber, and an accurate control of the refractive indices of the core and cladding, only one ray (i.e. a single mode of light) enters this single mode fiber and is the only one guided wave in the fiber guide. Though slightly harder to splice and couple to the source, single mode fibers can carry pulses of very high-speed data. There is no modal spread due to the singular presence of only one mode through the fiber (only when it is perfectly cylindrical, see Section 16.2.2.1).

16.2.2.6 Multiple Modes through the Fiber

Now consider a different geometry with a much larger core diameter, typically 50 μm, $n_2 / n_1 = 0.99$, and the same source of light with a typical wavelength of 0.66 μm in the core. The value of D/λ is 1/75 radian or 0.8 degrees. The value of θ_{max} (for a fiber with cosine $\theta_{max} = n_2 / n_1 = 0.99$) is about 8.1 degrees, and there are 10 rays (i.e., 8.1/0.8 since there are no fractional modes) that can coexist in each direction away from the axis, thus leading to 20 rays in two dimensions. For the three dimensional fiber, the total number becomes about 20x20 or 400. With this relation between the parameters for the fiber, many hundreds of modes can exist, giving rise to multi-modality in the fiber; and every one of these modes has some finite extent of propagation delay through the fiber.

Effects of Optical Energy Distribution in a Waveband: The second major reason for the pulse spreading is due to the presence of optical energy at numerous wavelengths in the light emitted from LEDs and perhaps a chirp in the laser source. Different optical wavelengths travel at different speeds. The range of light from LEDs and other noncoherent light sources may become quite broad (over 5 percent of its nominal value). The optical density or the index of refraction of silica glass is sensitive to the wavelength and is evident as the prism effect upon white light. This effect shows up because the propagation speed will vary with wavelength if the second derivative of the index of refraction with respect to wavelength is not zero. This effect, known as material dispersion, is measured in ns/(km-nm) and is characterized by the variation in delay (measured as ns/km of fiber length) with respect to the change in wavelength (nm). In Figure 16.7, the typical material dispersion effect is plotted for silica glass over a range of wavelengths. It is easy to see that the material dispersion effect can be sizable (in fact exceeding the modal delay spreading in graded index fibers) under certain conditions. At 0.85 μm wavelength, the dispersion is about 0.1 ns/km-nm. If the optical band spread

from an incoherent LED is 50 nm, then the pulse spread between the two
extremes of the wavelength is 5 ns/km. When the pulse width is only 20 ns
at 50 Mbps, then pulse spread due to material dispersion becomes significant.

The spectral spread of coherent light sources (laser) is small and the
effects of material dispersion can be low. However, silica glass offers one
other attractive feature; at 1.3 μm, the material dispersion crosses zero value
and the optical attenuation is low (0.5 dB/ km; a low enough value, though
not the least). For this reason, many lightwave systems are built with
components around a nominal wavelength of 1.3 μm.

It is possible to design single mode fibers in which the waveguide
propagation effects can counter the material dispersion effects. The net effect
is that the frequency of net zero dispersion is shifted. Also, single mode
fibers with uniformly low dispersion effects over a significant band of
wavelengths can be fabricated. This type of fiber design permits the concept
of multiplexing numerous channels over preassigned optical wavelengths
through the same fiber, thus leading to wavelength division multiplexing (or
WDM) of the wavelength parameter through the fiber with low dispersion of
the pulses in the many wavelength bands.

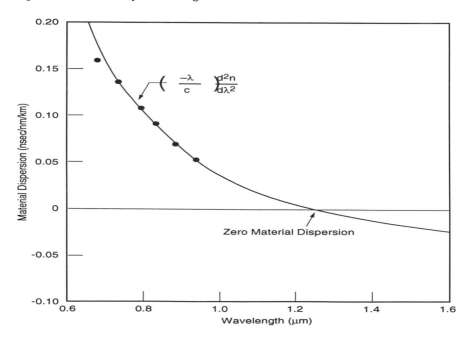

**FIGURE 16.7. Typical Plot for the Material Dispersion for Silica
Glass Fibers.**

The spectral spread of coherent light sources (laser) is small and the effects of material dispersion can be low. However, silica glass offers one other attractive feature; at 1.3 μm, the material dispersion crosses zero value and the optical attenuation is low (0.5 dB/ km; a low enough value, though not the least). For this reason, many of the earlier lightwave systems were built with components around a nominal wavelength of 1.3 μm.

It is possible to design single mode fibers in which the waveguide propagation effects can counter the material dispersion effects. The net effect is that the frequency of net zero dispersion is shifted. Also, single mode fibers with uniformly low dispersion effects over a significant band of wavelengths can be fabricated. This type of fiber design permits the concept of multiplexing numerous channels over preassigned optical wavelengths through the same fiber, thus leading to wavelength division multiplexing (or WDM) of the wavelength parameter through the fiber with low dispersion of the pulses in the many wavelength bands.

As the transmission capabilities are pushed higher and higher, the optical properties other than transmission loss become increasingly important for the fiber. Representative of these properties are the dispersion (chromatic and polarization) effects and scattering (stimulated Raman, stimulated Brillouin, cross-phase modulation, and four photon mixing). These limitations effect the various optical systems differently and each system need to simulated, studied and experimentally verified.

Bit rates, distance, transmitted power and the number of channels through the fiber all influence the ultimate capacity of an optical system. Once again the optical system is much more than the fiber and the system components (the sources, the multiplexers, the detectors the demultiplexers, etc.) exert their own influence. Good quality fibers offer attenuation loss of about 0.16 to 0.2 dB/km at λ =1.55 μm and 0.32 to 0.35 dB/km at λ =1.3 μm. For this reason the fabrication of complex lightwave system is an elaborate computer assisted iterative process that minimizes the BER (typically <10^{-9} to 2×10^{-10}) of up to 20 Gbps over 80 to 100 km. The eye closure penalty is still a good measure for these optical data transmission systems(see Chapter 18).

While it is possible to limit the effects of dispersion, modal delay lingers on and this effect cannot be canceled by the *two* (i.e., *waveguide and material*) dispersion effects. The total delay distortion (when modal delay and dispersion are both present) is the square root of the sum of the squares (of the individual spreads in ns/km).

The transmission capacity of appropriately designed fiber far exceeds the capacity of other media. Even if the fiber is used to a capacity that is four orders of magnitude less than it ultimate capacity, it surpasses other media for high volume communication systems. Existing communication systems that have been deployed by national and international communications networks have other restrictions, such as switching and intelligence, to channel the fiber capacity. On the national scene, fiber have been routinely introduced for interoffice high data capacity channels in the United States, Canada, Europe and Japan. Optical rates of 565 Mbps and OC-12 rates of 622.080 Mbps have been achieved by commercial systems.

16.2.3 OPTICAL DEVICES FOR EMERGING NETWORKS

The newest of networks blend the communication and processing at the very seminal level. In the information processing environments, the five major components to consider are the processors, the memory, the input/output systems, the bus structures, and the switching devices. Conventional computing systems assumes the availability of electronic power sources, detectors, transistors switches and their integration, and go on towards the next level of system integration. In the optical systems, these devices have to be considered individually because of the infancy of this technology. Further, possible steps toward integration also need investigation. These aspects of optical components and their integration are presented here.

In the network environments, the switching and transmission aspects are exaggerated and processing is becoming more and more sophisticated, especially in intelligent networks, in packet systems, and in the CCISS backbone network. Currently, in most digital networks, almost all processing is electronic and almost all new trunk routing and computer networking is optical. Fiber-to-the-home is economical where the expected demand for high volume data can be clearly assured for newer services, HDTV, or the high-tech home office of the future. The current technological constraints on the devices, and the practices within the networks as they are emerging are reflected in the information presented next.

16.2.3.1 Optical Sources

Optical sources for optical communications need to offer high radiance, easy direct modulation of the data at multigigabits per second rates, small size, low cost and high efficiency. The two optical sources generally in use are the light-emitting diodes (LEDs) and laser diodes. Both these devices are fabricated from multilayered structures of compound semiconductor materials. These layers are grown by the epitaxial growth process, thus preserving the crystalline symmetry throughout the many layers within the structure. Such structures with precisely controlled epitaxial growth of microscopic layers of the semiconductor materials become the functioning elements of both the optical sources and detectors. Recent innovations in liquid and vapor phase epitaxial growth technologies make the accurate fabrication of these devices possible. Heteroepitaxial growth process permits the deposition of one crystalline layer of one material upon the crystalline substrate wafer of the other. For photonic devices to operate in 1.3 or the 1.55 μm wavelength bands, the choice for the substrate material is indium phosphide (InP). The substrate material permits the deposition of heteroepitaxial layers of indium gallium arsenide phosphide (InGaAsP), which becomes the active material within the device.

16.2.3.2 Light Emitting Diodes

Light emitting diodes are used as sources for optical data transmission links with digital rates to about half a gigabit per second (Gbps). The distance range is limited because of the waveband spread of the optical energy. LEDs in the 0.85 μm band use the aluminum gallium arsenide materials and are used for short distance optical links such as those used in the electronic switching systems (such as AT&T's 5ESS). The details of the surface-light emitting diode fabricated for 1.3 μm are shown in Figure 16.8.

In this band, the fiber attenuation (or loss) is much smaller and the chromatic dispersion is negligible. LEDs for this band use the InGaAsP materials (see Figure 16.8) and the device emits light in response to current pulse injected into the device. The physical geometry of the device with curved hemispherical surface permits effective coupling of the generated light directly into the fiber. The doping density and the active layer thickness can both be controlled to give a suitable balance between the bandwidth (and thus the response time and the maximum data rate) and the output power (and thus the range of transmission) of the LED.

The buried heterostructure in the LEDs is planar and of one single growth. A much thicker active layer is used here and the light emitting region is defined by the processing steps. These steps cause the drive current to be channeled through a strip (as in an edge-emitting LED, or an EELED) or through a dot (as in a surface emitting LED or a SELED). In the later device, SELED, the light is emitted through a window in the InP substrate rather than from a facet.

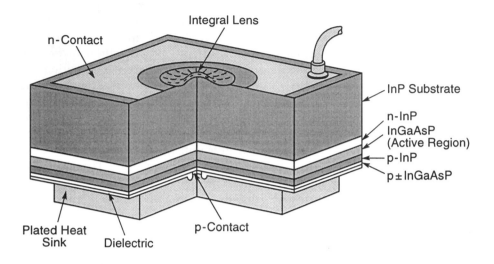

FIGURE 16.8. Details of a Surface Light-Emitting-Diode (LED). The InGaAsP (for $\lambda \approx$ 1300 nm) layer surrounded by p-InP and the n-InP is the light-emitting region. The lens refocuses the light emitted.

In contrast to the LEDs, lasers offer a much quicker response and thus the optical data rates are much higher. InGaAsP materials are generally used for the active region in the laser. The laser designs are generally optimized to yield optical linearity, temporal stability and prolonged operating life. The performance characteristics of the laser are controlled by its structure and index-guiding. The temporal stability is achieved by the optical confinement of its lasing mode. The operation of most laser sources is based upon burying the active light-emitting region in layers, which have lower index and higher bandgap which leads to index-guiding. This arrangement permits the two most important processes in the laser; electrical confinement of the primary and recombining electrons and holes resulting from the current injection, and optical confinement of the lasing mode. The electro-optical process is thus complete. Two possible laser structures are possible; CSBH lasers and CMBH lasers. These are discussed next.

16.2.3.3 CSBH Lasers

The light-emitting region is buried in a triangular groove within the channel-substrate buried heterostructure (CSBH) type of laser (Figure 16.9). The size of this active layer is about 1.5x1.5 μm in the groove, and the design can be optimized to suit the user specifications within reason. Typically, the optical laser. Cooling and operating temperatures also influence the life of the device. Lasers depend upon optical feedback to retain the lasing mode and this feedback is provided by the cleaved crystal facets at each end of the laser. Such lasers can be designed to emit around 1.3 μm wavelength and are dependable sources for lightwave applications when the bit rate to distance ratio is less than 80 Gb/km. The modality of these lasers is not strictly stable at 1.3 μm, and the chromatic dispersion can vary slightly causing pulse dispersion over very long distances, thus limiting its data rate. These multilasing lasers with multiple wavelength energy peaks, severely limit the data rates.

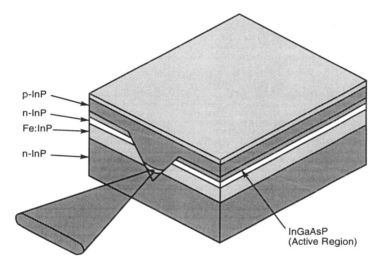

p-InP
n-InP
Fe:InP
n-InP
InGaAsP
(Active Region)

FIGURE 16.9. Schematic Representation of a Channel-Substrate Buried Heterostructure (CSBH). The light-emitting region in the small InGaAsP layer (≈1.5 μm X 0.15 μm) in the V-groove. This particular type of laser source has been used in the AT&T FT Series G transmission system (See Figure 17.1).

This is especially the case for lightwave systems operating around 1.55 μm wavelength, with conventional silica fibers offering high chromatic dispersion (about 15 ps/km-nm [picosecond per kilometer per nanometer]; see Section 16.2.2.4). Specially designed fibers designed for zero dispersion and minimum loss at 1.55 μm rather than at 1.3 μm, can be used with multiple wavelength lasers to enhance the data rate in specialized applications calling for 1.55 μm systems.

A special class of laser sources called the single wavelength distributed feedback laser (DFB) has also been developed to limit the effects of fiber dispersion. These lasers incorporate diffraction grating within the structure prior to the epitaxial growth of the numerous layers. The grating carried out by etching forces optical feedback, at a predetermined wavelength, controlled by the periodicity of the grating incorporated. Since the grating is an integral part of the lasers, these single wavelength lasers are more stable, operate longer, yield higher power, and operate faster than CSBH lasers. Enhanced data rate for applications in many digitally encoded lightwave transmission systems are thus feasible. This type of DFB laser is suitable for applications using wavelength-division-multiplexing (WDM) techniques for very high data rates and for submarine lightwave guide systems with fewer intermediate repeaters.

16.2.3.4 CMBH Lasers

The capped-mesa buried heterostructure (CMBH) type of laser (Figure 16.10) is a special version of the DFB laser. These laser sources emit high single frequency powers (in excess of 30 mW) with about 7 to 8 GHz bandwidth modulation.

These lasers can also operate over a life span comparable to that of the CSBH lasers in the 1.3 and 1.55 µm bands for submarine lightwave applications. Very high modulation bandwidths (over 10 GHz) have been achieved with very low capacitance and semi-insulating current-blocking layers. The metallization schemes enhance the high-speed modulation response and the power output from the laser. Figure 16.11 indicates the small signal frequency response of a typical CMBH DFB laser with about 20 mW power output and 17 GHz bandwidth.

The structures can also be fabricated for coherent optical transmission systems. Tunable optical sources are necessary to key the frequency. Two section DFB lasers (Figure 16.12) are used and the keying occurs as the refractive index of the light emitting region is changed by injecting a modulating current signal in either of the two sections. The emission wavelength (hence the frequency) is thus modulated. A gain in receiver sensitivity ranging from 6 to 8 dB has been achieved over the direct detection schemes.

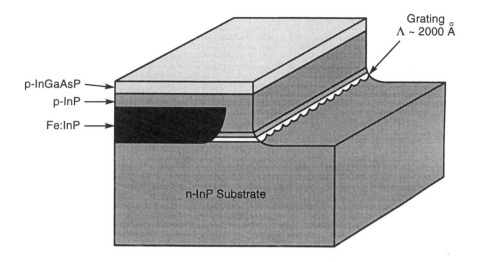

FIGURE 16.10. Schematic of a Capped-Mesa Buried Heterostructure (CMBH) Distributed-Feedback (DFB) Laser. The diffraction grating near the active region permits frequency-selective feedback and makes the laser emit at a single frequency.

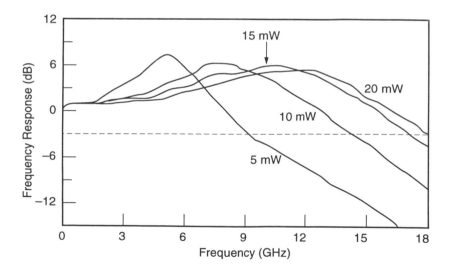

FIGURE 16.11. Small Signal Modulation Response of a Distributed Feedback (DFB) Laser with Different Biasing Conditions. The laser bandwidth increases with increasing output reaching a value of about 17 GHz at 20 mW.

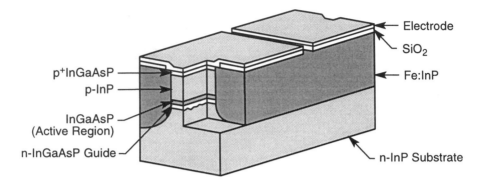

FIGURE 16.12. Single Wavelength Laser Structure for Coherent Lightwave Applications (see Figure 16.27). A two section distributed feedback structure is shown. Also see Figure 16.13.

The high performance requirement of the InGaAsP lasers requires tight lateral optical and electrical confinement of the buried heterostructure. *Two* sets of epitaxial growths may become essential. The *first* growth establishes the current blocking layers and etching of the channel through them (as in the case of the channeled substrate buried heterostructure or CSBH laser). The *second* growth establishes the double heterostructure in the channel, generally consisting of n-type InP, symbolized as n-InP, an undoped active InGaAsP layer, and finally a p-type InP symbolized as p-InP layer. The distributed feedback laser (DFB) also uses a buried heterostructure but the fabrication process is different. The initial growth of the double heterostructure contains the active layer. The first processing step etches the heterostructure to leave behind a narrow mesa. The second processing step grows a blocking layer around the mesa. These two epitaxial growths needed for the lasers are not necessary for the buried heterostructures in the LEDs. LEDs also have the non-reflective facets through which the light from the device is emitted. This facet is necessary in the lasers to sustain their intrinsic oscillation at the appropriate wavelength.

In the next configuration for the coherent optical transmission scheme (Figure 16.13), distributed Bragg reflector configuration has been used. The range of variation of this device is as high as 60 angstroms and the device can be used as a local oscillator for such coherent systems. The diffraction grating is directly etched into the 1.3 μm InGaAsP waveguide layer.

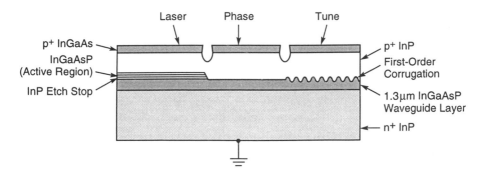

FIGURE 16.13. A Three Section Distributed Bragg Reflector Laser.
This type of laser is also for coherent Lightwave applications. Both these
laser sources (Figures 16.12 and 16.13) have adequate line width and
tunability for coherent frequency shift-keying (FSK) system.

This layer couples directly with the light emitting region of the laser. The
current through the phase and frequency tuning changes the waveguide
index, and the device can be used for continuously changing the wavelength
emitted.

16.2.3.5 Photodetectors

The *three* common types of solid-state optical detectors are the *(a)* back-
biased PN junction, *(b)* positive-intrinsic-negative (PIN) photodiode; and *(c)*
avalanche photodetectors (APD). The PIN and APD diodes are variations of
the more simple photo sensitive PN junction diode. The photo-cathode of a
vacuum device can also be considered as a photodetector, but is not viable
(because of the size, expense and poor response) in the present context of
fiber optic systems. An overview of the solid-state devices is presented in
the following sections.

Two of the basic concepts in dealing with photo detection are: quantum
efficiency and responsivity. Quantum efficiency of the detector is the fraction
of the incident photons that results in the electrons liberated in the device.
The maximum quantum efficiency is unity and generally quite low (typically
between 0.1 to 10 %) in the photo cathodes, and as high as 100% (in special
cases), typically about 84% in high-speed photodiodes. Quantum efficiency
depends dramatically on the materials used in the fabrication of the
photodetectors, and also upon the wavelength. Both have a profound effect
upon the output current of the device.

Another measure for the detection devices is the responsivity or the efficiency with which incident light is converted to the photo-current. Responsivity is defined as the ratio of photo current in amperes, or the product of number of electrons/second and charge/electron = 1.6 × 10^{-19} Coulombs = to the power in watts, or the product of number of photons/second and energy /photon = 2 × 10^{-19} Joules at wavelength of 1 µm. By the way in which these two parameters are defined, the responsivity at 1 µm is about 0.8 times the quantum efficiency (defined as amperes/watt), and it can change with the wavelength. For the basic relations between the units of charge, q and current, i, or energy, E and power, p see the footnote*.

The Back-Biased PN Junction: A voltage source and a load resistor and PN junction constitute this type of a archaic rudimentary detector, shown in Figure 16.14a. The back bias forces the electrons and holes to have moved away from the junction, and a depletion region (Figure 16.14b) around the junction is created without any positive donor ions nor any negative receptor ions. Now if the incident photons give rise to electron-hole pairs in the absorption region (Figure 6.14c), then the movement of the holes (towards the negatively biased P region) and the electrons (towards the positively biased N region) gives rise to a displacement current.

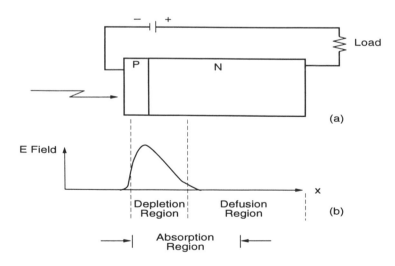

FIGURE 16.14a and b. PN Photodiode Structure and its Three Regions.

---

* q (in Coulombs) = $\int i \, .dt$ (in ampere-seconds) and E in watts = $\int^t p \, .dt$ (in Joule-seconds)

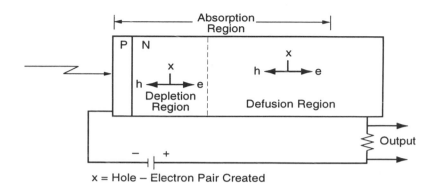

x = Hole − Electron Pair Created

FIGURE 16.14c. Carrier Regeneration Effects in Photodiodes.

The magnitude of this displacement current is proportional to the product of the carrier velocity and the local field strength. The carrier velocities are relatively high in the depletion region and slow in the diffusion region due to the lack of a strong electric field. The carrier movement in the depletion region produces an immediate response from the device and the movement in the diffusion layer becomes random and contributes to a slowly decaying current. The first effect gives rise to the a quick surge of current to follow the incident photons, and the later gives rise to the diffusion tail.

The collision process giving rise to the electron-hole pair is a random process, since any incident photon may or may not be on a collision path, and even so, the resulting carriers may be close or distant from the contacts sensing the photo current. Thus, the actual number of electrons responsible for the diode current at any given instant of time is a random number with a mean and a variance. This effect shows in the photo current generated from any pulse of light to contain *two* types of noise (a) the thermal noise and (b) the shot noise in addition to its average response due to the incident light. In the PN and PIN diodes the thermal noise dominates the shot noise and is generally ignored. The presence of dark current (no signal leakage current through the device) adds an offset to the output of the photodiode.

Positive-Intrinsic-Negative (PIN) Photodiode: PIN diodes arise due to the practical restrictions on the contact resistance on the N side of the device (see Figure 16.15a).

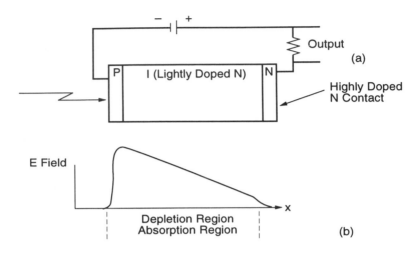

FIGURE 16.15a and b. PIN Photodiode Structure and its Depletion/Absorption Region.

A highly doped N device is essential to facilitate the flow of current, and this calls for a reduction in the doping in order to make a lightly doped N-region intrinsic (I type rather than the heavily doped N type). This doping profile of the three regions makes the device positive-intrinsic-negative or the PIN diode.   The depletion region is shown in Figure 16.15b and the fabrication layers of the earlier types of typical PIN diodes are shown in Figure 16.15c.

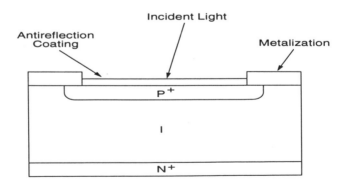

FIGURE 16.15c. Geometry of a Typical PIN Photodiode.

The desired levels of quantum efficiency and the speed of response influence the geometry of the PIN diodes. In practical devices, both need to be at their highest levels. A wider I-region slows the response and too narrow an I region limits the number photon that get absorbed to give rise to the electrons responsible for the photo current. Silicon exhibits excellent absorption of light at wavelengths < 0.9 μm, and hence, the diodes can be fabricated with almost a 100% quantum efficiency (of the light that does penetrate the device after the anti-reflection coating; see Figure 16.15c) and with a 0.5 ns response time. As the wavelength increases, the absorption in silicon falls rapidly and the quantum efficiency suffers. Other materials such as InGaAsP offer more desirable combinations of speed and efficiency for longer wavelengths.

Photodetectors operate under very low signal level constraints. Very sensitive digital receivers have to operate with only a few nanowatts (nW) of power since every bit can be encoded with a discrete number of photons. Even under conditions when the quantum efficiency approaches unity, only a few nA of photo-current results. Thus, the PIN devices are inherently limited to work with very low output current, since they can generate no more than one electron-hole for each photon. PIN diodes suffer from the same type of noise found in the PN diodes.

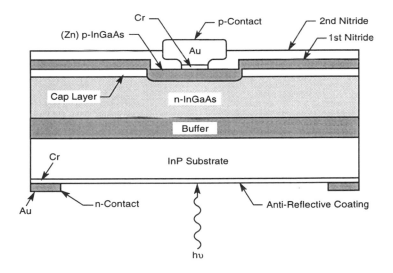

FIGURE 16.16. Schematic of a Modern PIN Diode. The light is absorbed in the low doped n-InGaAs Layer.

In the more recent version of the PIN diode shown in Figure 16.16 the incident light is absorbed in the low-doped n-InGaAs layer where the carriers are generated and supply the detector current sensed as electrical signals to the external sensing circuitry. Being simple to fabricate, they are used extensively in earlier low to moderate data rate optical transmission systems. The very low failure rate (about one failure in 10^9 hours of operation) makes them desirable in many applications such as the TAT-8 system operating at 295.60 Mbps.

Avalanche Photodetectors (APD): The APD devices, also being compound semiconductors like the PIN devices, append the electron-hole pair generation process of the PIN device with a multiplicative (or an avalanche) process that generates numerous electron-hole pairs for almost every incident photon.

Physically this is done by including a high field multiplicative region next to the lightly doped depletion region where the electron-hole pairs are originally formed. The doping profile is indicated in Figure 16.17a and the electric intensity E field distribution is shown in Figure 16.17b. This multiplicative process takes place because this region is more highly doped; the effect of the ionizing collision is shown in Figure 16.17c.

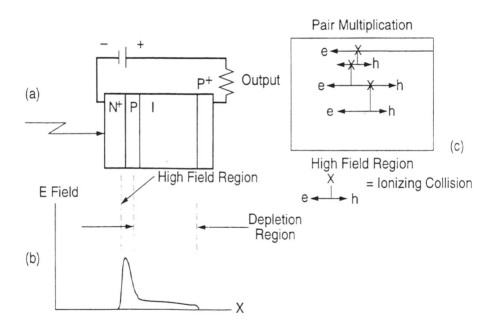

FIGURE 16.17a, b, and c. Structure, Function and Field Intensity of an Avalanche Photodiode (APD).

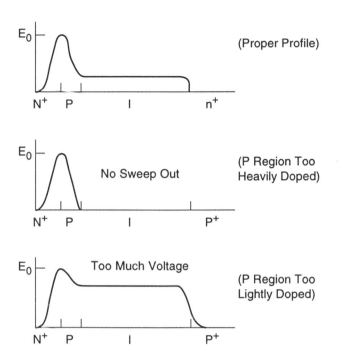

FIGURE 16.17d. Effect of Doping up on the Electric Intensity E over the N+, P, I and P+ Regions of the APD. APD - Avalanche Photo Diode

The doping profile of this device needs to maintain the required E field distribution. Excessive doping of the P region results in no E field in the I region able to sweep out the carriers, and very light doping results in over depletion of the depletion layer. Accurate reverse biasing of this device (see Figure 16.17d) is also essential to maintain the proper E field distribution in the depletion layer.

Under proper operating conditions, each hole-electron pair created by the photon absorption process can result in a number (ranging from tens to hundreds) of additional electron-hole process and the effective responsivity of the APD can be significantly greater than that of the PIN device. But the process of secondary generation has to respond to the incident photonic excitation of the diode (see Figure 16.17d for the geometry of the APD) and be self-terminating to be stable. Two depictions of the APD geometry are shown in Figures 16.18a and b.

However, the multiplicative process is a statistical and an unpredictable phenomenon, since the exact number of ionizing collision depends on the random events within the various layers of the device (also Figures 16.18a and b). This randomness (of the ionizing collisions) is the main cause of an additional noise characteristic (akin to randomly multiplied shot noise) of the APD and limits the sensitivity of the device.

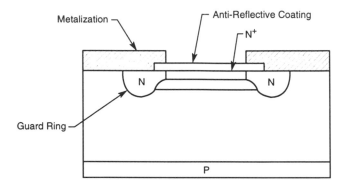

FIGURE 16.18a. Typical Structure of an Avalanche Photodiode

Within the operating region, higher gain is accompanied with higher shot-noise or quantum noise, and this dependence needs to be appropriately adjusted in the simulation (see Chapter 18) of the fiber optic systems. The device geometry and the operating conditions can limit the extent of the randomness of the multiplicative process, but not totally eliminate the randomness of the amplification within the APD diodes. Thus, the output of the APD will appear noisier than that of the PIN diodes, which exhibit predominantly thermal noise characteristics. The APD and PIN diodes both have the thermal noise and the shot noise. However, in the APDs the shot noise can become comparable, if not greater than, the thermal noise.

Both PIN diodes and APDs can be adequately represented by their equivalent circuits and their transfer functions. Accurate estimates of the thermal noise (and shot noise for APDs) for various types of diodes can be derived from the quantum mechanics relations, which govern the probability of the electron-hole generation process (and the probability of the ionizing collisions for the APDs). At the next level of designs the detector, the amplifier should be matched into next stage of signal amplification in the receiver. As is usually the case, this procedure turns into a computer-aided optimization process for any given application.

The APD (Figure 16.18b) absorbs the incident light in the low-doped InGaAs layer and generates carriers. However, these photo-generated carriers undergo a multiplication due to the impact ionization (secondary emission) in the n-InP multiplication layer before they are collected by the external sensing circuitry. The probabilistic avalanche process within the device amplifies the electrical signals generated by the reception of the photonic energy. The avalanching causes a multiplication of the photo-generated carriers. Lower level optical signals can thus trigger the APD, making the APD more sensitive than the PIN photodiode. In the high data rate systems, the number of photons per bit start to decrease. Thus, for a given incident optical power at the transmitter, the sensitivity required from the APDs at the receiver becomes higher.

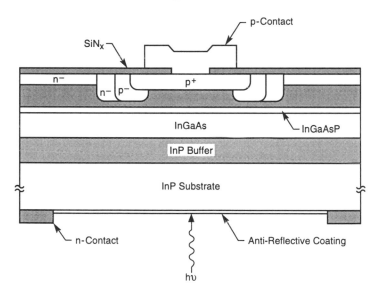

FIGURE 16.18b. Typical Structure of a Modern APD. The light absorbing region is the low doped n-InGaAs layer. This APD structure has separate absorption and multiplication (SAM-APD) regions. The multiplication occurs in the n-InP layer.

The typical gain-bandwidth product could peak at about 70 GHz and be over 60 GHz with a bandwidth over 6 GHz and a gain of 10. The high performance of these APDs makes them suitable for data rates of 3.4 GHz (i.e., the AT & TFT series G system deploying two channels, each carrying 1.7 Gbps.

16.2.3.6 Quantum Well Devices

Photodetectors, which utilize intraband transitions in a superlattice structure, can also be fabricated. These devices have much less noise than that in the PIN and the APD diodes, which have conventional single layer absorbing layer. PIN and conventional APD devices, using bulk non-layered material depend upon the interband transitions. In the quantum well devices, very thin light absorbing layers are separated by higher bandgap semiconductors serving as quantum wells (See Figure   16.19).   The kinetic energy of the electrons and holes is restricted to discrete values for velocities normal to the thin layer by the laws of quantum mechanics. The principle of operation is simply based upon the increased probability of escape of the electron (which has indeed made the transition from the low energy level to the higher energy level in the conduction band) from the potential energy barrier.   In Figure 16.19 the structure of a superlattice photodiode with GaAs and AlGaAs layers is shown. The corresponding responsivity of the structure (in kV/W) is plotted as a function of the wavelength (in µm) and its photon energy (in cm^{-1}).

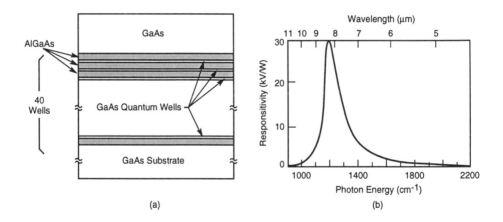

(a)   (b)

FIGURE 16.19a and b. Structure and Responsivity as a Function of Wavelength for a GaAs-AlGaAs Superlattice Photodiode.

Quantum well lasers can also be fabricated. These devices offer a change in emission wavelength with a change in the thickness of the active layer. Lower spectral width under modulation is typical of these lasers, thus making them suitable as sources for direct detection.

In the superlattice structures, the conduction-valance band edge discontinuities differ significantly from those in the bulk materials. A larger difference in the ionization rates of electrons and holes is thus realized, reducing the device noise. The reduced noise makes these devices preferable over the PIN and APD diodes.

16.2.3.7 Optical Amplifier and Repeaters

Direct restoration of the optical signal is desirable in most optical systems spanning long distances (such as the high-speed AT&T FT series G, or the TAT-8 systems). Optical detection, timing recovery, decision circuitry and an optical transmitter are all necessary for repeaters. Optical amplifiers are ideal candidates for direct optical signal restoration. However, these amplifiers can approximately restore the optical pulse. Thus, only ideal amplifiers (thus far nonexistent) can be total regenerators. However, optical amplification of the order of 30 dB has been achieved under laboratory conditions by using laser-like structures. The more recent approach the repeater problem is the erbium doping within the fiber material.

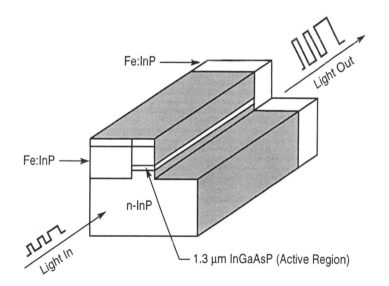

FIGURE 16.20. Schematic of a Buried Facet Optical Amplifier.

Figure 16.20 depicts the principle of the buried facet optical amplifier. In this device, both the facets have anti-reflection coatings to minimize laser oscillations and enhance the optical gain. Recent technology constraints can yield a facet reflectivity of about (about 10^{-2}) and thus a 30 dB or more optical amplification. The amplifying layer in this device is terminated before reaching the output facet. The coupling between the reflected light from the facet and the amplifying layer is reduced. The device simplicity makes such amplifiers serve as pseudo (and imperfect) regenerators. Even though imperfect, they are likely to satisfactorily meet the requirements for some of the applications.

16.2.3.8 Integration of Optical Devices

Photo electric and electronic functions are both handled by some of the devices discussed in this Chapter. A spectrum of hybrid functions thus evolves. The photonic functions are altered, controlled and optimized by the recently evolved InGaAsP technology and the electronic functions are handled by the well established GaAs or the silicon integrated circuit technology. The optimal composition of the functions with the material characteristics within the constraint of the two technologies (InGaAsP and GaAs/Si), is still to be realized in detail. For this reason, it is still not clear whether newer devices will completely overshadow the devices presented in this Section. However, several integrated structures (discussed next) have been fabricated.

The integrated transmitter combines the laser with a field effect transistor to generate light pulses from optical sources. The PIN diode and a field effect transistor may be integrated to serve as a receiver. Finally, numerous indium phosphide field effect transistors can be combined to give amplification. In an effort to fabricate a single chip repeater, an integrated DFB laser transmitter, a PIN photodiode FET receiver and a PIN-FET amplifier have been combined (see Figure 16.21). The sensitivity of the laser element to the input signal is in the order of a few dB, and the optical signal operates with a high bandwidth in excess of 12 GHz and furnishes single mode optical power, also in excess of 30 mW. The metal insulator semiconductor field effect transistor (MISFET) is fabricated atop an n-InP layer with low doping density. Metal-organic chemical vapor deposition (MOCVD) and the more recent metal-organic molecular beam epitaxy (MOMBE) both contributed to the synthetic materials known as heterostructures used in these devices.

16.2.3.9 Photonic Switches

Integration of signal processing (as is necessary in regeneration) and switching (as is necessary in switching systems) can also be attempted. Specialized configurations that use photonic or photo-electronic devices become essential. The approach is especially attractive since the lightwave

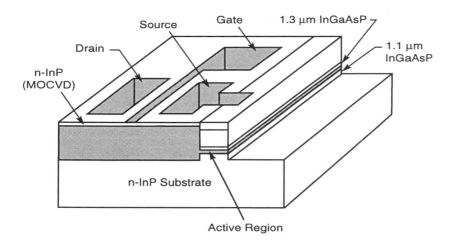

FIGURE 16.21. A Monolithically Integrated DFB Laser and Metal Insulator Semiconductor Field Effect Transistor (MISFET).

transmission systems are optical. Bistability is essential in most switches to enable them to serve as storage elements.

In the traditional integrated circuits, the flip-flop combination was one of the earliest switches. Then came the simple FET and CMOS technologies. Storage elements in CMOS can also be switched from a charged to discharged state and vice-versa. In most cases, this element has served as a storage device. Photonic switching is feasible with optical bistability of the III-V devices. The physical phenomenon behind the switching function is based upon quantum-confined Stark effect. A device using this effect is referred as a SEED (a term coined by AT&T Bell Laboratories to stand for self-electro-optic effect device). Light triggers the switch with a PIN configuration. The reverse bias in the undoped multi-quantum well region (see Figure 16.22) produces an electric field.

The multi-quantum well consists of very thin layers with a deposition of about 100 angstrom or less of absorbing and non absorbing regions. Carriers induced in the absorbing regions change the impedance of the device and thus alter the electrical field. This field affects the optical transmission if the original and incident light is near the band edge of the material used in the quantum well. The amount of light through the device increases monotonically with light into the device. However after a certain level the optical property of the device is altered and the output is reduced, as shown in Figure 16.23. On the reverse course, as the input light is decreased, the output light stays low until a lower value of the input light is reached, thus causing an optical hysteresis within the device.

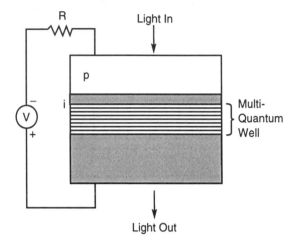

FIGURE 16.22. Schematic of a Basic Self Electro-Optic Effect Device (SEED).

This effect is similar to the magnetic hysteresis in which the flux density of the magnetic material retains (or remembers) the effects of the magnetic field and its reversal in either direction, thus permitting the device to be used as a memory element of computer systems. The optical switch can be made more robust and relatively insensitive to the bias level and the optical power and its variation.

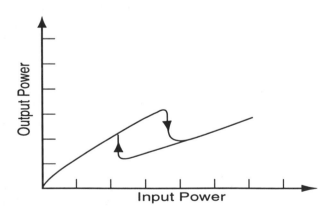

FIGURE 16.23. Typical Performance of a Multi-Quantum Well SEED Device.

For the symmetric self electro-optical effect device (S-SEED a term coined at AT&T Bell Laboratories), two such SEEDs, each with its own quantum well PIN diodes, are serially built on the same substrate (Figure 16.24) to function as very rapidly (few nanoseconds) switching memory elements and logic amplifiers.

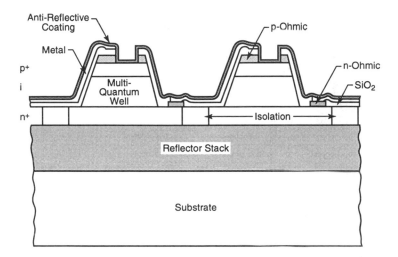

FIGURE 16.24. Cross Section of a Symmetric SEED or S-SEED device.

The current use of such devices has been demonstrated by building a very rudimentary optical processor, which is roughly the equivalent of a 4 bit microprocessor. The extent of use is confined to experimental and laboratory systems of insignificant computational value. However, the impact of these devices to the switching fabric of any telecommunication system, and eventually to the computing system, is still to be evaluated. Such devices can only influence the next generation of telecommunication and computing systems. The semiconductor technology, now mature and deeply ingrained in the telecommunication and computing systems, will need very substantial cost advantages and economic persuasion before it can be replaced. Speed alone is hardly enough since the modern networks can deliver enough gigabits over dedicated fiber optic channels. The business and social needs of modern cultures does not yet call for circuit switched multiple gigabit channels to corporations and private homes. But it can happen!

16.3

WDM FOR ENHANCED RATES

Wavelength division multiplexing (WDM) is a well accepted technique to enhance the bit rate. The fiber quality is high enough to offer extremely low dispersion of the optical signal.   The relative crosstalk between the information bearing wave bands is extremely low, and a wide band of low attenuation surrounds the water peak in any typical fiber. This offers a range of channels to coexist and carry high rate data through a single fiber.

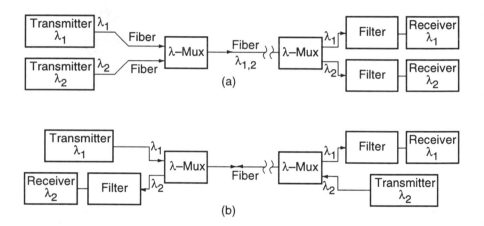

FIGURE 16.25. (a) Multiple Channels for Unidirectional Transmission with the Wavelength Division Multiplexing (WDM) Principle; (b) WDM Applications for Bidirectional Transmission.

In principle, wavelength division multiplexing offers two or more independent channels for communication. There is some minute spillover of optical energy from one band of wavelength or optical frequency to the other. However, the optical media by itself shows little cross coupling of energy from one band to another.   These channels may be unidirectional, as depicted in Figure 16.25a, or bidirectional, as depicted in Figure 16.25b. The number of channels could be numerous and tightly cramped in the regions of minimum attenuation. The WDM system needs to be designed as an entity based on the energy distribution and the stability of the laser sources, the grade of fiber, the spacing of the channels and receiver characteristics. Simulation results (of the type discussed in Chapter 18) of the entire system become essential.

16.4

DIRECT DETECTION AND COHERENT OPTICAL SYSTEMS

When the optical signals are directly converted to the electrical current by the photodetector, the process is called direct detection (see Figure 16.26). A local oscillator is necessary for coherent detection (see Figure 16.27). Most of the simulation and then the design consideration of the direct detection methods are discussed in Chapter 18. The entire simulation and design aspects have been streamlined, much like those in the simulation and design studies of the Digital Subscriber Lines (see Chapters 7 through 10). The coherent detection techniques and their significant results are presented in this Chapter.

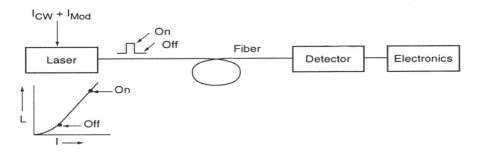

FIGURE 16.26. Block Diagram for a Direct Detection Lightwave System.

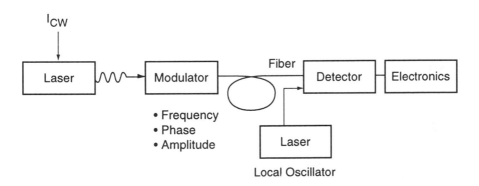

FIGURE 16.27. Block Diagram for a Coherent Detection Lightwave System.

At both the transceivers, coherent lightwave systems use a local signal from a local (laser source) oscillator (LO). Like other coherent systems, the transmitting modulator modifies the output signal, and the receiving demodulator detects the modification to the optical signal during detection. Coherent lightwave systems encompasses the techniques employing nonlinear mixing of two optical waves. Typically, one of these is the modulated wave (A_{sig} at an radian frequency of ω_{sig}), and the other (A_{LO} at an radian frequency of ω_{LO}) is locally generated by a local oscillator. At the receiver, the mixing is done by the photodetector. The heterodyning (nonlinear admixing) leads to the modulated photodetector current I $_{photo}$, can be written as

$$I_{photo} = A^2_{sig} + A^2_{LO} + 2 A_{sig} A_{LO} \cdot \cos (\omega_{sig} - \omega_{LO}) t \qquad (16.16)$$

Some advantages result in working with the intermediate signal. This intermediate signal carries the original optical signal as amplitude, frequency or phase that is then recovered. These coherent systems have good sensitivity for detection and increased selectivity at the receiver. The large bit rate capacity offered by single mode fibers can be harnessed by these systems. In context to the lightwave technology, coherence implies a nonlinear mixing of two signals.

The photodetector current is modulated at difference frequency of the two original signals (A_{sig} and A_{LO} ; see the cosine term in the Equation (16.16)). The local signal may be generated by a local oscillator, and its frequency is called the intermediate frequency or IF (as it is called in most of the radio literature). The detector current is generally measured as the photon count rate and the photonic processes are thus more accurately evaluated and controlled.

In homodyne systems, the instantaneous local oscillator signal frequency is made identical to the signal frequency, and the detector current frequency tends to become zero, leading to baseband signal recovery. The mode of operation is called the homodyne system, as opposed to the heterodyne system, where the LO frequency is distinct from the signal frequency. In principle, the homodyne process becomes similar to conventional direct detection where the photodiode current follows the optical signal on the photodetector. However, this process is severely restricted because of inherent thermal noise, amplifier noise and the detector's own noise. For this reason, it is advantageous to have a local oscillator at a high enough amplitude to drown the other noises, and make the admixed signal (carrying the original information) as large as possible. However, the limit is easily approached, since quantum noise of the detected photo current also gets amplified by the higher value of the local oscillator amplitude. This shot noise becomes the limiting factor in the performance of heterodyne systems with large local oscillator signal levels. Detailed theoretical analysis has shown that if a bit error rate (BER) of 1.0^{-09} is specified, the detector current (in terms of number of photons per bit) has to range between 9 to over 80, depending upon the type of modulation used. The quantum limited direct detection sensitivity has been determined at about 10 photons per bit.

Systems limited by the electronic noise considerations (rather than the shot noise considerations) need about 400 photons per bit while operating in the 1.3 to 1.5 μm wavelength for lightwave systems. On the other hand, well designed and efficient coherent systems can offer the required sensitivities with about 20 photons per bit. At 400 Mbps, the relative difference in sensitivity of the two systems (coherent versus direct detection) has been demonstrated at 45 photons/bit versus 450 photons per bit. In 1988, the coherent systems experiments reported rates to about 4 Gbps (NTT's 155 km [96 miles] experiment with 1445 photons per bit with DFB laser for local oscillators) and direct detection system experiments reported rates of up to 10 Gbps (8 Gbps over 68 km [42.2 miles] with 1 dB of dispersion penalty using external modulators in 1986). Also in 1988, computer simulations [16.6] indicated that for a 2 dB dispersion penalty, the maximum modulation rate of 5 to 9 Gbps is feasible over 100 km (62.1 miles) of fiber at 1.55 μM wavelength by using the on-off keying (OOK) modulation technique (with no laser chirp noise) (also see Section 16.10). This study [16.6] focuses only on the limitations caused by chromatic dispersion.

16.5

DIGITAL CAPACITY OF OPTICAL SYSTEMS

The rates utilized in the commercial systems are still much lower than the experimentally demonstrated rates achieved over fibers. Single channel fiber experiments have demonstrated bit rates of up to 8 Gbps, and multi-channel systems have achieved throughputs of about 30 Gbps. The maximum bit rate is still limited by the laser modulation speed, modulator capabilities, fiber optic detector and the receiver spectral dynamic range. The supporting electronic circuits can also become a serious limitation at these bit rates. While dealing with fiber optic systems, the rate-distance measure offers a realistic parameter, since the information has to traverse a minimum distance through a network. Through the late eighties, most system designers have pushed this rate-distance product to over 2.0^{12} bit km/s (or 2 terabit-km/s) over a single fiber.

The outer limits of the rate and range product for fiber optic transmission of data can be visualized by *two* fundamental characteristics inherent to the fiber: *(a)* its center band attenuation; and *(b)* its chromatic dispersion, polarization mode dispersion and various nonlinear effects. The constraints caused by component limitations (such as the limitations of modulators, pin-diodes, detectors and wide band signal amplifiers) impose the range of tolerance for the overall performance of the data transmission system. At a very basic level, there are *four* components within the system. The limitations inherent within each of the *four* components *(a)* the laser devices for photonic modulation, *(b)* the detectors and receivers, *(c)* the data encoding and modulators, and *(d)* the fiber glass optical medium that essential for the optical systems, are discussed in Sections 16.6 through 16.9.

16.6

LASER SOURCES

The photonic modulation of laser sources can be performed with high precision. The upper modulation bandwidth is bounded by the relaxation resonance frequency of these sources. This resonance frequency varies [16.7]. approximately inversely (in the region of interest) with the square root of the laser output power. The relationship may be approximated as:

$$f_{res} = 3 \sim 5 \text{ GHz} / \text{mW}^{1/2} \tag{16.17}$$

Some of these devices offer resonance frequency as high as 24 GHz. The frequency decreases as the power output is increased by square root relation. Multi-quantum well lasers offer an outer theoretical limit of about 50 GHz. Specially designed laser sources, which limit the parasitic capacitance, current leakage and nonlinearities, can offer modulation frequencies as high as 20 GHz with a small-signal 3 dB bandwidth. The optical bit stream (as early as 1987) has been documented at 16 Gbps, shown in a constricted 1.3 μm mesa laser (see Figure 16.28) by direct modulation techniques.

Alternatively, if the laser operates in a continuous mode, its light may be externally modulated. Thus, more external power may be available, the laser chirp may be substantially reduced and increase the modulation rate. Electro-optic effects in lithium niobate (with a molecular representation of $LiNbO_3$) have been harnessed in the guided wave directional coupler for approximately 20 GHz modulation bandwidth. However, for this high a bandwidth drive voltage of about 30 volts is necessary. With the electronic circuitry limitation for the voltage swings restricted to 12 volts, the modulation rate of about 8 to 10 GHz is realistic.

20 ps/div   125 ps/div

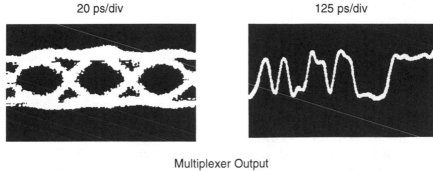

Multiplexer Output
0.16 V/div

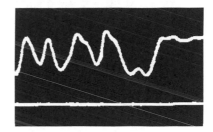

Detected Optical Signal
50 mV/div

FIGURE 16.28. Early Experimental Results (1987) for a Data Transmission Rate at 16 Gbps with Direct Modulation of a Constricted Mesa Laser System (Gnauck and Bowers).

16.7

LIMITATIONS OF THE DETECTION TECHNIQUES

The receiver imposes its own speed restriction. At the receiver a good signal-to-noise ratio is critical in an error-free optical system. Field effect transistors operating on the PIN-diode principle suffer a 4.5 dB loss in receiver sensitivity for doubling the bit rate. The avalanche photodiode (APD) offers slight improvement because of the inherent gain in these devices that overcomes the amplifier noise. However, the bottleneck is the electronic amplifier circuitry that has to provide about 100 dB of gain in a typical optical receiver. Linearity and constant gain over the extended bandwidth are the most severe limitations.

The detectors in the optical environment suffer from *three* additional effects: *(a)* dark current noise, *(b)* thermal noise and *(c)* inter-symbol interference. In coherent systems, the dark current effect is removed by increasing the local oscillator power. Typically, the dark current is less than 1 nA, and enhancing the LO power to be over 10 nW would significantly reduce the dark current effects. With direct detection OOK, the low level dark current can cause concern in the detector performance. The inter-symbol interference in the detectors can become serious as the bit rate is increased significantly. While dark current and thermal noise effects can be reduced by increasing the LO power the electronic distortion within the diode that ultimately shows up as inter-symbol interference lingers on. This limitation precludes the performance of coherent detection from reaching shot noise limited systems at highest frequencies with either direct or coherent detection. These considerations are discussed in greater detail next.

In an ideal noise-free environment, both the theoretical and experimental results [16.8] and [16.9] have confirmed that the direct detection techniques outperform the homodyne detection systems (10 versus 18 photons/bit); but the presence of any dark current (as low as one photoelectron per bit) makes the homodyne systems better than the direct detection systems. Thus, a certain minimum local oscillator power in the coherent systems is necessary and this value depends on the values of the shot-noise in the system.. Typically, this criterion is trivial in most coherent systems, since the dark currents are < 1 nA and the LO power requirement > 10 nW. Thermal noise is a more serious cause for degradation in the coherent systems raising the typical levels of the LO power to about 0.7 mW under special conditions as discussed in Reference 16.9. The requirement on the minimum LO power tends to grow as the square of the bit rate and becomes difficult to sustain at very high rates, thus limiting the coherent systems from competing effectively with the direct detection systems at the very highest rates. The intersymbol interference influences both the direct detection and coherent systems. Increased LO power helps the coherent systems, but not at a cost justifiable with the gains associated with these systems, especially as the bit rate increases. When the system comparison does become very detailed, only simulation studies (see Chapter 18) can provide the specific results that are applicable to the particular design problem.

16.8

SIGNAL MODULATION TECHNIQUES

The modulation techniques encode data onto the fiber optical carrier system. The *three* more common formats for modulation are: phase-shift-keying

(PSK), frequency-shift-keying (FSK), and amplitude-shift-keying (ASK). For the purely binary systems, on-off keying (OOK) is also used, and offers specific advantage over the other techniques (see Section 16.9). The optical sensitivity at the receiver is ultimately influenced by the original modulation format and detection technique necessary to recover the original information. Phase-shift-keying presents *three* minor variations: continuous phase-shift-keying (CPFSK), minimum-shift-keying (MSK), and differential-phase shift-keying (DPSK).

The input to the fiber may be represented for the different modulation techniques by *two* simple equations. For the various modulation systems [16.6] for the CPFSK and MSK, the time-domain fiber input may be represented as

$$X(t) = e^{j2\pi\Delta f} \int_{-\infty}^{t} \sum_{k} a_k\, P(t' - kT)\, dt' \tag{16.18}$$

where the output of the PR code is defined as $a_k = \pm 1$, k is an integer, T is the bit time, P(t) is the pulse function with unity amplitude within the bit time and zero elsewhere and Δf is 0.5/T for CPFSK and 0.25/T for MSK. For the PSK, DPSK, ASK and OOK systems the time-domain fiber input can be written as

$$X(t) = \sum_{k} a_k\, P(t - kT) \tag{16.19}$$

where $a_k = \pm 1$ is for PSK or DPSK, and $a_k = 0$ or 1 is for ASK and OOK. The effect of these modulation techniques in conjunction with the optical transmission through the fiber are discussed in Chapter 18.

16.9

FIBER LIMITATIONS

Finally, the fiber has its own outer limits [16.8 and 16.9] in carrying the data. The loss of signal strength through modern fibers is no longer of concern, except in single-channel coherent systems. In most cases, chromatic dispersion, polarization, mode dispersion (and other nonlinearities), and the more inherent properties of the fiber material cause more serious limitations as bit rate, distance, optical power and the number of channels are increased. The ASK systems are susceptible to nonlinearities on power changes and fluctuations. The FSK and the PSK systems offer some immunity to these nonlinearities. The high quality commercial fibers offer <0.2 dB/km loss at

1.55 μM and about 0.35 dB loss at 1.3 μM, thus shifting the attention to eye closures (as discussed in Section Chapter 18) due to chromatic dispersion arising from group delay effects.

Chromatic dispersion can be compensated by building a dispersion equalizer (similar to the phase equalizers for copper transmission mediums) or by the design of specially fibers with low chromatic dispersion in the 1.55 μm wavelength region. Low-loss dispersion-shifted fiber with a zero dispersion at 1.55 μm may be used. These techniques will become viable for very high capacity wavelength division multiplexed (WDM) systems. Presently, this approach has not been investigated because of the enormous capacity of the single mode fiber already available.

Polarization mode dispersion corresponds to the difference in propagation time of two-orthogonal principal states of polarization. It results from birefringence in the fiber and the difference in the propagation times for the two orthogonal principal states of polarization. For direct detection systems, this polarization effect is evidenced as pulse broadening. For the coherent systems, the arrival of different spectral components at different times and with different polarization causes performance degradation because of receiver admixes all these components and retains the time and polarization shifts of the spectral components. The mean differential propagation delay increases as the square root of the fiber length and can vary between 0.1 to 0.2 ps / ûkm for single mode fiber at 1.55 μM. The actual delay has a truncated Gaussian probability distribution. The maximum value corresponds to the case of no mode coupling. In poorly designed fibers, this maximum value can become as high as 30 ps / ûkm. The effect of the polarization mode dispersion is likely to become significant in single mode fibers at bit rates over 10 Gbps over fiber lengths exceeding 100 km (62.1 miles). For multi-channel fibers these limits are lower, but the *three* ways of compensation are: *(a)* polarization scrambling, *(b)* polarization diversity reception, or *(c)* the use of high quality fiber that reduces the differential polarization.

Nonlinear effects generally result from *four* innate characteristics within the fiber. The *first two* results are from stimulated Raman scattering (SRS) and stimulated Brillouin scattering (SBS). The *third* and *fourth* nonlinearities arise from cross-phase modulation and four-photon mixing. The effects of SRS are negligible for single channel lightwave systems until the launched power approaches one watt. In multi-channel systems, the crosstalk effects due to SRS can be a concern for widely spaced channels. Generally, this is the case because of the limited channel separation. In coherent systems, the channel spacing is close, and SRS is not likely to be a problem. The effects of SBS start to manifest themselves at lower power levels. These effects have been observed at 1.52 µM in fibers 30 km (18.6 miles) long with input power of about 2 mW. However, the gain is zero in the forward direction, and maximized in the reverse direction, thus the scattered beam traverses back to the transmitter. Hence, in the forward propagation of the signal, the incident wave is only slightly depleted, but does not travel to the receiver.

Optical power fluctuations cause slight changes in the fiber refractive index. The phase of the forward signal is thus affected. While this effect is known as self-phase modulation in single channel systems, its effect becomes cross-phase modulation in multi-channel systems. In most lasers, the power level is uniform, giving rise to negligible phase effects in single channel systems. In multi-channel systems, when the transmitter power for long fibers falls below 25 mW, unacceptable phase noise may result with about 50 or more channels. These considerations differ with the type of modulation techniques and the effect of this type of nonlinearity has to be investigated for each technique.

Four-photon mixing is likely to restrict the transmitter power in any multi-channel optical system. Fiber length, chromatic dispersion and frequency separation influence the extent of phase matching in optical waves. Thus, the crosstalk between the channels can become a severe restriction. The optical waves are well matched as the chromatic dispersion approaches zero. The channel separation should approach 400 GHz at 1.3 µM or about 300 GHz at 1.5 µM, before the mixing can be blocked. In the conventional fiber with dispersion of 15 ps/km-nm, the mixing efficiency reduces to about twenty percent at frequency separation of 15 GHz. The problem can become exaggerated as the frequency separation is reduced to 10 GHz, even with short fibers of about 10 km (6.21 miles), restricting the maximum power in each channel to just about 0.1 mW. It is foreseen that this restriction will be most severe for coherent multi-channel systems with close spacing between the channels.

16.10

CHROMATIC DISPERSION IN THE FIBER AND ITS LIMITATION

In the early years of fiber optic data transmission systems, signal attenuation was foreseen to limit the rate and range of such transmission systems. Greater control on the fiber quality and its clarity has shifted the focus to chromatic dispersion as the ultimate constraint. Fiber quality, rate, distance, modulation and methodology for detection, all influence the outer limits of the rate and distance over which data can be ultimately transmitted. One other factor also enters the picture for the eye diagram. The transmission quality is also considered. The measure for transmission quality is quantified by the bit error rate (BER), and this rate can be related to the eye closure due to chromatic dispersion. Data transmission systems display very cogent relationships between the signal-to-noise ratio and the probability of error. Assuming the noise in the fiber optic system follow the Gaussian distribution, a finite limit on the eye closure can be tolerated for one in $10.0E^{-09}$, or one in $10.0E^{-08}$, etc. Signal strength depends upon the transmission parameters discussed earlier. Noise depends upon the accumulation of a large number of noises (sources, photodetectors, couplers, circuit and other system components), and degradations (chromatic dispersion, polarization dispersion, stimulated Raman scattering, stimulated Brillouin scattering, cross-phase modulation and four-photon mixing). It can be seen that the precise simulation of the entire electro-optical-transmission optoelectrical process is monumental, if not impossible. For this reason, the initial studies have been limited to the effects of chromatic dispersion.

In reported simulations [16.6], only the dispersion criteria have been used under simplified conditions. The effects of fiber nonlinearities are not considered. Signal dependent noise such as source laser chirp and APD multiplication noise is also ignored to highlight the effect of chromatic dispersion. The first set of results is for one dB dispersion penalty (calculated as twenty times log, if ratio of an ideal eye opening is without fiber to the degraded eye opening). The second set of results is for a 2 dB dispersion penalty. Both sets assume a second order Butterworth baseband receiver filter with a 3 dB bandwidth at sixty-five percent frequency point. Six modulation techniques [16.6] studied are CPFSK, MSK, ASK nonsynchronous, PSK/ASK, DPSK and OOK. Results are shown in Figure 16.29 for coherent lightwave transmission systems. High quality fiber with 15 ps/km . nm at 1.55 μM is used (zero dispersion wavelength near 1.3 μM) in the simulations, and studies are carried out for 1.55 μm optical source wavelength.

In summary, a 2 dB dispersion penalty may be expected with rates in the range of 5 to 9 Gbps over a 100 km (62.1 miles) of fiber. The OOK systems hold a clear edge over continuous-phase frequency-shift-keying (CFFSK systems with delta frequency are the same as the bit rate frequency) coherent detection systems.

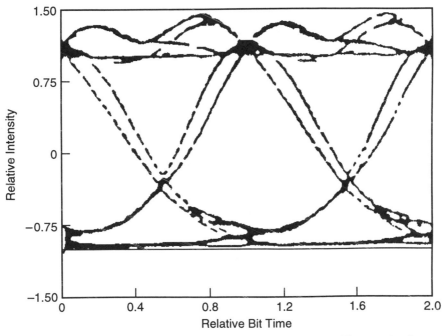

FIGURE 16.29. Simulated Eye Pattern for a 8 Gbps Transmission over 68.3 km of Single Mode Fiber. The closure is 0.5 dB due to chromatic dispersion for a coherent lightwave system. (Elrefaie, et al, 1988).

For the OOK systems, the simulated results offer a 0.5 dB eye closure for the 8 Gbps, 68.3 km (42.4 miles) fiber optic experimental results [16.10] reported in 1986. The eye closure measured with APD detector is reported at one dB. The allocated margin for the APD noise is 0.4 dB for an APD gain of 8, k factor of 0.4 and a dark current of 10 nA. In this computation, the thermal noise of the APD is assumed to be 75 pA/Hz$^{0.5}$. With this allocation for the APD noise, the simulated and the experimental results agree to within 0.1 dB.

16.11

NEWER STRATEGIES

Whereas rates of 8 Gbps were just being achieved in the late eighties with two [16.11 and 16.12] specially designed optical systems, data rates at 32 Gbps [16.13] over 90 km of fiber were achieved in 1992 by using specially designed optical amplifiers. Optical amplifiers are special segments of fiber tha` contain erbium (a rare earth element) to maintain minimum power levels to assure the very very low (0.1 failure in 10 billion hours) error rates. The system uses solitons rather than light pulses; but solitons are light pulses that retain their shape over long distances as they travel through fibers. Segments of fiber that refocus the optical energy back from its dispersed shape to its original shape are interspersed throughout the 90 km length of the transmission path. Stated alternatively, a part of the fiber rectifies the dispersion effects caused by the other part, and the integrity of the optical energy distribution is maintained within the bounds of the wavelength. It is expected that these fibers can carry solitons for distances up to 6,000 miles without the usual electronic regenerators.

Eight channels of information have been used; the signal pulses in each have been modulated, multiplexed, amplified, transmitted and then demultiplexed. For the demultiplexing, an all optical gate has been deployed. The gating device uses a strong gating pulse (to induce Kerr effect), which permits the information recovery to be more orderly and synchronized. The gate receives the probing pulse every 250 ps (or at the rate of 4 GHz), and thus is able to recover the eight 4 Gbps channels originally multiplexed on to the 32 Gbps data stream through the fiber.

16.12

REFERENCES

16.1   S.V. Ahamed and V.B. Lawrence 1996, *Intelligent Broadband Multimedia Systems*, Kluwer Publishers, Boston, MA.

16.2   Bell Laboratories 1982, *Transmission Systems for Communications*, Fifth Edition, Western Electric Company, Winston-Salem, NC.

16.3   J.A. Stratton 1941, *Electromagnetic Theory*, McGraw Hill Book Company, New York, NY.

16.4   D. Marcuse 1972, *Light Transmission Optics*, Van Nostrand Reinhold Company, New York, NY.

16.5   M.K. Barnoski, Editor 1989, *Fundamentals of Optical Fiber Communications*, Second Edition, Academic Press, London, England.

16.6   A.F. Elrefaie, et. al, 1988, "Chromatic Dispersion Limitations in Coherent Lightwave Transmission Systems," *IEEE Journal of Lightwave Technology*, Vol. 6, No. 5, May, 704-709.

16.7   R.A. Linke 1994, "Ultra-High-Speed Digital Transmission Systems,", AT&T Bell Laboratories, Crawford Hill Laboratory, *Private Communication*, Holmdel, NJ.

16.8 R.A. Linke and A.H. Gnauck 1988, "High Capacity Coherent Lightwave Systems," *Journal of Lightwave Technology*, November. Vol. 6, No. 11, 1750-1769.

16.9 U. Timor and R.A. Linke 1988, "A Comparison of Sensitivity Degradation for Optical Homodyne versus Direct Detection of On-Off Keyed Signals," *Journal of Lightwave Technology*, Vol. 6, No. 11, November.

16.10 A.H. Gnauck, S.K. Kortky, B.L. Kasper, J.C. Campbell, J.R. Taman, J.J. Veselka, and A.R. McCormick 1986, "Information-Bandwidth-Limited Transmission at 8 Gbit/s over 68.3 km of Optical Fiber," *Technical Digest of OFC '86*, February 24-26, Atlanta, GA.

16.11 A.H. Gnauch, J.E. Bowers, and J.C. Campbell 1986, *Electronic Letters*, Vol. 22, 600.

16.12 A.H. Gnauch, et. al. 1986, Paper number PDP-9, *OFC '86*, Atlanta, GA.

16.13 AT&T Bell Laboratories 1992, *Internal Communication*, September 3.

CHAPTER 17

OPTICAL LIGHTWAVE SYSTEMS IN EXISTING NETWORKS

To become economically feasible, broadband networks have to be optical, even though individual links may be microwave, coaxial, or even satellite based. The recent technological strides in making very high quality fibers and optical devices, and the availability of ATM switches, all point to the use of optical systems becoming more and more entrenched in current and future networks. In this Chapter, the existing networks where optical systems are an integral part are identified. The emphasis is on the telecommunication environment and the role these in which optical systems have significantly altered the economics of both broadband and wideband data communication. A recent study of a typical developing country in the Middle East has indicated that any delay in bringing optical systems into major telecommunication centers is a significant source of *negative* cash flow to the telephone business, and more than that, the lingering demand of services will not be met (in any economically viable fashion) except by establishing OC-3 or at least OC-1 links between the population centers of the country.

17.1

THE DEPLOYMENT OF FIBER

In the commercial network environments (Parts I and II of Reference 17.1), the major impact has been in the direction of lightwave transmission systems. Optical processors and optical switches have not been introduced even though the feasibility of building very rudimentary optical processors has been documented (see Chapter 15). Optical integrated circuits (OIC) have a long way to go before they can compete with ULSI (ultra large scale integration) circuits because of the infancy of economically viable integrated optical technology. However, the individual components (source diodes and lasers, detector diodes and avalanche devices, discussed in Chapter 16) have been very successfully employed in the optical lightwave technology for LANs, WANs, trunk, interoffice, intra-computer, inter- Central Office, computer networks, fiber-to-home [17.2] applications and even medical technology. In the rest of this Section, we concentrate upon the optical systems pertinent to existing telecommunication networks and emerging intelligent networks.

At the present time, optical switching is not yet implemented yet in conjunction with optical fibers to offer truly commercial optical networks even though the use of the optical switch for demultiplexing data at 32 Gbps is documented [17.3]. In a nutshell, the transmission is optical, encoding, signal generation, signal recovery and processing is in the VLSI or the DSP domain. However, fibers carry multi-gigabit data quite dependably. Hence, the public domain networks of the late twentieth century are very likely to deploy semiconductor switching under stored program control and deploy fiber for high capacity transmission. Optical channel switches (like the D1 or D3 channel switches currently implemented with conventional semiconductor technology) and complex logic circuits such as those used in computers need to evolve before the networks can assimilate optical switches (see Section 16.11) of the future. This scenario is similar to the delayed impact of semiconductor digital computers on switching systems even though it was realized before the mid sixties, that digitally controlled channel switching could be accomplished. The computer industry was already well into its prime before the first electronic switching system (#1ESS) was introduced in 1965.

The major driving force behind most broadband digital networks is the availability of high grade fiber for transmission. The secondary force is the availability of add-drop multiplexers for constructing high capacity OC-N rings (see Chapter 16). The transmission issue was very successfully resolved during the seventies and early eighties. Presently, the data rates and ranges are well beyond the projected needs of even the most data intensive society. The new national and trunk communication is invariably optical in networks where fiber is easily available. In the United States, Canada, Europe, Japan and Australia, fiber has been deployed for most trunk and shared transmission facilities. High-speed transmission on copper facilities is an economic alternative to homes and locations where the copper already exists for transmission. With efficient coding methods, (2B1Q, trellis, QAM, etc.) methods data to the T1 rate of about 1.544 Mbps is feasible to the individual customers on the HDSL mode and data to 6.34 Mbps in the ADSL mode. However, the capacity of even the most rudimentary fiber (at 51.64 Mbps) that reaches individual homes in a star or bus topology from a centralized data distribution point is about 30 times greater than the T1 rate at 1.544 Mbps. The information age and computer revolution has not prompted individual households to currently receive, manage, and deploy currently such bandwidths in any significant way.

In the telephone networks, fibers have been deployed at the T3 rate of 44.736 Mbps in the FT3 fiber optic systems [17.4] to interconnect Central Offices in the metropolitan areas. The AT&T field trial of the lightwave transmission system took place in 1977 in Chicago. During 1980 to 1985, the cost per channel dropped by about a hundred and the integration of the fiber in the networks has proceeded gradually from 1981, when the FT3 system with 0.82 μm technology was introduced at a nominal capacity of 45 Mbps. In 1982, the FT3C lightwave system using 0.82/0.87 μm technology offered 90 Mbps on the trunk environment. Improved fiber quality made the FT3C system design feasible at 1.3 μm wavelength in 1983. The FTX 1.3 μm system doubled the capacity to 180 Mbps in 1984. The FT-G systems now currently in use offer data rates to 1.7 Gbps using the 1.3/1.5 μm technology even though some versions of the FT-G systems offered data rates of 417 Mbps in 1985 with 1.3/1.5 μm technology and again 1700 Mbps in 1987 with 1.3 μm technology. These cost per channel with the newer FT-G systems is less than 1.0 percent of the cost in 1982. Figure 17.1 summarizes the major strides in the deployment of fiber optic systems in the AT&T network environment through the 1980s.

The SONET transport rates (STS-N rates, see Chapter 15) are expected to become dominant in the future. On a national and international basis, the modification and verification of the system performance become comparatively easy with the SONET standards. Very high speed switching [17.5] for ATM applications (typically 256 I/O ports each at 622 Mbps) also makes these networks more and more realizable. Typically the capacity to add and drop optical channels at any of the OC-N rates, to monitor transmission quality, to multiplex and to distribute the optical channel with standard multi-vendor hardware appears very attractive in the emerging high capacity optical networks. Some of the networks (such as AT&T's Metroplus) introduced before the standardization of SONET are not compatible with the emerging standard and compatibility issues need to be addressed.

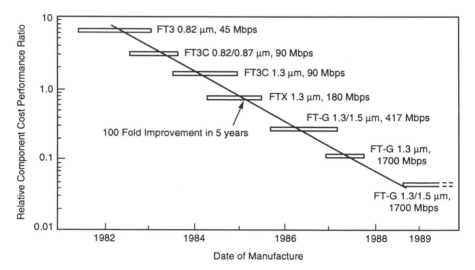

FIGURE 17.1. Performance and Cost of Typical Lightwave Transmission Systems over a Seven Year (1982-1989) Period.

Fiber has been actively used in numerous local area, wide area, metropolitan area and campus networks. Most of these networks use the fiber distributed data interface (FDDI). Most of the interfacing is handled by the second generation chip set now available for these networks. The *five* major functions in these networks are basic media access, physical layer control, clock recovery, clock distribution and basic media access system interface. The burst rate capacity can be as high as 800 Mbps with 32 bit parallel interface within the bus of the individual chips accomplish the individual functions. The data rates in the network is 100 Mbps on the fiber media. The transmission is ten times faster than the Ethernet LAN operating at 10 Mbps. Token access and ring topology are used in the FDDI environment. Up to 500 nodes (on an extended ring covering up to 125 miles in circumference) are expected to operate on these networks without significant degradation of the performance. These networks can operate as a backbone to connect independent islands of Ethernet LANs (operating under

IEEE 802.3 standards) and Token Ring LANs (operating under IEEE 802.5 standards). Even though the immediate use of these high capacity FDDI networks is for workstations and distributed computer systems, they are eventually earmarked to be able to interconnect the more powerful PCs and desktop workstations.

17.2

FIBER DISTRIBUTED DATA INTERFACE (FDDI)

The FDDI concept and its role in fast packet technologies were introduced in Section 4.8. Even though these technologies are robust and viable in their own right, they are overshadowed by the eminent and universal ATM networks lurking in the telecommunications arena. The older plain old FDDI had started to make some inroads before ATM started to gain universal acceptance. FDDI is covered because it may linger on for a few more years.

Metropolitan area networks may serve the public or remain private. In the public domain environment, the IEEE 802.6/DQDB standard is most likely to dominate the implementation. One such topological arrangement is shown in Figure 17.2.

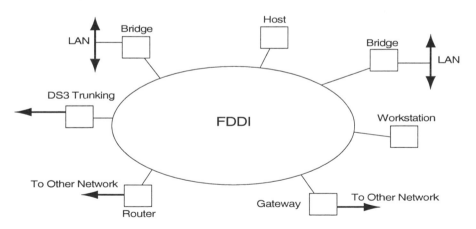

FIGURE 17.2. Typical FDDI Applications for Processor to Processor Workstation to Processor and LAN Interconnections. The future of such FDDI rings is in serious jeopardy because of the recent ATM infrastructure evolving through the 1990s.

In the private MAN environment, the FDDI defines a means of interconnecting data processing equipment to high-speed data networks operating at 100 Mbps. Very high speeds, excellent noise immunity and a timed-token passing protocol, guarantees the access of each node to the network. The maximum packet size is 4500 bytes and up to 1000 nodes may exist over 200 km fiber path. Each node regenerates and repeats the packets sent to it. This also serves to attach and identify devices on the network. Some rudimentary network management capability is built into the FDDI specification and provides enough capability to detect network failure and then automatically reconfigure it. The American National Standards Institute (ANSI) Interface Committee X3T9.5 has addressed *four* topics in the FDDI architecture.

First, the physical media dependent (FDDI/PMD) issue deals with optical transceivers and the media characteristics for point-to-point communication between various nodes.

Second, the physical layer (FDDI/PHY) issue deals with encoding, decoding, clock recovery and synchronization, and hand shake with adjoining nodes to test link integrity.

Third, the media access control (FDDI/MAC) issue addresses and governs the access to the medium. This issue is concerned with the frame format, interpretation of the frame content, generation and repetition of the frames, issuance and capture of the tokens, controlling the timers, monitoring the ring, and finally interface with the station management.

Fourth, station management (FDDI/SMT) deals with controls required for managing the node or station functions for proper operation and cooperation with adjoining nodes on the fiber optic ring. The SMT function is complete by the signals and interaction with PMD, PHY, and MAC issues and layers discussed earlier. Connections, configurations and interfaces are addressed in the context of SMT. Ring and station initialization, fault isolation/ recovery, statistics gathering, address administration and ring partitioning issues are addressed here.

The first three elements in the FDDI smoothly mesh into the logical link layer of the data link (DL) and the physical layer (PL) of the open system interconnect (OSI) model (see Figure 17.3). The SMT layer acts as the specialized manager for this particular application, and configures and facilitates the cohesive functions of the first three layers.

The nodes or station on the fiber optic ring may consist of data processing equipment (typically, a mainframe, a minicomputer, or even a peripheral such as disk drives or printers). *Three* configurations are possible. Single attached stations tap into one of the two fibers in the ring carrying data in the opposite directions.   Dual attached stations tap into both the fibers. Concentrators may be placed on the ring to share the node with a number of lower capacity devices. Tree structure of devices may thus be built with one fiber ring or with both the fibers. While constructing single-fiber-ring-tree structures, the other fiber is generally not tapped, such that at least one ring is less prone to being broken.

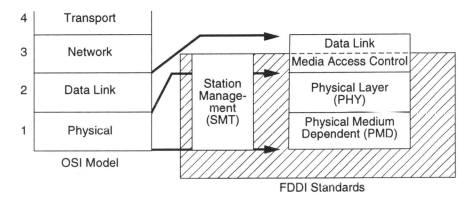

FIGURE 17.3. Functional Mapping of the Two Lowest Layers of the OSI Model into the FDDI Standards. ATM - Asynchronous Transfer Mode; DS - Digital Signal; FDDI - Fiber Distributed Data Interface; LAN - Local Area Network

The devices at the nodes interface the network via specially designed VLSI interface chips. One such architecture (from advanced micro devices [17.6]) is shown in Figure 17.4. This equipment, called the host, may undertake the network interfacing functions or utilize a slave processor (called the node processor) to carry out the interfacing functions. In simpler systems, the node processor responds to system level and packet level interrupts. In simple cases, it serves as a minimum state machine, while in the larger and more complex exchanges of data between computer systems, it can be designed to serve some of the data link functions of the standard OSI model. Buffer memory (static RAM) is also essential to hold the data from/to the host as it finds access in/out of the fiber optic ring media network.

Three controllers are also essential to monitor the flow of data. The *first* controller, called the RAM buffer controller generates the address in the RAM buffer so that the received and transmitted data frames may be temporarily stored. The main function of this controller is to provide DMA (direct memory access) to read/write from/to the buffer memory via the control and address lines. The node processor may request DMA access using instruction and bus interface lines. In essence it acts as a memory controller responding to the node processor or the host, thus permitting the flow of information to/from buffer memory from/to the host via buffer memory data bus. Since the buffer memory can be accessed by a host, the node processor or the DPC (see *second* controller), the buffer memory controller permits the access of only one of the three to access the memory at one time.

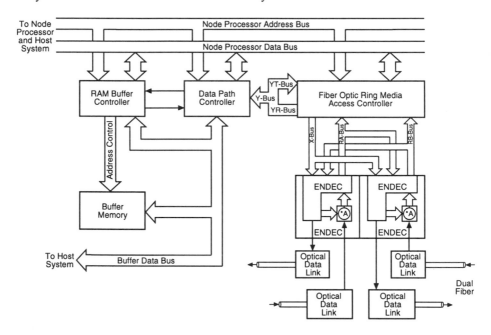

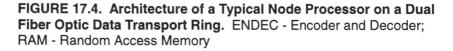

FIGURE 17.4. Architecture of a Typical Node Processor on a Dual Fiber Optic Data Transport Ring. ENDEC - Encoder and Decoder; RAM - Random Access Memory

The *second* controller called the data path controller (DPC) serves to control the flow data to/from the buffer memory (generally 32 or 64 bits wide) to 8 bit bytes that finds access in/out of the ring. Frame and node status are generated and parity checks are also performed in this controller. Its access to the RAM buffer memory is also direct via the data bus of the RAM.

The *third* controller called the media access controller (MAC or FORMAC for fiber optic ring media access controller) determines when the node may actually access the ring and performs the logical functions for token handling, address recognition and cyclic redundancy check (CRC) generation and checking. The physical layer headers and any preambles or start of frame delimiters are stripped in the MAC before the dispatch to data path controller. Likewise, the end of frame characters or postambles are also removed. The MAC also checks the frames for destination address and notifies the data path controller if a match does not exist. Node and frame bits identifying node conditions and frame status are also generated here. The interface between the data path controller and the MAC is generally half-duplex 8 bits wide. MAC chips also interface with the encoder/decoder (ENDEC) chip(s) via three 11-bit buses; two handle the received data and the third handles the transmitted data. The information flow on the three buses is synchronized with the byte rate clock.

The MAC chip transmits the data as 8 bit bytes with two control bits and one parity bit into the ENDEC chip(s). The ENDEC then performs a 4 bit/5 bit encoding, which maintains a DC balance in the output waveform to ensure that no more than 3 consecutive zeros are ever present in an encoded pattern. This data is converted from parallel to serial format and sends NRZI (non-return to zero, invert on ones) bit streams to the fiber optic transmitter. FDDI specified bus selection and media access (to receive or to transmit) is provided by the ENDEC; the byte clock is also generated in the ENDEC.

The use of fiber in the very long distance intercontinental and the relatively short subscriber loop, environments are presented next.

17.3

FIBER OPTIC SYSTEMS IN TRUNK SYSTEMS

The capacity and potential of fiber optic systems can vary dramatically. Primarily it depends on the source (LED, laser, quantum well, MQW, etc.) and its wavelength, on the fiber and its optical properties (step, multimode, graded) and the detector, (pin, APD, laser, quantum well, MQW, etc.). At the lowest extreme of performance lie the archaic LED multimode/graded fiber and PIN devices barely yielding about 30 Mbps over 4 km of fiber. The rate-distance product stands at 1.2×10^8 bit-km. At the highest extreme of performance lie the chirpless, wanderless lasers, self-focusing fibers, optical Kerr effect demultiplexer switches and MQW detectors yielding 32 Gbps over 90 km of specially fabricated fiber the rate-distance product is close to 3×10^{12} (about 3 terabit-km, even though it is a laboratory environment; see Section

15.11) and is three times two orders of magnitude greater than some of some of the older rate-distance products, which can be as low as 10^{10} (10 Gigabit-km) for any commercial system. The more realistic standard of the current value (early 1990s) of this product stands at 10^{12} (one terabit-km) for the high quality fiber optic systems.

17.3.1 EARLY SYSTEMS

In the early applications, numerous T1 (or DS1) signals were multiplexed to generate a high bit rate fiber optic signals at typical rates of 44.7 Mbps, 135 Mbps, etc. In the United States, the older fiber optic trunking used multimode fibers with short wavelength (0.8 to 0.9 μm) signal energy from light emitting diodes. PIN (p-doped, intrinsic, n-doped) design for semiconductor detectors was also used in some early fiber optic trunk facilities. Typically, GaAlAs LED directly coupled into a 50 μm core delivered to the optical power of about 50 μW (microwatts). The band of optical wavelength is about 50 nm and the pulse spreading due to material dispersion is about 5 nm/km assuming that the material dispersion in this region of wavelength is about 100 ps/nm-km. The loss in the fiber cable is significant at 3-6 dB depending upon the grade of the fiber (and thus the cost) and the temperature range of operation (micro-bending). The response of the silicon pin detector is satisfactory to about 100 MHz. The system, though very unsatisfactory according to current standards, offered a vast improvement in performance by increasing the range of distance to 4.8-6.2 km (15.75 kft. to 20.3 kft.) at 1.5 Mbps, and to 2.4-3.8 km (7.87 to 12.46 kft.) at 25 Mbps.

17.3.2 RATE AND DISTANCE CONSIDERATIONS

On a simplified basis, *four* factors affect the rate and distance considerations: *first*, the maximum rate at which the LED or any optical source can be modulated, and its energy be coupled into the fiber for transmission; *second*, the loss that the optical signal experiences as it travels down the fiber; *third*, the dispersion of the pulses such that the pulse spreading does not cause excessive intersymbol interference (ISI); and *finally*, the receiver sensitivity at which the receiving amplifier and photodetector can demodulate the photonic energy from the fiber and recover the data with an acceptable bit error rate (BER). Each of the *four* factors imposes its own limitation (see * footnote) on the rate or the distance. The practical design is based upon having some margin of safety before a optic, electrical or photonic limits are reached. These limits are investigated further here.

First, consider the LED source with a simple (pin) photo-detector in this most archaic application. Data rates around 30 Mbps for the GaAlAs LED are near the modulation limits of this type of optical source. Further, the grading of this 50 μm multimode fiber and the output limitation of the LED can together yield about -15 dBm (about 31 μW) of power if the on/off probability is the same (or the duty cycle is 50%)

$$P_{trans} = -15 \text{ dBm} \tag{17.1}$$

Second, consider the attenuation of the optical signal in the fiber. The attenuation of power is based upon the length of the fiber in relation to the minimum required power at the receiver, which is first computed. The energy in a photon at the present wavelength (0.8-0.9 μm) is about 2.0^{-19} Joules. If the typical amplifier circuits and the photodiode need 20,000 photons to detect a pulse with a 1.0^{-09} bit error rate, the power required at the receiver is:

$$P_{reqd} = 0.5 * 20000 * 2.^{-19} * B \quad \text{Watts}$$

or

$$P_{reqd} = 2.^{-12} * B \quad \text{mW} \tag{17.2}$$

where B is the bit rate in bits per second. The equal probability of "ones" and "zeros" in the bit stream brings in the factor of 0.5 in the power requirement. When the bit rate is doubled, the power required increases by 3 dB. The actual power required can be written down in dBm as:

$$P_{reqd} = -117 + 10.0 * \text{LOG (B)} \text{ dBm.} \tag{17.3}$$

From this equation, it is seen that the required power gets increased by 3 dB every time the bit rate is doubled (LOG 2 = 0.3010). If the loss of the cable (and the splices) is approximated as 5 dB per km (at a very high side), then the increased loss due to doubling the bit rate can be compensated by reducing the distance by 0.6 km.

The actual distances may also be calculated at any given bit by the power budget (in dBm) consideration from the transmitter to the receiver.

$$P_{trans} = \{ P_{receive} = P_{reqd} \} + \text{Losses} \tag{17.4}$$

---

* It is to be appreciated that the information presented indicates a design limitations rather than the actual numbers in Figures 17.5 and 17.6 which depicts the pulse spreading limitations and mode partition noise. The recent experiments of the 1990s indicate that the rate-distance product limit is in the order of 3-4 terabit*km (see Sections 17.3 and 17.5.4). The recent innovations are literally light years ahead of the limitations in Figures 17.5 and 17.6.

The losses consist of fiber, connector, and at splices, etc., that can be considered. If it is assumed that the connector loss and splice loss (for the older systems) can vary between 3 dB (optimistic value) and 10 dB (pessimistic value) and the connector loss is between 5 to 6 dB, then the length of the fiber and the distance can be correlated as follows: The losses in this relatively simple system are about 16 dB in the worst case and about 8 dB in the best case. Since the bit rate loss alone at 1 MHz is 60 dB, the fiber loss due to distance alone amounts to 26 dB (worst case) or 34 dB in the best case. Once more if the loss of the cable and splices (if any) is 5 dB/km, the distance in the best case is 6.8 km (3 dB margin) or 5.4 km in the worst case (10 dB margin). Now, if the bit rate is changed, new distance limits can be established. This type of a simple power budget diagram is depicted by *the straight lines* shown in Figure 17.5 for 3 dB or 10 dB margin.

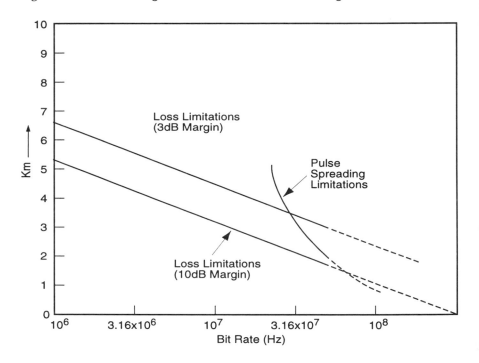

FIGURE 17.5. Limitation Imposed by Pulse Spreading. (See Footnote in Section 17.3.2).

The pulse spread considerations (discussed next) may lead to different design rules for other cables and rates and components. For this reason, the distance values are checked against the pulse spread considerations.

Third, consider the pulse spreading constraint, which stipulates that a pulse with fifty percent duty cycle may spread to half its maximum width in no more than one half the pulse duration. This stipulation constrains the intersymbol interference from becoming an additional reason for degradation. Thus, if the pulse spreading is assumed at 5 ns/km (see Chapter 16), then the maximum length at a data rate of B bits per second has to satisfy the relation

$$5.0 * 10^{-09} * \ell < 0.5/B \tag{17.5}$$

where ℓ is the cable length in km and B is the bit rate in bits per second (or 1/B is the pulse spacing). This pulse spreading constraint for this (poorly configured) system implies that the rate-distance product is 10^8. It also implies that the rate can be doubled by reducing the distance by half. The relation is shown by the curve shown in Figure 17.5.

At 1 Mbps, the distance constraint due to this limitation is far in excess of 6.8 km (22.3 kft.) or 5.4 km (17.7 kft.) due to power budget considerations. However, at about 31.6 Mbps, the distance due to pulse spreading (or dispersion) is about 3.7 km (12.13 kft.), which is also in the region of the power budget considerations. This also happens to be in the region of the maximum rate at which some LEDs can effectively modulate the optical energy that they can emit. The rate-distance product is 10^8.

17.3.3 COMPARISON WITH T1 AND COAXIAL L5E SYSTEMS

For the sake of comparison, it is noteworthy that even under the worst conditions (LED source, multimode poor quality fiber, connectors and very poor detector), the fiber capacity and range are considerably better than that of conditioned and specially selected copper twisted wire-pairs. The T1 carrier operating over AWG 22 wire-pairs with no gauge discontinuities nor any bridged taps, at a bit rate of 1.544 Mbps, needs repeating every 1.8 km (or 6,000 feet). Generally, four wires or two wire-pairs (one for each direction) are necessary for bidirectionality. The reason for the short distance between repeaters is to suppress the effects of crosstalk and the effect of Central Office impulse noise. The capacitive component C_0 i.e., one of the four primary constants which make up the characteristic impedance (Z_0), tends to attenuate the signal dramatically and near-end crosstalk (especially within-unit NEXT, i.e., the Self-NEXT from the binder group in the cable) starts to restrict the repeater spacing. As a comparison with the fiber media in the subscriber environment, the field trials (as far back as 1981) in Japan have been built to offer T1 rates for spans up to 12 km (39.4 kft.).

Coaxial cables offer a different environment for comparison. For example, the L5E analog carrier system carrying 13,200 voice circuits is used in Central Office trunk facilities (see Section 2.7.5) rather than in subscriber loops. For the 274.176 Mbps T4M coaxial system on 0.375 inches OD (outside diameter) cables, the restriction on the repeater spacing is caused by the thermal noise and intersymbol interference. This system is also used in the trunk rather than in the subscriber loop environment. The maximum half-bit rate loss is 54 dB at about 1.73 km (or 5,700 feet). Binary (two-level) T4 systems offer a good margin (about thirty-five percent) in the performance at this repeater spacing. The coaxial cable pairs (one for each direction) function well as dependable T4 transmission media (see Section 1.5.2). As a comparison with the fiber media in the long-haul digital environment, the TAT-8 (1989) fiber optic digital carrier systems, carrying 40,000 voice channels, operate at 295.6 Mbps with about 70 km (230 kft.) spacing.

17.3.4 EVEN AN ARCHAIC FIBER OPTIC SYSTEM

When the design is altered and components are selected from a set of standard, but very inexpensive, lasers, fibers, APDs and the rate-distance product can be enhanced considerably. For example, if the laser output of minus 6 dBm is broadcast into a single mode fiber with fiber and splice loss of 0.5 dB/km and a total connector loss of 6 dB, and a APD with a receiver sensitivity of [-119 + 10.0 * LOG B] dBm, then the product rises to about 1.25 × 10^{10} with a 3 dB margin for 10^{-9} BER. The power budget diagram is shown in Figure 17.6. These power budget constraints are shown as straight lines (3 dB and 10 dB margins as shown in Figure 17.5) and the pulse spread constraint is shown as the curve in the figure. Recent well designed fiber optic systems perform with a rate-distance product of about 10^{12}.

17.4

FIBER AND COPPER IN THE SUBSCRIBER LOOP PLANT

The inexpensive fiber optic systems can be gracefully merged into the loop plant for the current implementation of the very high-speed asymmetric digital line (ADSL) providing network access to the customer from the Central Office at DS3 rates (6.34 Mbps).

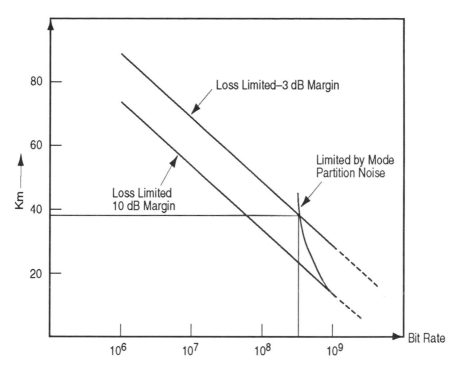

Figure 17.6. Limitation Imposed by Mode Partitioning Noise. (See Footnote in Section 17.3.2)

The possibility of providing data at 51.34 Mbps to the desk top from fiber distribution nodes in campus and building networks exists over Grade 5 wire-pairs or over fiber to the desk top. Copper based premises distribution systems (see Section 5.3.1) using the "copper distributed data interface" or CDDI, can handle data at this rate using MLT-3 codes with specially designed equalizers, or the 2-D cluster codes with specially designed transceivers (see Chapters 9 and 10). Fiber based distribution systems using the FDDI (see Sections, 4.8 and 17.4) can handle data through 100 Mbps.

The analogy of fiber entering the data distribution plant in the 1990s carrying OC-3 (or higher rate) data is much like the entry of conditioned wire-pairs in the 1960s that carried 1.544 Mbps T1-rate. The need for T1 rate data by the individual customers in 1960s was very low, if not non-existent. However, the advent of the T1 carrier systems led to the development of the subscriber loop carrier systems that brought 64 kbps speech and its superior quality closer to the customers, right into the remote terminals of these subscriber loop carriers or SLC systems (see Section 16.1). The projection is that the fiber in the loop plant can take the digital multimedia technology and its services to the termination of the information highways at the routers that make the branching to the individual customers via the VHDSL to homes (via copper) or desk tops (via fiber and/or copper). The FTTH architecture takes these services to the customers without any copper media and the HFC bring these high bandwidth services via the CATV facilities that already reaches a majority of households in the United States.

There are *two* closely intertwined approaches to providing data links and network access to individual customers: local area networks and ISDN. While the LAN, WAN and MAN context is used for dedicated private and computer networks, ISDN will provide unlimited open access to public domain networks, which in turn may be connected to private networks. The ISDN access thus becomes far more global than dedicated LAN, WAN, or MAN access. Copper and fiber are likely to prevail in the local loop access to these two types of networks. The *two* most seriously contenders for the carrying of the high(er) speed data are fiber optics and twisted wire-pairs. Fiber optic systems have proved to be very viable in most network environments. Numerous vendors have made the transceivers available to the system designers to multiplex, transmit, recover, demultiplex and redistribute circuit switched and packet switched data in the local networks via ATM switches. Sometimes the last link as a copper connection invariably exists in most cases and is likely to be far more prevalent. Hence, the high-speed digital path can be more easily established over the HDSL or the VHDSL wire-pairs. However, this phase is likely to be transitional until the fiber links are as distributed as wire-pairs and until the ATM interfaces are as inexpensive as the telephone jacks.

Optical media for the transmission of high-speed digital information and the cheap add drop multiplexers (see Chapter 15) are two major forces that shape the high-speed access possible in the public network environment. Optical fiber and optical communication systems are being introduced in most digital networks. The flexibility, the economy, and the rate of transmission (up to 1.7 Gbps with 20 mile repeaters, and 295 Mbps overseas intercontinental ranges) make these systems "move" gigabits with extreme flexibility and accuracy. Coupled with optical switching facility and optical computing techniques, intelligent networks will enter their next generation.

17.4.1 EARLY FIBER SUBSCRIBER LOOP SYSTEMS

One of the more dominant and widely accessible fiber optic systems in the subscriber loop plant in the United States is the Fiber SLC-96 (designed and developed by the pre-divestiture Bell System) during the early eighties. *Three architectural variations of the system are shown in Figure 17.7.*

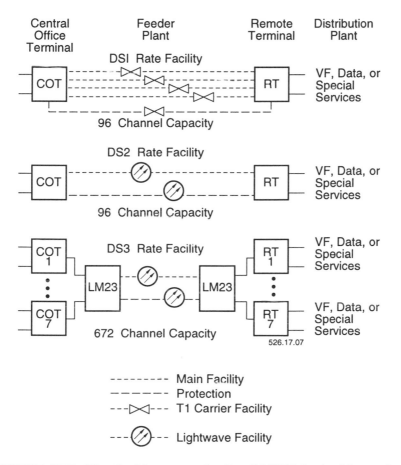

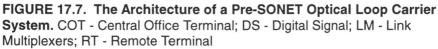

FIGURE 17.7. The Architecture of a Pre-SONET Optical Loop Carrier System. COT - Central Office Terminal; DS - Digital Signal; LM - Link Multiplexers; RT - Remote Terminal

This system has been designed to carry data at 6.314 Mbps between the Central Office and a remote terminal, which serves as a miniature switch to multiplex and demultiplex voice channels. The twisted copper wire-pairs connect the remote terminal and the individual customer in a star configuration.

This particular system links up to 96 voice channels to/from the remote terminals from/to a Central Office. Reduced power and size requirements with inherent immunity to crosstalk and electromagnetic inductive interference make the repeaterless fiber optic systems desirable in the subscriber and in the distribution plants. This particular system operates in the 1.3 μm wavelength region The dispersion is minimal and permits the use of light emitting diodes (LED) as opposed to the injection laser diodes (ILD), which tend to have temperature sensitive characteristics. At the receiver side positive intrinsic negative diodes (PIN) opposed to the avalanche photodiodes (APD) offer wider temperature stability.

During the early eighties, a range of about 20 km (66 kft.) was envisioned with the lightguide cable having a graded refractive index profile. The core diameter of 62.5 μm and an outside cladding diameter of 125 μm was selected. A modest data rate at DS2 (6.314 Mbps) was attempted. With N grade fiber, the loss was estimated at 0.70 dB/km and 1.06 dB/km or 1.8 dB/km with M or L grades of fiber.

17.4.2 SWEDISH ENVIRONMENT

In 1987, Ericsson participated in a European collaboration on studies of transmission at 2.4 Gbps, using an optical fiber transmitter contributed by Ericsson with a receiver developed by CNET (Centre National D'Études des Télécommunications). The tests demonstrated the use of direct modulation and externally modulated data transmit/receive facilities at 2.4 (to 4) Gbps over a relatively short distance of about 40 km with a bit error rate of 10^9. Some of the integrated optical devices that were explored include modulators, laser amplifiers, switching matrix and passive power dividers, in addition to laser and photodiodes. During that period, total integration of entire subsystems was not complete. However, the WDM techniques and integrated optoelectronic components had taken a firm hold in optical communication technology. As early as 1988, lithium niobate (LiNbO3) with diffused titanium waveguides, was being explored for a small (8x8) switching matrix. The switching elements in the matrix consisted of voltage controlled directional couplers. These subsystems were used for optical fiber bus at 2.4 Gbps in local LANs (building to building), rather than in subscriber systems, with a view to make them full fledged communication networks.

17.4.3 JAPANESE ENVIRONMENT

In Japan, field trials using fiber for subscriber loop carrier systems were initiated as far back as 1981. Wavelength division multiplexing concepts with graded index multimode fiber were employed to achieve a rate-km product of about 150 (Mbps.km). The maximum transmission length of 12 km in the 0.8 μm band at the T1 rate and about 2 km for the CATV and for 4 MHz analog TV signals was reported. High definition TV signal range was limited to about 5 km, and eight VHF TV channels are WDM-multiplexed over 2 km fiber length. However, the lack of single mode fiber technology during, the early eighties limited the rate and range of this trial with a transmission loss of about 4 dB/km.

In late 1988, the broadband optical ISDN requirements (spelled out by ITU [formerly CCITT] in draft form) were satisfactorily implemented at 150 Mbps. In fact, the Fujitsu system at 1.8 Gbps had ample bandwidth to accommodate twelve separate 155.52 Mbps channels over the fiber in the subscriber loop plant. Optical cross connects between different nodes is proposed and considerable topological flexibility is foreseen to implement different architectural variations. An enormous variety of services including the asynchronous transfer mode (ATM) of handling voice and high bandwidth data are available with this facility in the subscriber loop plant. With bulk switching facility at intelligent network nodes, the SONET specified optical signal format and interfaces can be implemented for the ATM packet switching system.

17.4.4 COMPARISON WITH COPPER BASED SUBSCRIBER LOOP CARRIER SYSTEMS

Loop carrier systems are not prevalent throughout the world. In the Western Hemisphere, the United States and Canada have incorporated the T1 carrier systems in the subscriber loops in the subscriber loop carrier (SLC) system for 40 voice channels (SLC-40®) with adaptive delta modulation (ADM) encoding of the voice since 1974. The two-bit companding of the step size uses a simple RC network to achieve a reasonable quality of speech at 37.6 kbps rather than the conventional PCM rate of 64 kbps. Experimental codecs with 3 bit, 3 and 4 bit companding have produced equally satisfactory speech quality at 32 kbps. Other adaptive delta pulse code modulation (ADPCM) techniques also exist for telephone quality speech to and from the customer. The SLC-40 system has also been successfully deployed in the United States to carry voice traffic to densely populated clusters via a T1 line.

The success of T1 carrier systems since 1962, and its growing popularity since, has prompted investigation of other systems that combine digital transmission, programmability, and possible switching. Optical transmission has been highly successful and semiconductor switching is well established. Fiber optic subscriber loop carrier systems (especially SLC-96) combine the fiber and switch to meet the distribution requirement in the subscriber end of the telecommunication networks (see Figure 17.7). The long-haul systems are presented in Section 17.3. The short-haul high speed data by fiber as it is applicable to public domain networks was considered in detail during the early 1980s. It is to be appreciated that private networks, such as LANs and MANs, can be tailored to meet any requirement within the constraints of the transmission and switching technology. However, more recently, the design of the public domain networks is becoming more and more standardized. The design principles and their feasibility of implementation are well established in the 1990s. Further, even at the very modest rates of optical transmission (6.3 Mbps), these systems proved to be economically viable to meet the standards and economic practices of the telecommunication industry of the seventies and eighties.

The SLC-96® system used a fiber medium to carry encoded voice at 64 kbps toll quality speech to the subscribers. The digital rate on the fiber is unusually low at a DS2 rate of about 6.3 Mbps. Typically, two fiber pairs are used with one pair as a standby or spare for the designated 96 voice channel capacity. The SLC-96 system was rapidly accepted in the United States market in spite of the gross under-utilization of the fiber capacity. The high quality fiber now available carries data at rates in multigigabits per second. The SLC-2000 (see Section 1.2) systems provide a good example to provide high-speed access to the users, or carry PCM quality voice and 64 kbps data services to remote users.

The SLC-96 system uses 1.3 μm wavelength LED/PIN diode technology. Low power repeaterless transmission is deployed to reach commercial and residential complexes. Remote terminals are essential to house the multiplexers and demultiplexers for redistributing the channels to customers, and also to house ringing supplies, battery plants, and power distribution panels. Some of the terminals have been designed to house up to 4,000 lines, thus creating satellite Central Offices (within the remote terminals), diverging from one main high capacity Central Office.

17.4.5 A RECENT FIBER OPTIC SUBSCRIBER CARRIER SYSTEM

The more recent designs promise significantly improved commercial systems. For example, the optoelectronic integrated circuit (IC) devices have been successfully tested at 1.2 Gbps over 52 km (170 kft.) with InGaAsP laser diodes and three FET drivers. This system uses InGaAs PIN photodiodes and low-noise amplifiers with 4 GaAs FETs and 5 GaAs Schottky diodes and a resistor on a 1x1 mm chip. Far higher rates are foreseen in the near future. More recently, the thrust towards extending the fiber to home of subscribers has prompted significant and viable designs. The Broadband local access architecture envisioned by the Bell Operating Companies has a capacity up to 560 Mbps to the home via remote electronic facilities capable of video distribution and broadband packet handling capability. The remote electronics facility is serviced by fiber feeders operating around 2.4 Gbps from broadband Central Offices distributed throughout metropolitan areas. Lower rates of 240/140 Mbps are obtained by demultiplexing the higher bit rates to individualized customers. Other smaller systems for carrying high-speed data can be assembled by the components manufactured by vendors of fiber optic diodes and laser sources, fibers, detectors, and pin diodes.

Public domain broadband digital networks can evolve only because of the massive capital expense necessary to standardize, design and construct national and international networks. Small private networks (such as dedicated LANs, WANs and MANs) with any reasonable amount of sophistication can be built with the existing technology. Between these two extremes lie a group of specialized, high capacity, sophisticated networks that can be classified as highly customized in their own right. Typical examples of these are military, industrial, computer, campus, and private networks. However, if one specifies the network environment (their customers, services, specialized features), then their components and architecture may be strategically implemented for optimal performance.

The impact of optical switching will have its own effect on the switching functions of the evolving all digital (and perhaps passively intelligent) networks. Optical switching holds enormous potential for speed and access of the time division multiplexing of numerous high-speed channels. Integrated optoelectronics offer features that the SPC (stored program control) and switching systems now accomplish with programmable software modules. Typical functions that can be performed are customized services; selected data base access; multimedia educational, financial and medical services; interface with packet switched networks, common channel control interface; and interface with a large number of digital carrier systems that are already in use.

Fiber has been used in the loop plant environment as a carrier system connecting Central Offices to remote terminals. The viability, pricing, very high rates, error free transmission, and potential growth characteristics of fiber optic transmission systems become more and more attractive in the subscriber loop plant. The recent digital carrier systems, SLC (mostly in North American countries), are fiber optic systems. When the data rate approaches the 45 Mbps region (North American DS-3), fiber is the only viable choice. In these SLC systems, fiber carries both data and signals. At rates higher than 400 Mbps, microwave links can also become attractive if the terrain or region is hostile to land lines. These systems do not adhere to any stringent standards, since the operating companies cater to particular needs individually, and conform to local operating environments, such as the type of signaling and switching capability of the Central Office. The Central Office access penetration of all the digital loop carrier systems in 1990 was about ten percent, with about four percent being optical access and six percent being traditional copper interface. The other ninety percent was still analog-voice frequency access in 1990. The data transmission within the distribution environment starts out with a detailed study of the existing loop plant. Fiber starts to enter the picture as the need for digital capacity increases. Typical reasons are the increased demand of conventional services (usually in developing countries) and/or the needs of businesses, hospitals, high-technology industries, and universities seeking higher and higher data rates or more and more modern services such as PRISDN, BISDN, DS-3, or OC-3 data links. In most of these cases, the solution is a phased expansion of the current systems rather than an outright replacement.

However, the universal truth is that for the present, fiber and copper coexist in the loop plant quite harmoniously. New installation of all fiber optic distribution systems (when existing customers have a copper distribution facility in place) appears uneconomical at this time. New facilities that can be shared between numerous buildings, customers and floors of modern information-based offices (such as banks, stock brokerage houses, knowledge and data banks, campuses, etc.) need special considerations. Even under these conditions, a composite facility (fiber optic ring and/or star combination) for the backbone communication facility with short copper drops (to the individual customer and/or work-stations) may satisfy the cost and switching requirements depending upon the individual network conditions. As the costs of fiber optic switches, modems, and other components start to decrease substantially, the proportion of the mix of fiber and copper may start to shift in favor of all fiber networks.

Customer need plays a crucial role in the choice of the medium for carrying high-speed data. Numerous types of customers have been identified. The data needs of businesses and subscribers are different. The data needs of governmental agencies and educational campuses are different, and so on. For this reason, the transmission aspects at the distribution level need to be evaluated individually. These customer needs are declared at initiation of the service but it may be modified from time to time.

It is foreseen that the SONET rates (OC-1 and OC-3) will penetrate the subscriber environment from Central Offices to sophisticated remote terminals. This penetration is projected to be modest during the mid-nineties, since only the new demand for digital services are likely to be met with all digital SONET service. However, by the end of the nineties, if the demand for all digital services is sustained, then OC-12 (622 Mbps), and in some instances OC-48 (2.48 Gbps), will become essential. This type of new (subscriber loop plant) architecture will slowly replace (by the year 2025) all copper loop plants, which prevails now.

17.4.6 FIBER TO THE HOME

Fiber to the home is also an emerging area of investigation. Fiber to the home (FTTH) has been proposed for all fiber very broadband network architectures for the twenty-first century. The Central Office is seen as the focus. Numerous homes (as many as 24) can be served on a single fiber pair serving as a bus (bus topology). Two fibers are used to isolate the directions of communication. When a single fiber is used to the individual home in a star topology, wavelength division multiplexing (WDM) is used to isolate the two directions of communication. When a remote terminal (RT) is used, the individual home is served by WDM fiber, but the RT is served by at least two fibers for the two directions of communication. Such RTs generally have a star topology serving many homes. The Central Office also generally has a star topology serving many RTs. A double star with two nodes (for the two directions of communication) has also been proposed. The ring configuration is generally not promoted in the public domain environments for security and for possible isolation of individual customers unless it becomes necessary.

Whereas, FTTH is seen as an ultimate for moving modern society towards information society and homes towards knowledge towers, fiber hybrid coax(ial) systems (HFC, see Section 3.6) are seen as plausible intermediate solutions for digital TV, HDTV, near HDTV, video-on-demand, multimedia services, etc., services. In current (mid 1990s) commercial systems, fibers have proven to be more economical (over copper) only when the capacity of the multiple RTs is required in isolated areas such as large office buildings or isolated packets of high technology industries. In the low capacity and conventional POTS environment, the digital loop carriers (DLC) based upon conventional copper media (coaxial and twisted wire-pairs) was less expensive during the late eighties. The DLC technology for the later systems is now mature and established. For this reason, it is foreseen that over a period of time, fiber SLC and HFC will compete at the higher bit rates when the demand for video services is to be met by the networks. The definite advantage is the customer programmability of the services that the network will perform.

17.5

INTEGRATION OF EXISTING COMMUNICATIONS NETWORKS

There are *three* architectural variations in the networks used in the public domain: *Circuit* Switched, *Packet* Switched, *Message* Switched networks. Private networks can vary dramatically and assume specialized topologies to serve specialized application needs. Whereas the switched and packet network constitutes the major variations, a third minor variation in the architecture also exists. This network that is called the message switched network uses large memory or data bases at each of the switching nodes. The entire message is collected at the receiver and transmitted in its entirety to the next node and the process repeats. These networks are presented next in view of their integration into the optical environments

17.5.1 THE SWITCHED OR CIRCUIT SWITCHED NETWORK

First, in the switched or the circuit switched architectures (connection orientation), there are *five* stages for data transfer through the older switched channel. The sequence may be summarized as follows: *(a)* an idle channel is identified and tagged, *(b)* a connection is established, *(c)* the data is transferred, *(d)* the channel is released, *(e)* the channel resumes to be idle again. This sequence of stages closely parallels the well established network steps in a typical voice call that follows the network functions *call setup, alert, connect, disconnect and release.* In context of the POTS network, the channel could be typically a voice channel, and the signaling for these distinct steps is accomplished in bands within the same channel. The functions remain same in all networks, but the protocol and the steps in the implementation vary slightly from the older copper based networks versus the newer fiber optics based networks.

When a new network (such as the HFC) becomes viable, the whole process of defining the protocol for the functions and the subfunctions optimized in the implementation in a particular technology become necessary, and new standards evolve. In context to the ISDN, the channel is typically a bearer (B) channel and the information necessary to accomplish these individual steps is incorporated in the CCITT (now ITU-T) Q.931/931 protocol at the network layer used over the D (delta) channel to set up the B (bearer) channel. Similar variations and standards exist for the ATM networks and are discussed in Section 7.8 of the companion volume, *"Intelligent Broadband Multimedia Networks"* [17.1].

Switched networks are perhaps the most common and universal networks around the globe. Almost all telephone networks utilize the basic principle of making the channel available to the user until the call termination is initiated by the caller or the called party. The older telephone environments have provided dedicated (or shared, as in multi-party) telephone lines to the subscriber(s) and used shared trunk lines for intra and interoffice connections. The principle works because all the subscribers do not use their telephones at the same time and numerous lines could be "concentrated" into fewer lines or channels over longer distances. Thus, subscriber lines are concentrated onto trunk lines at the local Central Office or a remote terminal. The ensuing trunk lines are then distributed through a geographical (or a logical) expanse, and at the distant Central Office (or its own remote terminal), the trunk lines are expanded (unconcentrated) to make the final connection. All the addressing information in the older networks is embedded in the dialed number.

17.5.2 THE PACKET OR PACKET SWITCHED NETWORK

Second, in the packet switched architectures (connectionless orientation), a packet of information is assembled with its own address and forward destination. ATM, the most recent of the networks, is indeed implemented within the framework (see Figure 15.8) of the packet technologies. The header plays the crucial role in making sure the packet, the frame, or the cell exhibit the features germane to the requirement of the application on hand. In a truly elegant protocol structure of the frame and the cell headers, the ATM systems (which may be fiber based and/or copper based) work well as packet networks, and in true sense, fiber (via ATM networks) has already impacted the packet switched network in a slow, sure and steady manner over the last decade or two.

The Aloha network was one of the first networks to make communications possible between certain digital systems without allocating a channel to the user during its use. The reasoning behind this concept was that computers would exchange information in a bursty mode and dedicating a physical channel over the entire time would be wasteful. Hence, at the packet assembly/disassembly interface of the network, the packet switching system would establish a logical channel, and collect data, but send out independent packets of data as they arrived.

The packet may transit through any number of nodes before it reaches its destination. The exact physical routing is not known. The innate functioning of the participating nodes through the network assures the correct delivery of the packet to the correct node. In context to packet switching, ITU (formerly CCITT) has specified the widely accepted X.25, X.75, X.28, and X.29 protocol at the network layer (see Section 4.5 of the companion volume, *"Intelligent Broadband Multimedia Networks"* [17.1].

The two architectures (CS and PS) can work cooperatively. One of the objectives of the ISDN network switches and the nodes is to assure the compatibility of two major network ideologies and hence their difference in functioning. The packet handling facility of the ISDN nodes makes the basic rate subscriber use bearer channel 1 (i.e., B1) for circuit switched network functions, bearer channel 2 (i.e., B2) for circuit/packet switched network functions via LAPB protocol, and the delta channel (i.e., D) for network and packet switched functions. The details of the ISDN bearer and signaling capabilities are presented in Chapter 7 of the companion volume, *"Intelligent Broadband Multimedia Networks"* [17.1]. Concepts culminating in the present standardization and implementation of the ISDN have initiated the evolution of sophisticated intelligent networks that take full advantage of the ISDN to link the various intelligent peripherals (IPs) within the architecture of Bellcore's IN/2 and the ITU's (formerly CCITT) Generic AIN Platform, discussed in Chapters 11 and 12 of the companion volume, *"Intelligent Broadband Multimedia Networks"* [17.1]. The generic ideas integrating most of the existing networks lingers from 1987, if not earlier. Even under the auspices of ITU-T (formerly CCITT) 6, an architecture encompassing switched networks, packet switched networks, international networks, and dedicated private networks has been proposed. Such architectures (see Figure 17.8) are discussed in Chapter 8 of Reference 17.1.

17.5.3 MESSAGE SWITCHED NETWORK

The *third* network in the environment is the message switched network. The functionality of this network is based upon the memory/data base being incorporated at each node. The entire message is relayed from one node to the other until its final destination. The entire message is carried in one packet and the packet size is adapted to the message size. Intermediate transmissions are suppressed and the network can be half-duplex (i.e., only one party is communicating to/from the network at one time). Low cost teletypewriters can be used in this network for customer premises equipment and the complete message arrives as one block. The bit rates in typical message switching (used in Western Union's Telex II service) can be very low (such as 110 bits/second) and use standard ASCII code. This type of network is perhaps the least expensive and least sophisticated.

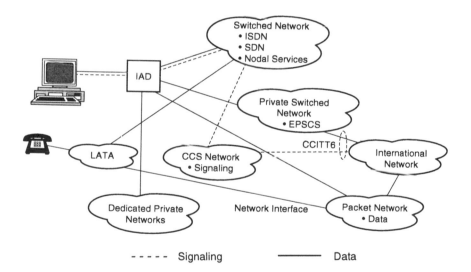

FIGURE 17.8. A Typical Version of Early (1987) Data Networks in the United States before SONET and ATM. Notice the independent signaling network as a precursor to the full fledged SS7 network. ATM - Asynchronous Transfer Mode; CCITT- International Telegraph and Telephone Consultative Committee (now ITU-T); CCS - Common Channel Signaling; IAD - Integrated Access Distributor; ISDN - Integrated Services Digital Network; EPSCS - European Packet Switched and Circuit Switched network access; LATA - Local Access and Transport Area; SDN - Software Defined Network.

17.5.4 PRIVATE AND SEMI-PRIVATE NETWORKS

Private and semi-private networks (sometimes called channel switched networks) may also use standard components and associated interfaces. Generally, private branch exchanges (PBXs), private and dedicated lines, and digital distribution facilities with computerized information handling capacity may be encountered in private (intelligent) networks. These networks are emerging with alarming amounts of sophistication and intelligence. Typically, such networks are found in the scientific community (e.g., National Science Foundation's NSFNET, Defense Advanced Research Projects Agency's ARPANET, Carnegie Mellon & Bell of Pennsylvania's Metropolitan Campus Network (MCN), etc.). Other examples from industry (e.g., VLSI vendors networks) and national laboratories (e.g., Lawrence Livermore National Laboratory's private network for their 12,000 employees spanning 500 buildings) also prevail. INTERNET (see Chapter 4) is another example of networks where the ownership is not proclaimed. Even though the INTERNET roots can be traced back to ARPANET, the INTERNET functions as a collaboration of many individual bodies, organizations and even nationally and internationally operated nodes and gateways.

More recently, the private networks have extended into the international scene on a grand scale. During the early 1980s, the transoceanic cable, using repeatered fiber optic systems were popular. As many as seven cable systems (TAT-8, TAT-9, TAT-10, TAT-11, TAT-12, TAT-13 and PTAT-1) span the Atlantic. Three major systems (NPC, TPC-4, TPC-5) span the entire Pacific. Three systems (HAW-4, HAW-5 and TPC-4) access the Hawaiian Islands and two more systems (TPC-3 and TPC-4) access Guam and then the Pacific rim (PCRIMEAST and PCRIMWEST) and the Far East.

In a recent collaboration (mid 1990s), AT&T and Alcatel will design, manufacture and install an international all optical fiber transmission system for Africa (Africa ONE). This optical fiber ring topology (see Figure 17.9) around Africa will facilitate advanced telecommunication in the Pan-African region by spanning about 40 countries and covering about 20,000 miles (32,000 km).

Segments of this topology reaches Europe, the Americas, Turkey, Cyprus, India, Sri Lanka, Indonesia and Singapore. To be completely functional by 1999, the fiber optic ring deploys optical amplification (see Section 16.11) and wavelength division multiplexing (see Sections 16.3 and 18.5.3) by multiplexing up to eight different signals in eight tightly controlled wave bands. Segments of this stretch of the lightwave system are operational at 40 Gbps over 1,420 km (rate-distance product of 56.8 terabit-km, but with 16 WDM channels of 2.5 Gbps each) without the need for regeneration in laboratory environments at the Bell Laboratories.

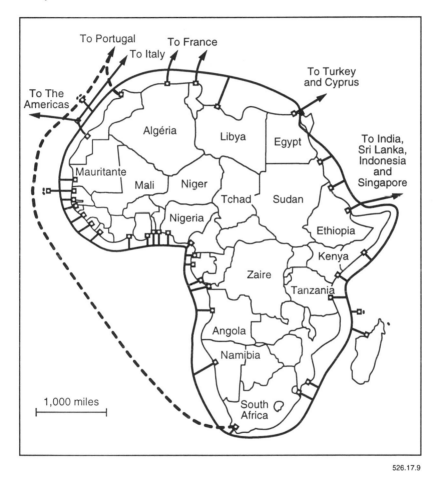

526.17.9

FIGURE 17.9. Topology of the Proposed African Undersea Fiber Optic Network. This network links 40 countries and deploys 16 WDM channels (see Section 18.5.3) each at 2.5 Gbps, placed at 100 GHz frequency intervals. Thirteen fiber spans of erbium doped fiber (see Sections 16.11 and 17.5.4) are used for optical amplification (from raw laser energy to modulated information carrier) throughout the optical ring.

Wavelength division multiplexing concept and implementation (simulated in Section 18.5) is particularly viable with tightly coupled lightwave sources, and the intermodulation between wave bands is minuscule, if laser sources and receptors are used.

Another extensive fiber optic link has taken place around the globe as FLAG (fiber optic link around the globe). It accommodates 120,000 DS0 rate (bearer rate or 64 kbps) channels on two fiber pairs. This link spans Europe, the Middle East and Asia with 27,300 km (about 17,000 miles) of fiber, and also provides for the projected growth in basic 2 Mbps and ATM services. Initially, the deployment of the 64 kbps channels will conform with the SDH mode of the functionality to easily support the demand for baseband synchronous facilities. Its closest competitor from Europe (Marseilles) to Singapore operates under the 565 Mbps plesiochronous digital hierarchy (a digital transmission hierarchy, which permits the transmitters and receivers to operate under different rates).

The routing of this global information highway follows twelve countries at 10 or 5 Gbps from the United Kingdom to Japan. The routes, the cable landing point(s), the local carrier, and the rates are presented in Table 17-1.

TABLE 17-1 Details of the FLAG Information Highway

Country	Landing Point(s)	Local Carrier	Rate
UK	Porthcurno	Mercury	10.6 Gbps
Spain	Estepona	Telefonica Espana	5.3 Gbps
Italy	Palermo	Telecom Italy	10.6 Gbps
Egypt	Alexandria, Port Said, Suez	Arento	10.6 Gbps
UAE	Fujairah	Etisalat	5.3 Gbps
India	Bombay	Videsh Sanchar Nigam Limited	10.6 Gbps
Malaysia	Penang	Telekom Malaysia	5.3 Gbps

TABLE 17-1 Details of the FLAG Information Highway, continued

Country	Landing Point(s)	Local Carrier	Rate
Thailand	Sonkhla, Satun	Communications Authority of Thailand	10.6 Gbps
Hong Kong (UK)	Lan Tau Island	Hong Kong Telecommunications International	10.6 Gbps
People's Republic of China	Shanghai	China Telecom	10.6 Gbps
Republic of Korea	Keoeje Do Island	Korea Telecom	10.6 Gbps
Japan	Miura	International Digital Communications	10.6 Gbps

This network also deploys the optical amplifier technology, which is made a part of the fiber optic link that is embedded every 45 to 85 km, and the received optical signal which is regenerated and amplified at the landing sites for the next stretch of fiber optic cable. The cable has twin pairs of optical fibers, housed in the circular unit fiber structure, surrounded with two sets of circumferencial strength wires housed inside a copper sheath. The copper sheath carries 0.9 A dc current at 7,500 V for repeater power with its return path through the sea water. These micro repeaters are made an integral part of the fiber optic cable as the fiber is assembled into various cable sections. This sheath has a insulation jacket and then armored protection with an total outside diameter of 22 mm.

The all optical technology for FLAG uses optical amplification by 10 m long segments of erbium doped fiber, discussed in Section 16.11. The doping is done by replacing a small fraction (up to 0.1 %) of the atoms within the fiber core with the rare earth element erbium. The segment now acquires properties of optical amplification. Low level laser energy at 1,480 nm is also pumped into the core with the signal at 1,558 nm. The shorter wavelength signal excites the erbium atoms and they, in their excited state, emit light at the signal wavelength, thus forcing a very dependable and reliable amplification process within the segment. They are placed at intervals between 45 to 85 km.

Each fiber has a 0.125 mm diameter. The two twin pairs handle 10.6 Gbps with 5.3 Gbps for each fiber pair, and 3.84 Gbps (i.e., 60,000 times 64 kbps) capacity is for information and 1.46 Gbps for overhead, signaling, framing, and error correction. Each fiber pair carries two ITU (formerly CCITT) standard STM-16 SDH signals (see Chapter 15) at 2.53 Gbps. At the operational level, each of the STM-16 signals are demultiplexed to the STM-1 signals at 155.52 Mbps. Add-drop multiplexers (ADMs; see Sections 15.4 and 16.2) channel the data into the landing points, or the digital cross connects throughout the network. These signals may be interfaced with other non-synchronous or pleisochronous cables or satellite links at 34, 45, and 140 Mbps.

Private dedicated networks can be localized or extended nationally. Generally, local area networks (LANs) are geographically limited and network speeds are restricted by the media and switching. The applications can vary from voice, data, facsimile, video, computer data base exchange, word/graphic processors, etc. Numerous terminal devices can be incorporated in the network. When these LANs do not use telephone lines, the Federal Communication Commission (FCC) regulations do not apply to their operations. Typical transmission media used in LANs can range from twisted wire-pairs, coaxial cable, optical fiber, digital radio, or even to laser beams. The network topology is a consideration in LAN design. Typically, the star, ring, and bus architectures are preferred, even though dropped bus, gateway oriented multiple rings, and bus ring structures can be built and operated. Such rudimentary architectures are discussed in numerous books [17.7, 17.8]

17.5.5 IMPACT OF FIBER ON ISDN

The transmission aspects in ISDN start out with a detailed study of the loop environment that has to carry the high-speed data between the digital Central Offices (or data switching centers) of the emerging ISDN network and customers. The question posed through the 1990s is not the availability of BRISDN, but the provisioning of PRISDN, if not broadband ISDN. The data needs of businesses can be significantly different from the data needs of individual subscribers. The data needs of governmental agencies can be quite different from those of educational campuses, and so on. For this reason, the transmission aspects in ISDN have been addressed by the data rates and the specialized needs of the customers. Enhancement of service becomes a necessity in the ISDN environment.

The provisioning of services at high speeds is not foreseen as a problem since the telephone network handles high-speed data communication effectively between Central Offices. The trunk transmission facilities are highly evolved in the United States. Further, the evolving fiber nets offer ample bandwidth for very high-speed communication ranging from a few megabits per second to a few hundred gigabits per second. However, the transmission of data at intermediate rates, ranging from 80 (B+D) kbps to 384 (5B+D) kbps, or even to 1.544 Mbps (23B+D), or the primary rate over the widely distributed telephone network, which provides access to every home, is a unique challenge.

Systems have been designed and currently tested to verify conformity to the ITU (formerly CCITT) standards. The systems and component vendors in the telecommunications industry have developed the VLSI chips that accomplish subfunctions satisfactorily, even though extensive network testing is not complete. Some of the Regional Bell Operating Companies in the United States are offering ISDN-like services more and more frequently. The ISDN uses the entire existing telephone network topology, media, switches, and data bases to serve the customers on a local, national and global basis. The evolution is gradual and the numerous phases of ISDN stage, (ISDN-1 [island], ISDN-2 [backbone] and the ISDN-3 [national and standardized ISDN throughout the world]) are documented in Chapter 7 of the companion volume, *"Intelligent Broadband Multimedia Networks"* [17.1]. The evolving ATM network architectures will permit all the networks to cooperate and coexist as one seamless entity. The principles behind networks are common to most multimedia communication systems via broadband networks (see Chapter 15) even though the applications, vendors, and expanse can differ considerably from customer to customer, company to company, or even nation to nation.

17.6

THE SIGNALING NETWORK

A supporting network that has emerged because of the need to control and monitor the flow of information is called the common channel signaling (CCS) network. Signaling is essential to the functionality of any network. In context to ISDN, the signaling is carried out by the standard CCS7 network adopted in the United States and Europe. This network uses the out-of-band information (i.e., the information on the D channel), to control and signal the various switches to complete and monitor the B channels. Figure 17.10 depicts the role of signaling and control of the information bearing B

channels.   This network becomes essential in the circuit switched context
because the B channels provide transparent end-to-end digital connectivity
for the network users. If and when a transition to the packet mode is to occur,
the CCS information is used to provide this transition and vice versa.

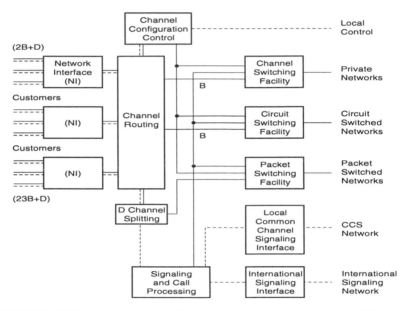

**FIGURE 17.10. Integration of the Signaling and Control within the
Traditional ISDN and BISDN Networks.** For ATM based BISDN
networks see Chapter 15.

In essence, the CCS network is a private and dedicated high-speed packet
network used essentially by the newer INs and ISDNs that provide services
ranging from IN services and functions to POTS. The customer uses this
network indirectly via the IN signaling protocol. The user has no direct access
to this rather illusive core signaling network.   User requests for special
services, such as CLASS, (see Chapter 11 of the   companion volume,
"Intelligent Broadband Multimedia Networks" [17.1]), 700, 800, 900, 911, network
services, etc., activate the signaling via the CCS7 network.   However, these
requests are first validated, and if considered legitimate, the signaling is
authorized.

The capacity of the signaling network (which is itself-packet switched)
appears adequate for a decade or so depending on the growth of the ISDN
and IN services.   This network is secure, duplicated and under utilized by

design for quickly transporting network signals and thus invokes quick network response. There are some steps, (such as enhancing the network rate to 56 kbps), that have been taken to increase the throughput in this network in congested areas. Optical signaling networks may never be necessary, but they can be fabricated by network designers if and when the multimedia and information revolution starts to reach every pocket transceiver.

17.7

REFERENCES

17.1 S.V. Ahamed and V.B. Lawrence 1996, *Intelligent Broadband Multimedia Networks*, Kluwer Academic Publishers, Boston, MA.

17.2 Lucent Technologies 1996, Bell Laboratories Innovation, *Private Communication*.

17.3 AT&T Bell Laboratories 1992, *Internal Communication*, September 3. Also see: "Undersea Communication Technology," *AT&T Technical Journal*, January-February 1995, Vol. 74, No. 1. 75-82.; "Undersea Lightwave Systems," *AT&T Technical Journal*, February-March 1992, Vol. 71, No. 1. 5-13 and "Erbium-Doped Fibers and the Next Generation of Lightwave Systems" *AT&T Technical Journal*, February-March 1992, Vol. 71, No. 1.

17.4 N.K. Dutta 1989, "III-V Device Technologies for Lightwave Applications," *AT&T Technical Journal*, January/February, 5-18.

17.5 S.J. Hinterlong and H.M. Hall, Jr. 1994, "Bringing Photonics to Broadband Switching", *AT&T Technical Journal*, November/December, 71-80.

17.6 Advanced Micro Devices 1989, "The SUPERNET™ Family for FDDI" *Publication # 09734, Revision C*, February.

17.7 J.P. Hayes 1986, *Computer Architecture and Organization*, Second Edition, McGraw-Hill, New York, NY, or
A.S. Tanenbaum 1981, *Computer Networks*, Prentice Hall, Englewood, NJ.

CHAPTER 18

A PC BASED FIBER OPTIC CAD ENVIRONMENT

There are *three* main objectives in writing this concluding Chapter.

The *first* objective is to demonstrate the ease of simulation studies of the fiber optic systems. If fiber optic systems are going to become more and more abundant, then their understanding and design should become equally easy. Here, the simulation and design methodology is generalized and implemented on PC based microcomputer systems.

The simulation methodology is quite akin to that of the digital subscriber lines presented in Chapters 7, 8 and 9. The system transfer functions are computed by using the basic equations germane to the propagation of energy through the fiber medium. The *second* objective is to show an architecture of a CAD platform which permits the selection of an appropriate set of components and displays the system characteristics based upon any selected topology of the data distribution through the network.

The *third* objective is to indicate the nature of simulation results. *Three* major categories of results are included: *(a)* time-domain results when the fiber is used as a single mode fiber carrying data under arbitrary conditions, *(b)* the performance of the fiber when it carries multi-level codes and *(c)* the quality of wave-length inter-modulations that can exist when wavelength division multiplexing (WDM) techniques are employed. Judicious use of the simulation and CAD facility permits lightwave system designers to fine tune complex optical communication systems rather than just experiment with different photonic sources, fiber optic links, and photodetectors to make up an optical network.

18.1

INTRODUCTION

The modeling of the various fiber optic components and of the fiber optic systems has been in vogue since the late 1970s. There is rich and complete documentation on almost all aspects of models and computational approaches. The interested reader is urged to refer to the Additional References listed at the end of the Chapter. This Chapter only provides a fleeting overview of a very well studied topic in fiber optic data communications.

Fiber optic simulations used to be processor time intensive in the computational environment. The integration of energy over the finite spectrum of wavelength active in carrying the optical power through the fiber leads to substantial demands on the central processor. The math co-processor in the computer is also taxed heavily. In addition, the conventional simulation techniques have used the time integration methodology thus making the simulations tedious and time consuming. However, the Pentium and dual/quad processors have progressed significantly through the early 1990s, thus reducing the overall simulation, design, and optimization of fiber optic systems faster than the corresponding DSL designs in the early eighties [18.1, 18.2].

In this Chapter, an enhancement of the HDSL simulation platform [18.3] is feasible, and is depicted in Figure 18.1. The numerous functional modules within this software are formulated, developed and explained throughout the rest of Chapter. In these simulations, the numerical solution employs a long repetitive pulse sequence to excite the fiber optic system. The Fourier components of the excitation sequence are evaluated and the attenuation of the optical signal is computed by the transfer function of the optic lines at various wavelengths. These techniques are discussed in some detail under various coding alternatives for coherent systems in Section 16.8. The alternative simulation technique for system simulation is based on the time step integration method. In this method, the system transients are first computed till these transients settle down to arbitrarily low values (i.e.,

within the accuracy of the numerical processor unit).   Such transient
responses are then cascaded for each step function exciting the optical
system. As far as the system is completely linear, both approaches produce
the same results.

The approach utilized in this Chapter reduces the long computations
required by the conventional approach of the time-step integration method. It
also eliminates errors that can cumulate due to slight inaccuracies in the
system characterization and computational errors.

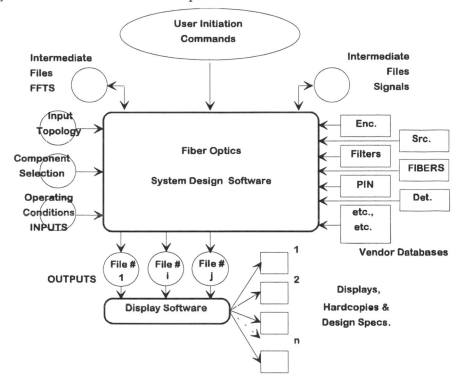

**FIGURE 18.1. Software Organization for the Design of Lightwave
Systems.**

The intense need for high accuracy is thus removed. However, the
proposed method may also require very long repetitive excitation cycles if
extreme accuracy in the simulation is desired. This approach is proposed for
optic system designers interested in installation of the fiber optic links rather
than for scientific researchers. In the local area fiber networks and in the high
capacity fiber optic data paths, the system acceptability does not critically

hinge upon extreme accuracy in simulation. Instead it depends upon the system viability and upon the cost of the system and of system components.

Some of the fiber optics simulation studies [18.4, 18.5] were traditionally carried out by main frame or extended mini computers. The simulation software system proposed here is oriented to a microcomputer environment. This software integrates the data base manipulation capability with active simulation of the critical components to offer the lightwave designer the capacity to modify the system, change the design parameters, and select components from vendor data bases to fulfill the system constraints. In the body of the Chapter, the capacity to carry out the simulation of complex lightwave systems with wavelength division multiplexing techniques is reported. Fiber splices generally cause attenuation and little, if any, phase distortion. The simulation software yields graphic distortion of the optical signal as it progresses through any user definable lightwave system. It is capable of generating plots that permit the designer to scan problem components and configurations in the system.

18.2

THE FIBER OPTIC SIMULATION PACKAGE

Like any other communication system, there are at least *five* components to make up a direct detection lightwave system: *(a)* the data source and the encoder, *(b)* the photonic source, *(c)* the fiber *(d)* the photodetector, and *(e)* the demodulator. Coherent detection systems need local oscillators at both ends. The modems are for modifying the frequency, phase or amplitude of the lightwave carrier. A variety of feasible encoders are presented in Section 16.8 for coherent systems. Two dimensional codes are also used in some cases [18.6]. Multilevel encoders are also feasible. If the information is being multiplexed by WDM techniques, then wavelength multiplexers (see Figure 16.25) are also necessary. The various types of sources (LEDs and lasers) are presented in Section 16.2.3.1 and the different types of frequently used detectors (PIN diodes and APDs) are presented in Section 16.2.3.5. Modems are usually unimportant (except at very high rates, [see Section 16.11]) and do not cause any serious concern in most broadband and wideband commercial lightwave communication systems.

18.2.1 SOURCES IN THE SIMULATION ENVIRONMENT

For the simulation studies, it becomes necessary to accurately characterize the performance of each of these components in order to evaluate the overall system performance. Typically, the sources (LEDs in particular) have a slow response. The faster devices will be able to track the encoded signal with little or no lag and the quality of the source is characterized by the time constant in emitting the optical energy into the fiber. Laser devices respond faster, and the pulse shape for the optical energy follows the encoded signal more accurately. The second characteristic of the source is indicated by the spread of the wavelength over which the optical energy is distributed. Generally this is a Gaussian distribution for LEDs represented as

$$S(\lambda) = \frac{1}{\sigma_s \sqrt{2\pi}} \cdot e^{-\frac{(\lambda - \lambda_s)^2}{2\sigma_s^2}} \tag{18.1}$$

where $S(\lambda)$ is the energy distribution from the source around the nominal wavelength of λ_0, λ_s is the nominal wavelength of the source and σ_s is the standard deviation of the device.

The energy distribution exhibits a much lower variance for laser sources. Once again the components from different vendors may be characterized by the different constants that the simulation package will deploy. Typically, the data bases for these components will contain all the relevant information as the simulation and/or design progresses. If the laser devices are not totally stable, or are subject to insidious noise, such as "chirp" or "mode hop and/or back" effects, then the computational models should reflect the possible behavior. Data bases, which characterize the sources, can be made elaborate to reflect these impairments.

18.2.2 PHOTODETECTOR IN THE SIMULATION ENVIRONMENT

The detectors also have their respective data bases and the characteristic constants are appropriately stored. Detectors generally add electronic noise in the system as a number, and distribution of photons gets transformed to the electrons and holes in the semiconductor devices are uncertain. *Two* types of noises are inherent in photodetectors; *(a)* the thermal Gaussian noise and *(b)* the impulsive or shot noise that depends upon the power level from the photodiode. Both PIN devices and APDs (see Section 16.2.3.5) are affected by these two types of noises; the blend and the extent differs. Mostly in the APDs, the shot noise can become a significant proportion of the total noise in the device. Both of these types of noise have their own characteristic constants, and based on the randomness of the photonic collision within the semiconductor, electrons are released and migrate to the edge of the device (giving rise to the very low detected currents). Both of these types of noises get combined in the detectors, and are the primary cause of limitation in single mode optical fiber systems. The exact model of these noises and their computations differ from PIN devices to the APD and are discussed next.

The PIN Diode Signal Process: The average number of the electron-hole pairs N_{sig} generated within an interval of time Δt, if the device receives an incident optical power of $P_{inc}(t)$ at a time, 't' is given as

$$N_{sig} = \frac{\eta}{hv} \cdot P_{inc}(t) \cdot \Delta t \tag{18.2}$$

where η is the quantum efficiency, h is Plank's constant and v is the optical frequency. The sum of N_{sig} and the electron-hole created by the dark current in the device, gives rise to the average (total) number of electron-hole pairs N in Δt seconds is given by

$$N = \frac{\eta}{hv} \cdot P_{inc}(t) \cdot \Delta t + I_d \cdot \Delta t \tag{18.3}$$

where I_d is the dark current in number of electron-holes pairs per second. The number of electron-hole pairs generated, n is a random time-varying Poissonian process, the mean and variance for this process is N and hence the probability that exactly k pairs are generated in Δt can be written as

$$P(n=k) = \frac{(N)^k \cdot e^{-N}}{k!}. \tag{18.4}$$

The total current $i(t)$ for the PIN device is simply the rate of flow of the electronic charge (q = n.e, where e is the charge per electron) and $i(t)$ is written as

$$i(t) = \frac{n.e}{\Delta t} \tag{18.5}$$

The APD Diode Signal Process: For the APD, the device current can be modeled as

$$i(t) = \frac{e}{\Delta t} \sum_{\ell=1}^{n} g_\ell \tag{18.6}$$

where g_ℓ is the number of resulting electron-hole pairs (i.e., secondary plus the primary pairs) for every primary pair 'n' in Δt seconds. Typically g_ℓ is a random variable with its own probability density function [18.6,18.7 and 18.8]

$$P(g_\ell = m) = \frac{(1-k)^{m-1} \; \Gamma\left[\dfrac{m}{1-k}\right]\left[\dfrac{1+k(G-1)}{G}\right]^{\frac{1-k(m-1)}{1-k}}}{\left[1+k(m-1)\right](m-1)! \; \Gamma\left[\dfrac{1+k(m-1)}{1-k}\right]}\left[\dfrac{G-1}{G}\right]^{m-1} \tag{18.7}$$

where G is the average gain of the APD and k is the ionization ratio given in the equation

$$\langle g^2 \rangle = \left[kG + \left[2 - \frac{1}{G}\right](1-k)\right]G^2 \tag{18.8}$$

The characteristic constants of the variable g_ℓ are presented in Reference 18.7, its approximate distribution is given in Reference 18.8 and a computational representation (for speed during simulations) is presented in Reference 18.9. The Equations (18.2) to (18.8) need to be adapted for the computational environment; an excellent methodology is presented in Reference 18.6. To save on execution time, most of the computations that are not signal dependent are stored in the data bases for each of the devices.

18.2.3 FIBER IN THE SIMULATION ENVIRONMENT

The basic equation that is instrumental in the simulation is the transfer function of a *single mode* fiber. It assumes *two* slightly different expressions based upon the (a) the electric field intensity model or (b) the power intensity model. If the light source is coherent and the information is communicated by modulating the wavelength or phase of the encoded signal, then laser or single (very very narrow) wavelength devices are used as local oscillators. This calls for the electric intensity model and here the equations given in Section 18.2.3.1 are applicable. On the other hand, if the information is communicated by modulating the intensity of the light of an LED, for instance, then it is the power over an entire spectrum of wavelengths that becomes crucial. This calls for the power intensity model and here the equations in Section 18.2.3.2 are applicable.

18.2.3.1 Electric Intensity Model

For the single mode optical fiber, the assumption that it can be represented as a flat bandpass filter with a flat amplitude response and linear group delay (within the bandwidth of the data) is justified, because of the very narrow data bandwidth with respect to the absolute frequency. The group delay (see Reference 18.10) is determined from the usual equation for the chromatic dispersion in the fiber (see Chapter 16 and Section 18.5 for details), and can be written as

$$\frac{d\tau}{dv} = \frac{d\tau}{d\lambda} \cdot \frac{d\lambda}{dv} = \left(\frac{-1d\tau}{\ell\,d\lambda}\right)\left(\frac{\lambda^2}{c}\right)\ell \tag{18.9}$$

and the chromatic dispersion $D(\lambda)$ is defined as

$$D(\lambda) = \left(\frac{-1d\,\tau}{\ell\,d\lambda}\right). \tag{18.10}$$

The fiber length is ℓ, the operating wavelength is λ, the optical frequency is v and the velocity of light is c. When the optical carrier frequency V_c is present, the group delay in the fiber can be written as:

$$\tau(v) = \tau(v_c) + (v - v_c)\frac{d\tau}{dv}\bigg|_{v=v_c} \tag{18.11}$$

The associated phase for this or any function is given by

$$\phi(v) = 2\pi \int \tau(v')\,dv' \tag{18.12}$$

$$\phi(v) = 2\pi v\,\tau(v_c) + 2\pi\left[\frac{(v-v_c)^2}{2} - \frac{v_c^2}{2}\right]\frac{d\tau}{dv}\bigg|_{v=v_c} \tag{18.13}$$

The transfer function of the equivalent model of the low-pass single-mode can now be evaluated as

$$H(f) = e^{-j\phi(f)} = e^{-j\alpha f^2} = e^{-j\left[\alpha B^2 \left(\frac{f}{B}\right)^2\right]}$$ (18.14)

with B being the bit rate and

$$\alpha = \pi D(\lambda) \frac{\lambda^2}{c} \ell$$ (18.15)

and

$$f = v - v_c.$$ (18.16)

Since only the chromatic distortion is considered, the constant phase term and phase terms linear with respect to f are not included since they do not introduce any distortion in the received signal. The computed spectral domain components of the fiber output signals from the fiber are derived from Equation (18.17). In case of coherent detection lightwave systems, the expression for X_{in} (f) is the Fourier component (at each harmonic f) of the input signal. The nature of coherent lightwave system signals is discussed in References 18. 10 and 18.11 and summarized in Section 16.8.

$$X_{out}(f) = X_{in}(f) \cdot H(f)$$ (18.17)

where

$$X_{in}(f) = \int_{-\infty}^{\infty} X_{in}(t) \cdot e^{-j2\pi f t} \cdot dt.$$ (18.18)

The inverse Fourier transform of the components X_{out} (f) leads to the time domain representation of the waveform output from the fiber.

Two major sets of results have been published. *First,* the eye pattern and the signal degradation due to chromatic dispersion [18.10] is simulated for a two dB penalty (i.e., the eye closure in the active region) and depicted in Figure 18.2 (5(d)) for ASK (amplitude shift-keying) or PSK (phase shift-keying) modulations and in Figure 18.3 (5(e)) for the DPSK (differential phase shift-keying). In the same vein, the dispersion limit charts may also be generated (see Figure 18.4) for the various types of signal modulation techniques (see Section 16.8) with a 2 dB penalty at 1.55 μm and a fiber chromatic dispersion D(λ) of about 15 ps/km nm.

For both Figures 18.2 and 18.3, the value of the chromatic dispersion index γ defined in Equation (18.19) is adjusted to produce an eye closure of 2 dB. The level of intersymbol interference depends on γ at a data rate of B. Additional penalties, such as imperfections of the lasers (chirping and phase drifts), are not included.

$$\gamma = \frac{1}{\pi} . B^2 \, \ell . D(\lambda) . \frac{\lambda^2}{c} = \frac{\alpha}{\pi^2} . B^2 \qquad (18.19)$$

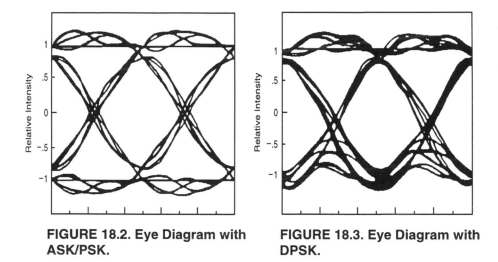

FIGURE 18.2. Eye Diagram with ASK/PSK.   **FIGURE 18.3. Eye Diagram with DPSK.**

Second, if the polarization dispersion is considered as the fiber limitation in coherent lightwave systems, then it is possible to evaluate the limit that is imposed by the propagation delay difference between the two orthogonal polarization modes. For this delay difference to approach 0.3 times the bit duration, the receiver penalty can be as high as 1 dB [18.12] with a fiber length of about 100 km at 10 Gbps. Simulation results are presented in Figure 18.5 for the different signal modulation algorithms (OOK, ASK/PSK, DPSK, MSK and CPFSK discussed in Section 16.8. Generally, this limit is beyond the chromatic dispersion limits presented earlier.

The approach for the baseband or direct detection fiber optic systems, which depends on the intensity of light (rather than the wavelength or the phase), is presented in the next section. The simulated results for the eye openings and for a typical case of fiber transmission are presented in Section 18.5.

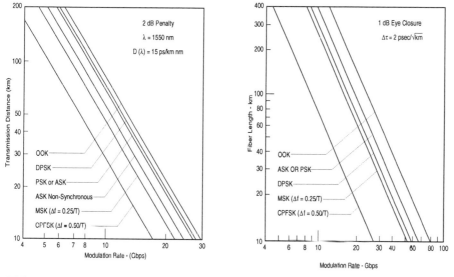

FIGURE 18.4. *Dispersion* Limitation at 1550 nm & 15 ps/km nm Fiber; 2 dB Eye Closure Penalty.

FIGURE 18.5. *Polarization* Dispersion Limitation at $\Delta\tau =$ 2 ps / (km)$^{0.5}$; 1 dB Eye Closure Penalty.

18.2.3.2 Power Intensity Model

The basic equation that is instrumental for this mode of operation in the simulation is the transfer function. For a *single mode* fiber, it can be written in a simplified form as

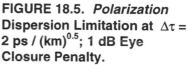

$$H(f) = \int_{-\infty}^{+\infty} S(\lambda).L(\lambda).e^{-j\,\omega.\,T(\lambda).\,\ell}\,.d\lambda \qquad (18.20)$$

where λ = wavelength
ω = $2\pi f$
$S(\lambda)$ = optical source spectrum as a function of λ
$L(\lambda)$ = attenuation of the entire fiber as a function of λ
$T(\lambda)$ = group delay of the fiber (per unit length) as a function of λ
and ℓ = fiber length.

The group delay of the fiber is sometimes called the antiderivative of the dispersion characteristic. In silicon fibers, the chromatic dispersion characteristic has been documented [18.12] as

$$\frac{d\,T(\lambda)}{d\,\lambda} = \frac{S}{c} \cdot \frac{\lambda - \lambda_0}{\lambda^2} \tag{18.21}$$

where S = a dimensionless constant
c = velocity of light
λ_0 = zero-dispersion wavelength.

Integrating this equation leads to T(λ), which is written as

$$T(\lambda) = \frac{S}{c} \cdot \left(\ln\frac{\lambda}{\lambda_0} + \frac{\lambda_0 - \lambda}{\lambda} \right) \tag{18.22}$$

used in the transfer function. S(λ) in Equation (18.20) is defined as the source spectrum and typically represented as Equation (18.1).

In elaborate simulation facilities, imperfections of the devices are accurately modeled in the data bases (either by algebraic equations or by look-up tables) and the influence of the device limitations can be accurately predicted. In addition, the variety of the devices (the sources, and detectors) to fine tune and optimize the system performance can be extensive. The data bases are also extensive and indicative of every limitation inherent in these devices.

The Fourier components of the excitation function are first computed by a series of fast Fourier transform (FFT) programs. The effect of the fiber on each of the harmonics is computed together with the transfer functions of the other components in the system. The entire series in the spectral domain is re-inverted in the time domain by inverse FFT programs to get the time domain response of the fiber optic systems. Eye diagrams are obtained by superposing the entire repetitive pulse sequence in one time slot. The noise sources through the fiber optic system consist of the photo-sensitive PIN-diode and the final detector with its own thermal noise component.

18.2.4 SIMULATION PROCESS

Simulation of complex systems can become time intensive. The APD model, if it is used, becomes most time intensive in the simulations. For the PC environment, it is more desirable to generate the APD distributed numbers in its distribution and store it away on the disk for every APD in the data base. For this reason, the preliminary runs can be made with standardized parameters and inputs, such as standard/average PIN-diode noise, or standard/average coupler characteristics, etc. The unpromising designs can be ruled out without excessive toil. However, the user has an option to force a full simulation of promising and or marginal design. The approach permits the usual designer practice of concentrating on designs that uniquely satisfy his own criteria without wasting much time on unproved and unsuccessful designs. The system can be easily modified to incorporate the rules for suggesting components and/or their combinations for an inexperienced designer. At the user's discretion, this option can be removed for the innovative designer experimenting with new features, and a full simulation can take place.

18.2.5 EXECUTION TIMES

The system works on any PC environment with about 40 MB of disk storage. For the simulation runs, the turnaround time is less than 50 seconds on a 20 MHz Intel 486 based PC. The execution time for a 100 MHz Pentium machine is about 20 seconds or less for a medium sized simulation. The APD diodes can be a source of most delay during simulation. Disk access and elaborate data base functions, if any, can also become tedious and time consuming. System optimization, where the machine picks the components and keeps track of the various steps through the optimization process, can be time intensive. The graphic post processors and display influence the final depiction of the simulation results. Perhaps this is the most elaborate part (and human aspect) in the simulation studies of fiber optic systems.

18.3

THE CAD ASPECTS

The CAD procedure can be divided into *three* phases of interaction with the designer. The *first* phase consists of defining the system configuration. The components are assembled by a set of interactive graphic routines. These routines explain to the designer all types of components available in the model library. The system configuration is then specified as the user selects the components from the library according to the network function. The system then presents the user with vendors in the data bases for each component selected. The vendor choice for each component is user defined input. This procedure is also carried out in an interactive graphics mode. The system permits the user to backtrack at any stage and change the components or the vendors selected. Once the system is configured and vendor selection is complete, the system offers the user the choice to generate a hard copy of the final system before simulation starts.

During the *second* phase dealing with the simulation of the system, the user is actively informed of status simulation procedures and any errors resulting from user inputs or data base access are posted. Recoverable errors are tackled by the system and unrecoverable errors terminate the simulation system.

The *final* phase permits display (and generation of the hard copy) of the system performance. This phase also operates in an interactive mode. The system offers the designer to see the time domain results as wave shapes at any of the nodes in the fiber optic network. It can also display the eye diagrams as repetitive cyclic excitation occurs at the transmit end of the network. Finally, the designer can track down the signal strengths at different nodes in the network as a function of the wavelength. This feature makes the simulation system extremely useful as different vendors of light emitting diodes offer different distribution of energies at various wavelengths. Likewise, the different fiber vendors offer different grades of fibers with varying attenuation at these wavelengths. The user is capable of determining precisely the benefit to the cost ratio for different vendor choices. The system performance can be studied and ascertained in ant amount of detail.

The user can exercise ample flexibility throughout the design procedure. Coupled to price and inventory data bases, the system yields cost and timing information for a typical fiber optic installation.

Some of the preliminary results on the eye openings (see Section 7.6, or Reference 18.13) obtained during data transmission on a single mode fiber with binary and multilevel code have been presented. For this reason, only a brief overview of this system (in context to the series of simulations) is presented in this Chapter.

Central to the CAD facility is a basic simulation program capable of computing the loss and dispersion (see Section 18.2) that optical energy experiences as it propagates through a single mode fiber. In view of the computational limitations of a microcomputer, the incremental time domain analysis presented in the SYSTID simulation program [18.4,18.5] is not used here. Instead, long repetitive periodic excitations for the fiber optic system are utilized. Any random or arbitrary pulse pattern (generally between 2,048 and 16,384 pulses) is selected. The performance is evaluated by studying the signal wave shapes at any point through the lightwave system. This approach has been also used for evaluating the ISDN performance for digital subscriber lines (see Chapter 7).

Intelligent and integrated computer aided design (CAD) environments are especially desirable [18.13,18.14] for network design and their components. The need for every component to fulfill its role many billions of times every second of its use over extended periods of time is non trivial. The traditional approach towards over-design of components and under-utilization of the channel capacity is not viable in competitive network environments. For these reasons, the CAD environments are becoming sophisticated, accurate, economical and highly interactive to make them optimal. In this Chapter, an on-line PC based simulation software for the analysis and design of fiber optic data transmission links is presented. The system integrates the data base access and manipulation capability with active simulation of components and their functions. The numerical solution employs a very long repetitive pulse sequence to excite the system. The Fourier components of a long excitation pulse sequence are evaluated and the transmitted signal undergoes systematic degradation as it progresses through the systems. The transfer functions are accurately simulated to determine signal levels through the entire fiber optic system.

The system designer has the flexibility to modify the system, change design parameters, and select components from vendor data bases to fulfill system objectives. The capacity to carry out the simulation of complex lightwave systems with wave division multiplexing techniques and fiber splices is reported. The simulation software offers the designer the opportunity to examine the distortion of the optical signal and the noise contribution as the signal progresses through any given lightwave system. In addition, if so instructed, it generates the time domain plots that permit the designer to identify the components and their interactions that cause most performance degradation.

An on-line microcomputer based simulation software for the analysis and design of fiber optic data transmission links in large data communication networks is used. The system integrates the data base manipulation capability with active simulation of the critical components. The numerical solution employs a very long repetitive pulse sequence to excite the system. The Fourier components of a long excitation pulse sequence are evaluated and the transmitted signal undergoes systematic degradation as it progresses through the system. The transfer functions are accurately simulated to determine the signal levels at any node. The simulation software also yields graphic distortion of the optical signal as it progresses through any user definable lightwave system. It is capable of generating plots that permit the designer to scan problem components and configurations in the system.

The fiber optic CAD facility operates in conjunction with a large number of vendor data bases. These data bases hold the optical characteristics of the vendor products for the simulation of the system. The data bases also hold the cost and availability of the components to meet the economic and installation goals of the network operating companies.

18.4

THE FIBER CAD PROCEDURE

The CAD process (see Figure 18.1) involves simulation. The man-machine feedback occurs when the designer defines the system, and studies the simulated results to evaluate system performance. The input and output aspects are presented in Sections 18.4.1, 2 and 3. System optimization is carried out by the iterative procedure of readjusting the input parameters till a satisfactory design is negotiated.

The simulation software consists of *four* major segments as depicted in Figure 18.6. These segments correspond to the *four* functions discussed earlier. The entire simulation process consists of providing enough information for the system as follows: (*a*) defining the configuration of the lightwave system; (*b*) selecting the components of the system from vendor data bases; (*c*) approving the final configuration and component selection; and (*d*) defining which of the parameters need to be displayed.

Graphics is used extensively to facilitate the user interface. Accordingly, the following four subsections present the four phases (a) through (d) listed above. At the user's request, hard copies are generated during the entire CAD procedure for user records. This run data may also be made an integral part of the design procedure which is a finite step in the overall design procedure. The system is also capable of keeping personalized records of the individual users such that preferred and successful designs may be reused. Highly successful designs may be moved into a common knowledge base of the designers at users' requests.

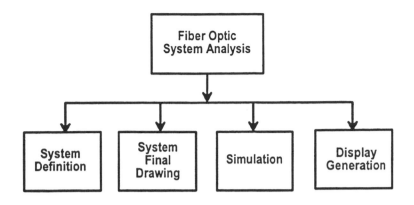

FIGURE 18.6. Software Segments for the Lightwave Design System.

18.4.1 SYSTEM CONFIGURATION

The input, function and the output of this software segment is depicted in Figure 18.7.

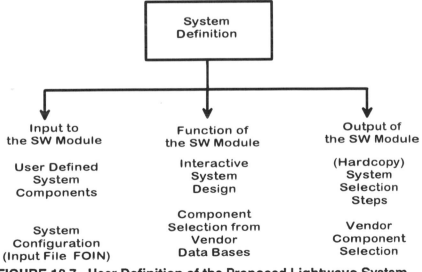

FIGURE 18.7. User Definition of the Proposed Lightwave System.

At the entry to the CAD software platform, the user defines the number of nodes and components, the generic names of the components selected, and the interconnection between the generic components. For simplicity, only simulations from a serially connected system are presented, even though multiple optical taps can be incorporated and their performance be evaluated. Distinct commonality between VLSI, HDSL and fiber optic simulation and CAD software exists.

The Lightwave Design System
Step 1

System is:

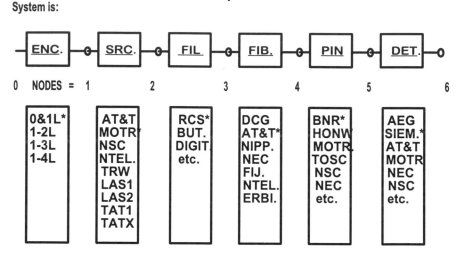

* Selected Components for the Simulations to follow. The System Configuration is selected by the user (<u>Underlined</u>). The selection (*) is made by the user.

FIGURE 18.8. Configuration and Component Selection Procedure in the CAD Facility. These data bases are extensive and only a subset of the typical components are shown.

The system generates a block diagram of the system that the user has defined. At present, the user has the capability to select components from the any number of generic components (encoders, sources, fibers, filters, PIN-diodes, detectors, optical couplers, optical taps, optical signal splitters, etc.) commonly used in optical systems. Next, it scans the vendor data bases to offer the user particular components by vendor serial numbers. In Figure 18.8, the major vendor for each generic component is displayed.

18.4.2 COMPONENT SELECTION

Component selection is also facilitated by graphic routines. The user selects the components (see the asterisk (*) in Figure 18.8) in an interactive mode as the component menu is offered to the user. These selected components are tagged by the system and a user defined system configuration is redrawn (Figure 18.9) with the appropriate components placed in the block diagram.

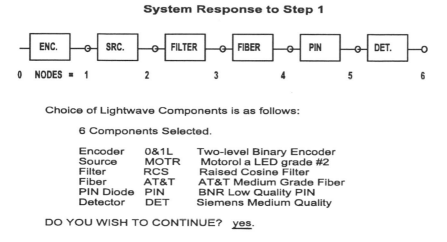

System Response to Step 1

Choice of Lightwave Components is as follows:

6 Components Selected.

Encoder	0&1L	Two-level Binary Encoder
Source	MOTR	Motorol a LED grade #2
Filter	RCS	Raised Cosine Filter
Fiber	AT&T	AT&T Medium Grade Fiber
PIN Diode	PIN	BNR Low Quality PIN
Detector	DET	Siemens Medium Quality

DO YOU WISH TO CONTINUE? yes.

FIGURE 18.9. System Response and Verification. User response is underlined.

The user reaffirms the configuration and components that constitute the simulation to follow. After validation, the component identification is stored away for the simulation run.

> UTS-Fiber Optic Simulation Program Sequence FS*.
> 128 Pulses with 128 Points per Pulse.
> FFT for 16,384 Points.
> The Program takes about 2.0 minutes on
> A 125 MHz Pentium Microcomputer.
>
> FIBER LENGTH: in kilometers = 100 km
> DATA RATE: in Megabits per second = 622 Mbps
>
> Fiber Loss Interpolation.
> Fiber Delay Interpolation.
> Harmonics have been Computed.
> Fiber Characterization Completed.
> Time Domain Values Computed at *SIX* nodes.
>
> I AM DONE. Thank you for waiting.
> Execution Terminated.

FIGURE 18.10. System Updates During the Simulation Phase.

The system generates ample updates throughout the simulation process and uses an intricate file management methodology to provide the user with the necessary information.   The simulated data is condensed during the simulation, and is stored for further analysis or for more intricate displays.

18.4.3 GRAPHICAL DISPLAYS AND HARD COPIES

For the results presented in this Chapter, a 640 × 400 resolution color monitor is used for the graphical output.   Standard displays depict the simulated results and screen buffers are used to dump the displays to a laser printer. Windowing options for displaying and screen dumping is built in the last segment of the software. The user also has the option of defining the nodes at which the displays will be depicted. Higher picture quality and resolution have also been achieved with the high resolution graphics boards and 1,280x1,024 color displays.

18.5

SIMULATION RESULTS

18.5.1 TWO LEVEL POINT TO POINT TRANSMISSION

In Figures 18.11a through f, the signal levels and their shapes are depicted for six nodes of the system shown in Figure 18.6. The length of the fiber link is 100 km and the rate is a modest 622 Mbps. The simulation and design procedure provides enough updates through the simulation phase (see Figure 18.10). The components selected perform with a wide enough eye-opening, as depicted in Figures 18.11e and f. At the receiver node (Node 6, Figure 18.11f), one level threshold detector is necessary, and the eye is closed about 25 percent.

   The series of these six figures (18.11a through f) indicate the effect of each of the components selected from the six data bases.   The extent of the degradation of the performance can be tied to the cost of the components from the cost data base (not shown in the Figures), which is a subset of the vendor data bases. The system designer has an extremely fine control on the extent of the degradation in performance and the cost penalty for its improvement. In Figure 18.11g, a two cycle eye-diagram is plotted at Node Number 6. The extent of amplitude jitter due to PIN diode and detector noise can be identified at each of the 64 sampling instants for the pulses and the 0-0, 0-1, 1-0, 1-1 transition paths can be identified.

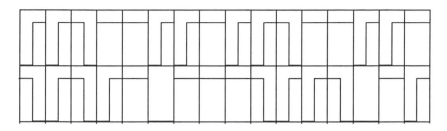

FIGURE 18.11a. Binary Data from Source.

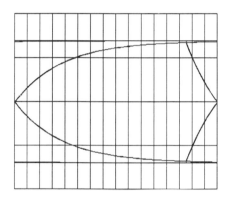

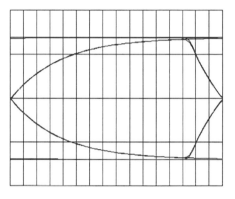

FIGURE 18.11b. ED at LED.

FIGURE 18.11c. ED after Filter.

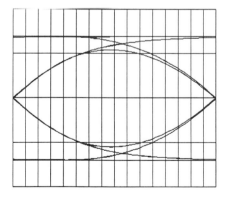

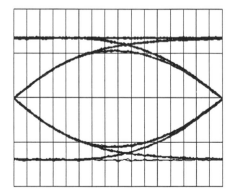

FIGURE 18.11d. ED after Fiber.

FIGURE 18.11e. ED at PIN Diode.

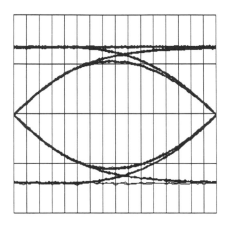

FIGURE 18.11f. ED after Detector

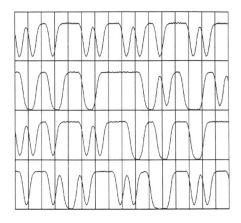

FIGURE 18.11g. Waveshape (Det.)

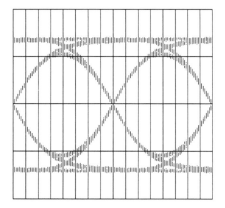

**FIGURE 18.11h. Effect of
Amplitude Jitter.**

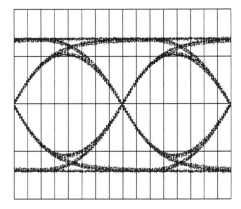

**FIGURE 18.11i. Effect of Amplitude
and Phase Jitter.**

In Figure 18.11h, the extent of the timing jitter in the clock recovery circuit is also shown. The plotting routines are also interactive and provide the user only the information requested from the CAD software.

The system also plots the individual characteristics of the components. In Figure 18.12 a through d, the source energy distribution, the fiber loss, the received signal energy distribution, the group delay are also plotted.

Typically every product has its own influence on the system performance and this can be accurately tracked in any simulation and design procedure.

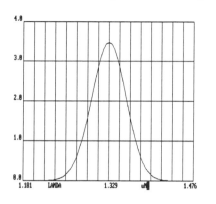

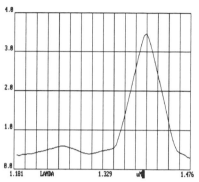

FIGURE 18.12a. Source Energy.   **FIGURE 18.12b. Fiber Loss.**

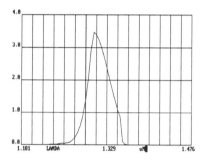

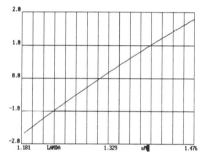

FIGURE 18.12c. Received Signal Energy Distribution.   **FIGURE 18.12d. Fiber Delay Distribution.**

18.5.2 MULTILEVEL TRANSMISSION THROUGH THE FIBER

Multilevel signals can be used to encode data at the input to the fibers. The effects of these were of some interest when the limits of the poorer quality fiber was being investigated in the early 1980s. The results of four level coding are shown in Figures 18.13a through f. The differences between Figures 18.13c, d, and e become apparent when they are viewed on large (12"x15") display computer terminals. The system topology is the same as shown in Figure 18.6 and the outputs are depicted in two cycle eye diagrams. It is to be appreciated that the capacity of single mode fibers is very large with the newer systems, and multilevel coding may or not be necessary. Even though the capacity is 1.24 Gbps, there are three threshold levels for the detection, and each of the three eye diagrams is closed by over 25 percent.

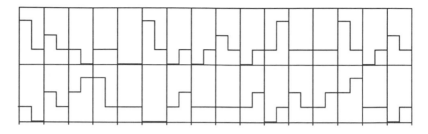

FIGURE 18.13a. Multilevel Data into the FO System.

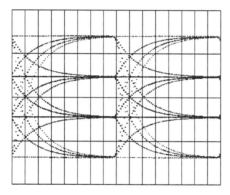

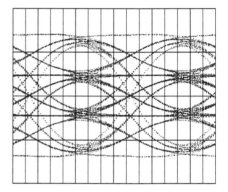

FIGURE 18.13b. Signal into the
Fiber.

FIGURE 18.13c. Signal out of the
Fiber.

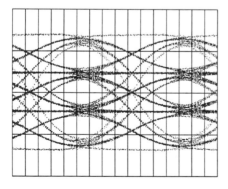

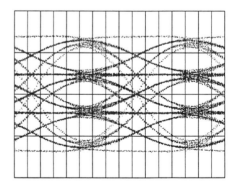

FIGURE 18.13d. PIN Diode
Output.

FIGURE 18.13e. Detector   Output.

18.5.3 WAVELENGTH DIVISION MULTIPLEXING

Wavelength division multiplexing (WDM) through the fiber can be simulated easily in the CAD environment. Different attenuation's and dispersions are encountered in the same fiber for the signals at various wavelengths. Crosstalk between the channels has been identified and correlated to the separation of the wavebands, fiber characteristics, signal levels, etc. The WDM simulations need about seventy percent more execution time.

For typical results that can be simulated, a simple two channel WDM concept shown in Figure 16.25a is chosen. Two sources are active at the same time and broadcast optical energy in the same fiber. The optical energy distributions are plotted in Figure 18.14a. The fiber attenuation and group delay distribution curves are depicted in Figures 18.14b and c.

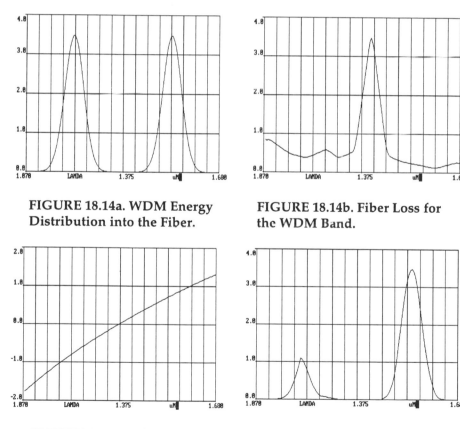

FIGURE 18.14a. WDM Energy
Distribution into the Fiber.

FIGURE 18.14b. Fiber Loss for
the WDM Band.

FIGURE 18.14c. Fiber Group
Delay Distribution.

FIGURE 18.14d. Receive Energy
Distribution.

At the receive end of the fiber, the signal and the crosstalk energy distributions are as shown in Figure 18.14d.   The relatively different attenuation at different wavelengths is the reasons for the signal levels.   In Figures 18.15a and b, the time domain signals at the PIN diodes and the detector for optical waveband channel 1 at 1.226 μm are depicted as eye diagrams.   Corresponding results for channel 2 at 1.55 μm are shown in Figures 18.15c and d. The channel separation is wide in this case, and there is relatively no interference and crosstalk between the channels.   When the sources are allocated very narrow wavebands and placed closer and closer to each other, the interference becomes more and more excessive, and the significant leakage of optical energy occurs through the fiber.

These effects have been accurately quantified and depicted as the degradation of the received signal. With very closely spaced channels, the simulation studies with extremely narrowband laser sources have yielded very attractive results.

18.6

THE USE OF DATA BASES

The organization and classification of the data base for fiber optic simulations is not unlike that presented for the DSL studies presented in Chapter 8.   In fact, the CAD platform for the DSL, HDSL, ADSL, etc. and that for the lightwave systems is essentially the same.   It is also well suited for transferring signals and wave shapes to any other signal processor design package.

18.6.1 VENDOR DATA BASES

Vendor data bases follow a hierarchical structure.   The *first* level of classification is the component type.   The *second* level is according to the quality of its performance (e.g., rate x distance product or the attenuation and delay characteristics for fiber, noise level for PIN-diodes, etc.). Since files and data structures are user definable, there is ample flexibility to store and retrieve data according to any pre-selected attribute of the device.   The storage of noise data for photodiodes and preamplifiers can become quite challenging.   If the data base holds only the distribution characteristics, computations can become time intensive.   In order to enhance execution speeds, the signal independent to noise values are precomputed and stored in appropriate files. The signal dependent noise is then computed as the actual simulation proceeds, and the final summation occurs at the device output.

FIGURE 18.15a. Channel 1 Eye Diagram at PIN Diode.

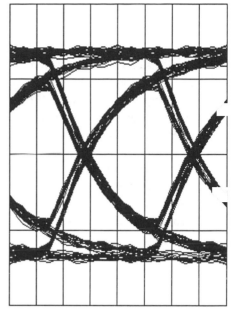

FIGURE 18.15b. Channel 2 Eye Diagram at PIN Diode.

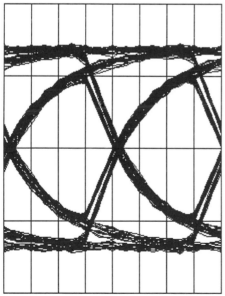

FIGURE 18.15c. Channel 1 System Output at Detector.

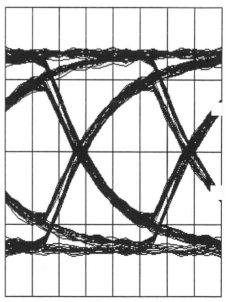

FIGURE 18.15d. Channel 2 System Output at Detector.

Vendor data bases and their organizations are user and applications dependent. At an initial level, it is appropriate to organize these data bases as VLSI component data bases. From a designers' perspective, the selection of the various components for a microcomputer system is akin to the selection of the optical components for a lightwave system. In many cases, the designers' preferences may also be incorporated in the vendor data bases by including the various systems in which the components have been used, and their relative performance levels. When a wide range of designs and their components are available, then the CAD facility can be prompted to suggest components in context to the particular design in progress. When the correlation between the simulation results and the components is complete, the system will be able to identify the components that enhance or degrade the performance objectives.

18.6.2 PROGRAM DATA BASES

The simulation programs may be classified as *four* categories. *First,* the generic algorithms for FFTs, transfer functions generators, filters, convolutional techniques, etc., and their code that is common to most simulators resides in the program library. *Second,* the routines for numerical evaluations such as integration, partial differentials, trignometric approximations, and statistical routines etc., reside in the routines library. Sometimes the computer vendor supplied routines are not be adequate or accurate for specialized applications and they need to be modified or enhanced for the DSL or fiber optic design studies. *Third,* the graphics programs for display and post processing of the data also become crucial if detailed optimization is necessary. *Finally,* the programs for cross comparisons of numerous designs is also necessary if various vendors and their components should be analyzed in any given lightwave system. Routine data base management techniques may or may not be adequate depending on the designers objectives.

The system design flexibility is tailored in the software organization. This software is designed to document and monitor events in the design procedure. The switch between optical signal processing and silicon signal processing environments can take place at any step with a complete option to retract any false or undesirable steps. Entry and/or exit from one environment to the other is completely discretionary.

18.7

INCORPORATION OF FIBER OPTIC SIMULATION WITH OTHER SIGNAL PROCESSOR DESIGN SOFTWARE

The data base methodology is well suited for transferring signals and wave shapes to other signal processing design packages. The simulation phase generates *three* categories of files.

In the *first* category, input files containing data to the simulation run are stored. The system does not automatically save all input files unless directed by the user. Thus the current input is likely to overwrite previous input files.

The *second* category of the file contains transitory information between one segment of the program to the next. The system does not automatically save these files to conserve disk space, but it can save these files for future reference.

The *third* category of the file contains actual simulation results that may be essential to the signal processing environment. Systematic saving of these files is a user option. The user may elect to optimize the design of the lightwave system before entering the signal processing environment or optimize the integrated fiber-IC design.

Extreme flexibility exists, since the system is designed to track the history of the procedural steps in the fiber or in VLSI environments.

18.8

CONCLUSIONS

The simulation and CAD facility can be made quite elaborate, but efficient and inexpensive. By forcing repeated runs with automated storage of design parameters for any given combination of design criteria, the machine yields an optimum design. The facility can be effectively used in the design of typical local, subscriber, and metropolitan area fiber optic links and systems. A wide variety of results can also be obtained.

The vendor data base for fiber optic components constitutes an entire library. When the fiber optic devices function with peripheral IC devices, the two signal processing environments can be tightly coupled in the same CAD environment. Under these conditions, IC design routines constitute an entire and distinct library. Since IC technology is more highly evolved than lightwave technology, the system also interfaces IC design and layout routines rather than IC simulation packages. This facilitates the choice of the IC chip/design ideally suited to work in conjunction with the lightwave system. The results presented in this Chapter are a small subset of the results that can be obtained from the simulation facility.

The data base for fiber optic components organized by the limits of the components constitutes a library in its own right. This library can be created if the design requirements demand extremely stringent tolerances. Sophisticated data base search techniques can reduce the extent of duplication of raw component and vendor data.

18.9

REFERENCES

18.1   S.V. Ahamed, P.P. Bohn, N.L. Gottfried 1981, "A Tutorial on Two-Wire Digital Transmission in the Loop Plant," *IEEE Transactions on Communication*, Vol. COM 30, 1554-1564.

18.2   M.J. Miller and S.V. Ahamed 1988, *Digital Transmission and Networks, Volume II, Applications*, Computer Science Press, Rockville, MD,. Also see: M. K. Barnoski 1981, Editor *Fundamentals of Optical Fiber Communications*, Second Edition, Academic Press, New York, NY.

18.3. S.V. Ahamed and V.B. Lawrence 1990, "An Intelligent CAD Environment for Integrated Services Digital Network (ISDN) Components", *Proceedings of the IEEE International Workshop on Industrial Applications of Machine Vision and Intelligence*, Paper 02, February 2-4, Roppongi, Tokyo, Japan.

18.4. K.S. Shanmugan, et. al. 1985, "Simulation of Digital Lightwave Communication Links Using SYSTID", *Globecom '85*, December. Also see: M. Fashano, A.L. Strodtbeck 1984, "Communication System Simulation and Analysis with SYSTID", *IEEE Journal of Selected Areas in Communications*, Vol. J-SAC-2, No. 1, 8-28, January.

18.5   W.I. Way and A.F. Elrefaie 1987, "Transmission of Microwave 16-QAM Signal over Single Mode Fiber in the 1.3 µm Wavelength Region", *Record of International Conference on Communications*, June 7-10, 177-182. Also see: A. Elrefaie, M. Romeiser 1986, "Computer Simulation of Single-mode Fiber Systems", *Optical Fiber Communications Conference*, February, 54-56.

18.6 S.D. Personik 1971, "Statistics of a General Class of Avalanche Detectors with Application to Optical Communications," *Bell System Technical Journal*, Vol. 50, 3075-3095.

18.7 P.P. Webb, R.J. McIntyre and J. Cornadi 1974, "Properties of Avalanche Photodiodes," *RCA Review*, Vol. 35, 236-276.

18.8 P. Balaban 1976, "Statistical Evaluation of the Error Rate of the Fiberguide Repeater Using Importance Sampling," *Bell System Technical Journal*, Vol. 55, 745-766.

18.9 A.F. Elrefaie, J.K. Townsend, M.B. Romeiser and K.S. Shanmugan 1988, "Computer Simulation of Digital Lightwave Links," *IEEE Journal on Selected Areas in Communications*, Vol. 6, No. 1, January.

18.10 S.V. Ahamed and V.B. Lawrence 1989, "A PC Based CAD Environment for Fiber Optic Simulations", *Proceedings of Globecom '89*, Vol. 2, 19.1.5.1-6, Dallas, TX.

18.11 A.F. Elrefaie, R.E. Wagner, D.A. Atlas, and D.G. Daut 1988, "Chromatic Dispersion Limitations in Coherent Lightwave Transmission Systems," *IEEE Journal of Lightwave Technology*, Vol. 6, No. 5, May, 704-709.

18.12 R.E. Wagner and A.F. Elrefaie 1988, "Polarization Dispersion Limitations in Lightwave Systems," *Proceedings of the OFC*.

18.13 S.V. Ahamed 1988, "The Integration of Fiber Optic Simulations With Integrated Circuit Design", *Proceedings of MILCOM '88*, Paper No. 2.3, October 23- 26, San Diego, CA.

18.14 S.V. Ahamed and V.B. Lawrence 1987, "A Computer Aided Design Environment for the Local Lightwave Communication Systems", *IEEE Communication Society's Fifth International Workshop on Integrated Electronics and Photonics in Communications*, October 21-23, North Carolina. Also see: D.G. Duff 1984, "Computer-Aided Design of Digital Lightwave Systems", *IEEE Journal of Selected Areas in Communications*, Vol. J-SAC-2, No. 1, 171-185, January.

GLOSSARY

#

1B type of switching processor

2B1Q two binary to one quartinary code for BRISDN

3B20 a type of specialized switching processor

1ESS, 1A ESS, 2B ESS, 3ESS precursor ESS systems introduced before the current 4ESS and 5ESS

4ESS Number 4 electronic switching system for toll and trunk switching

4A ETS 4A electronic translator system for toll switching

5ESS a generalized electronic switching system for most types of CS, PS, ISDN, suscriber loop carrier and trunk switching

5ESS-2000 a new 5ESS platform and concept to contend with wireless, intelligent network functions and almost all of the routine 5ESS functions

---

A

A and B class service types of ATM services

A/D analog to digital converter

A6 sixth generation of the A-Type channel bank

AAL ATM adaptation layer

ADM add-drop multiplexers

ADM adaptive delta modulation

ADPCM adaptive differential pulse code modulation

ADSL asymmetric digital subscriber line

AEC adaptive echo cancellation, or adaptive echo canceler

AI artificial intelligence

AIN advanced intelligent network

AIU access interface units

A-law, μ-law two laws for companding for speech signals

AL application layer

AlGaAs aluminum gallium arsenide

ALOHA network the earliest packet network developed at the University of Hawaii at Monoa

ALU arithmetic logic unit within the CPU of a computer

AM administrative module of a modern ESS.

AM/PM amplitude modulation and/or phase modulation

AMAC asynchronous media access control

AMI alternate mark inversion

AMI 2B1Q, 3B2T, 4B3T four of the line codes for copper medium

ANSI American National Standards Institute

APD avalanche photodiode

AR-6A 6 GHz radio SSBAM analog carrier system

ASK amplitude shift keying

AT&T American Telephone and Telegraph

ATM asynchronous transfer mode

ATM/BISDN ATM to broadband ISDN

AWG American wire gauge

AXC ATM cross-connect

AXE a telecommunications switching system built by Ericsson

B

B1 channel the number one B channel of BRISDN service that is used to carry 64 kbps PCM voice data for telephone service.

B2 channel the number two B channel of BRISDN service that can be used to carry packet-switched data

BB baseband

BDCS broadband digital cross connect switch or system

BER bit error rate

BHCA busy hour call attempts

BISDN broadband integrated services digital network

B-NT network terminator for broadband ISDN

BPF bandpass filter

bps bits per second

BRISDN basic rate integrated services digital network

BRT broadband radio terminal

BSS broadband switching system

BT bridged taps

B-TA terminal adapter for broadband ISDN

B-TE terminal equipment for broadband ISDN

BW bandwidth

C

C control (and signaling)

C and D class service type of ATM services

CAD computer aided design

CAD/CAM computer aided design/computer assisted manufacture

CAL customer access line

CAP carrierless amplitude and phase (modulation)

CATV cable television

CBR constant bit rate

CCITT International Telegraph and Telephone Consultative Committee (now ITU-T)

CCS Common Channel Signaling

CCS7 Common Channel Signaling 7

CDDI copper distributed data interface

CENTREX a widespread telephone company switching service that uses Central Office switching equipment and to which customers connect via individual extension lines

CEPT Conference of European Posts and Telecommunications

CLASS custom local area signaling services

Class A, B, C, D classes of ATM services

CLP cell loss priority

CMBH capped mesa buried heterostructure

CMOS complementary metal oxide semiconductor

CNET Centre National D'Etudes des Telecommunications

CNI common network interface

CO Central Office

COCF connection oriented convergence function

CO-LAN AT&T's LAN with CENTREX

COT Central Office terminal

CPCS common part convergence sublayer

CPE customer premises equipment

CPFSK continuous phase shift-keying

CPU central processing unit

CRC cyclic redundancy check

CS circuit switched

CS convergence sublayer

CSA carrier serving area

CSBH channel substrate buried heterostructure

CSDC circuit switched digital capability

CSMA/CD carrier sense multiple access/collision detect

CS-MUX circuit switching multiplexer

CT cordless telephone

D

D4 digital channel bank for switching each of 24 voice band signals into 8-bit PCM words

DACS digital access cross connect system

dB decibel

dBrnC a power level in dB relative to noise reference of -90 dBm, as measured with a noise meter, weighted by a special frequency function called C-message weighting that expresses average subjective reaction to interference as a function of frequency

dBrnCO noise measured in dBrnC and referred to as the 0 transmission level point

DC direct current

D-channel delta channel; in ISDN a 16 Kbps signaling channel for basic rate access or a 64 kbps signaling channel with other access rates

DCLU digital channel line unit

DCS digital cellular switch

DCS digital cross connect switch

DDS digital data service

DFB distributed feedback

DFE decision feedback equalizer

DLC digital loop carrier

DLL data link layer

DMS switching system

DMT discrete multi-tone

DNU digital network units

DPSK differential phase shift-keying

DQDB distributed queue dual bus

DS digital signal (North America, Taiwan and South Korea)

DS0 universal 64 kbps channel (or rate); same in the CEPT and Japanese digital hiearchies

DS0A, DS0B digital signal of a specified rate (64 kbps)

DS1, DS-1 digital signal 1, a formatted digital signal, transmitted at 1.544 Mbps, usually referred to as T1 (same in the Japanese digital hiearchy)

DS1C digital signal level 1C rate of 3.152 Mbps (same in the Japanese digital hiearchy)

DS2, DS-2 digital signal 2, a formatted digital signal, transmitted at 6.312 Mbps, usually referred to as T2 signal (same in the Japanese digital hiearchy)

DS3 standard third level digital carrier system used in the Noth America at 44.736 Mbps; also known as T3 signal

DS3C digital signal level 3C rate of 91.053 Mbps (Japanese equivalency at 97.728 Mbps)

DS4 fourth level digital carrier system used in the Noth America at 274.176 Mbps; also known as T4 signal (Japanese equivalency at 397.2 Mbps)

DSBSC double sideband suppressed carrier (in AM-FM modulation systems)

DSBTC double sideband transmitted carrier (in AM-FM modulation systems)

digital subscriber line

SP digital signal processing

DSX digital signal cross connect

DT distribution terminal

DTE data terminal equipment

E

E field electric field vector in the electromagnetic propagation of signals

E1 or E_1 European transmission link with bandwidth capacity of 2.048 Mbps

EC echo canceler, echo cancellation

ECSA exchange carrier standards association

ED eye diagram

EELED edge-emitting LED

ELU End Office line unit

ENDEC encoder - decoder

EO End Office

EPSCS European packet-switched and circuit-switched (network access)

ERLE echo return loss enhancement

ESS electronic switching system

ET exchange termination

ETSI European Telecommunications Standards Institute

EWSD An electronic switching system built by Siemens

F

FDDI fiber distributed data interface

FDI feeder distribution interface

FDM frequency-division multiplexing

FE far end

FEP front end processors

FET field effect transistor

FEXT far end crosstalk

FFOL FDDI follow-on LAN

FFT fast Fourier transform

FIR finite impulse response

FLAG fiber optic link around the globe

FM frequency modulation

FO fiber optic

FR frame relay

FR/ATM frame relay/asynchronous transfer mode

FSE fractionally spaced equalizer

FSK frequency shift keying

FSLC fiber subscriber loop carrier

FT3C fiber optic lightwave system

FTG an version of transoceanic fiber optic cable system

FTTC fiber to the curb

FTTH fiber to the house

G

GaAlAs gallium aluminum arsenide

GaAs gallium arsenide

GaAs/Si gallium arsenide with silicon substrate

Gpbs Gigabits per second

GFC generic flow control

GTE General Telephone and Electronics

H

H_0 (5B+D) service at 384 kpbs; also know at C6

H_1 **or** H_{11} (23 B+D) service at 1.536 Mbps; also know as C24

HAW4, HAW5 and TPC4 Hawaiian Island cable access

HBT heterojunction bipolar transistor

HDLC high level data link control

HDSL high-speed digital subscriber line

HDTV high definition TV

HE_{21} one of the modes for optical propagation in fiber

HEC header error check

HFC hybrid fiber coax

HL higher layers of the OSI model

HMC hybrid multiplexer control

HSM high-speed multiplexer

HTU-C HDSL termination unit

Hz Hertz

I

I/O input/output

IAD installation and accounting data

IC integrated circuits

ICF isochronous convergence function

IDLC integrated digital loop carrier

IDT integrated digital terminal

IEC interexchange carrier

IEEE Institute of Electrical & Electronics Engineers

ILD injection laser diodes

IMAC isochronous media access control

IN intelligent network

INAP intelligent network access points

InGaAs indium gallium arsenide

InGaAs/InP indium gallium arsenide/indium phosphide

InP indium phosphide

INS impulse noise data base

IP intelligent peripheral

ISDN integrated services digital network

ISDN-1 island ISDN

ISDN-2 backbone ISDN

ISDN-3 worldwide ISDN

ISO international organization for standardization

ITU-T International Telecommunications Union - Telecommunications Standardization Sector (formerly CCITT)

IXT interaction crosstalk

K

KB knowledge base

kbps kilobits per second

kHz kiloHertz

L

L analog carrier system (North American)

L5, L5E wideband analog carrier systems

LAN local area network

LAP link access procedure/protocol

LAPB link access protocol, balanced

LAPD link access protocol for the D channel

LATA local access and transport area

LCO local Central Office

LDS local digital switch

LED light emitting diode

LFLC loop data base

LiNbO$_3$ lithium niobate for optical materials

LIU line interface unit

LLC logical link control

LM link multiplexer

LME layer management entities

LO local oscillator (optical coherent systems)

LPD loop data base

LPF low pass filter

LPLC loop data base for subscriber loops

LPLCF enhanced loop data base for limit cycle repetitive excitations

LPSVAC enriched loop data base blended by loops of different geographical areas

LSB lower side band

LT line termination

M

MAC media access control

MAC CF media access control convergence function

MAN metropolitan area network

MAU medium attachment unit

Mbps megabits per second

MDF main distribution frame

MHz megaHertz

MISFET metal insulator semiconductor field effect transistor

MLT-3 multilevel ternary level code

M*n* multiplexer *number*

MOCVD metal organic chemical vapor deposition

MOMBE metal organic molecular beam epitaxy

MPEG Motion Picture Experts Group

MSE mean square error (algorithm)

MSK minimum shift-keying

MTSO mobile telephone switching office

MUX multiplexer

N

n, or N hierarchical number appended to T, D, STS or optical carrier systems

NCP network control point

NE near end

NEXT near end crosstalk

NI network interface

NISDN narrowband ISDN

NIU network interface unit

NL network layer

NNI network-to-network interface

NT network termination

NTSC national television standards committee

NTU network termination unit

NYNEX New York telephone operating company (now merged with Bell Atlantic)

O

OA&M operations administration and management

OC optical carrier

OCn optical carrier nth level of multiplex

OCU office channel unit

OH overhead; also hydra-oxyle (in fiber optic systems)

ONU off network unit

OOK on-off keying

OP output files (for simulation and design studies)

OSDS operating system for digital switching

OSI open system interconnect

OSI-RM open system interconnection - reference model

P

PA physical access

PAR project authorization request provided by IEEE

PBX private branch exchange

PCM pulse code modulation

PCRIMEAST, PCRIMWEST Pacific Rim and Far East access

PCS personal communication services

PDS premises distribution systems

PE paging entity in UMTS

PE polyethylene

PEIQL, PEUIT Australian brands of TWP

PHY physical layer protocol

PIN positive-intrinsic-negative (photodiode)

PIUT Australian brand of TWP

PL plotting (programs); also playload; also physical layer

PLCF physical layer convergence function

PLCP physical layer convergence protocol

PLL phase locked loop

PLT playload type

PM physical medium

PMD physical layer medium dependent

Glossary

PN type of semiconductor junction diode

POTS plain old telephone service

PRI primary class

PRISDN primary rate integrated services digital network

PS packet service

PSK phase shift-keying

PSTN public switched telephone network

PTT Postal, Telegraph and Telephone (European government carriers)

PVC polyvinylcholride

PVC permanent virtual service

Q

QAM quadrature amplitude modulation

R

R receiver

RAM random access memory

RC regional center

RD remote destination

RDT remote digital terminal

REGEN regenerator

RF radio frequency

RG cable type

U radio port control unit

T remote terminal

S

SAAL signaling

SAR segmentation and reassembly

SBS stimulated Brillouin scattering

SCP signal control point of intelligent networks

S/D signal/distortion ratio

SD standard deviation

SDDN software defined data network

SDH synchronous digital hierarchy

SDLC synchronous data link control

SDM standard deviation of the mean

SDN switched digital network

SDN software defined network

SDSL symmetric digital subscriber line

SECT Sectional Center

SEED self electro-optic effect device

SELED surface emitting led

SL session layer

SL sublayer

Glossary

SLC subscriber loop carrier

SLC-96 a subscriber loop carrier system over copper wire

SLC-2000 a subscriber loop carrier system over fiber

SM single mode

SMDS switched multimegabit data service

SMS service management system

SMT station management

S-MUX service multiplexer

SNA system network architecture (IBM)

SNA/ATM systems network architecture/ATM cell relay

SNA/FR systems network architecture/frame relay

SNR signal-to-noise-ratio

SONET synchronous optical network

SRS stimulated Raman scattering

SS7 Signaling System 7

SSBAM single side band amplitude modification

SSCS service specific convergence sublayer

S-SEED symmetric SEED

SSP service switching point

STM-i synchronous transport module level i ($i = N/3 = 1, 4, 16, 64$)

STP signal transfer point

synchronous transport signal/system

TS-N synchronous transport signal level *N* (*N*= 1, 3,.12, 48 or 192)

SU Syndes unit

SUB subscriber

SVC switched virtual service

SYN synchronization

SYSTID set of simulation programs for fiber environments

T

T transmitter

T1 a digital carrier facility used to transmit a DS-1 formatted digital signal at 1.544Mbps, also called T1 carrier system

T1 or E1 primary rates for North American or European environment (see DS-*n* levels)

TA terminal adapter

TASI time assignment speech interpolation

TAT transatlantic telephone

TC transmission convergence

TCM time-compression multiplexing

TCS transmission convergence sublayer

TD type of analog radio carrier system

TDM time-division multiplexing

TDM-MTN time division multiplex maintenance field

TE terminal equipment

TE$_{01}$ one of the modes for optical propagation in fiber

TIRKS Bellcore developed trunk signaling system

TL transport layer

TLP transmit level power

T-n standard North American carrier lines used to transmit a formatted digital signal

TOC television operating center

TOLL toll center

TP toll point

TPC3 and TPC4 transmission control protocols

TR-303 transmit/receive interface

TRANS transmitter

Tx transmit fiber

TWP twisted wire-pair

U

U user

ULSI ultra large scale integration

UNI user network interface

USB upper sideband

UTP unshielded twisted wire-pair

V

VBR variable bit rate

VCI virtual channel interface

VCO voltage controlled oscillator

VCR video cassette recorder

VF voice frequency

VHDSL very high-speed digital subscriber line

VPI virtual path interface

VSAT very small aperture terminal

VT virtual terminal

W

WAN wide area network

WDCS sideband digital cross connect switches

WDM wavelength division multiplexing

X

X-talk crosstalk

X.25 a CCITT (now ITU) recommendation that specifies the interface between user data terminal equipment (DTE) and packet-switching data circuit-terminating equipment (DCE)

INDEX

INDEX

INDEX

INDEX

-M-

-U-

Syed V. Ahamed received his Ph.D. and D.Sc. (E.E.) degrees from the University of Manchester and his MBA (Econ.) from the New York University. He taught at the University of Colorado for 2 years before joining AT&T Bell Laboratories in 1966. In 1981 he taught Electrical Engineering at the University of Hawaii at Monoa and in 1982 he became a Professor of Computer Science at the City University of New York and a member of the Doctoral Faculty in 1985. He has taught at New York Polytechnic at Brooklyn as a visiting Professor of Computer Science from 1982 to 1986. Professor Ahamed has been a Telecommunications consultant to Bell Communications Research, AT&T Bell Laboratories, Lucent Technologies, Timplex Corporation, Telecom Australia, ETH Zurich, Schweizerischer Nationalfonds zur Forderung, (Swiss NSF), Berne, and Schmid Switzerland. He is the co-recipient of the 1981 IEEE Communications Society, Leonard G. Abraham Prize Paper Award and the co-recipient of the 1964 Honorable Citation for the IEEE Aerospace Society paper. With Professor Michael J. Miller he has written two other books "Digital Transmission Systems and Networks," - Volumes I & II in 1987 and 1988. In 1991 he became a Fellow of the IEEE and listed in Marquis 1995, Who's Who in the World. He holds over 16 American and European patents and has published over 150 papers, numerous chapters in books and works with an excellent team of Ph.D. students at the Graduate Center of the City University of New York.

Victor B. Lawrence received his Ph.D. from the University of London. After teaching at Kumasi University of Science & Technology in Ghana, he joined AT&T Bell Laboratories in 1974. Currently, he is the Director of the Advanced Communications Technologies Center with responsibility for technology transfer, systems engineering and exploratory development of multimedia products and services, including the processing, storage and transport of multimedia systems over wire and wireless networks. He has taught at Columbia, Princeton, and Rutgers Universities as Adjunct Professor. He delivered the 1986 Distinguished Chancellor's lecture series at the University of California at Berkeley. He has been the Repporteur on coding for CCITT in 1984, Editor-in-chief of the IEEE Transactions on Communications from 1987 to 1991, Member of IEEE Fellow Committee from 1988-1991, the Chairman of IEEE Awards Board from 1994-1995. He is Fellow of both the IEEE and AT&T Bell Laboratories. He was co-recipient of 1981 Gullemin-Caauer prize paper award, and a best paper award at Interface 1984. He holds fourteen patents and has published over 40 technical papers. He is author of a chapter in a book "Introduction to Digital Filtering", co-editor of a book "Tutorials in Modern Communications".